1

028259

WATER
AND THE
ENVIRONMENT

WATER

AND THE
ENVIRONMENT

INNOVATION
ISSUES
IN
IRRIGATION
AND DRAINAGE

Edited by

Luis S. Pereira
Institute of Agronomy,
Technical University of Lisbon, Portugal

and

John W. Gowing
University of Newcastle, UK

Portuguese National Committee of the International Commission on
Irrigation & Drainage
Selected papers of the 1st Inter-Regional Conference
"Environment-Water: Innovative Issues in Irrigation and Drainage"
held in Lisbon, Portugal, September 1998

ICID·CIID

This edition published 1998
by E & FN Spon.
11 New Fetter Lane, London EC4P 4EE

Simultaneously published in the USA and Canada
by Routledge
29 West 35th Street, New York, NY 10001

Reprinted 2000, 2001

Spon Press is an imprint of the Taylor & Francis Group

© 1998 Luis S. Pereira and John W. Gowing

Printed and bound in Great Britain by
T.J.I. Digital, Padstow, Cornwall

Transferred to Digital Printing 2002

Publisher's note
This book has been produced from camera-ready copy supplied by the authors.

British Library Cataloguing in Publication Data
A catalogue record for this book is available from the British Library

Library of Congress Cataloging-in-Publication Data
Inter-Regional Conference "Environment-Water: Innovative Issues in
        Irrigation and Drainage" (1st : 1998 : Lisbon, Portugal)
        Water and the environment : innovation issues in irrigation and
    drainage / edited by Luis S. Pereira and John W. Gowing ; Portuguese
    National Committee of the International Commission on Irrigation &
    Drainage ; selected papers of the 1st Inter-Regional Conference
    "Environment-Water: Innovative Issues in Irrigation and Drainage",
    held in Lisbon, Portugal, September 98.
            p. cm.
    Includes bibliographical references and index.
    ISBN 0-419-23710-0 (hardcover)
    1. Irrigation engineering--Congresses. 3. Drainage--Congresses.
    4. Water quality management--Congresses. I. Pereira, Luis S.
    II. Gowing, John W. III. International Commission on Irrigation and
    Drainage. Portuguese National Committee. IV. Title.
    TC803.I54   1998
    627'.52--dc21   98-26951
                                                                 CIP

ISBN 0 419 23710 0

CONTENTS

PREFACE

As we approach the end of the millennium it is clear that sustainable management of water resources represents a major challenge for the future. Imbalances between availability and demand coupled with degradation of water quality must be viewed in the context of growing pressures for environmental protection and increasing concern over inter-sectoral and inter-regional conflict over access to resources. The problem is particularly acute for the agricultural sector, which dominates demand in many countries, especially those in arid, semi-arid and drought-prone regions.

Water is required for life as well as for economic activity. Its dual role as an environmental asset as well as an economic good poses a particular challenge to us all. We must ensure that in exploiting this valuable resource for our own benefit we do not unwittingly compromise the lives of others now and in the future. To meet this challenge will require the full application of human creativity and innovation.

The contributions to this volume provide insights into recent developments in theory and practice in irrigation and drainage engineering. New approaches are proposed which aim to promote environmentally sustainable development and management of irrigation and drainage facilities. These include research into environmental processes, management tools and models, technological solutions, social and institutional arrangements and insights into policy. They are arranged into five sections:

- managing water quality
- managing on-farm demand
- managing supply in irrigation systems
- water conservation and hydrologic behaviour
- coping with water-scarcity and droughts

Of course, there are significant gaps in our existing knowledge. Identification of needs for new research and technology development is therefore, in itself, an important outcome. There are many such needs, but it is apparent that particular attention needs to be given to promoting long-term investigations, which alone can provide validation of predictions of the sustainability of new or existing practices. In the short-term we must rely upon extrapolations and simulations, but in time these must be tested against measurements if we are to continue to improve our understanding of our complex environment. Future innovations will be built upon knowledge of successes and failures which comes from case study analysis together with field monitoring and evaluation at a range of scales.

For the present the editors are grateful to:

- the Portuguese National Committee on Irrigation and Drainage, who promoted the conference;
- the many experts from a large number of countries, who contributed their ideas and experiences;
- the scientific panel, who reviewed all papers and therefore helped greatly in maintaining the high standard of the individual contributions to this volume.

May 1998   Luis Santos Pereira   John W. Gowing
                              Lisboa   Newcastle upon Tyne

ACKNOWLEDGEMENT

The first Inter-Regional Conference on Environment-Water: Innovative Issues in Irrigation and Drainage is organized by the Portuguese National Committee of the International Commission on Irrigation and Drainage (ICID) and is sponsored by the ICID, namely through the European Regional Working Group, and by the CIGR (International Commission on Agricultural Engineering) Land and Water Section.

The Conference, mainly the publications and the attendance by participants from developing countries, is supported by several international and national organizations, in particular the following ones:

CIHEAM (Paris) and IAM-B (Bari)
CTA (Wageningen)
Instituto da Água (Lisboa)
Instituto de Hidráulica, Engenharia Rural e Ambiente (Lisboa)
Instituto Superior de Agronomia and Centro de Estudos de Engenharia Rural (Lisboa)

The Conference was also co-sponsored by other scientific and technical organizations: IAIA, IWSA and EWRA.

The conference papers have been reviewed and selected after the advice of the following members of the Scientific Committee:

M. Ait Kadi, Rabat, Morocco
R. G. Allen, Logan, Utah, USA
A. B. Almeida, Lisbon, Portugal
J. F. Baptista, Lisbon, Portugal
T. Brandyk, Warsaw, Poland
R. A. Brito, Sete Lagoas, MG, Brasil
L. G. Cai, Beijing, China
J. Castro Caldas, Lisbon, Portugal
J. Chambouleyron, Mendoza, Argentina
W. Dirksen, ERWG-ICID, Bonn, Germany
D. El-Quosy, Cairo, Egypt
J. A. Fernandes, Lisbon, Portugal
J. Feyen, Leuven, Belgium
J. Gowing, Newcastle, UK
J. Henessy, ICID, Watsave, Reading, UK
I. Ijjas, Budapest, Hungary
R. Kanwar, Ames, Iowa, USA
P. Kerkides, Athens, Greece
N. Lamaddalena, CIHEAM, Bari, Italy
J. Lobo Ferreira, Lisbon, Portugal
H. M. Malano, Victoria, Australia
Z. Mao, Wuhan, China

J. Martinez-Beltran, FAO, Rome, Italy
B. Maticic, Ljubljana, Slovenia
A. Musy, Lausanne, Switzerland
I. Nicolaiescu, Bucharest, Romania
B. Nicolardot, Reims, France
L. S. Pereira, Lisbon, Portugal
R. Ragab, ICID, Wallingford, UK
G. Rossi, Catania, Italy
B. Schultz, Delft, The Netherlands
R. Segura, Madrid, Spain
E. M. Sequeira, Lisbon, Portugal
R. Serralheiro, Évora, Portugal
Y. Shevah, Tel-Aviv, Israel
A. A. Soares, Viçosa, MG, Brasil
P. L. Sousa, Lisbon, Portugal
V. S. Sousa, Vila Real, Portugal
J. L. Teixeira, Lisbon, Portugal
N. Tyagi, Karnal, India
I. Varlev, Sofia, Bulgaria
A. Zairi, Tunis, Tunísia
D. Zimmer, Antony, France

SECTION I

MANAGING WATER QUALITY

APPLICATION OF THE ROOT ZONE WATER QUALITY MODEL FOR ENVIRONMENT-WATER MANAGEMENT IN AGRICULTURAL SYSTEMS

L. R. AHUJA, L. MA, and J. D. HANSON
Great Plains Systems Research, USDA-ARS, Fort Collins, Colorado, USA
R. S. KANWAR
Iowa State University, Ames, Iowa, USA

Abstract
Increasing public concern for agricultural resources and environmental pollutants have led to more use of agricultural systems models to assess water management practices under field conditions. The Root Zone Water Quality Model (RZWQM) was developed as a tool for agricultural management and tested under various conditions around the world. This paper summarizes results of model application and evaluation for environment-water management.
Keywords: Agricultural management, crop production, drainage, modelling, nitrate, pesticide, surface runoff, water management, water quality.

1 Introduction

The Root Zone Water Quality Model (RZWQM) was developed by USDA-ARS scientists of the Great Plains Systems Research Unit in Fort Collins, USA in cooperation with other scientists. The model integrates state-of-the-science knowledge of agricultural systems into a tool for managing and conserving agricultural water, environmental quality, crop production, and global change [1],[2]. RZWQM has the capability of simulating soil water storage and movement [3], [4], subsurface drainage [5],[6],[7], surface runoff [3],[4], evapotranspiration (ET) [8],[9], and crop production [10],[11]. It can also accommodate various agricultural management practices, including irrigation, fertilization, tillage, and manure application [6],[7],[11],[12],[13]. The transport of pesticides and nitrate is also simulated [12],[13],[14],[15],[16]. The purpose of the paper is to review results from field applications of the model.

Water and the Environment: Innovative Issues in Irrigation and Drainage. Edited by Luis S. Pereira and John W. Gowing. Published in 1998 by E & FN Spon. ISBN 0 419 23710 0

2 Field application and results

2.1 Soil water movement and tile drainage flow

Ahuja et al. [3] found that, with proper determination of soil hydraulic properties, the model gave good results for soil water, runoff, and macropore flow (Fig. 1). RZWQM optionally provides estimates of soil hydraulic properties from soil texture and, if available, 1/3 bar water content. Model-estimated parameters have shown to be adequate in several studies [12],[14],[17]. Although soil water content (storage) is a widely available experimental measurement and has been used as an indicator for water movement in soils, goodness of model prediction is very sensitive to crop transpiration estimation [18],[19].

Ghidey et al. [20] calibrated soil hydraulic conductivity based on surface runoff measurements from a Missouri MESA (Management Systems Evaluation Areas) site. With consideration of soil macropore flow and surface crusting, model predictions were slightly higher than measured runoff. However, they pointed out that the simulation of soil surface cracking was needed to correctly predict runoff under different drying and wetting cycles.

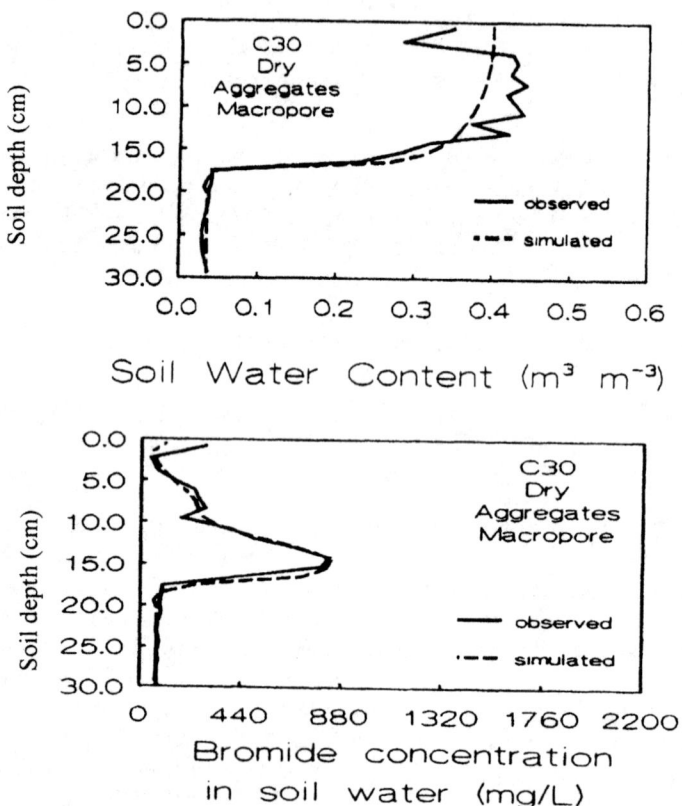

Fig. 1. Comparison of observed and simulated soil water and Br distributions in soils with macropores and surface aggregates [3]

With partially calibrated soil hydraulic parameters, Jaynes and Miller [21] found that the model underestimated drainage, because runoff water infiltrated into the drainage tile at the experimental site. Farahani et al. [22] found that estimated hydraulic properties could not produce surface runoff from a topographic sequence although evidence of runoff was observed for the field This most likely resulted because rainfall intensity was averaged for the storm period.

Johnsen et al. [5] tested the drainage component of RZWQM for three drain spacings (7.5, 15, 30 m) at 0.9 depth and obtained a fairly good prediction of water-table fluctuation in a North Carolina study. However, they found that RZWQM slightly over predicted water-table depth. Singh and Kanwar [6],[7] evaluated RZWQM for tile drainage flow under four different tillage systems (moldboard plow, chisel plow, no-tillage, and ridge-tillage). They found the model predicted timing and amount of drainage within one standard error when calibrated from 1990 data (Fig. 2). The model correctly predicted tillage effects on subsurface drainage with maximum tile-flow under no-tillage and minimum tile-flow under moldboard plow. They also identified the important contribution of macropore flow to tile drains under no-tillage systems.

Bromide (Br) has been used as field tracer for water movement because it is exogenous to the soil environment. Ahuja et al. [14] obtained good predictions of Br in the soil profile except for 14 and 34 days after Br application, which was explained by possible errors in ET estimation. Ahuja et al. [3] found that RZWQM simulated better Br distributions in the soil profile in a column study than in runoff or macropore flow (Fig. 1). Poor understanding of the role of soil aggregates on partitioning surface applied chemicals into macropores may be responsible for deviations in Br simulation in runoff.

2.2 Nitrate in soil and water
RZWQM was used to predict nitrate fates without special calibration procedure. In a manure study in eastern Colorado, Ma et al. [12] found the model correctly predicted nitrate concentration in the soil profile for the manured and non-manured plots ($r^2 = 0.83$ and 0.86, respectively). In a later study, Ma et al. [13] were able to predict the response

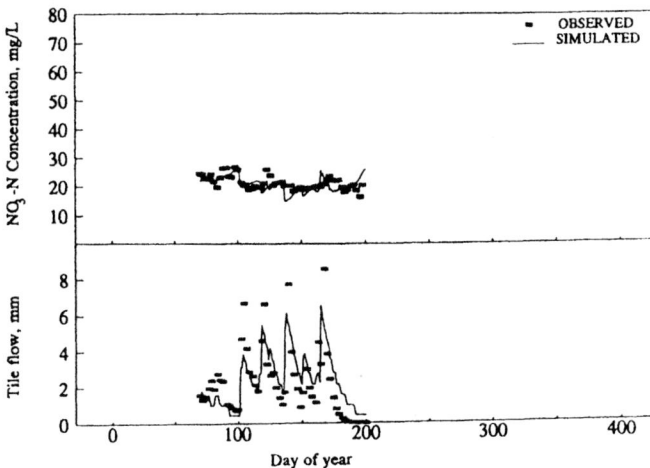

Fig. 2. Simulated and observed nitrate-N concentrations in tile flow for ridge-tillage [7]

of soil nitrate to manure applications in a tall fescue field in Arkansas. However, the prediction of nitrate concentrations in soil water samples taken from suction lysimeters was not successful. They attributed this failure to failing to characterizing the fragipan in the model. In Portugal, field nitrate movement under level-basin fertigated corn compared well with modeling results [17].

Ghidey et al. [20] tested RZWQM for nitrate in the soil profile, in soil solution, and in surface runoff. Calibrated soil nitrate was within 5% of measured values before planting and almost 50% lower after harvesting, which was attributed to poor initialization of the soil humus pools. Given the large variability of nitrate concentration in suction lysimeters, the model prediction of nitrate in soil solution was reasonable. However, nitrate in surface runoff was greatly over predicted for all rainfall events. Landa et al. [19] also attributed lower soil nitrate prediction on a soybean plot at the Ohio MSEA site to inadequate initialization of soil humus pools and to lower estimation of nitrogen fixation of soybean. At the Iowa MSEA site, RZWQM adequately predicted total profile nitrate and nitrate in the tile drains, but the model failed to reproduce nitrate distribution in the soil profile [21]. Singh and Kanwar [7] correctly predicted the tillage effects on nitrate concentrations in tile drainage water with higher concentrations under moldboard plow and chisel plow and lower concentrations under no-tillage and ridge-tillage (Fig. 2).

2.3 Pesticide transport

Several pesticides have been studied using the RZWQM, including atrazine, alachlor, metribuzin, prometryn, fenamiphos, and cyanazine. Pesticide adsorption can be instantaneous (equilibrium) or equilibrium-kinetic in nature, and degradation can be one-stage decomposition or two-stage decomposition. Ahuja et al. [14] applied RZWQM to cyanazine and metribuzin transport by estimating pesticide parameters from laboratory experiment. They found the equilibrium model underpredicted pesticide persistence and overpredicted pesticide movement. The equilibrium-kinetic model, on the other hand, improved model prediction. Better predictions of pesticide leaching and runoff with equilibrium-kinetic model were also obtained in a study by Ma et al. [15] (Fig. 3). In a field study in Georgia, Ma et al. [16] found a two-stage decomposition model was superior to the one-stage decomposition model in predicting atrazine fates in the soil. Predicted water runoff, atrazine runoff, atrazine persistence, and atrazine distribution were highly correlated to experimental observations (r^2 = 0.87, 0.92, 0.97, and 0.73, respectively). Similar results were obtained for atrazine and metribuzin at the Iowa MSEA site [21].

At the Missouri MSEA site, Ghidey et al. [20] found RZWQM underestimated atrazine and alachlor concentrations near the soil surface for samples taken two days and one week after application and overpredicted the amounts for samples thereafter. Predicted atrazine concentrations in seepage water were within the range of the measured values, given the large variability of experimental measurements. Atrazine and alachlor concentrations in surface runoff compared well with experimental values except for the first runoff event where they were under estimated by more than 50%. Poor predictions of pesticide concentration were explained by the high amount of runoff water simulated.

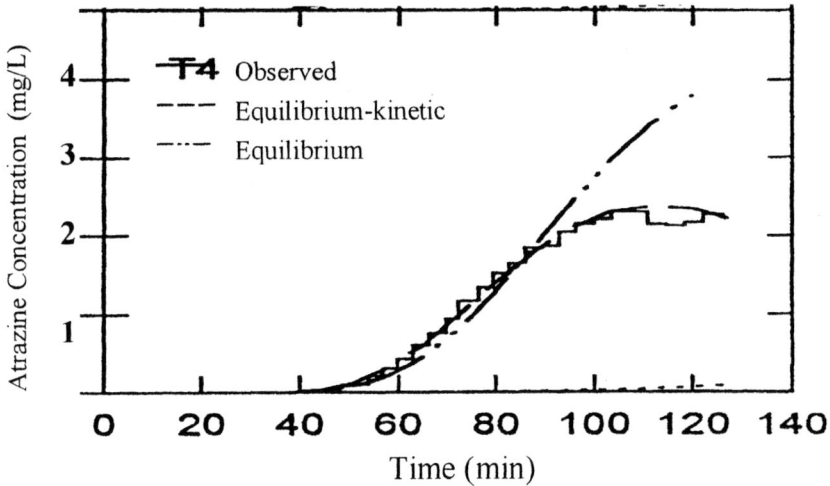

Fig. 3. Measured and predicted atrazine solution concentration in leaching water [16]

2.4 Evapotranspiration (ET) and crop production

Farahani and Bausch [9] and Farahani and Ahuja [8] tested the existing Shuttleworth and Wallace ET module [23] and enhanced it for surface residue effects. Figure 4 shows a comparison between S-W model predictions and measurements in a corn growing-season using the Bowen ratio energy balance (BREB) method. The regression slope is not significantly different from one (r^2 =0.84). Similar success in ET estimation was obtained in a corn-soybean cropping system [21], a silage corn field [12], and a tall fescue field [13]. Farahani et al. [22] later tested the RZWQM for corn production in eastern Colorado under dryland no-till systems, and found that the model under predicted corn yield and nitrogen uptake at toe slope positions and over predicted corn yield and nitrogen uptake at summit positions. The pattern of these prediction errors eluded to the difficulty of simulating runoff in the soil catenary. Other applications of RZWQM show the model provided better predictions of crop yield than leaf area index and leaf and stem biomass [10],[18],[19].

2.5 Varying agricultural management practices

One of the most distinguished features of agricultural systems models is their ability to predict once the model is calibrated. Buchleiter et al. [24] evaluated the effects of overirrigation by 40% on crop yield and nitrate leaching after the model was calibrated for a center-pivot irrigated-corn system. They observed an increase of 6.4 cm in water percolate and an increase of 110 kg N ha^{-1} of nitrate leaching with a reduction of 5% in corn biomass. In an irrigated corn study in eastern Colorado, Ma et al. [12] found that reducing water application to 50% significantly decreased water and nitrate leaching beyond the root zone without reducing corn yield. Ahuja et al. [4] studied macropore transport of nitrate and pesticides (atrazine and prometryn) under two different rainfall intensities (2.54 and 5.08 cm h^{-1}), two different initial soil moisture conditions (-33 and – 1500 kPa), and two different macropore sizes (0.1 and 0.0125 cm). They found high rainfall intensity increased macropore flow, especially for the initially wet soil. However, macropore size had little effect on macropore transport of water and chemicals, which was

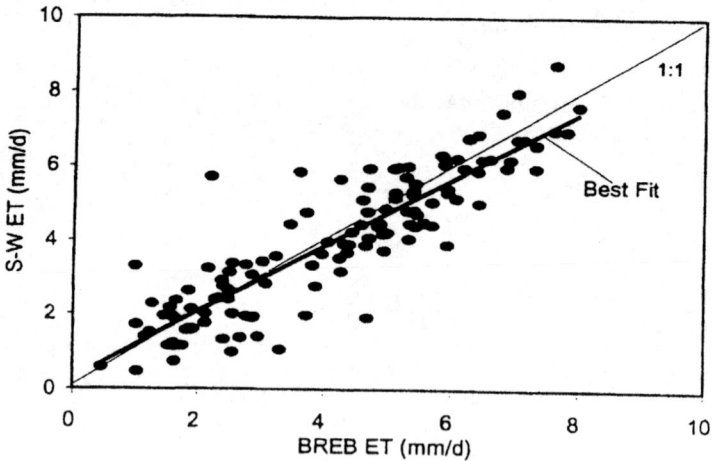

Fig. 4. BREB ET vs. S-W predictions for alternative furrow-irrigated maize [9]

favorable for the purpose of model application. The moderately adsorbed atrazine was more subject to macropore transport, followed by strongly adsorbed prometryn and nonadsorbed nitrate. Water seepage was higher under wet initial conditions than under dry initial conditions. Therefore, besides adjusting soil hydraulic conductivity, modifying initial soil water content can affect timing and amount of tile drain flow under field conditions [6],[7].

Sensitivity analysis was also performed by Ahuja et al. [4] to compare the role of evaporation and transpiration on chemical transport through macropores and the soil matrix. They found evaporation increased the amount of chemicals transported in macropores, but decreased their downward movement through the soil matrix. On the other hand, transpiration (water uptake by roots) decreased both the amounts of chemicals entering the macropores and the movement in the soil matrix (Fig. 5). Therefore, cropping should reduce transport of environmental pollutants into groundwater.

3 Conclusion and perspectives

RZWQM has been tested with various degrees of success at several locations around the world. Estimating model parameters was the most difficult issue concerning the use of RZWQM. This is, however, an unavoidable fact regarding the state-of-the-science process-level modeling of a cropping system. Some of these parameters were available from measurements, but several of them were not. Many of the model input parameters vary across the field [18]; therefore, parameters should be treated as a distribution rather than a single value [13]. In addition, since some of the parameters are correlated, future efforts should emphasize the relationship between those parameters [24]. The inconsistency of macropore effects on chemical transport and surface runoff warrants further study on soil macroporosity, such as introducing a variable soil-cracking approach

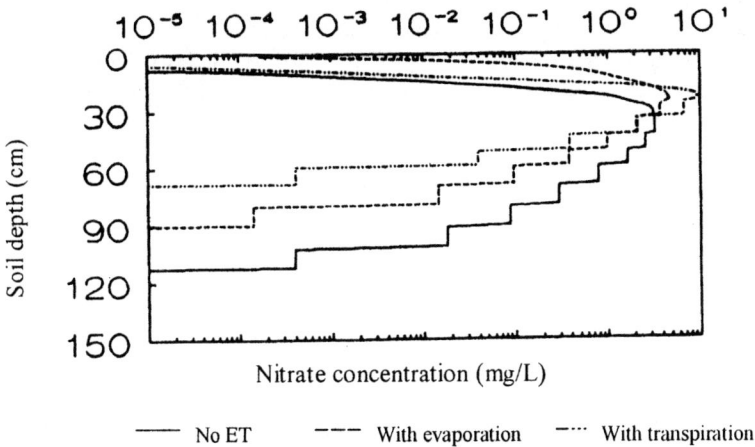

Fig. 5. Simulated nitrate concentrations with/without evaporation and transpiration [4]

[20],[21]. Although RZWQM documentation has provided some guidelines on how to apply the model, the calibration/validation procedure changes depending on data availability and indicator variables [12],[13]. More work is needed in standardizing model calibration and database development [11]. Furthermore, since the criteria for model calibration are not rigorous (5%-20% error), there may be several combinations of model parameters to obtain similar accuracy of an indicator variable [21]. Such a non-unique problem may be minimized when the model is tested for a wide range of conditions [22].

In summary, RZWQM has shown promise in integrating state-of-the-science knowledge into a potential tool for environment-water management in highly complicated and dynamic agricultural systems. On the other hand, water management is only practically useful and environmental safe if it is simulated in an open system such as the agricultural field. Not all physical, chemical and biological processes in RZWQM are fully understood and improvements are still forthcoming. Yet application to field research has provided confidence in RZWQM and helped us to better understand the processes and interactions within agricultural systems. Therefore, model application is a self-improving process and is contributing new knowledge through systems analysis. At this stage, RZWQM can be used to help experimental design, analysis and quantification of results, and technology transfer. These roles will be improved and enhanced through continuous interplay between modelers, experimentalists, and agricultural resources managers.

4 References

1. RZWQM Team. (1998) RZWQM: Simulating the effects of management on water quality and crop production. *Agricultural Systems*, (In press).
2. Ahuja, L. R., J. D. Hanson, K. W. Rojas, and M. J. Shaffer (eds). 1998. The Root Zone Water Quality Model. Water Resources Publications LLC. Highlands Ranch, CO.
3. Ahuja, L. R., Johnsen, K. E. and Heathman, G. C. (1995) Macropore transport of

a surface-applied bromide tracer: model evaluation and refinement. *Soil Science Society of America Journal*, Vol. 59, pp.1234-1241.

4.   Ahuja, L. R., DeCoursey, D. G., Barnes, B. B. and Rojas, K. W. (1993) Characteristics of macropore transport studied with the ARS root zone water quality model. *Transactions of the American Society of Agricultural Engineers*, Vol. 36, pp.369-380.

5.   Johnsen, K. E., Liu, H. H., Dane, J. H., Ahuja, L. R., and Workman, S. R. (1995) Simulation fluctuating water tables and tile drainage with a modified root zone water quality model and a new model WAFLOWM. *Transactions of the American Society of Agricultural Engineers*, Vol.38, pp.75-83.

6.   Singh, P. and Kanwar, R. S. (1995) Modification of RZWQM for simulating subsurface drainage by adding a tile flow component. *Transactions of the American Society of Agricultural Engineers*, Vol.38, pp.489-498.

7.   Singh, P. and Kanwar, R. S. (1995) Simulating NO3-N transport to subsurface drain flows as affected by tillage under continuous corn using modified RZWQM. *Transactions of the American Society of Agricultural Engineers*, Vol.38, pp.499-506.

8.   Farahani, H. J. and Ahuja, L. R. (1996) Evapotranspiration modeling of partial canopy/residue-covered fields. *Transactions of the American Society of Agricultural Engineers*, Vol.39, pp.2051-2064.

9.   Farahani, H. J. and Bausch, W. C. (1995) Performance of evapotranspiration models for maize - bare soil to closed canopy. *Transactions of the American Society of Agricultural Engineers*, Vol.38, pp.1049-1059.

10.   Nokes, S. E., Landa, F. M., and Hanson, J. D. (1996) Evaluation of the crop component of the root zone water quality model for corn in Ohio. *Transactions of the American Society of Agricultural Engineers*, Vol.39, pp.1177-1184.

11.   Hanson, J. D., Rojas, K. W., and Shaffer, M. J. (1998). Calibration and evaluation of the root zone water quality model. *Agronomy Journal*, (in press).

12.   Ma, L., Shaffer, M. J., Boyd, J. K., Waskom, R., Ahuja, L. R., Rojas, K. W., and Xu, C. (1998) Manure Management in an Irrigated Silage Corn Field: Experiment and Modeling *Soil Science Society of America Journal*, (In press).

13.   Ma, L., Scott, H. D., Shaffer, M. J., and Ahuja, L. R. (1998) RZWQM simulations of water and nitrate movement in a manured tall fescue field. *Soil Science*, (in press).

14.   Ahuja, L. R., Ma, Q. L., Rojas, K. W., Boesten, J. J. T. I. and Farahani, H. J. (1996) A field test of root zone water quality model – pesticide and bromide behavior. *Pesticide Science*, Vol.48, pp.101-108.

15.   Ma, Q. L., Ahuja, L. R., Wauchope, R. D., Benjamin, J. G., and Burgoa, B. (1996) Comparison of instantaneous equilibrium and equlibrium-kinetic sorption models for simulating simultaneous leaching and runoff of pesticides. *Soil Science*, Vol.161, pp.646-655.

16.   Ma, Q. L., Ahuja, L. R., Rojas, K. W., Ferreira, V. F., and DeCoursey, D. G. (1995) Measured and RZWQM predicted atrazine dissipation and movement in a field soil. *Transactions of the American Society of Agricultural Engineers*, Vol.38, pp.471-479.

17.   Cameira, M. R., Sousa, P. L. Farahani, H. J., Ahuja, L. R. and Pereira, L. S. (1998) Simulation of water and nitrate movement in level basin fertigated maize in Portugal. *Journal of Agricultural Engineering Research*, (in press)

18.   Martin, D. L. and Watts, D. G. (1998) Application of the root zone water quality model in central Nebraska. *Agronomy Journal*, (in press).

19.   Landa, F. M., Fausey, N. R., Nokes, S. E., and Hanson, J. D. (1998) Evaluation of the root zone water quality model (RZWQM3.2) at the Ohio MSEA. *Agronomy Journal*, (in press).

20.   Ghidey, F., Alberts, E. E., and Kitchen, N. R. (1998) Evaluation of RZWQM using field measured data from the Missouri MSEA. *Agronomy Journal*, (in press).

21.   Jaynes, D. B. and Miller, J. G. (1998) Evaluation of RZWQM using field measured data from Iowa MSEA. *Agronomy Journal*, (in press).

22.   Farahani, H. J., Buchleiter, G. W., Ahuja, L. R., and Sherrod, L. A. (1998) Model evaluation of dryland and irrigated cropping systems in Colorado. *Agronomy Journal*, (in press).

23.   Shuttleworth, W. J. and Wallace, J. S. (1985) Evaporation from sparse crops – An energy combination theory. *Quarterly Journal of The Royal Meteorological Society*, Vol. 111, pp.839-855.

24.   Buchleiter, G. W., Farahani, H. J., and Ahuja, L. R. (1995) Model evaluation of groundwater contamination under center pivot irrigated corn in eastern Colorado. *Proceedings of the International Symposium on Water Quality Modeling*. Orlando, Florida. pp 41-50.

ASSESSING ATRAZINE IN IRRIGATED SOIL PROFILES

A.S. AZEVEDO and R.S. KANWAR
Agricultural & Biosystems Engineering Dept, Iowa State University, Ames, USA
L.S. PEREIRA
Department of Agricultural Engineering, Institute of Agronomy, Technical University of Lisbon, Portugal

Abstract
A study conducted during the corn cropping season for a silty loam alluvial soil and a sandy eutric soil in Sorraia Valley, Portugal is reported. Soil, water, and atrazine concentrations through the soil profiles were observed. Result indicate that hydraulic properties of the alluvial soil do not favor atrazine transport, without evidence of heavy leaching by irrigation. For the coarse soils, where overirrigation is practiced irrigation water plays a role in leaching atrazine.
Keywords: Atrazine transport, herbicide, irrigated soils, leaching, soil water movement

1 Introduction

Groundwater pollution from non point sources is an important concern. The often excessive rates of fertilizers and pesticides applied to crops contribute to surface and subsurface water contamination, affecting drinking water supplies and the natural habitats. Herbicides represent the main class of pesticides for groundwater contamination. A review on pesticide occurrence in groundwater [1] shows that 32 herbicides have been reported to be present in groundwater. Triazines are the most frequently detected, the atrazine having the highest frequency in contamination of groundwater (38%).

In the SorraiaValley, irrigated agriculture plays a very important role in the region's economy. Near half of the cropped land is irrigated. Corn is the main irrigated crop in this region. A study in the area, concluded that the atrazine used in the corn crop affects the groundwater quality in the region [2]. Atrazine was detected in all sampled irrigation wells. For the same region [3] reported that atrazine's dynamic in the soil is

Water and the Environment: Innovative Issues in Irrigation and Drainage. Edited by Luis S. Pereira and John W. Gowing. Published in 1998 by E & FN Spon. ISBN 0 419 23710 0

influenced by irrigation.

Although many studies were conducted worldwide on atrazine movement through the soil profile [4] [5] [6] [7], atrazine's dynamic for Portuguese conditions is not well known. Only few studies were conducted to investigate atrazine movement in soils and in water [2] [3] [8].

The objectives of this study are to investigate the movement of atrazine through the soil profile, in two different soils, representatives in this region, under irrigated conditions. The results presented in this paper are preliminary, using data from 1996.

2 Materials and methods

Field data was collected at the António Teixeira's Experimental Station, located at the Sorraia Valley, near Coruche, Portugal, during the summer of 1996. The latitude and longitude at the experimental station are 38° 57' N, and 8° 32' W, respectively. Two plots were used, one located in a silty loam soil, of recent alluvial origin (Plot 5) and the other a sandy soil, of ancient alluvial formation (Plot 32). The silty loamy soil is mainly constituted by fine sand and silt, with near 20% clay and only 1% coarse sand. On the contrary the sandy soil has near 75% of coarse sand and only 3.5 % of clay. The hydrodynamic characterization of this soils is shown in [9], and [10]. Each plot was subdivided in 3 sub-plots of approximately 100 m². All soils were cropped with furrow irrigated corn. Atrazine was surface applied at a rate of 1.2 kg a.i./ha in Plot 5, and at a rate of 1.0 kg a.i./ha in Plot 32, 2 days before planting. The rainfall and irrigation events, and farming operations are presented in Figure 1.

Soil samples were taken at five different depths: 0-10, 10-20, 20-30, 30-45, and 45-60 cm depth. Three sets of soil samples were collected from the center of each sub-plot. Sampling was done with a soil probe using a zero contamination tube made of polyethylene terephthalate, glycol-modified (PETG) plastic. Soil cores were frozen promptly after collection. The soil samples relative to each depth and sub-plot were then combined giving five composite samples for the five depths for each area in each sub-plot. Each composite sample was wrapped in aluminum foil and at the time os extraction was homogenized and divided in three parts, one to determine the soil water content, and the other two for atrazine extraction.

In the Plot 5 soil samples were taken before and after every irrigation event. For Plot 32, due to the high frequency of irrigations, samples were taken every month. At the same time soil water measurements were made with a neutron probe. Access tubes were installed in the center of each sub-plot. Measurements were made at 20, 30, 45, 60, and 80 cm depths. A sample was taken at the soil surface to calculate the soil water content using the gravimetric method.

For atrazine extraction, approximately 40 g of soil were weighted and placed in a bottle, and 20 g of water and 20 ml of toluene were added to the soil samples. The bottles were covered and shacked during two and a half hours at a speed of 150 rpm. After this period they were again weighted and the toluene was decanted to a test tube. The samples were kept in the freezer until gas chromatography analysis were performed. Atrazine extraction was performed at Instituto Superior de Agronomia, Lisbon, and gas chromatography was done at the Water Quality Laboratory of the Agricultural and Biosystems Engineering Department, Iowa State University. Samples were analyzed using a gas cromatograph equipped with a N-P thermoionic detector. Column temperature was 170°C with helium carrier gas at a flow rate of 25 mL/min. Both inlet and detector temperatures were 250°C with hydrogen and air reaction gas flows of 2.5 mL/min and 100 mL/min, respectively.

3 Results and discussion

3.1 Alluvial soil
Figure 2 presents the soil water content profiles averaged for the 3 sub-plots from planting to third irrigation. After the second irrigation, the soil water content below 40 cm depth remains unchanged, the soil being wet below this layer. Observations in a similar soil in the area [11] indicate that macropore's contribution to water movement reduce by 70% at 30 cm depth. Data indicates that flow conditions do not favor solute transport below the root zone.

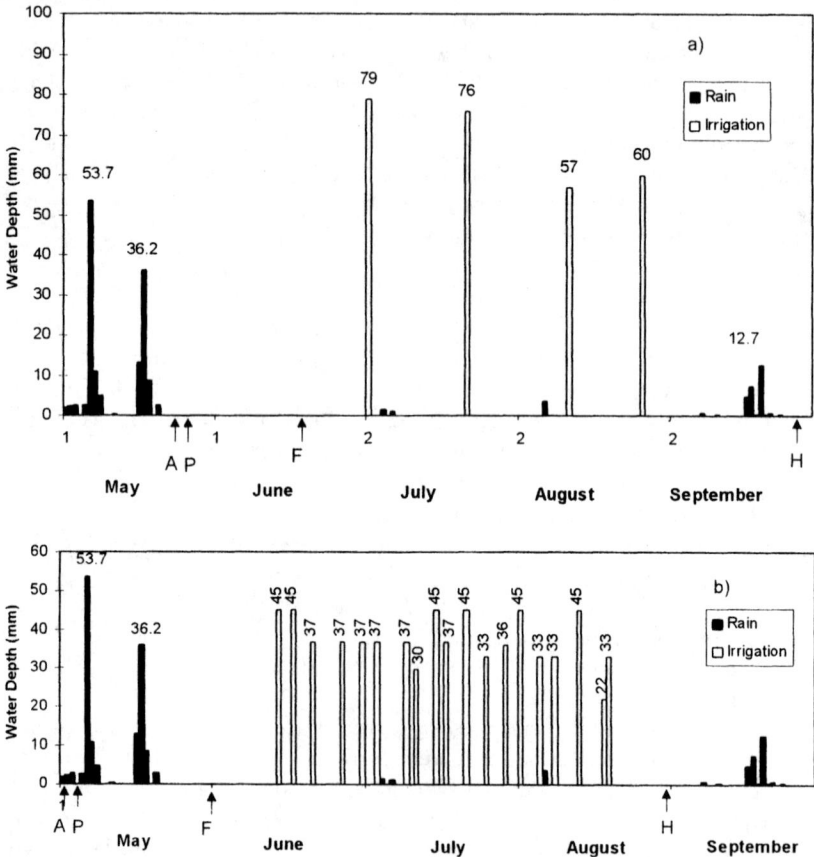

Fig. 1. Rainfall and irrigation events, and main farm operations (A- atrazine application; P-planting; F- furrow opening; H- harvesting) for a) alluvial soil in Plot 5; b) sandy soil in Plot 32.

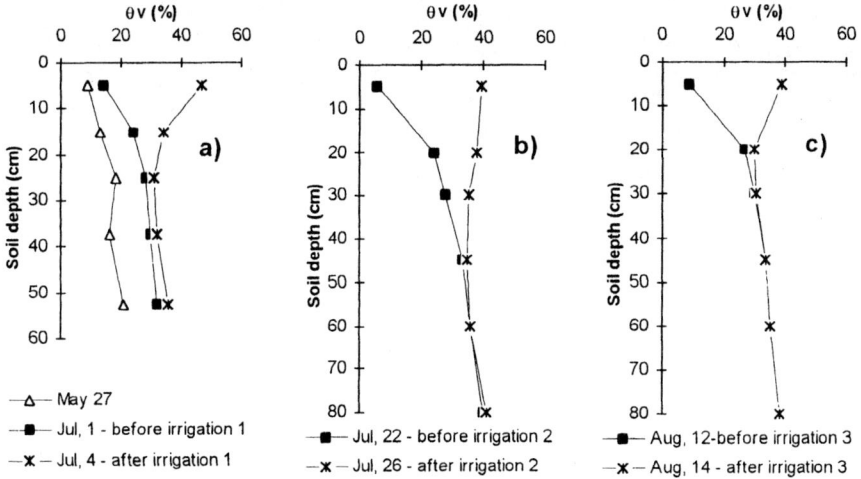

Fig. 2. Soil water content profiles for Plot 5 (alluvial soil): a) from planting to 1st irrigation; b) before and after the 2nd irrigation; c) before and after the third irrigation

Figure 3 presents the atrazine concentration profiles averaged for the 3 sub-plots, during the growing season of 1996. Table 4 shows the atrazine concentration averages and the respective standard deviations. From May 27 until the first irrigation (Fig. 3a), a significant reduction was verified on the atrazine concentrations in the top layers of the soil. Following the subsequent irrigations (Fig. 3b and 3c) part of the remaining atrazine moved deeper in the soil profile. However, during the entire period atrazine concentrations below 15-20 cm depth were always low (below 0.04 ppm). Rainfall did not play any role in atrazine movement. The variation of the atrazine mass in the soil during the crop season in given in Table 2.

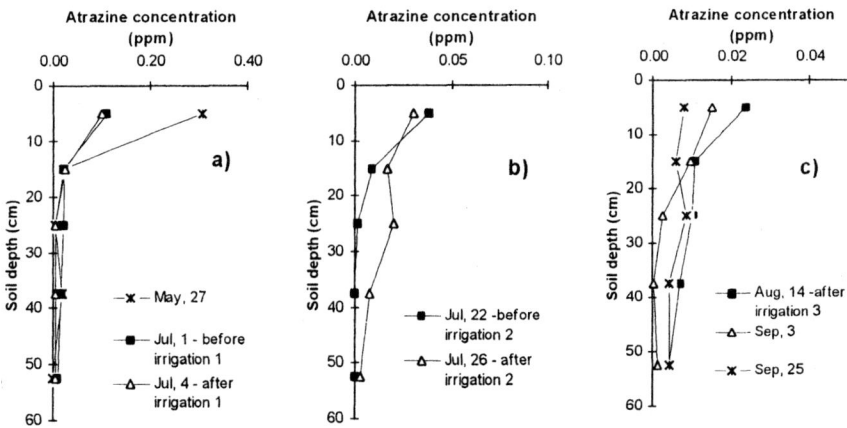

Fig. 3. Atrazine concentrations through the alluvial soil profile (Plot 5), during the corn crop season

Results show a great variation and high standard deviation (as for the concentrations in Table 1), which may be explained by sampling. Results show that the major portion of atrazine was degraded in the upper soil layer, for about the first month after application. Later degradation is slower and part of the atrazine remaining in the soil moves when irrigation water is applied. However, the amount of atrazine involved in this movement is relatively small.

Table 1. Atrazine concentrations (ppm) through the alluvial soil profile (Plot 5) during the corn growth season

Date	Depth (cm)	Average	SD	Date	Depth (cm)	Average	SD
May 27	0 - 10	0.3077	0.2399	Jul 26	0 - 10	0.0306	0.0043
	10 -20	0.0224	0.0126		10 -20	0.0173	0.0152
	20 - 30	0.0065	0.0084		20 - 30	0.0206	0.0228
	30 - 45	0.0210	0.0096		30 - 45	0.0078	0.0113
	45 - 60	0.0006	0.0011		45 - 60	0.0025	0.0025
Jul 1	0 - 10	0.1115	0.1521	Aug 14	0 - 10	0.0235	0.0116
	10 -20	0.0213	0.0176		10 - 20	0.0107	0.0105
	20 - 30	0.0234	0.0357		20 - 30	0.0102	0.0076
	30 - 45	0.0164	0.0074		30 - 45	0.0072	0.0078
	45 - 60	0.0099	0.0106		45 - 60	0.0045	0.0039
Jul 4	0 - 10	0.1021	0.0555	Sep 3	0 - 10	0.0151	0.0043
	10 -20	0.0261	0.0055		10 - 20	0.0099	0.0013
	20 - 30	0.0064	0.0110		20 - 30	0.0026	0.0024
	30 - 45	0.0042	0.0038		30 - 45	0.0005	0.0008
	45 - 60	0.0064	0.0111		45 - 60	0.0015	0.0025
Jul 22	0 - 10	0.0385	0.0216	Sep 25	0 - 10	0.0080	0.0043
	10 -20	0.0093	0.0082		10 - 20	0.0060	0.0060
	20 - 30	0.0011	0.0020		20 - 30	0.0080	0.0044
	30 - 45	0.0000	0.0000		30 - 45	0.0044	0.0076
	45 - 60	0.0000	0.0000		45 - 60	0.0043	0.0074

Average - average of observations from the 3 sub-plots; SD - standard deviation

Table 2. Variation of the atrazine mass (g/ha) for the entire profile (0-60 cm depth), for the alluvial soil (Plot 5), when 1200 g/ha were applied May 23

Plot 5 - Atrazine mass (g/ha)								
	May 27	Jul 1	Jul 4	Jul 22	Jul 26	Aug 14	Sep 3	Sep 25
Area 1	724.2	471.7	202.1	99.0	168.3	18.7	36.6	21.9
Area 2	128.9	170.0	299.0	67.2	74.5	123.7	31.2	17.6
Area 3	586.8	148.7	91.6	22.5	101.3	112.7	52.6	111.3
Avg	480.0	263.5	197.6	62.9	114.7	85.0	40.1	50.3
SD	311.7	180.7	103.8	38.4	48.3	57.7	11.1	52.9

Avg - average; SD - standard deviation

3.2 Sandy soil

When compared to the plot in the alluvial soil, the irrigation depths applied to this plot were smaller but frequency was much higher (cf. Fig. 1). Despite the water holding capacity of this soil being small overirrigation has been practiced. Figure 4a shows, for one irrigation event, that the soil water content only changes at the top 20 cm. Excess water quickly percolates out of the root zone.

Figure 4b presents the atrazine concentration profiles averaged for the 3 sub-plots. Table 3 presents those average atrazine concentrations and the respective standard deviations. A great variability among values is observed. Table 4 shows the atrazine mass for the entire soil profile. The low mass and concentration of atrazine at May 27, just 25 days after application, representing only 16% of that applied, may be explained by leaching due to the heavy rains occurring few days after planting (Fig. 1b). The very low values at harvesting may be explained by additional leaching due to the very frequent irrigations. Figure 4b illustrates that transport of atrazine occurs during the irrigation season.

Table 3. Atrazine concentrations (ppm) and respective standard deviations through the sandy soil profile (Plot 32) during the corn growth season

Date	Depth (cm)	Average	SD	Date	Depth (cm)	Average	SD
May 27	0 - 10	0.0555	0.0322	Jul 30	0 - 10	0.0035	0.0057
	10 -20	0.0142	0.0101		10 -20	0.0022	0.0012
	20 - 30	0.0056	0.0058		20 - 30	0.0232	0.0279
	30 - 45	0.0127	0.0037		30 - 45	0.0050	0.0066
	45 - 60	0.0054	0.0010		45 - 60	0.0235	0.0236
Jul 10	0 - 10	0.0176	0.0000	Aug 29	0 - 10	0.0047	0.0033
	10 -20	0.0139	0.0000		10 - 20	0.0025	0.0042
	20 - 30	0.0024	0.0013		20 - 30	0.0048	0.0054
	30 - 45	0.0060	0.0046		30 - 45	0.0011	0.0014
	45 - 60	0.0037	0.0045		45 - 60	0.0000	0.0001

Average - average of observations from the 3 sub-plots; SD - standard deviation

Table 4. Variation of the atrazine mass (g/ha) for the entire profile (0-60 cm depth) in the sandy soil (Plot 32) when 1000g/ha were applied May 2

Plot 32 - Atrazine mass (g/ha)				
	May 27	Jul 10	Jul 30	Aug 29
Area 1	209.1	ND	157.3	12.5
Area 2	84.4	65.6	89.9	25.2
Area 3	194.3	ND	88.7	27.0
Avg	162.6	-	112.0	21.6
SD	68.1	-	39.2	7.9

ND - no data available; Avg - average; SD - standard deviation

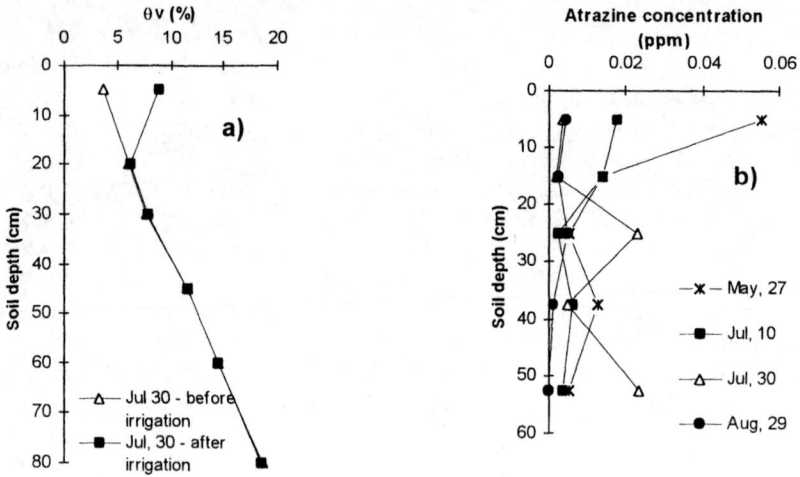

Fig. 4. Water and atrazine profiles in the sandy soil (Plot 32): a) typical soil water profile before and after the irrigation at July 30; b) atrazine concentrations from May27 until harvesting.

4 Conclusions

Observations performed for water and atrazine concentrations in a corn cropped silty loam alluvial soil do not show evidence that the transport of atrazine by the irrigation water is important. Transport occurs but to a limited extend only. Similar observations in a sandy soil indicate that the main decrease in atrazine occurred due to heavy rains few days after pesticide application. Results also indicate that overiirigation could have contributed to a relatively important leaching in this coarse soil. These preliminary conclusions have to be confirmed with new field observations and the use of an appropriated simulation model.

5 Acknowledgments

The authors thank the funding support by the Program PRAXIS XXI, Lisbon, and the "Centro de Estudos de Engenharia Rural", Lisbon.

References

1. Funari, E., Donati, L., Sandroni., D. and Vighi, M. 1995. Pesticide levels in groundwater: value and limitations of monitoring, in: *Pesticide risk in groundwater* (ed. M.Vighi and E. Funari). Lewis Publishers, Boca Raton, FL:3-43.

2. Batista, S.B., Cerejeira, M.J., Trancoso, A., Centeno, M.S.L., and Fernandes, A.M.S. 1998. Pesticidas e nitratos em águas subterrâneas na região do Ribatejo e Oeste em 1996, in 4º *Congresso da Água*, APRH, Lisboa.

3. Cerejeira, M.J. 1993. Estudo da destribuição e destino final dos pesticidas no ambiente numa abordagem integrada. Caso da atrazina na Zona Agrária da Chamusca. Ph.D. Dissertation. ISA. Universidade Técnica de Lisboa.

4. Azevedo, A.S., Kanwar, R.S., Singh, P., Ahuja, L., and Pereira, L.S. 1997. Simulating atrazine transport through soil profile using the RZWQM for an Iowa soil. *Journal of Environmental Quality* 26(1): 153-164.

5. Heatwole, C.D., Zacharias, S., Mostaghimi, S., and Dillaha, T.A.. 1997. Movement of field-applied atrazine, metolachlor, and bromide in a sandy loam soil. *Transactions of the ASAE* 40(5):1267-1276.

6. Kanwar, R.S. 1991. Preferential movement of nitrate and herbicides to shallow groundwater as affected by tillage and crop rotation, in Proceedings of National Symposium on Preferential Flow. (ed. T.J.Gish and A.Shirmohammadi), ASAE, St. Joseph, MI: 328-337.

7. Weed, D.A.J., Kanwar, R.S., Stoltenberg, D.E. and Pfeiffer, R.L. 1995. Dissipation and distribution of herbicides in the soil profile. *Journal of Environmental Quality* 24:68-79.

8. Rocha, M.F. 1989.Comportamento dos resíduos de atrazina em solos Portugueses. Ph.D. Dissertation. ISA. Universidade Técnica de Lisboa.

9. Tabuada, M.A., Rego, Z.J.C., Vachaud, G. and Pereira, L.S. 1995. Two-dimensional infiltration under furrow irrigation: modelling, its validation and applications. *Agricultural Water Management* 27: 105-123.

10. Gonçalves, M.C., Pereira, L.S. and Leij, F.J. 1997. Pedo-transfer functions for estimating unsaturated hydraulic properties of Portuguese soils. *European Journal of Soil Science* 48:387-400.

11. Cameira, M.R., Fernando, R.M. and Pereira, L.S. 1997. Modeling infiltration and redistribution in a plowed soil with a two domain flow model, in *Water Quality and Pollution Control* (Proceedings of International Conference), IAM, Bari: 287-306.

INFLUENCE OF SUB-SURFACE DRAINAGE ON HERBICIDE MANAGEMENT IN LOW-LAND AREAS

D.E. CRUCIANI, G.C. BAPTISTA, P.J. CHRISTOFFOLETI and K. MINAMI
Escola Superior de Agricultura "Luiz de Queiroz", University of São Paulo, Brazil

Abstract
The fate of soil herbicides and their impact on the environment plays an important role in the assessment of the agricultural efficiency of these chemicals. Soil drainage may influence movement, infiltration and mobility through the soil profile and, consequently their presence in the groundwater. Therefore, the purpose of this research was to ascertain the effect of a soil drainage system on the surface and lateral movement, leaching and groundwater contamination by the herbicide trifluralin sprayed on corn crop in pre-emergence conditions. The experiment was conducted from January to April of 1995 at the drainage experimental field of the Department of Rural Engineering, ESALQ/USP. Surface removal of herbicides by run-off in the well drained blocks was found to be much less compared to the check block (no drainage). Herbicide infiltration in the soil was found to be directly related to drainage as indicated by a retention four times higher in drainage blocks than in the no-drainage block. Herbicide residues in drainage water were insignificant considering that the water table was at an average depth of 0.60 cm below surface. Residuals found on the soil surface, in the 0-15 cm soil layer and in groundwater allowed for the conclusion that subsurface drainage reduces movement and leaching of the herbicide and thus the risks of environmental contamination.
Keywords: Drainage, environmental impacts, herbicide, leaching, low-land areas, sub-surface drainage.

1 Introduction

Herbicides used in the agriculture are mostly synthetic organic compounds sprayed on agricultural fields to control weeds. These compounds are applied according to certain

Water and the Environment: Innovative Issues in Irrigation and Drainage. Edited by Luis S. Pereira and John W. Gowing. Published in 1998 by E & FN Spon. ISBN 0 419 23710 0

regulations in order to cause minimum environmental impact to the agroecosistems. However, when the herbicide reaches the soil, the fate and mobility of these compound depend on several edaphic and environmental factors and on the herbicide characteristics.

Among the edaphic factors that influence the herbicide behavior in the soil after application, soil drainage is certainly of great importance. The herbicides atrazine and metolachlor sprayed in post-emergence conditions in corn [1]. The water table in the experimental area was very superficial, and the treatments were applied to the soil with superficial and deep drainage for comparison. The results after two year of research revealed that subsurface drainage system reduced the losses of both herbicides, mainly through superficial water, at 55% and 51% respectively. The herbicide losses were up to 75% after 30 days of application to the soil.

Several other papers report the relationship between pesticide movement in the soil, and the influence of good soil drainage, [2]; [3]; [4]; [5]; [6]; [7]; [8]; [9], [10] and [11]. However, there is little or no research developed under the Brazilian conditions.

The major objective of this research was to determine the effects of a drainage system installed in low-land area, on herbicide movement by superficial water (run off) and through soil profile (leaching), being sprayed in pre-emergence conditions of a corn crop. It was evaluated how the different drainage conditions affect the herbicide mobility and retention. The research was also set in order to get conclusions about the possible reduction of the herbicide mobility and environmental contamination, by decresing the incidence of herbicide in the run-off water and increasing its retention in the superficial layer of the soil. The other goal of the research was to analyze if the herbicide reaches the water table in low-land areas.

2 Material and methods

The experiment was conducted on experimental field from January to July of 1995, in a low-land area of the Rural Engineering Department of ESALQ/USP, Piracicaba-SP, Brazil. After soil preparation it was planted corn, variety BRASKALB-XL 380, in a density of 5 seeds/meter row and row spacing of 90 cm. The herbicide sprayed was trifluralin, in pre-emergence conditions, immediately after crop planting, using a commercial formulation, containing 600 g of a.i./L, without incorporation to the soil, since this formulation is recommended to be sprayed in pre-emergence. The application was made by a tractor mounted sprayer, using tips Teejet 11003, spaced of 0.5 m, resulting in a spray volume of 300 L/ha. The corn was seeded prior to herbicide application, 5 cm deep, in order to avoid any direct contact of the seeds to the herbicide sprayed.

The experimental area was prepared keeping six blocks with subsurface drainage tubes 10 m apart (blocks A and B), 20 m apart (blocks C and D) and 30 m apart (blocks E and F), besides a check plot with no drainage system, as shown in figure 1. The tubes discharged the drained water in a main channel outside of the experimental area. The drainage tubes were corrugated PVC of 100 mm in diameter, wrapped in a Bidim geotextile mantle.

Each blocks was divided into six plots and each plot correspond to the points of collection (S=surface and P=sub-surface). The plots were 20 m in extension, with three

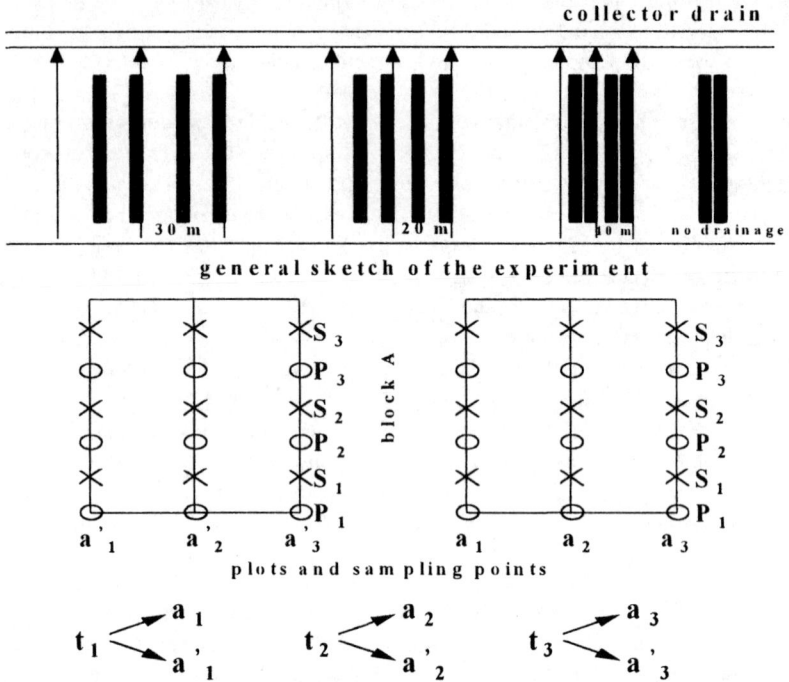

Fig. 1. Experimental system with collecting points in plots. S = soil samples from surface. P = soil sample in profile. a, a', time of sampling along the drain line and between drain lines, rispectively

drain tubes, so three soil samples were collected on the tube line and three right in the center between the tubes. These samples were made in three diferent time after herbicide application on the surface (S) and subsurface (P) (figure 1).

The soil samples P were collected from the soil layer of 0 to 15 cm, using small soil sampler, and additionally it was collected water from the drainage tubes. The sampling was made at t_1 = 4 days (01/22/95), t_2 = 15 days (02/01/95) and t_3 = 30 days (02/21/95) after herbicide treatment (DAT). The soil and water samples were processed in soil drainage laboratory of Rural Engineering Department, and kept in aluminum foil containers and glass jars of 200 ml, and at temperature of 2°C for later analysis in gas chromatograph of Entomolgy Department at ESALQ/USP.

The quantitative analysis was made by using gas chromatograph equipped with an electron capture detector (H^3) and cromatographic column 5% QF1, made of glass, and cromosorb W, AW-DMCS. The analytical method used showed a limit of quantification of 0.01 ppm, with average recovery of 87 +/- 4%.

In the southern part of Brazil it is very common the occurrence of heavy rain from December to march, since it is a tropical area in south hemisphere. In February, the amount of precipitation in 16 days was up to 415 mm, causing certain difficulties in the coduction of the experiment. However, it is important to emphasize that the occurrence of heavy rains indicated that even under critical conditions of soil water saturation, there is a clear evidence of beneficial effects of a drainage system in the environmental management.

3 Results and discussion

The initial trifluralin concentrations measured in the soil samples, four days (t_1) after herbicide application, are listed in table 1. Ten totally randomized samples were collected from the whole area. Subsequent soil samples, fifteen days (t_2) and thirteen days (t_3), respectively after herbicide spray, were collected according the sketch of figure 1. Results are shown in tables 2 and 3. Residual concentration of trifluralin as shown in these tables are average values of samples collected along a drain line and between two drain lines, as shown in figure 1.

These results clearly indicate that the herbicide concentration on soil surface is much greater on the check plot without drains. However, there is not a clear distinction among the results found in drains spaced 10, 20 and 30m, respectively. Similar results were observed at (t_2) and (t_3), as indicated in tables 2 and 3.

Table 1 - Trifluralin residuals on soil surface at (t_1), four days after spray

Samples	Residual in ppm
1	0.6
2	0.5
3	0.4
4	0.6
5	0.5
6	0.4
7	0.5
8	0.4
9	0.4
10	0.5

On the other hand, the same tables indicate the percentage variation of trifluralin concentration along the soil surface. Percentages are relative to the variation between the smallest and the greatest concentration found in the plots, comparing each drain spacing in the plots with the check plot.

Table 2 - Average values of trifluralin residual on soil surface, at (t2) 15 days after spray and % of surface dispersion along a 20m strip in each plot

Blocks	Drain spacing	Trifluralin residue (ppm)	Concentration variation
A+B	10 m	0.104	28%
C+D	20 m	0.281	95%
E+F	30 m	0.179	85%
Check	no drainage	0.533	117%

There is a sharp increase in that percentage as the drain spacing is increased. The greatest range was observed in the check plot without drains. Same feature was observed in both monitoring areas or soil strips, i.e., over the drain line and laterally, between two drain lines.

Table 3 - Average values of trifluralin residual on soil surface, at (t3) 30 days after spray and % of surface dispersion along a 20m strip in each plot

Blocks	Drain spacing	Trifluralin residue (ppm)	Concentration variation
A+B	10 m	0.051	14%
C+D	20 m	0.046	36%
E+F	30 m	0.032	46%
Check	no drainage	0.123	114%

Consequently, the worse the drain system efficiency is, the greater the herbicide movement overland and also the greater the potential risk to environment are. The question is that a pesticide should never leave the limits of the area of its application. At t_3 residual concentrations found on soil surface were always smaller than at t_2 in all drained blocks, except for the check plot where residual concentrations are almost constant and higher. It is evident that the potential risk of leaching along the surface of clay soil by rains is also higher. It can be said that subsurface drainage reduced surface movement of trifluralin with time.

Tables 4 and 5 show average concentration of the herbicide in the 0-15 cm soil layer, at t_2 and t_3. Both tables indicate that blocks A + B were able to retain more trifluralin than the remaining treatments. Consequently, surface movement was smaller in blocks A + B, with drains spaced at 10m.

At the same time a smaller overland movement of pesticide occurs, its retention within the top of soil profile is increased. This is also inversely proportional to drain spacing.

Table 4 - Trifluralin residuals within the top of soil profile at (t_2)

Blocks	Drain spacing	Trifluralin residue (ppm)
A+B	10 m	0.066
C+D	20 m	0.045
E+F	30 m	0.046
Check	no drainage	0.033

Table 5 - Trifluralin residuals within the top of soil profile at (t_3)

Blocks	Drain spacing	Trifluralin residue (ppm)
A+B	10 m	0.021
C+D	20 m	0.012
E+F	30 m	0.009
Check	no drainage	0.005

Table 6 shows trifluralin concentration measured in discharge water from drains. No significant residual was measured in all the blocks. This is important considering that the water table was at 60-80 cm below surface during the experiment.

Trifluralin has a very low solubility, consequently, a very low concentration should be expected in drainage water as it is easily and firmly retained in the colloidal system of soil. Provided that the water table is at safe depth.

Trifluralin is a very safe herbicide for soil spray with no potential risk to environment considering very few chances to find it in groundwater. These results agree with other authors, specially [1]. They demonstrated in a 2 year experiment, subsurface drainage

Table 6 - Trifluralin residuals in discharge water from drains.

Blocks	Drain spacing	Trifluralin concentration in the drained water (ppm)		
		$t_1 = 4$ days	$t_2 = 15$ days	$t_3 = 30$ days
A+B	10 m	0.000	0.004	0.000
C+D	20 m	0.000	0.000	0.001
E+F	30 m	0.000	0.001	0.001

benefits for herbicide management, even considering atrazine with very high solubility. Drainage sharply reduced herbicide losses beyond the limits of the application area.

4 Conclusions

a) Even considering a very low soluble herbicide as trifluralin, the methodology allowed to detect sharp differences among treatments.
b) Herbicide movement along the soil surface with time, was significantly smaller in plots with subsurface drainage than in check plot without drains.
c) Trifluralin infiltration and retention within the 0-15 horizon of soil profile was greater in plots with smaller drain spacing. Consequently, less movement of herbicide along soil surface.
d) Subsurface drainage benefit to pesticide management is significant, even in wet lands, because it minimizes the occurrence of surface runoff. As a consequence it reduces surface migration of herbicide and the risk to environment.

5 References

1. Bengston, R.L.; Southwick, L.M.; Willis, G.H.; Carter, C.E. (1990) The influence of subsurface drainage on herbicide losses. *Transactions of the ASAE.* 33(2), 415-518.
2. Kenimer, A.L.; Shanholtz, V.O. (1987) Effect of residue cover on pesticide losses from conventional and nontillage systems. *Transactions of the ASAE.* 30(4)953-959.
3. Helling, C.S.; Zhuary, W.; Gish,m T.J.; Coffman, M.D.; Woodward, M.D. (1988) Persistence and leaching of atrazine, alachlor, and cyanazine under no-tillage practices. *Chemosphere.* 17:175-187.
4. Jury, W.A.; Elabd, H.; Resketo, M. (1986) Field study of napropamide movement through unsaturated soil. *Water Res. Research.* 22(5),749-755.
5. Ritter, W.F. (1986) Pesticide contamination of groundwater. ASAE paper 86-2028. St. Joseph.
6. Southwick, L.M.; Willis, G.H.; Bengston, R.L.; Lormand, T.J. (1988) Atrazine in subsurface drain water in southern Louisiana. *Proc. on Planning Irrigation and Drainage.* ASCE. pp. 580-586.
7. Cheng, H.H.; Lehman, R.G. Characterization of herbicide degradation under field conditions. *Weed Sci.*, 33(2), 7-10, 1985.

8. Kenimer, A.L.; Mostaghimi, S.; Dillaha, T.A.; Shanholtz, V.O. (1989) Pesticide losses in erosion and runoff simulator. *Transactions of the ASAE.* 32(1) 127-136.

9. MC Smith, D.; Thomas, L.; Bottcher, A.B.; Campbell, K.L. (1990) Measurement of pesticide transport to shallow grounwater. *Transactions of the ASAE.* 33(5)1573-1582.

10. Gish, T.J.; Helling, C.S.; Mojasevic, M. (1991)   Preferential movement of atrazine and cyanazine under field conditions. *Transactions of the ASAE.* 34(4),1699-1705.

11. Ritter, W.F.; Scarborough, R.W.; Chirnside, E.M. (1994) Herbicide leaching in coastal plain soil. *J. Irrig. and Drain. Engineering,* 120(3) 634-649.

EFFECTS OF THE TRADITIONAL AND NO-TILLAGE SYSTEMS UPON RUNOFF AND HERBICIDE TRANSPORT

F.J. B. TEIXEIRA and G. BASCH
Department of Crop Science, University of Évora, Évora, Portugal

Abstract
Three field trials were carried out in order to assess the effects of the traditional and no-tillage system upon runoff and herbicide transport. Two were conducted in a wheat crop under rainfed conditions, during the winters of 1993/94 and 1994/95, at two different sites on a *Haplic Luvisol*, and the herbicide used was *isoproturon*. The third trial, corn irrigated with a *center pivot*, was carried out in the summer of 1994, on a *Gleyic Luvisol*, and the herbicides under study were *atrazine* and *metolachlor*. In all trials, the total runoff volume and sediment yield was higher under traditional tillage. The total amount of herbicides found in the runoff water was higher under the traditional tillage treatment, although the concentration of the herbicides in the runoff water from the first rainfall or irrigation/runoff event had been higher under the no-tillage treatment. The concentration of all the herbicides in the runoff water was considerably higher in the first and second rainfall or irrigation/runoff event, decreasing rapidly in subsequent events. The results show that care should be taken when adopting the no-tillage system since the residues and vegetation at the field surface may intercept much of the applied pesticides, where they are bound weakly and may be washed off easily.
Keywords:, Herbicides, no-tillage, runoff, sediments, tillage.

1 Introduction

Runoff water from farmland can carry both eroded sediments and agrochemicals. Conventional tillage (CT), that usually comprises the inversion of the top soil layer, and no-tillage (NT), when applied continuously, lead to marked differences in the physical, chemical and biological soil properties, which may affect the soil's hydrological behaviour (e.g. [1]). Although there are many studies showing that NT

Water and the Environment: Innovative Issues in Irrigation and Drainage. Edited by Luis S. Pereira and John W. Gowing. Published in 1998 by E & FN Spon. ISBN 0 419 23710 0

can be effective in reducing total runoff and the sediment load, this tendency does not seem to be universal, mainly in terms of total runoff volume, when infiltration rate is low due to poor internal drainage or compaction (e.g. [2]). Less understood is the interaction between tillage systems and the loss of pesticides in the runoff. Several studies show that pesticide losses in runoff are usually reduced when no-tillage is applied (e.g. [3]). Nevertheless, sometimes the reduction of total runoff volume goes along with an increase in the pesticide concentration that may equal or even surpass the total amount of pesticide losses of conventionally tilled plots (e.g. [4]).

The objectives that guided this work were the assessment of the effects of the conventional and no-tillage systems upon runoff volume, eroded sediment and herbicide transport in runoff, under rainfed and center pivot irrigated conditions, in order to contribute to a better understanding of the environmental impact of these tillage systems.

2 Materials and Methods

Three trials were conducted at 2 different locations. The trials under rainfed conditions with winter wheat, took place at "Herdade do Louseiro", Évora, in the winters of 1993/94 and 1994/95, in two different areas. The center pivot irrigated trial took place at "Herdade da Parreira", Ciborro, about 60 km Northwest from Évora, in the outmost area of the irrigated circle, during the summer of 1994, and corn as crop. Soil characteristics for the 3 sites are summarised in table 1.

Table 1. Soil Characteristics at Louseiro and Ciborro.

	Horizon	Louseiro 1993/94	Louseiro 1993/94	Ciborro
Depth (cm)	A	0-30	0-30	0-30
Clay (%)	A	8	4.3	4.3
Silt (%)	A	14.3	7.3	7.3
Sand (%)	A	77.7	88.4	88.4
Organic C (%)	A	0.60	0.56	0.63
Sat. Cond. (mm/h)	A	2.24	7.23	8.29

Each trial consisted in 6 plots, 2 tillage treatments – no-tillage and conventional tillage – with 3 replications per treatment. The layouts were a randomised block design. Plot features are summarised in table 2. At Louseiro, the CT plots were ploughed and harrowed to a depth of 0.25 m. At Parreira, CT consisted of disking twice at a depth of 0.15 m. Wheat sowing at Louseiro was done using the *Amazone combi-drill* sower for the CT treatment and a *Moore direct drill* seeder for the NT plots. At Parreira, for both tillage treatments, corn was sown with a *Gaspardo* pneumatic precision sower, equipped with a 7 cm large rotary harrow as furrow opener in the seeding line, and a distance of 0.75 m between lines. The plots were delimited by small earth dams, except for the lower edge where the runoff collecting device was installed (figure 1). Differences in the trials' design made inevitable the use of different strategies to measure runoff. For the 1993/94 winter trial, because of the size of the plots (100 m^2), barrels were connected in a way that only 10% of the runoff from the first barrel would be received by the second. For the second winter trial, the smaller plot size allowed the total runoff to be collected in the two barrels. At Parreira, irrigation obeyed to a schedule that allowed the use of runoff meters, attached to the barrels, making possible to record the runoff rate at different times of the runoff event.

Table 2. Plot features at the different locations.

Site	Crop	Plot size	Slope	Water	Reservoirs		
					Type	Amount	Sampling
Louseiro 1993/4	wheat	100 m² (20*5)	5.8 %	Rainfall	Barrel	Partial	after event
Louseiro 1994/5	wheat	15 m² (15*1)	5.0 %	Rainfall	Barrel	Total	after event
Ciborro	corn	60 m² (15*4)	4.7 %	Irrigation	Barrel	Fluxmeter outlet	after event

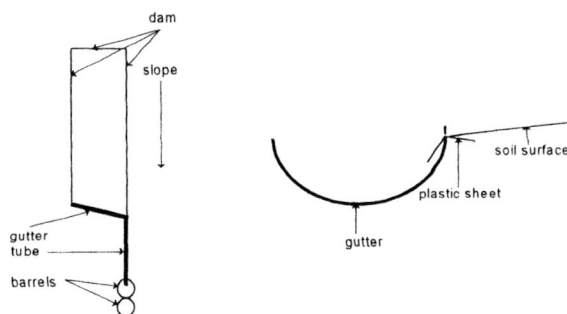

Fig. 1. Scheme of runoff plots and water collection devices.

For the three trials, herbicides were applied using a motorised backpack sprayer with a bar width of 3 m and a working pressure of 3 bar. The herbicides used were commercial formulations of *isoproturon*, applied in post-emergence in the wheat trials, at rates of 1.6 and 1.3 kg ha^{-1}, respectively for the 1993/94 and 1994/95 trials, and a mixture of *atrazine* and *metolachlor*, in pre-emergence in the corn crop, at a rate of 1.2 and 1.8 kg ha^{-1} respectively.

Two water-sampling techniques were used due to the different nature of the precipitation event -rain or irrigation water. At Louseiro, after a rainfall/runoff event, the water and sediments in the reservoirs were stirred thoroughly in order to obtain a complete sediment suspension, and two 1.5 l samples were then collected, one from each reservoir. Runoff water never stood more than 24 h in the reservoirs, and usually was collected within 12 h from the event. At Parreira, where the irrigated trial took place, 2 samples were usually collected from the connection pipe, the first one at approximately 1/3 of the expected runoff time and the second one soon after the peak discharge were reached. At Louseiro all the runoff events after the herbicide application were recorded and analysed for herbicide residues; in the first trial there were 6 rainfall/runoff events, and 2 in the second, analysed for herbicide residues. From the 28 irrigation/runoff events recorded at Parreira, only 10 were analysed for herbicide residues. Sediment mass load was determined by passing the runoff sample through a paper filter (Whatman). As the sediments were not analysed for herbicide residues (due to the texture of the topsoil) the filtered sediments were oven dried at

105 °C for 4 h and then weighed. 1.1 l of filtered water was stored at 4 °C for the following herbicide residue extraction.

The water reserved for the analyse of herbicide residues was first filtered through a 0.45 μm pore size filter, and then 20 ml of *methanol* and a known quantity of *metoxuron* (for isoproturon and *monomethyl-isoproturon*) or *terbuthylazine* (for *atrazine* and *metolachlor*) were added to 1000 ml of the water sample in order to obtain a concentration of 1 ppb of internal standard. Extractions were done by the solid phase extraction with C_{18} cartridges. Elution of the herbicides from the SPE cartridges was done using methanol (after elution of unwanted components with water), and collected in vials for high performance liquid chromatography (HPLC).

Herbicide concentrations were determined by HPLC. Technical features can be found in [5].

A full description of the sites, materials and methods can be found in [6].

Variance analysis was performed with the MSTAT program.

3 Results and discussion

3.1 Runoff water and sediments

For the 1993/94 winter trial, tillage had little effect on surface runoff volume and sediment transport (table 3). Nevertheless, for the set of rainfall/runoff events, CT generated 1.4 times more sediment.

The second winter trial, at a different spot, brought significant differences between tillage treatments (table 3). NT produced less runoff volume and a much lower sediment load than CT. For the 6 rainfall/runoff events recorded, total sediment yield and runoff volume was respectively 4 and 1.76 times higher under CT.

Table 3. Precipitation features; runoff and sediment load under NT and CT at Louseiro, winter of 1993/94 and 1994/5 (after herbicide application).

		Louseiro 1993/94						Louseiro 1994/95	
Days after herb. application		2	6	30	34	43	55	9	12
Precipitation (mm)		37.5	16	16	20.5	24	34	15	21
Runoff (mm)	NT	7.4	5.6	1	2.7	6.3	9.3	0.4	0.8
	CT	7.1	7.2	0.9	2.4	6.9	10.5	0.8	1.4
% of precipitation	NT	19.8	35	6	13.2	26.1	27.5	2.8	3.9
	CT	18.8	45.1	5.7	11.7	28.7	31.1	5.5	6.7
Eroded material ($g*m^{-2}$)	NT	21.8	13.6	11.1	15	17.1	23.7	0.2	0.5
	CT	21.5	17.1	29.1	23.9	26.3	25.8	1	1.3

The differences of tillage effect on runoff volume found between the two years are attributed to different hydraulic conductivity of the top soil layer (table 1), which was 3 times higher at the site of the 2nd trial. These findings are in agreement with those obtained by other authors e.g. [2], pointing out that NT is especially effective in reducing runoff, when compared to CT systems, in soils with a higher permeability.

At Parreira, in the summer of 1994, the first 3 irrigation events, approximately 8 mm each, did not produce runoff with both treatments. Except for the first irrigation/runoff event, CT always produced a higher runoff volume and sediment load. However, differences between tillage systems were only significant in terms of eroded material. On average, runoff volume and eroded material were, respectively, 1.3 and 4.7 times higher under CT. Assuming that for such small plots the only erosion processes involved are water drop detachment and redetachment, and that runoff acts only through the passive transport of detached sediment, the differences found between tillage treatment are most probably due to different soil cover levels (2.35 and 0.77 tons ha^{-1}, for NT and CT, respectively).

Table 4. Irrigation features; runoff and sediment load under NT and CT at Ciborro, summer of 1994. (These data corresponds to those events that were analysed for herbicide residues. Further data can be found in [7]).

Days after application		3	4	10	17	24	30	36	43	50	64
Irrigation (mm)		9.5	7	8	6.6	6.5	16.7	6.4	18.7	8	18.4
Intensity (mm/h)		57.3	55.7	64.8	48.4	47.2	59	47.3	64.5	65.5	68.1
Runoff (mm)	NT	0.29	0.23	0.74	0.6	0.68	3.01	0.95	3.66	0.84	3.74
	CT	0.23	0.26	0.86	0.89	0.95	3.87	1.33	4.87	1.46	4.68
% of irrigation	NT	3	3.3	10	9.7	10.1	19.2	14.6	19.1	9.9	20.8
	CT	2.4	3.8	10.1	12.8	15	22	21.2	26.8	19.4	24.8
Eroded material (g·m^{-2})	NT	1	1.1	3.1	2.1	1.5	8.2	1.8	5.2	1	4.1
	CT	0.8	1.8	7.2	8.3	6.8	35.7	8.3	39.5	12	31.6

Other parameters such as the soil water content of the top layer and the runoff peak discharge, along with relationships found between precipitation/irrigation and runoff and sediment yields are discussed in detail by [7].

3.2 *Isoproturon* and *Monomethyl-Isoproturon* losses in runoff water

For the 1993/94 winter trial, *isoproturon* transport in solution in the runoff water showed no significant differences between tillage treatments, both in terms of total herbicide transport and herbicide concentration (figure 2). Overall, the 6 rainfall/runoff events transported 1.37 and 1.93% of the amount of the herbicide theoretically applied, respectively with NT and CT. In terms of *isoproturon* concentration in the runoff water, for the 1st runoff event concentration was 1.47 times higher under NT, and 6.25 times lower under this treatment for the second event. This results seems to be related to a much higher interception of the herbicide by the residues and vegetation at the surface of the NT plots, being washed off easily by the runoff producing rain. Whereas this explanation can be applied to the results of the first event, the complete inversion of the situation for the second one is probably due to a higher herbicide sorption at the soil surface under NT. *Isoproturon* distribution through the soil profile supports these assumptions [6].

Monomethyl-isoproturon, a main metabolite of *isoproturon*, in the runoff water, did not differ significantly between tillage treatments either in terms of total mass or concentration (figure 2). On the whole, the metabolite transported was 40 and 50 times less when compared with for the original molecule, respectively with NT and CT. In terms of concentration, the metabolite showed the same trend than the original molecule. The hypothesis that a higher microbial degradation in the CT plots might

32   *TEIXEIRA and BASCH*

have occurred between the 1st and the 2nd runoff event is not supported by the results of a laboratory trial on the dissipation of *isoproturon* using the same soil [8].

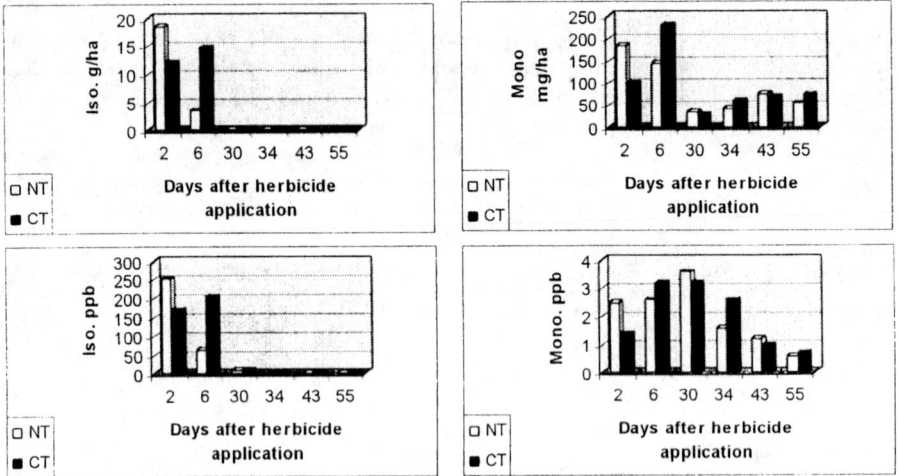

Fig. 2. *Isoproturon* and *monomethyl-isoproturon* concentration and total load in runoff water, at Louseiro, winter 1993/94.

For the second year, only two runoff events were recorded after the herbicide application. *Isoproturon* recovered in the runoff water (figure 3) did not differ significantly, in terms of total amount or in terms of concentration, with the tillage treatment.

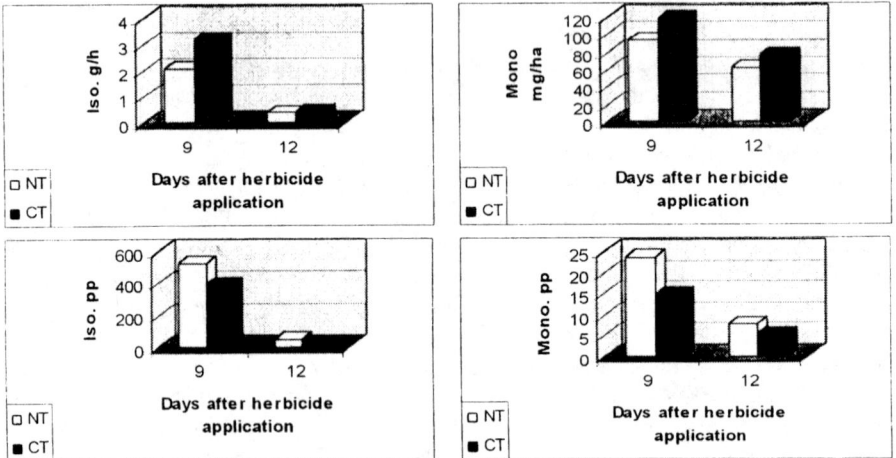

Fig. 3. *Isoproturon* and *monomethyl-iso* concentration and total load in runoff water, at Louseiro, winter 1994/95.

The total amount transported by the two events was 0.19 and 0.32% of the theoretically applied under NT and CT respectively. The results in terms of the herbicide concentration in the runoff water are likely to be related to a higher interception by the residues and vegetation at the surface of NT plots.

The presence of the metabolite *Monomethyl-isoproturon* did not show any significant difference between the tillage treatments (figure 3). Total losses were 16 and 21 times less than those observed for the original molecule under NT and CT respectively. The concentration of the metabolite in the runoff water was 1.45 and 1.95 times higher under NT for the 1st and 2nd event, respectively.

3.3 *Atrazine* and *metolachlor* losses in runoff water
In the irrigated trial in the summer of 1994 the presence of *atrazine* and *metolachlor* in the runoff water did not differ significantly between tillage treatments, both in terms of concentration or transported mass (figure 4).

For the 10 irrigation/runoff events analysed, 1.43 times more *atrazine* and 1.58 times more *metolachlor* were transported in solution in runoff water with the CT treatment. Though the concentration of both herbicides decreased consistently with time, the total amount of herbicide displaced followed a different pattern due to the increase in total water runoff (table 4).

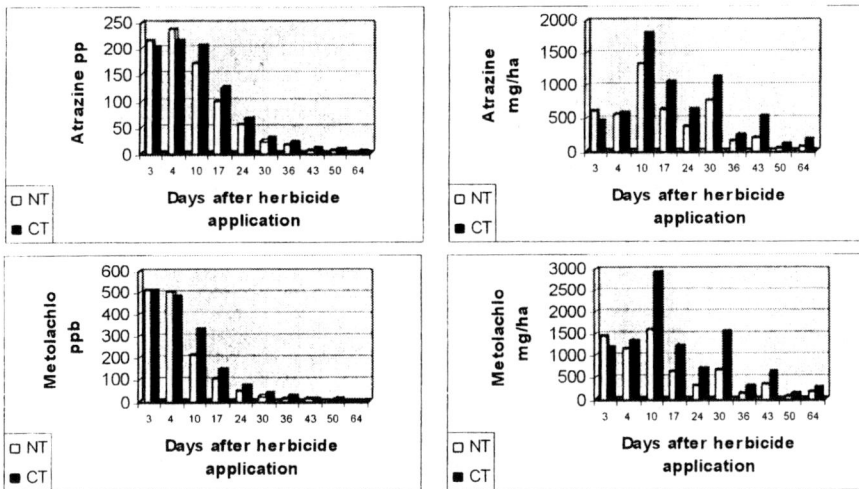

Fig. 4. *Atrazine* and *metolachlor* concentration and total load in runoff water, at Ciborro, summer 1994.

4 Conclusion

These results clearly show that NT is effective in reducing both the runoff volume and the sediment load, when compared to CT, though the soils where the trials took place were tilled less than 1 year before the trials were set up. This is a relevant fact if we consider that the soil physical, chemical and biological properties under NT improve considerably only if the soil is kept under this tillage system for longer periods.

Herbicide concentrations in the runoff water were higher for the first or second runoff event, decreasing rapidly for later events. NT showed a trend for a higher concentration of the herbicides for the first event of each trial, which is attributed to a higher interception of the herbicides by the plant residues and vegetation under NT, where they remain weakly bound, and thus, available for the wash-off and transport by water. Total amounts of herbicides transported in the runoff water were, for the 3 trials, always lower for the NT treatment.

Summarising, it can be concluded that NT, in comparison to CT, can increase pesticide concentration in the runoff water under certain circumstances. Those are given when the first rainfall event after application is of high intensity and the amount of pesticide intercepted by the vegetation or plant residues (normally higher under NT) is not infiltrating into the soil but washed-off. However, the possible higher pesticide concentration in the runoff water, is often compensated, regarding total pesticide transport, by a reduced runoff under NT, once the infiltration rate under this tillage treatment tends to be higher. Therefore it seems necessary to have in mind the probabilities of heavy rainfall after application and to determine the infiltration characteristics of NT and CT of a given soil in order to evaluate the potential of NT to reduce pesticide transport by runoff.

6 References

1. Carvalho, M. and Basch, G. (1995) Long term effects of two different soil tillage treatments on a vertisol in Alentejo region of Portugal. In: *Proceedings of the EC-Workshop - II - on no-tillage crop production in the West-European Countries*, 2, Silsoe, 15-17 May, 1995, ed. F. Tebrugge, A. Böhrnsen, Wissenschaftlicher Fachverlag.

2. Logan, T.J.; Eckert, D.J.; Beak,D.G.(1994) Tillage, crop and climatic effects on runoff and tile drainage losses of nitrate and four herbicides. *Soil & Tillage Research* Vol. 30, pp.75-103.

3. Hall, J.K.; Mumma, R.O.; Watts,D.W.(1991) Leaching and runoff losses of herbicides in a tilled and untilled field. In *Agriculture, Ecosystems and Environment* Vol. 37, pp.303-314.

4. Isensee, A.R., and Sadeghi, A.M.(1993) Impact of tillage practice on runoff and pesticide transport. *Journal of Soil and Water Conservation* Vol. 48, pp.523-527.

5. Quiao, X. (1992) Das Verhalten von Atrazin und Isoproturon in ausgewählten Böden aus Mittelhessen: Abnahme, Abbau und Verlagerung. PhD Thesis, University Giessen

6. Teixeira, F.J.B. (1995) Efeito de diferentes sistemas de mobilização do solo sobre a movimentação de herbicidas no solo e no escorrimento superficial. Universidade de Évora, Trabalho de Fim de Curso, Évora.

7. Basch, G. and Carvalho, M. (1995) Effect of soil tillage on runoff and erosion under dryland and irrigated conditions on Mediterranean soils. In: *Proceedings of the Conference on Erosion and Land Degradation in the Mediterranean*, Aveiro, June 1995.

8. Basch, G.; Düring, R.A.; Teixeira, F.J.B. (1998) Degradation of isoproturon as affected by soil tillage under controlled conditions. *Agriculture, Ecosystems and Environment* (to be submitted).

FIRST RESULTS ON CONTROLLED DRAINAGE TO REDUCE NITRATE LOSSES FROM AGRICULTURAL SITES

M. BORIN, G. BONAITI and L. GIARDINI
Dept. of Environment Agronomy and Vegetable Production, Università Bò, Padova, Italy

Abstract
To reduce pollutant loads from agricultural lands, especially NO_3-N, water table management practices such as controlled drainage and subirrigation, and constructed wetlands have recently been considered. An experimental facility was set up in 1996 on the University of Padova Agricultural Experimental Station's Legnaro farm near Padova, Italy, to make an original contribution to knowledge on operation and effectiveness of controlled drainage and wetlands for environmental purposes in North-East Italy. A hypothetical agricultural basin was simulated over a 6-hectare area. It was subdivided into 12 0.3-0.5 ha cultivated plots and one 0.65 ha wetland through which plots drainage water passes for a purification before flowing into an outer stream. Treatments include a combination of two drainage systems (subsurface pipe drainage and surface drainage) with two water table management systems (conventional drainage and controlled drainage-subirrigation) in a random-blocks experimental scheme with three replicates. The management of water table according to controlled drainage schemes allows to reduce the NO_3-N losses from agricultural fields compared to the absence of drainage control; the reduction of losses is related both to lower volumes of discharge and NO_3-N concentration.
Keywords: Controlled drainage, nitrate losses, subsurface drainage, surface drainage.

1 Introduction

Agricultural cropland has been pointed out as one of the non point sources of nitrogen and phosphorus contributing to environmental nutrient enrichment, and drainage is one of the agronomic practices that may contribute [1] [2]. Drainage has been linked to

Water and the Environment: Innovative Issues in Irrigation and Drainage. Edited by Luis S. Pereira and John W. Gowing. Published in 1998 by E & FN Spon. ISBN 0 419 23710 0

increased runoff, percolation, and sediment and pollutants losses [3] [4]. The quantity and quality of losses from drainage systems are related to the hydraulic layout considered. Subsurface drainage yields larger drainage volumes and greater losses of nitrates and soluble salts compared to surface drainage systems which are more connected with phosphorus and organic nitrogen losses. In the former, percolation is prevailing whereas in the latter run off is the main form of water flow [5] [6].

The effects of drainage on nutrients losses can be reduced with design and management suitable to specific site conditions and with the introduction of water table management practices such as controlled drainage and subirrigation. These practices, fairly new in Europe, were present in the U.S. since the begin of this century and have developed greatly in the last 20 years because of their environmental benefits [6]. In controlled drainage suitable devices allow flow of only the superficial water which may damage crops or limit machinery traffic. If subirrigation also occurs, water table depth is kept at a constant level throughout the year with the exception of periods when machinery access is needed. In this case, the water table is allowed to fall until field work is done.

Controlled drainage reduce nitrate losses primarily because of the reduction in outflow volume and increased denitrification in the soil. The latter process is promoted by the shallow water table which produces anaerobic conditions and faster development of denitrifying microorganism [7]. Many experiments reported that drainage control reduced the annual transport of total nitrogen (NO_3-N and TKN) at the field edge by 30-50%. It is harder to define subirrigation effects on nitrate losses because variable interactions occur between drainage water and rainfall depending on the subirrigation management. Subirrigation seems to increase denitrification and crop nitrate uptake, but outflow volume also increase [6].

The effectiveness of water table management practices on reduction of environmental pollution and on maintenance of good cultural conditions is strongly connected to system design and management. To optimise effectiveness, well trained staff and experimental indications and tests are required [3].

An experimental facility was set up in 1996 on the University of Padova Agricultural Experimental Station's Legnaro farm near Padova, Italy, to make an original contribution to knowledge on operation and effectiveness of controlled drainage and wetlands for environmental purposes in north-east Italy. An agricultural basin was simulated over a 6-hectare area. It was subdivided into twelve 0.3-0.5 ha cultivated plots and one 0.65 ha wetland. Drainage water from the plots passes through the wetland for purification before flowing into a receiving stream.

A long-term study started at this facility; in this paper, the features of the experimental facility and the first results regarding volume discharged, nitric nitrogen concentration and nitric nitrogen losses are presented.

2 Materials and methods

2.1 Experimental environment
The land where the study has been carried out was, in part, formerly used for an experiment on subsurface pipe drainage [8], that was dismantled to allow construction of the new facility.

The texture of the tilled layer (first 50 cm) is quite heterogeneous but it is always high in silt (30-48%). Sand contents range from 15 to 47%, clay contents range from 24 to 45%. On the average, a loam and a silt loam texture are prevailing, but sandy-loam and silty-clay textures are also present. Sand lenses with different thickness and directions run through the soil at depths equal to or greater than the drainage pipes. Soil is high in carbonate content (approximately 32% total calcium carbonate and 12% calcium carbonate active fraction) and has sub-basic reaction (pH 7.6-8.1). Soil moisture at field capacity (0.01 MPa) and at wilting point (1.5 MPa) is respectively 21.5-27.5% and 5-6.5% of dry matter. Bulk density is 1.3 t/m^3 and infiltration rates vary between 0.025 m/d, if soil is compacted with crust or silty deposits on the surface, and 2.4 m/d as evaluated with the double tube method with a constant head of 10 cm [9]. Hydraulic conductivity values determined from auger holes 1.0-1.2 m deep [10] were always larger than 40 mm/h.

Water table depth in undrained conditions ranges during the year between 1 m at the end of winter and 1.9 m in summer, but values of 0.5-0.6 m are not rare (Figure 1a). Annual rainfall reaches about 850 mm and is rather uniformly distributed throughout the year. The long term rainfall variability show larger differences in September, October, and November (Figure 1b).

2.2 Experimental layout traits

Treatments are a factorial combination between two drainage layout (subsurface pipe drainage denoted as P, and surface drainage denoted as S) and two water table management schemes (conventional drainage and controlled drainage-subirrigation, denoted as CD). The 12 plots have a longitudinal slope of 0.2%, and are arranged in a randomized complete block design with three replicates. Downstream of the basin a 0.65 ha area is used as a wetland.

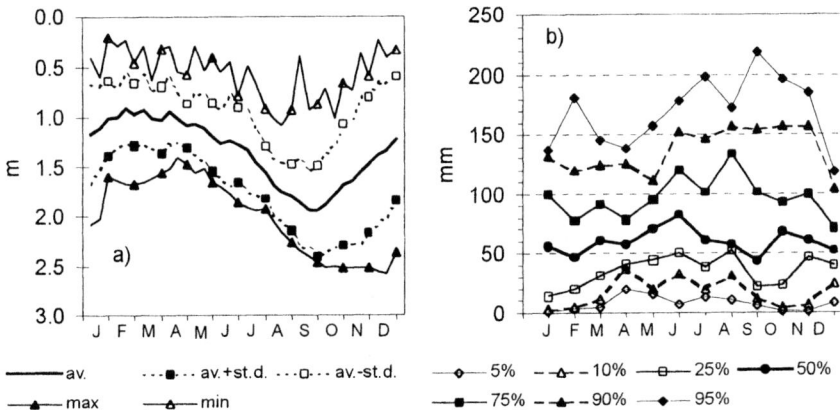

Fig. 1. a) Annual water table depth fluctuation observed at the Experimental Station's Legnaro farm on undrained soil (data recorded between 1980 and 1995). b) Monthly precipitation thresholds not exceeded at some critical probability levels at the Experimental Station's Legnaro farm. Values are the average of data observed between 1964 and 1990.

Each subsurface drainage plot has seven 65 mm diameter subsurface drains with coconut envelope and installed 8 m apart, 0.9 m below the soil surface, and at a 0.2% slope (parallel to the soil surface). Drains are connected at the upstream end to a 100 mm diameter unperforated PVC pipe that receives subirrigation water through a 0.8 by 0.8 by 0.8 m concrete sump. Surface drainage plots each have two 30 m-spaced ditches with a ridge in between that has a lateral slope of 1%. These ditches are 0.6 m deep, 0.3 m wide at the bottom and 0.6 m wide on the top. Subirrigation occurs upstream.

To limit lateral seepage between conventional and controlled drainage plots, a 0.15 mm PVC film has been buried at the plot boundary to a depth of 1.5 m. Where adjacent plots are both conventional or controlled drainage treatments, border effect is controlled by two 5 m-spaced drain lines.

2.3 Water table and drainage control systems

Downstream of each plot subsurface and surface outflows are collected in a main ditch which is hydraulically isolated from the others. Water can flow out of each ditch only via a 200 mm diameter PVC main buried at a depth of 1.3 m. Water is directed into a concrete collector sump (1.2 by 1.2 by 2.2 m) where flow is measured and water samples are collected. The full system is designed to allow water table control (Figure 2). PVC risers may be inserted in the main ditch allowing flow only at higher water levels.

Four concrete collector sumps have been buried in the basin and each collects drainage water from 3 plots. Flow is then diverted by gravity to another concrete sump via a 250 or 300 mm diameter PVC main buried between 1.5 and 1.6 m of depth. A 3 kW pump with a flow rate of 30 l/s is located in this latter sump and diverts water toward the wetland, through which water flows for a few days before discharging into a receiving stream. Wetland flow is controlled by means of PVC risers similar to those described for the concrete sumps. The experimental facility was installed in spring 1996.

2.4 Observations

Hydrologic observations refer to plot hydrologic balances (inlet and outlet), water table depth, and periodic soil water content. Water table depth is monitored by 65-mm diam. perforated PVC pipes 2.0 m deep. Samples of water table and drainage water are

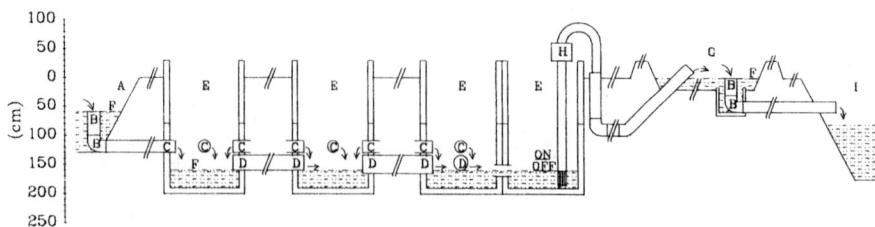

Fig. 2. Section of the water table control system. A: main ditch for plot outflows collection; B: PVC risers for water level control; C: 200 mm diameter PVC main for water transfer to the concrete collector sump; D: PVC main for water transfer to the pump sump; E: concrete collector sump; F: water level; G: wetland; H: concrete pump sump; I: receiving stream.

collected to measure the concentration of NO_3-N with a spectrophotometer.

All the elements concerning the crop management are regularly recorded and the main agrometeorological variables are collected.

The period of observation considered in this paper (June 1996 - January 1998) was characterised by high rainfall in October-December 1996 and November-December 1997; the spring and September-October 1997 were quite dry, while the summer 1997 was slightly more rainy than the median.

3 Results

3.1 NO_3-N concentration in the water table
The concentration of NO_3-N in the shallow water table showed the typical fluctuation during the year [11], in relation to rainfall distribution and N fertilisation. Peak values occurred in late spring and in November-December 1997, following N fertilisation in rainy periods. With rare exceptions, controlled drainage reduced the concentration in comparison to the conventional one, both with pipe and surface systems (Figure 3).

3.2 Drainage volumes and NO_3-N losses
Only few data were collected in the first winter season, because the instruments for measurement were not already available. In this period, only the discharge occurred between the 8th January and 21st February was measured and samples of water collected, showing that P treatment gave the highest losses (126 mm of discharged water and 3.7 kg ha^{-1} of NO_3-N) and that controlled drainage, both P-CD and S-CD treatments, reduced the volumes discharged and the nitrate losses (on the average, 40 mm of water and 1.4 kg ha^{-1} of NO_3-N).

In the following winter period, the first drainage events started in December. The NO_3-N concentration in the discharged water was initially higher in the controlled drainage plots, in particular P-CD, probably because the water table was shallower and received faster the nitrates leached from the upper soil layer. Later in the period, the

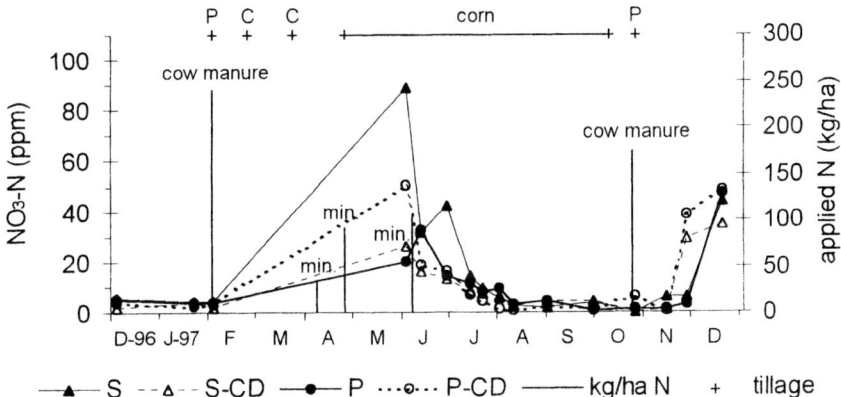

Fig. 3. NO_3-N concentration in the shallow water table during the monitoring period, in relation to drainage system and water table control; P: plowing, C: cultivation, min.: applied mineral N.

NO3-N concentration was higher in the P treatment, where almost all samples exceeded the EU safe drinking water standard of 11 ppm (Figure 4). The cumulative volume discharged at the end of the two months was dramatically reduced with controlled techniques: in P-CD, it decreased from almost 120 to very few mm, and, in S-CD, from 60 to some mm.

As a consequence of concentrations and volumes, the cumulative losses of NO_3-N were sharply influenced by drainage system and management of water table: maximum losses, close to 15 kg ha^{-1}, were obtained in the P treatment and about 3 kg ha^{-1} were lost in the S treatment. In both, S-CD and P-CD treatment, NO_3-N losses were almost absent (Figure 4).

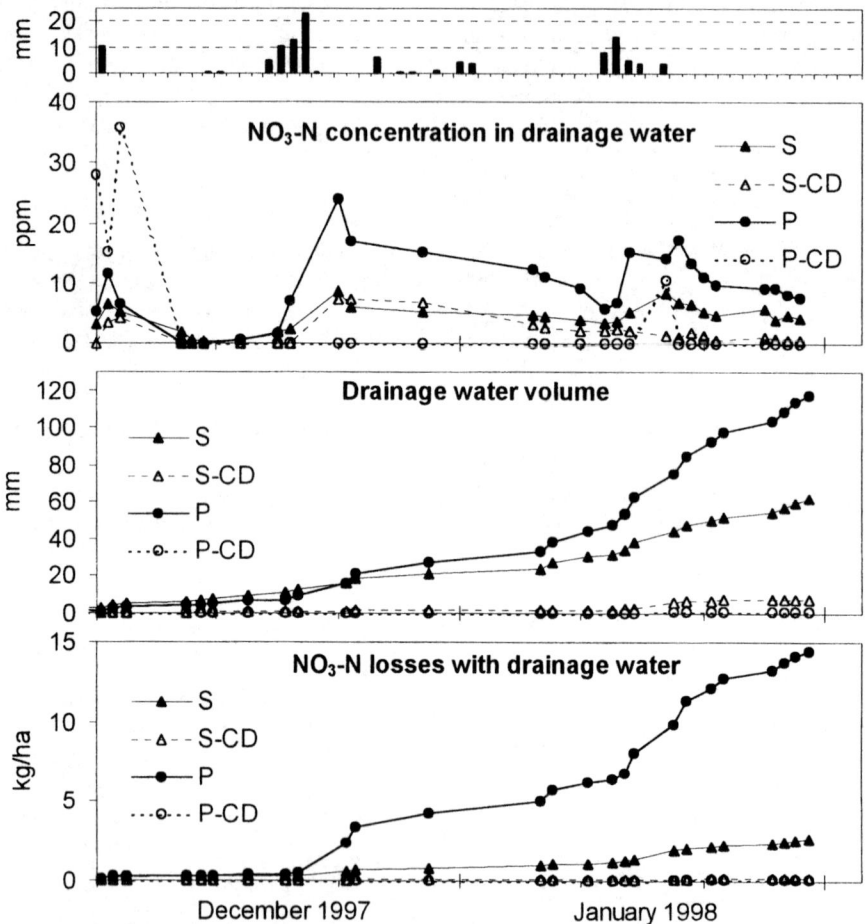

Fig. 4. Rainfall, NO_3-N concentration, volume of discharged water and NO_3-N losses during the period December 1997-January 1998 in relation to drainage system and water table control.

4 Conclusions

The results presented in this paper have to be considered as preliminary information produced by the experimental facility and the research programme.   In fact, the measurement system was incomplete during the first winter season and the second drainage season had a good rainfall only from November to the first half of January, but was preceded by a very dry period of two months. Nevertheless, some interesting first consideration can be drown: the management of water table according to controlled drainage schemes allow to reduce the NO_3-N losses fromagricultural fields compared to the absence of drainage control; the reduction of losses is related both to lower volumes of discharge and NO_3-N concentration; the control of water table techniques gives better results with pipe drainage systems than with surface systems, based on ditches and land modelling.

5 References

1. Thomas D.L., Hunt P.G. and Gilliam J.W. (1992) Water table management for water quality improvement. *Journal of Soil and Water Conservation*, Vol. 47, No 1, pp. 65-70.
2. Capone L.T., Izuno F.T., Bottcher A.B., Sanchez C.A., Coale F.J. and Jones D.B. (1995) Nitrogen Concentration in Agricultural Drainage Water in South Florida. *Transactions of the ASAE*, Vol. 38 No 4. Pp. 1089-1098.
3. Evans R.O., Gilliam J.W. and Skaggs R.W. (1990) *Controlled drainage management guidelines for improving drainage water quality.* Cooperative Extension Service North Carolina State University, Publication Number AG-443.
4. Arjoon D., Prasher S.O. and Gallichand J. (1995) Water table management systems and water quality, in *Subirrigation and controlled drainage*, (ed. H.W. Belcher and F. M D'Itri), Lewis , Boca Raton, pp. 327-352.
5. Bendoricchio G. and Giardini L. (1994). A controlled drainage demonstration project in Italy. In (eds. M. Borin and M. Sattin) *Proc. of Third Congress of the European Society for Agronomy.* Abano-Padova, pp. 768-769.
6. Skaggs R.W. and Breve M.A. (1995) Evironmental impacts of water table control, in *Subirrigation and controlled drainage*, (ed. H.W. Belcher and F. M D'Itri), Lewis , Boca Raton, pp.247-268.
7. Skaggs R.W., Evans R.O., Chescheir G.M., Parson J.E., Gilliam J.W., Sheets J.T. and Liedy R.B. (1993) *Water table management effects on water quality and productivity.* Tidewater Research Station, Plymouth North Carolina.
8. Giardini L., Giovanardi R., Borin M. (1983) Ricerche sul drenaggio tubolare sotterraneo e sulla sub-irrigazione nel Veneto (2° contributo). *L'irrigazione*, Vol. XXX, No 3-4, pp. 11-26.
9. Kessler J. and Oosterbaan R.J. (1974) Determining Hydraulic Conductivity of soils, in *Drainage principles and applications.* Vol. III ILRI, Wageningen, pp. 253-296.
10. Van Beers W.F.J. (1958) - *The Auger Hole Method, a field measurement of hydraulic conductivity of the soil below the water table.* IRLI, Wageningen, Bull. 1. 9-32.

11.   Borin M (1997) Effects of agricultural practices on nitrate concentration in groundwater in north east Italy. *Italian Journal of Agronomy*, Vol. 1, No. 1. pp 47-54.

6 Acknowledgement

Research supported by Finalized Project MIPA-PANDA, Subproject 2, Series 2

NITRATE MOVEMENT IN LEVEL BASINS: REDUCED *VERSUS* CONVENTIONAL SOIL TILLAGE

D.V. SANTOS, P.L. SOUSA and L.S. PEREIRA
Depart. of Rural Engineering - High Institute of Agronomy, Technical University of Lisbon, Portugal.

Abstract
The influence of hydrodynamic behaviour of the soil on water and nitrate movement in level basins with conventional (CT) and reduced tillage (RT) is discussed. Observations during the last five years have shown very low losses of nitrate by leaching in CT. Results indicated important upward fluxes from the water table during the irrigation season producing important contributions to the root zone. In CT this contribution was lower than RT because root development were limited by a compacted layer at tillage depth which reduces the fluxes. Losses of N to the atmosphere were important in CT and N mineralization assumed high values in RT.
Keywords: denitrification, level basin irrigation, nitrate leaching, soil hydraulic properties, tillage.

1 Introduction

The increments on fertilizers-N usage during last decades have conducted the nitrate leaching to the groundwater to become a main concern to public health. Soil tillage practices can play an important role on the control of N losses, either by leaching or to the atmosphere, improving the efficiency of N use by crops.

Under irrigation, the potential nitrate pollution could be high if irrigation, fertilizing and tillage practices are not in consonance. Several studies have been developed with the aim to establish the best management practices for these three yield factors - soil, water and nutrients - but results may differ according to the local conditions. Studies have shown that conservative tillage reduces runoff, increases infiltration, reduces soil evaporation and promotes macropores development [1,2]. Splitting fertilizer applications

Water and the Environment: Innovative Issues in Irrigation and Drainage. Edited by Luis S. Pereira and John W. Gowing. Published in 1998 by E & FN Spon. ISBN 0 419 23710 0

with irrigation water can reduce residual NO_3-N soil profile, reducing the potential for leaching during the rain season. Associating no-till and split fertilizer-N has given encouraging results. Varshney et al. [3] have reported lower amounts of residual NO_3-N in the soil profile when no-till with three split N applications (125 kg N.ha[-1] total) are adopted in comparison to a single N application of 175 kg N.ha[-1], without corn yield significant differences between treatments.

However, there is a concern that conservation tillage may increase the risk of groundwater pollution because higher infiltration leads to increase groundwater recharge. On the contrary, in poorly drained soils with high water table, the nitrate losses by leaching could be lower than under artificial drainage because this favours denitrification and nitrate movement [4].

Rice et al. [5] have suggested that the N availability is smaller under no-till systems, usually resulting in reduced N uptake by crops. This may be the result of immobilization of surface applied N fertilizer, reduced N fertilizer recovery and greater denitrification losses. The same authors have observed that N mineralization was at least as great in long-term no-till plots as in long-term conventional tillage plots and the lower availability of N frequently observed in no-till soils is a transient effect. Higher mineralization rates have been associated to conventional tillage [6,7]. Kanwar et al. [2] have found greater concentrations of mineral N under conventional tillage than for no-till due to increased mineralization with mouldboard plowing, and to reduced leaching with no-till. Varshney et al. [3] have observed more residual N accumulated in the top 60 cm of the soil profile under conventional tillage than under no-till system due to higher mineralization rates.

Despite several studies under different conditions have conducted to similar conclusions, for other particular conditions different or even opposite results could be obtained. This kind of studies may only be conclusive after some years of observations because the soil under reduced tillage requires time to establish the equilibrium. Results herein presented are preliminary because the data regards only one year of observations, despite this is the third year under reduced tillage and fertigation treatments. The observed differences are not conclusive but contribute to understand the most relevant differences between conventional and reduced tillage concerning the hydrodynamic behaviour of the soil and the N processes as influenced by the soil water regime in a poorly drained alluvial soil with shallow water table.

2 Material and methods

The experiments were conducted in two furrowed level basins in Coruche, with different soil tillage treatments: one, where reduced tillage (RT) is applied after 3 years and the crop residues are crumbled, removed from the furrows and accumulated on the ridges; the other under conventional mouldboard tillage (CT). The soil is a Fluvissol (modern alluvial not calcaric soil), poorly drained, which may remain flooded during the rainy season, and is in presence of a shallow water table that strongly influences the water and solute movement. Experimental facilities are described in [8].

The soil hydraulic potential was monitored through tensiometers installed at each 15 cm until 90 cm depth. Soil water content was observed with a neutron probe performing readings at the same depths. Observations were performed with 2 replications for both treatments on a weekly basis, and before and after each irrigation. Soil samples for nitrate and ammonium analysis were collected at the same depths with 3 replications before and after each irrigation, dried at air and analysed by specific ion electrode method. The depth of water table was observed before, after and between irrigation events.

A total of 200 kg N ha^{-1} was applied in each irrigation and in each treatment: 50 kg N ha^{-1} at seeding (fertilizer 7:21:21), 90 kg N ha^{-1} with the first irrigation and 60 kg N ha^{-1} with the second. The fertilizer used for the fertigations was solution UAN(32%) dosing 25% of NO_3-N, 25% of NH_4-N and 50% of urea. 4 and 5 irrigations were performed respectively in the CT and RT treatments. The average water depths were 110 mm for the two first irrigations and 80 mm for the next ones.

In Tables 1 and 2 some relevant physical and chemical characteristics of the soil under CT and RT treatments are presented.

The two level basins are contiguous and of the same size, 17 m x 210 m. The CT plot presents greater percentage of fine sand and less clay and silt than RT. The water content at field capacity and wilting point are similar. The bulk density in the root zone is greater for CT. In RT the BD increases beneath root zone.

Chemical characteristics differ due to crop residuals management, namely N-residues at the soil surface, C and organic matter content (OM) and C/N ratio through the entire soil profile.

Table 1 - Soil physical characteristics. Averages of 3 replications.

Depth	Conventional tillage (CT)						Reduced tillage (RT)					
	BD	Texture			θ_v		BD	Texture			θ_v	
		fine sand	silt	clay	FC	WP		fine sand	silt	clay	FC	WP
cm	kg m^{-3}	%			%		kg m^{-3}	%			%	
0-15	1.31	61.3	22.0	14.7	32.4	11.8	1.37	52.2	27.6	17.9	33.8	9.8
15-30	1.48	62.9	22.2	13.2	32.7	12.0	1.31	53.6	27.6	16.5	34.4	10.8
30-45	1.57	59.5	23.1	14.5	34.3	14.7	1.40	51.6	28.4	17.9	37.3	11.9
45-60	1.51	68.8	17.6	12.0	33.4	11.0	1.59	52.7	28.2	17.2	36.1	14.5
60-75	1.55	69.1	18.0	11.5	36.1	11.3	1.58	55.6	25.2	17.1	35.1	12.3
75-90	1.49	66.6	19.9	13.1	37.5	11.2	1.6	58.6	24.0	15.3	36.5	10.8

BD - bulk density; FC - field capacity; WP - wilting point.

Table 2 - Soil chemical characterization. Average of 3 replications.

Depth	Conventional tillage (CT)					Reduced tillage (RT)				
	N-residues	N	C	OM	C/N	N-residues	N	C	OM	C/N
(cm)	(kg.ha^{-1})	(Mg.ha^{-1})				(kg.ha^{-1})	(Mg.ha^{-1})			
0-15	1.26	2.03	9.86	17.02	4.9	3.69	1.66	12.02	20.72	7.3
15-30	2.85	2.09	3.66	6.32	1.8	8.83	1.55	12.38	21.34	8.0
30-45	0.605	2.09	14.48	24.97	6.9	-	1.13	16.53	28.51	14.6
45-60	-	2.12	2.54	4.39	1.2	-	0.86	14.48	24.98	16.8
60-75	-	2.13	3.65	6.31	1.7	-	0.87	7.01	12.10	8.1
75-90	-	2.00	5.36	9.25	2.7	-	0.86	8.53	14.71	10.0

During the observation period, from 15 April to 8 October, did not occur precipitation. The mean temperature ranged 17-23°C. The average maximum temperature ranged 23-31°C, the daily maximum being 36.5°C in June.

Figure 1 shows the soil water retention curves for the surface at RT and CT treatments. These curves were determined in laboratory on 100 cm^3 samples as described in [8,9]. The van Genuchten equation is used to describe the curves as it adjusts well [9].

The differences between treatments at layers more affected by tillage are not significant. However, at 15-30 cm depth the differences are more important. Treatment CT presents lower saturation water content for small log|h| where BD is greater (Table 1) as an effect of compaction caused by tillage at this depth.

The water retention curves for the RT plot were also determined in a monolite. The differences between laboratory and monolite values could regard to saturation method. At the laboratory, samples are saturated by bottom permitting the air escaping from the sample free surface. At the monolite, all soil surface is flooded so infiltration becomes difficult because of the entrapped air, impeding all the soil pores be completed by water. As the soil surface has the most of porosity, the differences between laboratory and monolite measurements are greater for these layers.

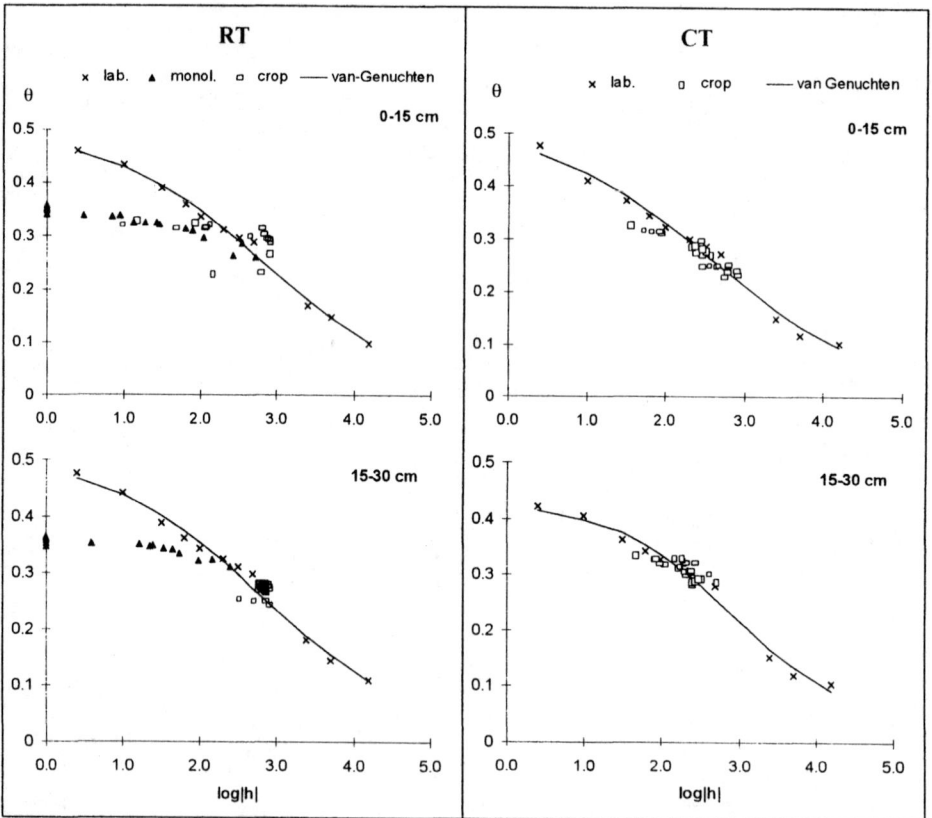

Fig. 1 - Water retention curves θ (m^3 m^{-3}) vs. log|h| (cm) for RT and CT. Laboratory and field measurements.

For both treatments the presence of roots slightly increase the soil effective pressure for the same water soil content at the interval $2 \leq \log|h| \leq 3$.

Figure 2 shows the hydraulic conductivity curves determined in laboratory [8,9] before irrigations. The unsaturated hydraulic conductivity is about the same for both plots when only the water in the soil matrix remains in the soil. Differences occur near saturation, with CT having a saturated hydraulic conductivity (Ks) of 11.2 cm d^{-1}, double then the RT treatment which is 5.7 cm d^{-1}. Below the layer affected by tillage, Ks for RT is about twice the Ks for CT. At the layer 60-75 cm, Ks is 47 cm d^{-1} and 106 cm d^{-1} for CT and RT, respectively.

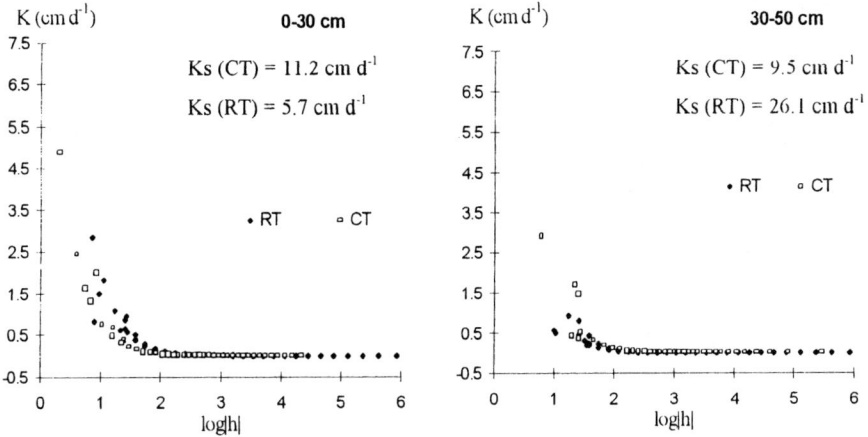

Fig. 2 - Unsaturated hydraulic conductivity for the RT and CT treatments.

3 Results and discussion

3.1 Water fluxes

The upward and downward fluxes from and to the water table are represented in Figure 3. Depth 52.5 cm represents the root depth for CT. At this layer fluxes are almost invariably negatives (upward fluxes) for the treatment CT. At the same level, fluxes are null or positives (downward fluxes) for RT. At a deeper level (75 cm), the fluxes are negative during all the observation period for both treatments.

Results indicate that in the CT treatment there was no potential for leaching to the deep soil layers and to the water table. For the RT treatment solutes could be transported from the upper soil layers but not directly to the water table.

3.2 Nitrate movement

Statistical analysis were performed to determine significant differences in NO$_3$-N concentration means with fertigations between both treatments using the analysis of variance procedure ANOVA and Student-Newman-Keuls method at significant level of 10%. According to literature, summarized in the introduction, one should expect higher mineralization in the CT treatment. However Figure 4 shows the opposite: NO$_3$-N concentration were most of the time significantly higher for the RT treatment. Before first fertigation (Fig. 4a) the NO_3-N concentration is very low in the upper layers, where root absorption is higher. For CT plot, the amount of nitrate is very low for the entire profile, but for the RT there is a large amount of nitrate below root depth. This may be explained by the upward flux from the water table carrying nitrates.

In fact, during the interval of highest NO_3-N concentrations the water table presented 13 ppm of NO_3-N. Considering the high fluxes occurred between 4 July and 16 July, the contribution from water table could be about 57 kg NO_3-N ha^{-1} during this interval. The water table fluctuations and NO_3-N concentrations are not directly related to performed irrigations in the corn fields but is a consequence of percolation losses from the rice paddy in the area. The same contribution does not occur in the CT plot because it has an elevation of 0.3 m higher than RT. This difference in elevation may justify the lower upward fluxes during the same interval for CT treatment. In addition, hydraulic gradients between 60-90 cm were higher in RT since root development was deeper extending up to 60 cm, promoting high soil capillary suction at this layer.

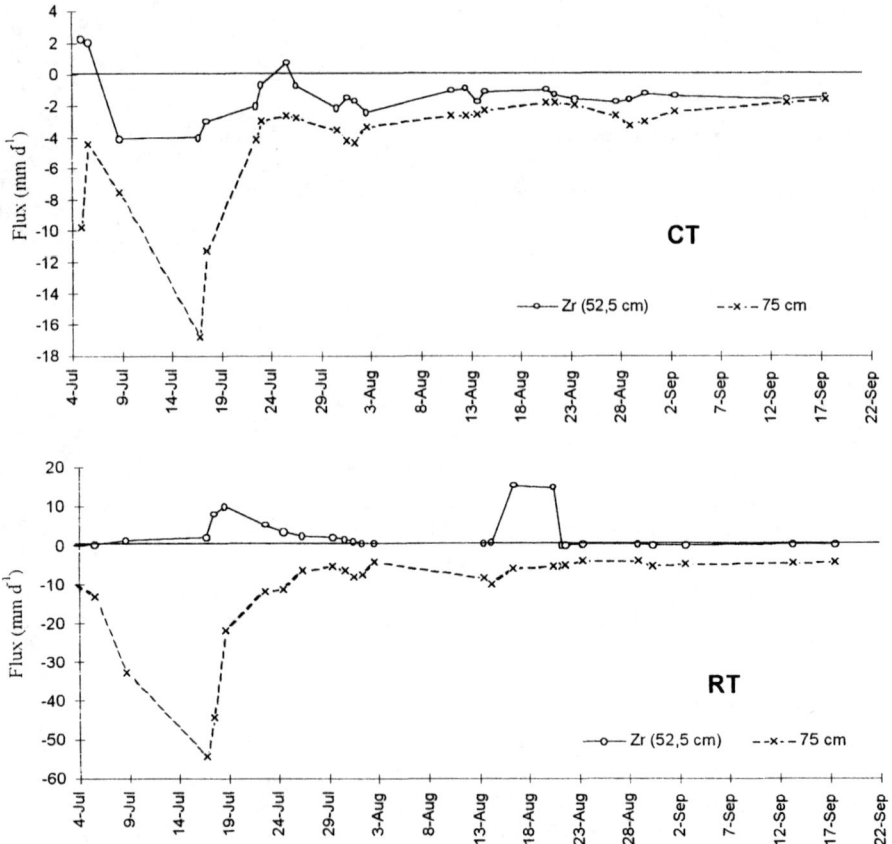

Fig. 3 - Upward and downward fluxes from or to water table during the irrigation season at 52.5 cm and 75 cm depth.

Infiltration during first irrigation lasted 12 hours to be completed in RT plot and 48 hours in the CT plot. This creates favourable conditions to N losses to the atmosphere, by denitrification and/or volatilization at CT. It is believed volatilization is not important because water and soil pH is neutral to slightly acid in studied case. From N budget carried out for CT, it was estimated losses of 30 kg N ha^{-1} and 9 kg N ha^{-1} for the first and second irrigation, respectively. This may help to justify the differences in Figure 4b.

During the drying period between irrigations (Figure 4c), the high temperatures associated to higher OM content and N-residues at soil surface in RT was highly favourable to mineralization. The gains were 183 kg NO$_3$-N ha^{-1} from which about 150 kg NO$_3$-N ha^{-1} was resulted from mineralization. This is only partially justified by the fact that crop residuals are all placed on the ridge. However, during the interval before first fertigation (Fig. 4a) and after second fertigation (Fig 4d) was registered gains of about 140 kg N.ha^{-1} more than the applied one and absorbed by crop. This value corresponds to an annual mineralization of 46% of total soil N. Stanford *et al.* [10] have studied several soils registering annual mineralization rates ranging 8.3 to 26% of total soil N.

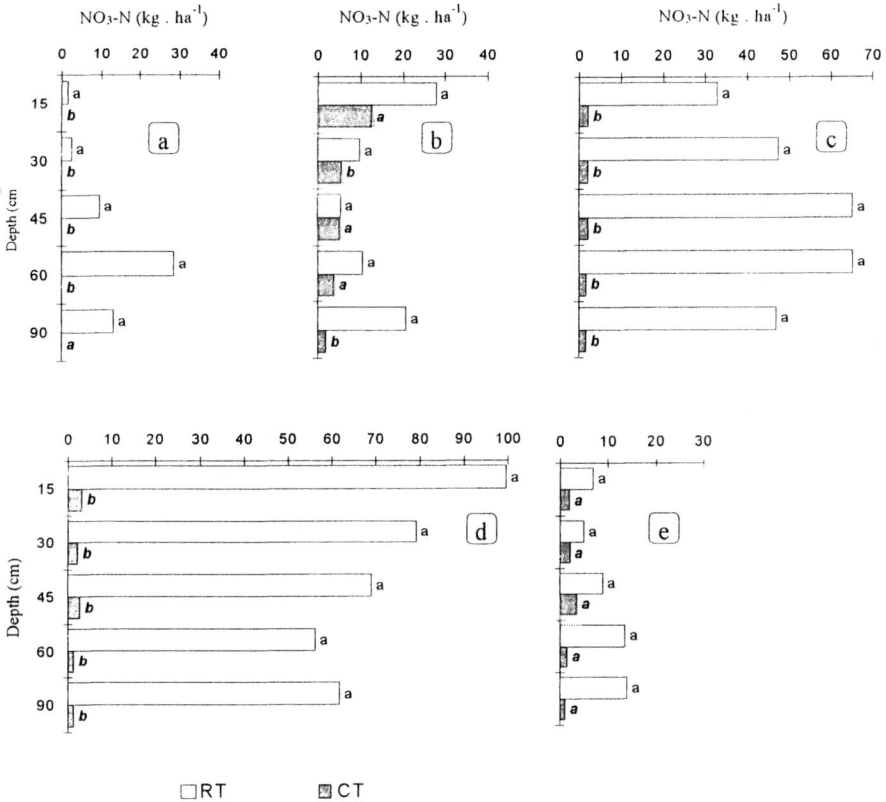

Fig. 4 - Comparative evolution of NO$_3$-N with depth in RT and CT . a) before 1st fertigation, b) after 1st fertigation, c) before 2nd fertigation, d) after 2nd fertigation, e) residual after harvesting. (same letter represents no significant statistic differences between means) .

Facing so high values is reasonable to put the hypothesis of a spatial variability in fertilizer distribution with irrigation has occurred but ammonium concentrations presented values in consonance with applied N-fertilizer. Therefore, it is believed the mineralization rate in RT is significantly high during irrigation season but reduced during the rest of the year and thus, the mean annual mineralization is certainly lower than 46% of total soil N.

The observed differences in Figure 4c are also partially coherent with the assumption that poorly drained soils have higher denitrification, explaining low values for the CT plot. Higher nitrate values in the deeper layers up to 60 cm depth for RT plot may be explained by aeration promoted by presence of roots which in RT reached deeper layers than in CT. High concentrations below to 60 cm remain as a continuous contribution from water table, since upward fluxes at these layers were quite high during the interval before first and after second irrigation (4 July to 16 July).

However, a definitive explanation is not found yet, even considering the differences in the chemical composition of both soils (Table 2). As a coherent result, the residual nitrate profile after harvesting, show to be different with lower concentrations for the CT plot, namely at deeper depths. Nevertheless, the differences between treatments are not statistical significant. At this date water table was placed at 1.5 m depth, so the contribution of nitrate is unmeaning and then RT profile shows low nitrate concentrations at deeper layers.

4 Conclusions

The comparison on the soil water and solutes movement in two plots with conventional tillage (CT) and reduced tillage (RT) produced expected results concerning water movement but not relative to nitrates. The CT plot has poor drainage and lower saturated hydraulic conductivity. It was observed decreasing of infiltration rates with irrigations as consequence of surface pores clogging by quite fine sand and silt sediments. Root development was restricted by compacted layer at tillage depth (30-45 cm). Therefore, it did not occur downward fluxes but upward fluxes towards to root zone at depth 45-60 cm.

The RT plot, placed at lower elevation presented very higher upward fluxes at deeper layers than CT, but null or downward fluxes at 45-60 cm depth since the root development extended up to 60 cm depth. At this depth suction capillary was very high promoting more important upward fluxes from water table to deeper layers when compared to CT. Such contribution from water table allowed the maintenance of water contents close to saturation over the entire season despite of the intense crop water absorption at these layers.

The CT plot presented favoured conditions to higher losses of N to atmosphere while the RT shows to have high mineralization rates and important contributions of nitrate from the shallow water table. Both treatments showed to have low potentiality to nitrate leaching through soil matrix.

The most relevant differences between treatments were consequences of the soil tillage effects over hydrodynamic soil behaviour. The presence of a shallow water table and fair soil structure with low OM and high fine sand and silt content confer a peculiar behaviour to this soil. In addition, the presence of furrows in RT leads to results much different to the expected one by literature.

5 Acknowledgements: Program Praxis XXI, Projects PAMAF 4050, PAMAF 5183, JNICT 2322.

6 References

1.   Ritter, W.F., Scarborough, R.W., and Chirnside, A.E.M. (1993) Nitrate leaching under irrigated corn. *Journal of Irrigation and Drainage Engineering.* Vol. 119, No. 3. pp. 545-553.

2.   Kanwar, R.S., Baker, J.L., Baker, D.G. (1988) Tillage and split N-fertilizer effects on subsurface drainage water quality and crop yields. *Transactions of the ASAE* Vol. 31 No. 2. pp. 453-460.

3.   Varshney, P., Kanwar, R.S., Baker, J.L., Anderson, C.E. (1993) Tillage and nitrogen management effects on nitrate-nitrogen in the soil profile. *Transactions of the ASAE.* Vol. 36. No. 3. Pp. 783-789.

4.   Kanwar, R.S., Baker, J.L., Johnson, H.P. (1984) Simulated effects of fertilizer management on nitrate loss with tile drainage water for continuous corn. *Transactions of the ASAE.* Vol. 27. pp. 1396-1399, 1404.

5.   Rice, C.W., Smith. M.S., and Blevins, R.L. (1986) Soil nitrogen availability after long term continuous no-tillage and conventional tillage corn production. *Soil Sci. Soc. Am. J.* Vol. 50. pp. 1206-1210.

6.   Thomas, G.W., Smith, M.S., and Phillips, R.E. (1989) Impact of soil management practices on nitrogen leaching. In: R.F. Follett (ed.) *Nitrogen Management and Ground Water Protection.* USDA-ARS. pp. 247-276.

7.   Doran, J.W., Power, J.F. (1983) The effects of tillage on the nitrogen cycle in corn and wheat production. In: R.R. Lowrance, et. al. (eds.) *Nutrient Cycling in Agricultural Ecosystems.* Univ. of Georgia. Spec. Pub. 23.

8.   Santos, D.V., Sousa, P.L., Pereira, L.S. (1997) Calibration of hydrodynamic and solute components of OPUS model for an alluvial soil in presence of a shallow water table. In: International Conference in *"Water Management, Salinity and Pollution Control Towards Sustainable Irrigation in the Mediterranean Region"* Vol. VI. pp. 91-110. Istituto Agronomico Mediterraneo, CIHEAM/MAI-B, Italy.

9.   Gonçalves, M.C., Pereira, L.S., Leij, F.J. (1997) Pedo-transfer functions for estimating unsaturated hydraulic properties of Portuguese soils. *European Journal of Soil Science,* Vol. 48. pp. 387-400.

10.   Stanford, G., and Smith, S.J. (1972) Nitrogen mineralization potentials in soils. *Soil Sci. Soc. Am. Proc.* Vol. 38. pp. 465-472.

IMPACT OF AGRICULTURAL DEVELOPMENTS ON WATER CONTAMINATION AND ENVIRONMENTAL SUSTAINABILITY

R.S. KANWAR
Department of Agricultural and Biosystems Engineering
Iowa State University, Ames, Iowa, U.S.A.

Abstract
Agricultural production systems have been found to have had negative impacts on the overall quality of water resources. Growing environmental awareness in the public has, however, led to improved management systems within agricultural watersheds of the nation to a certain degree. Contamination of surface and groundwater sources by fertilizers and pesticides has been well documented by various state and federal agencies in the United States as a result of agricultural practices in the watersheds. Nation's water resources include underground aquifers as well as lakes, rivers, and the ocean. Agriculture is viewed as a significant nonpoint source of groundwater contamination, presenting a difficult problem for the design of governmental methodologies to prevent pollution. Groundwater contamination is of increasing concern in the United States because about 50 percent of the drinking water comes from groundwater. To society at large and to producers, research and educational programs were initially aimed at water quality protection over compulsory regulations. Several studies were initiated at Iowa State University in the late eighties to evaluate the effects of various agricultural production systems on the environment. The objective of this paper is to review the results of some of these studies that were aimed at reducing the contamination of surface and groundwater sources. Some of the results of these studies indicate that better chemical management practices in conjunction with conservation tillage systems, can reduce contamination potential of nations water resources.
Keywords: environmental sustainability, herbicides, nitrate management, subsurface drainage, tillage, water pollution.

Water and the Environment: Innovative Issues in Irrigation and Drainage. Edited by Luis S. Pereira and John W. Gowing. Published in 1998 by E & FN Spon. ISBN 0 419 23710 0

1 Introduction

The quality of ecosystems and improved economic opportunities are twin elements of global stability. About twenty years ago, it was a popular belief that goals of economic development and environmental protection were mutually exclusive. Today, this view has largely given way to a belief that we need a better understanding between the economic development and the environment quality of our ecosystems. Therefore, it is necessary to develop a comprehensive environmental assessment policy that assures that developments are environmentally sound and sustainable.

The changing agricultural production systems in the world are adding more and more pressure on the limited natural resources. Agricultural land areas are finite on earth. We may have to intensify agricultural production on these limited land resources to feed the growing population in the world. But the key to success would be to maintain the sustainability of these intensive agricultural production systems.

The increased use of agricultural chemicals created another serious environmental threat for water quality as well as for human health. The application of modern pesticides, introduced in the mid-1940's, soon became a common practice. The industrial synthesis of ammonia in 1920s resulted in a cheap supply of nitrogen fertilizer; farmers responded by applying increasingly liberal amounts of the synthetic fertilizer, and crop yields increased dramatically [1]. In 1945 less than 0.2 kg/ha. of nitrogen fertilizer was applied to Iowa's croplands, but application rates rose to 160 kg/ha by 1985. In 1995, 4.6 million hectares of maize received nitrogen fertilizer. Further, 8.3 million hectares of maize and soybeans received herbicide treatment.

The quality of surface and groundwater suffered because of the off-site impact of field-applied agricultural chemicals. Beginning in the 1950s, research slowly defined the nature of water quality problems caused by agricultural chemicals. Runoff from agricultural land was shown to be a major cause of surface water contamination from pesticides and other agricultural chemicals. Runoff from agricultural land was shown to be a major cause of surface water contamination from pesticides and other agricultural chemicals [2]. Similarly, many studies conducted in Iowa have shown that subsurface drain water leaving agricultural watersheds carries agricultural chemicals into surface and groundwater sources [3].

Serious environmental concerns arise because of the adverse impact of agricultural chemicals and pesticides on the biodiversity of aquatic ecosystems. Toxic levels of some chemicals disrupt the flora and fauna of aquatic ecosystems. Excessive amounts of phosphorus in surface water bodies accelerate eutrophication. Water quality problems not only trigger biodiversity loss in aquatic ecosystems but also pose serious health problems for people, livestock, and wild animals [1].

2 Materials and methods

Field studies were conducted at the Iowa State University's Northeast Research Center near Nashua. Nine treatments involving no-till or chisel plow tillage treatments; differential fertilizer N rates based on the late-spring soil nitrate-nitrogen (NO_3-N) test, single fertilizer application rates of 112 or 135 kg-N/ha from UAN (urea ammonium nitrate), or swine manure; corn and soybean rotation or continuous corn cropping strategies; narrow strip cropping with a combination of corn, soybean, and oat plus a berseem cover crop; and alfalfa were evaluated with regard to their effects on the loss of N to shallow groundwater. The experimental site consisted of 40, 0.4 ha plots that were instrumented with subsurface drainage flow measuring and monitoring devices for continuous water quality assessments when tile lines are flowing. All management treatments were replicated three times except the alfalfa and narrow-strips which were replicated twice. Table 1 gives the N application rates for various treatments for five years (1993-97). Paired samples were collected from each crop grown in the narrow strips and from each alfalfa plot to accommodate the reduced replication. Subsurface drain water samples were analyzed for NO_3-N and herbicides.

3 Results

Fig. 1 gives the five year (1993-97) average NO_3-N concentrations in the surface drainage water as a function of different N management systems studied in this project. Highest five year average NO_3-N concentrations of 22.1 mg/l in the drainage water was observed from manure plots under continuous-corn production and the lowest average NO_3-N concentration of 6.3 mg/l was observed from plots under strip cropping. Manured plots under corn-soybean rotation resulted in four average NO_3-N concentration of 14.1 mg/l. Chisel plow plots resulted in higher NO_3-N concentrations in drainage water compared with no-till plots under similar treatments (late spring nitrate test, LSNT, and single N application of 112 kg/ha). Also, although no-till plots received higher amounts of N application under LSNT in comparison to chisel plow plots the average NO_3-N concentrations in drain water from no-till plots was 9.8 mg/l in comparison to 10.9 mg/l from chisel plow plots. These results indicate that several N management systems (LSNT, strip cropping, alfalfa crop, single N application at 112 kg/ha) could be used successfully to reduce the leaching of NO_3-N to shallow groundwater. This is one of the studies that is being conducted at Iowa State University to develop agricultural production systems that are likely to be sustainable.

Table 1. Applied nitrogen and corn yields by treatments for 1993 through 1997

Treatments	Applied Nitrogen (kg N/ha)					Corn Yield (Mg/ha)				
	1993	1994	1995	1996	1997	1993	1994	1995	1996	1997
NT-LSNT[1]	144	169	193	195	125	6.6	7.3	4.8	9.3	10.1
NT-single	110	110	110	110	110	4.1	6.3	4.5	8.5	9.5
CP-LSNT[1]	93	160	160	169	125	6.8	8.2	5.2	9.4	10.1
CP-single	110	110	110	110	110	5.0	8.0	5.2	9.0	9.6
CP-manure[2]	73	210	195	74	76	5.6	8.4	5.6	8.9	9.2
CC-manure[2]	61	234	270	91	92	2.8	7.3	4.8	8.4	7.4
CC-single	135	135	135	135	135	4.2	5.8	4.1	7.2	8.9

[1] Amount for late spring soil nitrate test includes 30 kg N/ha applied with planter.
[2] Assumed all ammonia and 50% of organic nitrogen was available in the first year.
NT - no-till, CP - chisel plow, CC - continuous corn, LSNT - late spring nitrate test.

Table 1 also gives seven years data on corn yields as a function of various N management systems. This yield data in Table 1 show that rotated corn yields have

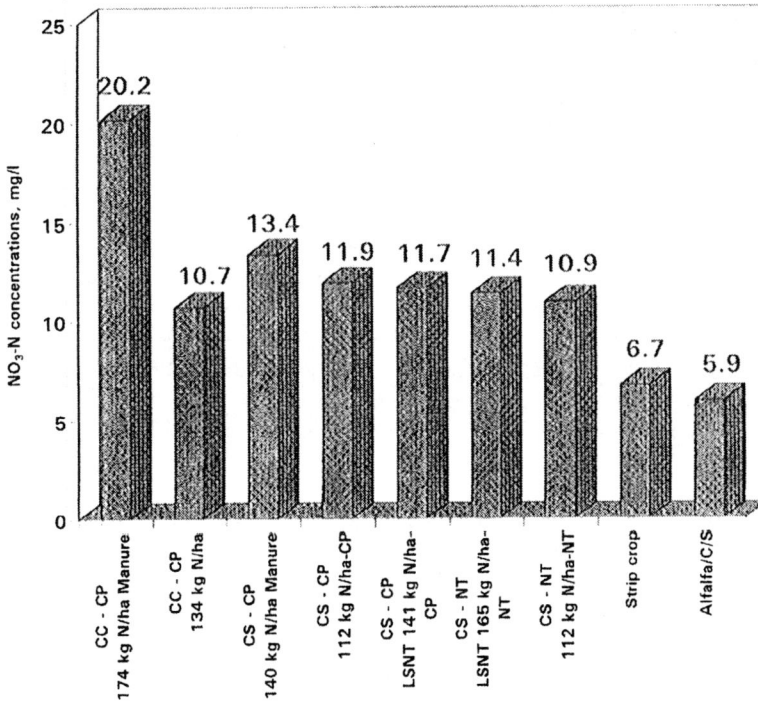

Fig. 1. Effect of different N-management systems on five year (1993-97) average concentrations in the subsurface drain water (CC = continuous corn; CS = corn soybean rotation; NT = no-till; CP = chisel plow).

greater than the previous long-term average of 7.3 to 8.4 Mg/ha for continuous corn. This shows that corn in rotation with soybean is a better management practice than a continuous-corn production system. Also, the average NO_3-N concentrations in tile water under corn-soybean production system are much lower in comparison to continuous corn production system. The five years of data clearly indicate that lower NO_3-N concentrations can be obtained in the shallow groundwater by reducing N application rates to 112 kg-N/ha. Use of the late-spring N test and differential N fertilization rates based on that test resulted in the lowest NO_3-N concentrations in subsurface drainage water under both no-till and chisel plow treatments when compared with manure application of single N application rates of 112 kg/ha. The alfalfa and narrow-strip crop plots had the lowest NO_3-N concentrations (< 10 ppm) in the subsurface drainage water at this research site. Corn following soybean on plots fertilized with swine manure had an average yield that was slightly higher than with the LSNT, but because of the difficulty in applying the intended amount of N with manure, NO_3-N concentrations in the tile drainage water were much higher. When averaged for the five years, no-till yield was slightly lower, but for both tillage practices, use of the LSNT resulted in equal or better yields than either preplant fertilizer or application of swine manure.

Another component of this research was to study the effect of banding on herbicide leaching to shallow groundwater. The effect of banding appears to be highly significant in reducing the overall yearly atrazine and metolachlor losses with the subsurface drainage water. These results also indicate that banding of herbicides could reduce herbicide leaching losses to less than one-tenth compared to the surface broadcast practice. Also, atrazine leaching losses were much lower from manure plots compared to the non-manured plots. This shows that swine manure may have a positive effect in reducing the leaching of atrazine to groundwater because of two reasons. One reason could be the increased microbial activity in the manure plots which may increase the microbial degradation of atrazine making it less available for leaching to groundwater. The second reason could be the greater degree of adsorption of atrazine to organic matter supplied by swine manure which also makes atrazine less available for leaching. We do not have much data in the literature to demonstrate the positive effects of manure application on pesticide contamination of groundwater. Therefore, long-term studies are extremely important to determine the role of swine manure in reducing pesticide leaching to groundwater.

4 Discussion

Growing environmental awareness among farmers and implementation of government conservation programs have resulted in some reduction in the environmental degradation associated with agricultural practices and the restoration of some natural habitats to a healthier state so that they are once again attractive to wildlife.

Serious conservation efforts began in the 1930s, with the introduction of such soil conservation practices such as contouring, terracing, and tree planting in sloughs and gullies . Legislation was introduced to study soil erosion in 1933, create the Soil Conservation Service in 1935, and create soil conservation districts. By the 1980s farmers had become increasingly aware of the need for conservation and adopted several ameliorative practices, such as conservation tillage, crop rotation, strip cropping, and reduced application of agrochemicals. More comprehensive government conservation programs were also implemented during the 1980s, including the Conservation Reserve Program (CRP), Conservation Compliance, Sodbuster, and Swampbuster. In 1990 the Food, Agriculture, Conservation, and Trade Act added the Wetlands Reserve Program (WRP), with a target of improving conservation on 400,000 hectares.

On a national scale soil erosion was reduced on about 22 percent of the 37 million hectares under CRP. Six percent of CRP land was under trees, while another 6 percent had been set aside specifically for wildlife [4]. Some 8,300 km of buffer strips have been created along waterways as a result of CRP. The WRP provides another option for enhancing biodiversity within agricultural landscapes. According to the U.S. Fish and Wildlife Service, because of various conservation practices of CRP and efforts under WRP, more than 37,000 hectares of wetland systems have been affected [5]. Wetland protection in agricultural areas has improved significantly in the last decade, although more work needs to be done [6].

Although crop and livestock productivity are still in the forefront of Iowa's economy, sustainability issues regarding agricultural production and recognition of the need to conserve biodiversity are beginning to emerge. Only recently have Iowans collectively come to appreciate the magnitude and importance of environmental losses that have been incurred by agricultural progress. The past two decades have been marked by increasing environmental and conservation programs introduced by federal and state agencies and more public awareness about environmental issues as related to agricultural development.

5 Conclusions

Nitrogen and pesticide management is the key factor for controlling groundwater contamination from pesticide use, nitrogen fertilizers, and manure applications. The potential negative water quality impacts due to the excessive use of chemicals have been recognized and accepted by farmers and the chemical and livestock industry. Increased emphasis is being placed on the use of soil and plant analyses to determine more appropriate N application rates and to give proper credit to N sources such as animal manure and crop residues. Manure and fertilizer applications should be based on manure nutrient test and soil N test information. Animal manure should be applied to avoid excessive amounts of nutrient based on manure analysis. This research will help to develop long-term plans for sustainable and environmentally friendly agriculture in the Midwestern states. The innovative N and herbicide management practices (such as late spring nitrate test, strip cropping and herbicide banding) are essential to the success of sustainable production systems.

6 References

1. Schaller, R., and G.W. Bailey, eds. (1983) *Agricultural Management and Water Quality.* Iowa State University Press, Ames, Iowa.
2. Nicholson, H.P. (1969) Occurrence and significance of pesticide residues in water *Journal of the Washington Academy of Science*, Vol. 59. pp.77-85.
3. Kanwar, R.S., T.S. Colvin, and D. Karlen. (1995) Tillage and crop rotation effects on drainage water quality, in *Proceedings of the Conference on Clean Water-Clean Environment, 21st Century*, American Society of Agricultural Engineers, St. Joseph, MI. Vol 3. pp.163-166.
4. Osborn, T. (1993) The Conservation Reserve Program: Status, Future, and Policy Options. *Journal of Soil and Water Conservation*, Vol. 48, No. 4. pp. 272-79.
5. Lant, C.L., S.E. Kraft, and K.R. Gillman (1995) The 1990 Farm Bill and Water Quality in Corn Belt Watersheds: Conserving Remaining Wetlands and Restoring Farmed Wetlands. *Journal of Soil and Water Conservation*, Vol. 50, No. 2. pp. 201-05.
6. Robinson, A.Y. (1993) Wetlands Protection: What Success? *Journal of Soil and Water Conservation*, Vol. 48, No. 4. pp. 268-70.

EFFECT OF IRRIGATION WITH SALINE WATER ON SOIL AND CROP IN SOUTH-WEST SPAIN

M.M.RIDAO, F. MORENO, F. CABRERA, J.E. FERNANDEZ and M.J. PALOMO
Instituto de Recursos Naturales y Agrobiología de Sevilla (IRNAS, CSIC), Seville, Spain
E. FERNANDEZ-BOY
Department of Agricultural Chemistry, University of Seville, Seville, Spain.

Abstract
The drained and irrigated marshes of the Guadalquivir river (SW Spain) are formed on soils of high clay content (about 70%), high salinity, and a shallow, extremely saline, water table. Irrigation is necessary for successful crop production in this region of low and variable rainfall. In some years, however, water supply for irrigation is limited due to drought periods, and farmers are obliged to irrigate with river water, which at this location is of high salinity due to tidal flow. The objective of this work was to evaluate the effects of irrigation with water of high salinity on soil properties and growth and yield of cotton crop. The experiments were carried out during 1997. Irrigation was applied by furrows. Water content profile, tensiometric profile, water table level, drainage water flow, soil salinity, leaf water potential, stomatal conductance, and crop development and crop yield were monitored. The results showed that after the irrigation with saline water, the soil salinity increased. This increase was more noticeable in the top layer (0-30 cm depth). After five irrigations with water of good quality, the salinity of the soil in the subplot irrigated with saline water reached values similar to those before the application of saline water. The irrigation with saline water affected crop development. Despite the negative effects on crop development and water status in plants, the crop yield was the same as on the subplot irrigated with non-saline water.
Keywords: Crop development, drainage, furrow irrigation, saline water, soil salinity.

1 Introduction

In semiarid and arid areas water is a limiting factor for the sustainability of irrigated agriculture, and the use of water of low quality is also a risk for the environment. The

Water and the Environment: Innovative Issues in Irrigation and Drainage. Edited by Luis S. Pereira and John W. Gowing. Published in 1998 by E & FN Spon. ISBN 0 419 23710 0

Guadalquivir river marshes in south-west Spain cover an area of 140000 ha. They were formed by the accumulation of fine material dragged by the river into the large estuary excavated in the Diluvial Era. The sediments, originated from Miocene and Triassic formations in the medium and high river basin, are mainly calcareous marls, with some chalk marls. The materials forming the Guadalquivir marshes are typical of sediments deposited at different levels. The most recent sediments, 0-2 m thick, were deposited on the lower parts, in depressions with water-logging or run-off phenomena, enriched with salts by evaporation. In most cases, soils developed in this area are of clayey nature (about 70% clay content), mainly illite type with a very advanced degree of alteration [1]. They are difficult to manage in agriculture due to their high clay and salt contents [2] and to the presence of a shallow, very saline water table.

In this region of low and variable rainfall, irrigation of these soils is necessary for successful crop growth. Drainage is also required to ensure that the highly saline water table does not encroach into the root zone. In some years, however, water supply for irrigation is limited due to drought. The scarcity of good quality water during the drought period 1993-1995 imposed on farmers the necessity to irrigate with river water, which at this location is of high salinity due to tidal flow.

The objective of this work was to evaluate the effects of irrigation with water of high salinity on soil properties and growth and yield of cotton crop. The results provide data for the management of irrigation with saline water and for sustainability of agriculture in periods of water scarcity in the area.

2 Materials and Methods

Experiments were carried out during 1997 in a farm plot of 12.5 ha (250 m x 500 m) situated in an area of marshes on the left bank of the Guadalquivir river, near Lebrija (south-west Spain). The soil of the plot is of clayey texture and its general characteristics are given in Table 1. The mineralogical composition of the clay fraction (70% altered illite; 15% smectite; 10% kaolinite; < 1% interstratified) is very homogeneous throughout the profile. The plot has been installed with a drainage system, consisting of cylindrical ceramic sections (30 cm long) forming pipes 250 m long, buried at a depth of 1 m and spaced at intervals of 10 m. These drains discharge into a collecting channel perpendicular to the drains. This drainage system controls the water table level, which remains at a depth of approximately 0.9 m. The electrical conductivity (EC) of the water table is > 80 dS m-1. It is imperative that this water table does not encroach into the root zone, hence irrigation must be complemented by drainage.

Two subplots of 0.5 ha (20 m x 250 m) each were selected. Cotton was growing on both subplots, and irrigation was applied by furrows. Cotton was sown on February 23,

Table 1. General characteristics of the soil

Depth (cm)	Soil particle size (% w/w µm)			CaCO₃ (%)	O.M. (%)
	> 50	50 - 2	< 2		
0-30	1.0	32.0	67.0	16.0	1.03
30-60	1.0	30.0	69.0	16.0	-
60-90	1.0	30.0	69.0	19.0	-

1997, and mulched with plastic film. One subplot was irrigated with good quality water (EC = 0.89 dS m^{-1}) during the whole season, while in the other subplot one of the irrigations (on 7 July 1997 at flowering stage) was with water of high salinity (EC = 22.7 dS m^{-1}). The saline water used for irrigation was taken from the collecting channels. The amount of water applied in each irrigation was 60 mm.

Four measurement sites were situated within each subplot, at which the water content profile, tensiometric profile, water table level, and salinity of soil were monitored. The drainage water discharge flow was also measured and the salinity of water analysed periodically.

A neutron probe was used to measure water content in the soil. Mercury tensiometers were used to measure the water tension at different depths. Drainage water flow was measured by means of a limnigraph with a V-notch weir. Soil samples (0-30, 30-60 and 60-90 cm depth) were taken periodically at the four sampling sites.

Electrical conductivity (EC), alkalinity (Alk.), Cl$^-$, SO$_4^=$, Na$^+$, K$^+$, Ca^{2+} and Mg^{2+} were determined in water samples, and in saturated paste extracts.

Plant height and leaf area index (LAI) were measured at several dates during the growing season. Leaf water potential was measured with a pressure chamber (Soilmoisture Equipment Corp., Santa Bárbara, California, USA), and measuremets of stomatal conductance were made with a steady-state porometer (LI-1600, LICOR) in leaves, which were sunny and healthy, of six plants in each subplot.

3 Results and Discussions

Figure 1 shows the drain outflow hydrographs corresponding to the irrigation on 7 July 1997. These results clearly show no differences between the drainage behaviour in the subplot irrigated with non-saline water and subplot irrigated with saline water. Similar results were obtained in the next irrigations. These hydrographs show the same pattern as those described in previous experiments with furrow irrigation in this area [3]. The drain discharge started between 30 min and one hour after the beginning of irrigation, which also agrees with the results of Moreno et al. [3].

Changes in EC$_{sp}$ of the saturated paste extract of soil samples, taken on various dates after the irrigation on 7 July 1997, are shown in Fig. 2. In the subplot irrigated with saline water, the EC$_{sp}$ increased immediately after irrigation in both soil layers (0-30 cm and 30-60 cm depths). In the top layer (0-30 cm), the EC$_{sp}$ started to decrease after the first irrigation with non-saline water, and reached similar values to those before the irrigation with saline water. In contrast, the decrease of EC$_{sp}$ in the 30-60 cm layer was lower than in the top layer. Significant differences (P < 0.05) between the EC$_{sp}$ of the two subplots for the 0-30 cm layer were found only for the period between the irrigation with saline water and the sampling date after the last irrigation with normal water. For the 30-60 cm layer, significant differences were maintained till the end of the growing season, and even after the rainy period.

Figure 3 shows the increase of soluble Na$_{sp}$ in the subplot irrigated with saline water. The change of Na$_{sp}$ in both soil layers follows a similar pattern to that of the electrical conductivity. Differences in Na$_{sp}$ between the two soil layers were significant (P <0.05) from 15 July 1997 till the end of the growing season, and also after the rainy period.

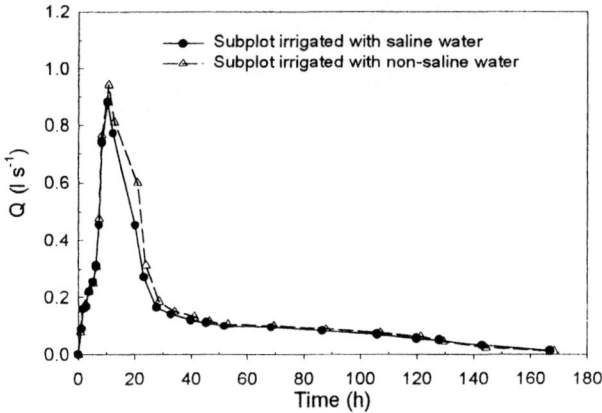

Fig. 1. Drain outflow (Q) hydrographs after irrigation on 7 July 1997 in both subplots

On 16 January 1998, Na_{sp} values were significantly different between the subplot irrigated with saline water and the subplot irrigated with non-saline water at both depths, but these values are of the same order as those reported by Cabrera et al. [4] and Moreno et al. [3] for the same commercial plot. As in the case of EC_{sp}, the soluble sodium concentration in the saturated paste extract decreased after irrigation with non-saline water, and tended to reach similar values to those before the irrigation with saline water. The pattern of sodium adsorption rate (SAR) was similar to that of Na_{sp} in both soil layers (data not shown). Significant differences in SAR were observed between the two soil layers. At the end of the rainy period, significant differences in SAR were maintained between the soil of the subplot irrigated with saline water and the subplot irrigated with normal water. SAR values of both subplots are similar to those reported by Moreno et al. [3] for the soil of the same commercial plot under cotton crop irrigated by furrow with non-saline water.

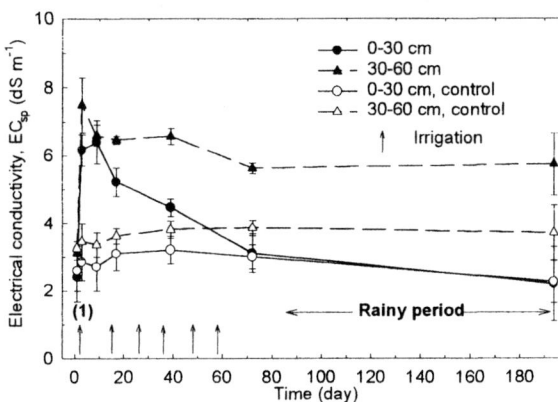

Fig. 2. Change of electrical conductivity of the saturated paste extracts (EC_{sp}) after irrigation on 7 July 1997 in subplot irrigated with saline water (filled symbols) and in subplot irrigated with non-saline water (empty symbols). (1) date of application of saline water. (Vertical bars are standard errors)

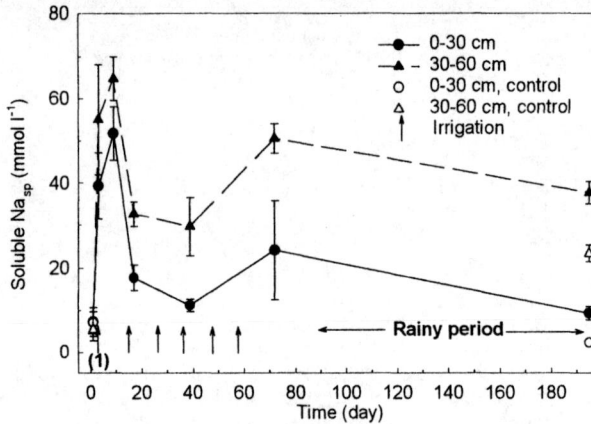

Fig. 3. Change of concentration of soluble sodium in the saturated paste extracts (Na_{sp}) after irrigation on 7 July 1997 in the subplot irrigated with saline water (filled symbols) and in the subplot irrigated with non-saline water (empty symbols). (1) date of application of saline water. (Vertical bars are standard errors).

Higher Na_{sp} and SAR values, after the rainy period (the total amount of rain was 480 mm), in the subplot irrigated with saline water than in the subplot irrigated with non-saline water seem to indicate that leaching was less effective in the former. However, as mentioned above, these values are similar to those found in these soils under irrigation with non-saline water, and do not represent a serious problem for the next crop.

Soil water potential at a depth of 40 cm was lower in the subplot irrigated with saline water than in the subplot irrigated with non-saline water during the period between 9 July 1997 and 14 August 1997. However, the differences were not significant (P < 0.05).

Changes of soil water content profile (Fig. 4) were similar in both subplots. This,

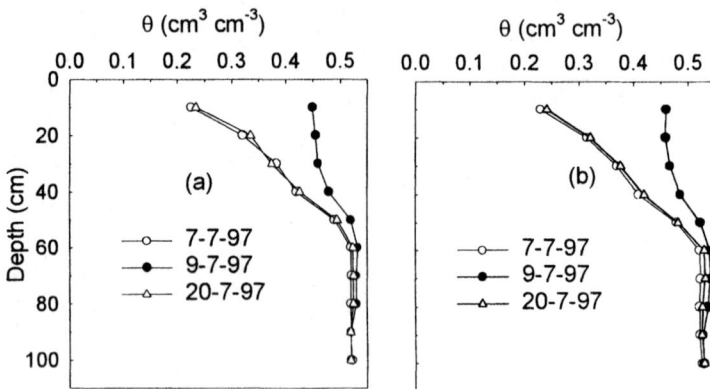

Fig. 4. Soil water content profiles before and after irrigation on 7 July 1997: (a) subplot irrigated with non-saline water; (b) subplot irrigated with saline water.

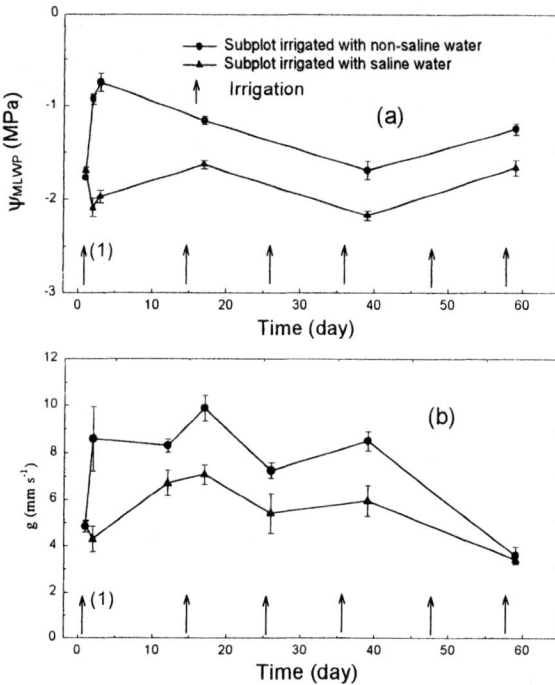

Fig. 5. Change of (a) the midday leaf water potential (Ψ_{MLWP}) and (b) stomatal conductance g, after irrigation on 7 July 1997 in the tow subplots.). (1) date of application of saline water. (Vertical bars are standard errors).

together with a similar drainage behaviour, seems to indicate that a single application of this very saline water did not affect the physical properties of the soil.

The irrigation on 7 July 1997 with saline water affected the crop development. Results of plant height and LAI (Table 2) show that a few days before the irrigation with saline water, plant height and LAI were practically the same in both subplots. On 27 July 1997, the plants of the subplot that received the application of saline water showed significantly lower plant height than the subplot irrigated with non-saline water. LAI value was also significantly lower. It seems that the saline water produced a severe stress and consequently reduced the crop development. Fig. 5 shows results of midday leaf water potential (ψ_{MLWP}) and stomatal conductance (g) in both subplots. Before application of saline water (on 7 July 1997) ψ_{MLWP} and g were practically the same in the two subplots. After this application ψ_{MLWP} and g were significantly lower in the subplot irrigated with saline water than in the subplot irrigated with non-saline water as reported by other authors, [5]. In contrast, fruit development was accelerated in these plants. The yield was slightly (but not significantly) higher than in the subplot irrigated with non-saline water (Table2). These results seem to indicate an increase in the water use efficiency by the crop irrigated with saline water.

Table 2. Plant height, leaf area index (LAI) and yield in the two subplots.

Subplot	Plant height (cm)		LAI		Yield (kg ha^{-1})
	3-7-97	27-7-97	3-7-97	27-7-97	
Irrigated with saline water on 7-7-97	64.4 a	69.7 a	3.50 a	3.71 a	4303 a
Irrigated with Non-saline water	63.5 a	101.5 b	3.68 a	4.86 b	4160 a

Values in the same column followed by the same letter are not significantly different (P< 0.05).

4 Conclusions

Results presented in this paper clearly show that the use of water of high salinity for irrigation in the reclaimed salt-affected soils of the Guadalquivir marshes increases the electrical conductivity, soluble sodium, and sodium adsorption rate of the soil profile. After several irrigations with non-saline water these parameters decreased to values similar to those before the irrigation with saline water.

The EC_{sp} reached a maximum value of 7.5 dS m^{-1} during the period between the application of saline water and the next irrigation with normal water. This EC_{sp} value was slightly lower than the EC_{sp} threshold (7.7 dS m^{-1}) given for cotton crop. The application of one irrigation with saline water negatively affected crop development. In contrast, yield was not affected.

These results indicate that it is possible to use saline water for supplementary irrigation in this area during years when the amount of of good quality water is limited, without serious problems for soil and crop.

5 Acknowledgements

The authors wish to thank J. Rodriguez and J.P. Calero for help with measurements in the field. This study was supported with funds of the Spanish CICYT, project HID96-1292, and the Junta de Andalucía (Research Group AGR-151).

6 References

1.   Moreno, F., Arrue, J.L., Murillo, J.M., Pérez, J.L. and Martín, J., 1980. Mineralogical composition of clay fraction in marsh soils of SW Spain. *Polish Journal of Soil Scicience*, Vol. 13. pp. 65-72.
2.   Moreno, F., Martín, J. and Mudarra, J.L., 1981. A soil sequence in the natural and reclaimed marshes of the Guadalquivir river, Seville (Spain). *Catena*, Vol. 8. pp. 201-221.
3.   Moreno, F., Cabrera, F., Andreu, L., Vaz, R., Martín-Aranda, J. and Vachaud, G., 1995. Water movement and salt leaching in drained and irrigated marsh soils

of southwest Spain. *Agricultural Water Management*, Vol. 27. pp. 25-44.

4. Cabrera, F., Vaz, R., Rieu, M. and Gaudet, J.P., 1992. Irrigation in a reclaimed salt affected soil of southwest Spain: II. Chemical properties. In: *Proceedings of the International Symposium on Strategies for Utilizing Salt Affected Lands,*. Bangkok, Thailand. pp. 179-189.

5. Henggeler, J. and Moore, J. (1995) The effect of salinity on cotton. In: *Proceedings Beltwide Cotton Conferences*, San Diego, CA. National Cotton Council of America, Memphis, TN, USA. pp. 1134-1136.

SOIL HYDRAULIC PARAMETERS FOR ENVIRONMENTAL AND LEACHING STUDIES

M. C. GONÇALVES
Department of Soil Sciences, National Agricultural Research Station, Oeiras, Portugal
R. M. FERNANDO and L. S. PEREIRA
Department of Agricultural Engineering, Institute of Agronomy, Technical University of Lisbon, Portugal

Abstract

Several laboratory and field methods are available for the direct measurement of the soil hydraulic properties consisting of the soil water retention curve and the hydraulic conductivity curve. However, the measuring techniques remain expensive and time-consuming, mainly for the hydraulic conductivity. Alternatively, indirect methods have been developed. Among the indirect approaches are the pedotransfer functions (PTF), which correlate the soil hydraulic properties with basic soil properties routinely available from soil surveys. In this study the unsaturated hydraulic properties estimated from PTF are compared with measured values in the laboratory, using data from 21 soil horizons or layers relative to 9 soil profiles. Soil hydraulic properties computed from PTF are compared with those obtained from field measurements when utilised in a mechanistic simulation model. Results show the potential for using these indirect methods in modelling.
Keywords: Hydraulic conductivity, modelling, pedotransfer functions, soil water retention curve.

1 Introduction

The quality of the soil and water resources is being adversely affected by the release of a variety of agricultural and industrial pollutants into the environment. A large number of computer models have been developed describing how water and chemicals move into and through the unsaturated zone [1]. They have become indispensable tools in research to simulate physical, chemical and biological processes in that zone [2]. The reliable application of computer models to field scale flow and transport problems implies the knowledge of a large number of model parameters. As the accuracy of the

Water and the Environment: Innovative Issues in Irrigation and Drainage. Edited by Luis S. Pereira and John W. Gowing. Published in 1998 by E & FN Spon. ISBN 0 419 23710 0

results of the simulations depend on the accuracy with which the model parameters are estimated, there is an increasing need for more efficient and accurate methods to estimate them. Among the more relevant parameters are those characterising the unsaturated hydraulic properties of soils.

The soil hydraulic properties consist of the soil water retention curve, $\theta(h)$, which relates the volumetric water content (θ) with the soil water pressure head (h), and the hydraulic conductivity curve, $K(h)$, relating the conductivity (K) to the soil water pressure head or to soil water content.

Even though several laboratory and field methods are available for the direct measurement of the hydraulic properties, the measuring techniques remain expensive and time-consuming, especially for the hydraulic conductivity [3] [4] [5]. Alternatively, indirect methods have been developed to estimate the hydraulic properties from more easily measured soil properties. Among the indirect approaches are the theoretical methods, which give predictions of the hydraulic conductivity from more easily measured soil water retention data (e. g. [6] [7]), and pedotransfer functions (PTF), which correlate the soil hydraulic properties with basic soil properties routinely available from soil surveys.

The term pedotransfer functions (PTF) was introduced by Bouma & van Lanen [8] for relating different land characteristics and soil properties to one another. Tietje & Tapkenhinrichs [9] defined a PTF as 'a function that has as arguments basic data describing the soil (e.g. particle-size distribution, bulk density, and organic C content) and yields as a result the water retention function or the unsaturated hydraulic conductivity function (including saturated hydraulic conductivity)'. In more recent approaches, PTF are used for the estimation of parameters in algebraic functions for describing $\theta(h)$ and $K(\theta)$ or $K(h)$ (e.g. [10] [11] [12] [13]). For Portuguese soils, PTF for estimating the parameters of the Mualem-van Genuchten equations [14], describing the $\theta(h)$ and $K(h)$ curves, from the basic soil properties [15] are used.

In this study, PTF are analysed using two processes: i) a statistical analysis, in which, using an independent data set, the estimated values of the different parameters are compared with the ones obtained from measured data; ii) a functional verification in the context of a specific model application. The model SWATRER [16] [17] was selected because the mechanistic simulation of the flow process in this model is the more often utilised in unsaturated zone leaching models [1]. Simulations are performed using measured hydraulic properties and those estimated with PTF. Results are compared with observed ones.

2 Material and methods

Soil samples have been collected in 21 soil horizons or layers relative to 9 soil profiles. Table 1 lists the locations, the classes of the soils according to the FAO/UNESCO system and the number of soil profiles that were sampled. Several basic soil properties have been determined for each sample. Laboratory methods are described in [13].

The measured hydraulic properties were parameterised using the Mualem-van Genuchten's equations [14] fitted simultaneously using the RETC code [18]:

$$\frac{\theta - \theta_r}{\theta_s - \theta_r} = \left[1 + (\alpha h)^n\right]^{-(1-1/n)} \tag{1}$$

$$K(h) = K_s \frac{((1 + (\alpha h)^n)^{1-1/n} - (\alpha h)^{n-1})^2}{(1 + (\alpha h)^n)^{(1-1/n)(\ell+2)}} \tag{2}$$

where θ is the observed volumetric water content (cm^3/cm^3), h is the imposed soil water pressure head (cm water), θ_r and θ_s are the residual and saturated water contents (cm^3/cm^3), K is the hydraulic conductivity (cm/day), K_s is the saturated hydraulic conductivity (cm/day) and α, n and ℓ are empirical shape factors. Table 2 shows the mean, maximum and minimum values of basic soil properties, measured hydraulic conductivity and Mualem-van Genuchten parameters for the data set.

Table 1. Classification of soils according to FAO/UNESCO, location and number of profiles

FAO Soil Unit	N° of profiles	Location
FLUVISOLS (FL)		
Eutric Fluvisols (FLe)	3	Coruche, Monte dos Alhos
VERTISOLS (VR)		
Pelic Calcic Vertisols (VRkp)	1	Beja
Chromic Calcic Vertisols (VRkx)	1	Beja
CAMBISOLS (CM)		
Chromic Calcaric Cambisols (CMcx)	3	Beja, Ferreira do Alentejo
LUVISOLS (LV)		
Vertic Luvisols (LVv)	1	Cuba

Table 2. Mean, maximum and minimum values of basic soil properties, K_{mes} and Mualem-van Genuchten's parameters for the data set.

	Mean	Minimum	Maximum
Soil Properties:			
Coarse Sand, (200-2000 μm), % (CS)	14.2	0.5	35.2
Fine Sand, (20-200 μm), % (FS)	34.9	15.6	70.0
Silt, (2-20 μm), % (S)	22.8	12.9	36.9
Clay, (< 2 μm), % (C)	28.1	10.3	56.0
Mean particle diameter, mm (GPD)	0.033	0.007	0.088
Geometrical standard deviation (GSD)	105.9	28.6	256.7
Bulk density, g.cm^{-3}(ρ_b)	1.50	1.27	1.77
Organic matter content, % (OM)	0.86	0.10	1.72
pH	7.1	5.4	8.6
Mean soil depth, cm (Z)	40.4	10.0	137.5
Measured hydraulic conductivity (K_{mes})	125.9	1.06	619.7
Model parameters:			
θ_r	0.1426	0	0.3522
θ_s	0.4440	0.3059	0.5742
α	0.1329	0.0120	0.3388
n	1.257	1.080	1.718
ℓ	-5.22	-11.15	-2.20
K_s	125.9	1.1	619.7

PTF from [15] are given by the following equations:

$$\theta_r = 0.33 - 0.0016\,S + 0.0071 OM - 0.20\,\rho_b \tag{3}$$

$$\theta_s = 1.00 - 0.00052\,FS + 0.040\,GPD - 0.35\,\rho_b \tag{4}$$

$$\ln \alpha = -6.84 + 0.63\ln CS - 0.45 \ln S - 0.65 \ln GPD - 0.74 \ln \rho_b + 1.38 \ln pH \tag{5}$$

$$n = 1.14 + 0.0025\,FS + 0.74\,GPD + 0.00037\,GSD - 0.020\,pH \tag{6}$$

$$\ell = 1.62 - 0.11\,S - 0.29\,C + 0.043\,GSD + 0.39\,OM - 0.43\,pH \tag{7}$$

$$\log_{10} K_s = 4.82 - 0.95\log_{10} FS - 1.61 \log_{10} S + 1.18 \log_{10} OM + 0.27 \log_{10} Z \tag{8}$$

3 Results and discussion

3.1 Statistical analysis

To determine how well each PTF estimates the Mualem-van Genuchten parameters, the estimated values of each parameter were plotted against the ones obtained from measured data so that the scatter around and the relationship of the data to the 1:1 line could be observed (Figure 1). In addition the mean error, the mean relative error and the root mean square error were determined (Table 3). The mean error (ME) was calculated by summing the absolute difference between the calculated and measured parameter values and dividing by the number of observations. The mean relative error (MRE) was determined by calculating the mean of the difference between the calculated value and the measured value divided by the measured value. The root mean square error (RMSE) was calculated by taking the square root of the sum of the squares of the differences between the calculated and measured parameter values divided by the number of the observations minus one [13] [15].

Table 3. Mean error, mean relative error and root mean square error resulting from use of PTF to estimate the Mualem-van Genuchten parameters ($n = 21$ observations).

Parameters (P)	Mean Error (ME)	Mean relative error (MRE)	Root mean square error (RMSE)
θ_r	0.132	*	0.170
θ_s	0.029	0.073	0.038
α	0.083	0.920	0.132
n	0.128	0.093	0.191
ℓ	2.817	0.682	3.420
K_s	118.7	2.230	210.5

* Because θ_r takes frequently the value zero, the MRE was not computed.

The sole statistical analysis of each parameter (Figure 1) "per se" is meaningless, because parameters act together to describe the curve. However, some interest exists in analysing parameters θ_r, θ_s and K_s because they have a physical significance and they condition the position of the curves.

3.2 Functional verification

The functional verification was performed in three steps. First, comparing the parameters of the Mualem-van Genuchten equations computed when using the PTF with those obtained from laboratory measurements in an Eutric Fluvisol in Coruche.

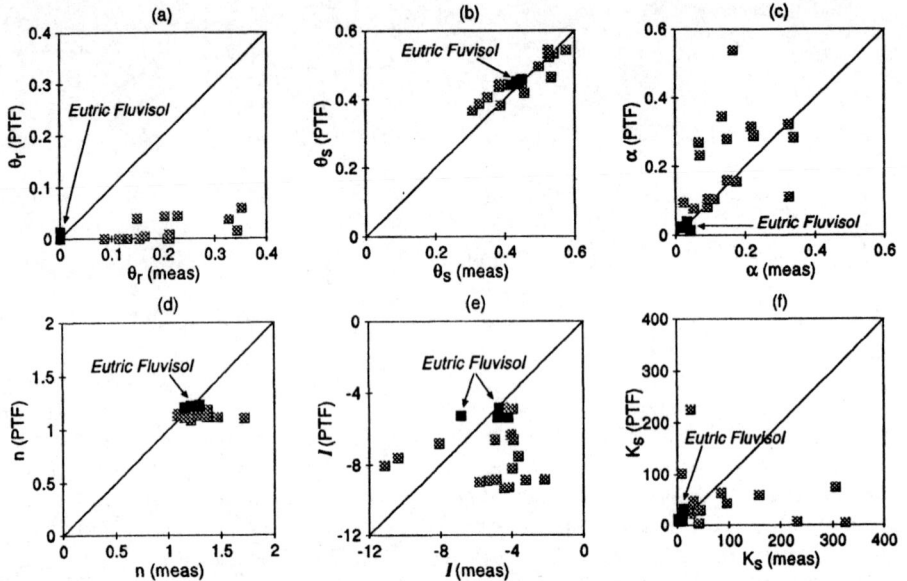

Fig. 1. Statistical analysis of Mualem-van Genuchten parameters when computed from measurements data or from the PTF: a) θ_r; b) θ_s; c) α; d) n; e) ℓ; f) K_s.

Results in Table 4 show that, for this soil, parameters estimated from PTF are close to those observed for the 4 soil layers.

When comparing the $h(\theta)$ and $K(\theta)$ curves obtained from using the parameters in Table 4, it was verified that the curves obtained from the PTF result very similar to those computed from laboratory measurements (Figure 2).

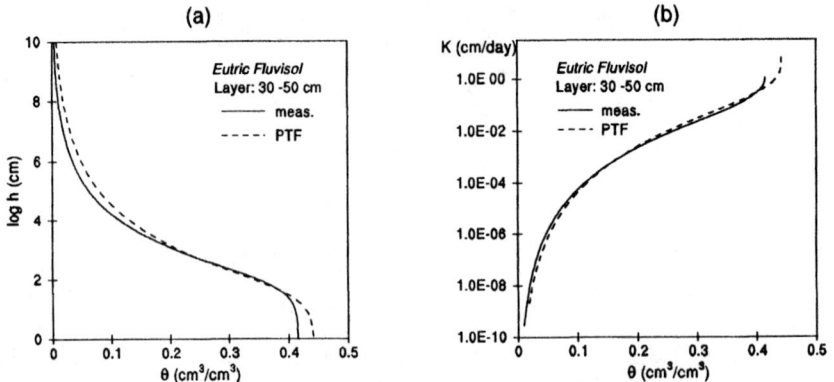

Fig. 2. Functional verification of PTF comparing: a) the $h(\theta)$ function obtained from experimental data (–) and from the PTF (--); b) the $K(\theta)$ function obtained from experimental data (–) and from the PTF (--) (curves refer to soil layer 30-50 cm in an Eutric Fluvisol).

Finally the two curves, for the 4 layers, were utilised in a model to simulate the water fluxes. SWATRER model is used to simulate soil water processes in an irrigated-cropped soil (silage maize, LG68-FAO600). The soil is an Eutric Fluvisols (Table 1) with a shallow watertable, that varies from 1.3 m, at crop emergence, to 1.75 m, at harvest. Observed values of soil water content, soil water fluxes and crop evapotranspiration are obtained from [17]. The crop was irrigated twice, respectively

Table 4. Values of the different parameters obtained with PTF for an Eutric Fluvisols in Coruche, used in the simulation. The parameters obtained from measured data and the measured hydraulic conductivity are also presented for comparison.

Prof. (cm)		θr	θs	α	*n*	ℓ	K_s	K_{mes} (cm/day)
0-25	Meas.	0	0.4329	0.0317	1.150	-6.88	12.4	18.0
	PTF	0.009	0.4525	0.0418	1.206	-5.25	32.9	
30-50	Meas.	0	0.4163	0.0125	1.266	-4.23	2.4	2.0
	PTF	0	0.4425	0.0267	1.217	-5.36	8.5	
80-100	Meas.	0	0.4507	0.0419	1.206	-4.81	10.2	12.0
	PTF	0.014	0.4589	0.0157	1.223	-5.64	11.4	
130-145	Meas.	0	0.4382	0.0120	1.286	-4.74	1.1	1.5
	PTF	0.009	0.4473	0.0222	1.231	-4.82	12.9	

with 100 and 54 mm. Simulations are performed using measured and estimated (with PTF) hydraulic properties (Figure 2 and Table 4).

Figure 3 presents the simulated and observed soil water storage from 0 to 35 cm, 35 to 70 cm and 70 to 120 cm. Observed soil water storage is calculated from profiles of water content measured with a neutron probe [17]. Simulation results with hydraulic properties estimated with PTF (sim ptf) show a very good agreement with observed values in the upper and intermediate layers, but the results are poor in the deeper layer. On the other hand, simulation results with measured hydraulic properties (sim meas.) show a poor agreement in the upper layer, and good results in both deeper layers.

Results from simulation of soil water fluxes at 110 cm depth compared with calculated values (observed), presented in Figure 4, show a good agreement with both measured and PTF hydraulic properties. Calculated values of soil water fluxes are obtained with Darcy equation, using measured hydraulic head profiles and *in situ* hydraulic conductivity determinations plus inverse parameter estimation [17]. Cumulative soil water flux at 110 cm was 76 mm for sim meas and 74 mm for sim ptf.

Observed cumulative crop evapotranspiration (ETR$_c$) during the simulation period was 347 mm. Simulation with measured hydraulic properties (ETR$_c$ = 289 mm) gave worst results than the simulation with hydraulic properties estimated with PTF (ETR$_c$ = 307 mm). However, both simulation underestimated ETR, which could be explained by the shapes of the h(θ) curve, with high water contents at a pressure head of 15 000 cm, below which most of the roots stop to extract water. This is also the reason of the poor agreement in the simulation of water storage with measured hydraulic properties in the upper soil layer (Figure 3).

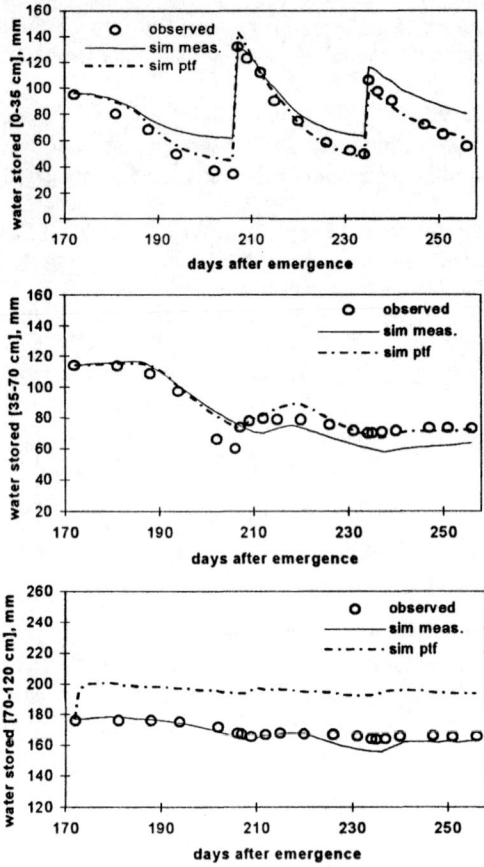

Fig. 3. Observed and simulated values of soil water storage at 0 to 35 cm, 35 to 70 cm and 70 to 120 cm layers.

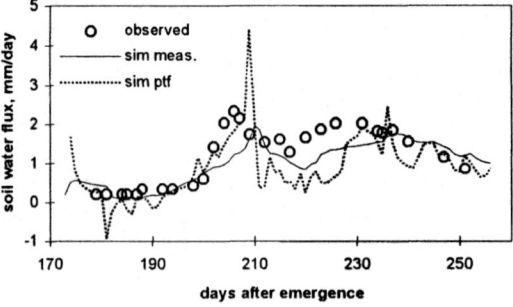

Fig. 4. Comparison between calculated (observed) and simulated values of soil water fluxes at 110 cm depth: i) sim meas. - simulation with measured soil hydraulic properties; ii) sim ptf - simulation with soil hydraulic properties estimated with PTF. Positive values represent upward fluxes.

4 Conclusions

Results show that, in the case of the Eutric Fluvisols, no significant differences exist between measured and PTF estimated soil properties. Deviations between simulated and observed values are of the same order of magnitude than differences between both simulations. Laboratory measurements are not sufficient to improve simulations. Field calibration and testing is advised when precision on the simulation is required. However, as it can be concluded from Figure 1 where the parameters for the Eutric Fluvisols are closed to the 1:1 line, for other soils showing a greater scatter the use of PTF could origin different results. Caution should be taken when using PTF and related improvements are needed.

5 References

1. Ma, L. and Selim, H. M. (1997) Physical non-equilibrium modelling approaches to solute transport in soils. *Advances in Agronomy* 58, pp 95-150.
2. Pereira, L. S. and Cameira, M. R. (1997) Unsaturated zone leaching models: An users view, in *Control of Agricultural Water Pollution* (ed. A. Kandiah and A. Hamdy), IAM-Bari, pp. 83-99.
3. Van Genuchten, M. Th. and Nielsen, D. R. (1985) On describing and predicting the hydraulic properties of unsaturated soils. *Annales Geophysicae*, 3, pp. 615-628.
4. Mualem, Y. (1992) Modeling the hydraulic conductivity of unsaturated porous media, in *Indirect Methods for Estimating the Hydraulic Properties of Unsaturated Soils* (eds M. Th. Van Genuchten *et al.*), University of California, Riverside, pp. 15-36.
5. van Genuchten, M. Th. and Leij, F. J. (1992) On estimating the hydraulic properties of unsaturated soils, in *Indirect Methods for Estimating the Hydraulic Properties of Unsaturated Soils* (eds M. Th. Van Genuchten *et al.*), University of California, Riverside, pp. 1-14.
6. Burdine, N. T. (1953) Relative permeability calculations from pore-size distribution data. *Petroleum Trans. Am. Inst. Mining Engng.* 198, pp.71-77.
7. Mualem, Y. and Dagan, G. (1978) Hydraulic conductivity of soils: unified approach to the statistical models. *Soil Sci. Soc. of America J.* 42, pp. 392-395.
8. Bouma, J. and van Lanen, H. A. J. (1987) Transfer functions and threshold values: from soil characteristics to land qualities, in *Quantified Land Evaluation* (eds K. J. Beek *et al.*), ITC Publications 6, Enschede, pp. 106-110.
9. Tietje, O. and Tapkenhinrichs, M. (1993) Evaluation of Pedo-Transfer Functions. *Soil Sc. Soc. of America J.* 57, pp. 1088-1095.
10. Wösten, J. H. M. and van Genuchten, M. Th. (1988) Using texture and other soil properties to predict the unsaturated soil hydraulic functions. *Soil Sci. Soc. of America J.* 52, pp. 1762-1770.
11. Rawls, W. J. and Brakensiek, D. L. (1989) Estimation of soil water retention and hydraulic properties, in *Unsaturated Flow in Hydrologic Modelling. Theory and Practice* (ed. H. J. Morel-Seytoux), Kluwer Academic Publishers, Dordrecht, pp. 275-300.

12. Vereecken, H., Maes, J., Feyen, J. and Darius, P. (1989) Estimating the soil moisture retention characteristic from texture, bulk density, and carbon content. *Soil Science*, 148, pp. 389-403.

13. Gonçalves, M. C., Pereira, L. S. and Leij, J. F. (1997) Pedo-transfer functions for estimating unsaturated hydraulic properties of Portuguese soils. *Europ. J. of Soil Sci.* 48, pp. 387-400.

14. Van Genuchten, M. Th. (1980) A closed form equation for predicting the hydraulic conductivity of unsaturated soils. . *Soil Sci. Soc. of America J.* 44, pp. 892-898.

15. Gonçalves, M. C., Almeida, V. V. and Pereira, L. S. (1998) Estimation of hydraulic parameters for Portuguese soils, in *Characterization and Measurement of the Hydraulic Properties of Unsaturated Porous Media.* (Proc. Int. Workshop, Riverside, CA, 1997) (in Press).

16. Dierckx, J., Belmans, C. and Pauwels, P. (1986) *SWATRER, a computer package for modelling the field water balance. Reference Manual 1*, Lab. of Soil and Water Engng., Kath. Univ., Leuven.

17. Fernando, R. M. (1993) Quantificação do balanço hídrico de um solo regado na presença de uma toalha freática. Simulação com o modelo SWATRER. Dissertação de doutoramento. Instituto Superior de Agronomia, UTL, Lisboa.

18. van Genuchten, M. Th., Leij, F. J. and Yates, S. R. (1991) *The RETC code for quantifying the hydraulic functions of unsaturated soils.* Environmental Protection Agency, USA.

REDUCING ACID POLLUTION FROM RECLAIMED ACID SULPHATE SOILS: EXPERIENCES FROM THE MEKONG DELTA, VIETNAM

T. P. TUONG
Soil and Water Sciences Division, International Rice Research Institute, Los Banos, The Philippines
L. Q. MINH and D. V. NI
Can Tho University, Can Tho, Viet Nam
M.E.F VAN MENSVOORT
Department of soil and geology, Wageningen Agricultural University, Wageningen, The Netherlands

Abstract
Using innovative land and water management methods, developed by trial and error, farmers have turned vast tracks of the traditionally abandoned acid sulphate soils (ASS) in Southeast Asia into reasonably productive lands. Leaching of ASS however creates environmental hazards. The amounts of acidity released are strongly influenced by the severity of the soil acidity, but also by the type of land use and farmers practices. In the reclaimed ASS of the Mekong Delta, Vietnam, on monthly basis, rice fields with good water management facilities release less than 2 kmol(+) ha^{-1} of aluminum even on severely acid soils. Corresponding amounts from fields provided with raised beds for upland crop cultivation, or from spoils of newly dug canals can be 8 times higher. Through diffusion and mass transport, the pollution can spread to the surrounding, affecting bio-diversity, the food chain... over a large area. Pollution hazard can be reduced by proper land uses, surface soil management, and adjusting the timing of leaching. Rehabilitating natural wetlands by impounding fresh or saline water may help prevent further oxidation and acidic export. *Melaleuca* forests have a potential as "sinks" for improving water quality drained from the reclaimed soils. The research implications of the above strategies and practices are discussed.
Keywords: Acid sulphate soils, aluminium, drainage, land reclamation, oxidation, rice irrigation, water pollution.

1 Introduction

Through oxidation of reduced S-compounds (mainly pyrite), acid sulphate soils (ASS) generate sulphuric acid that brings their pH below 4, leaks into the surface water, and

Water and the Environment: Innovative Issues in Irrigation and Drainage. Edited by Luis S. Pereira and John W. Gowing. Published in 1998 by E & FN Spon. ISBN 0 419 23710 0

attacks clay minerals, liberating soluble aluminum and other toxicities [1].   These adverse conditions have hindered the use of ASS for agriculture.

Nevertheless, under population pressure, ASS areas are increasingly being developed for agriculture, aquaculture, resorts and urban uses.   By trial and error farmers in Vietnam, Thailand, Indonesia developed innovative water management and agronomic practices enabling reclamation of large tracks of ASS for a variety of crops. Reclamation of ASS has contributed to the recent increased rice production and exportation from the Mekong Delta, Vietnam.

Acid sulphate soil reclamation processes however involve exporting the toxicities from the root zone to the surroundings, creating a serious conflict with environmental protection. Significant environmental damage, especially to fish and aquaculture, due to bad quality drainage water and changes in land use of reclaimed ASS has been reported [2, 3].

The conflict between production and environment protection poses a major challenge to managing the reclaimed ASS.   Can they be managed to maintain agricultural productivity while minimizing downstream impacts or are they best returned to wetlands? In order to address this question, it is necessary to understand the processes that lead to sulfide oxidation, the exportation of oxidation products and their subsequent distribution in the surrounding environment, and to appreciate how land use and hydrology affect these processes.

This paper reviews the key processes operating in reclaimed ASS, especially the interactions among soil chemistry, soil physics, hydrology and land use on the exportation of acidic materials to the surroundings.   We will examine management options to reduce the pollution hazards while maintaining the agricultural production. We will use examples from the Plain of Reeds in the Mekong Delta of Vietnam as case studies, but the principles are also applicable to other areas.

2 The study area

The Vietnamese part of the Mekong Delta covers a total area of 3.9 million hectares and is characterized by two contrasting seasons.   In the dry season, December to April evaporation far exceeds rainfall, which accounts for only 6-10% of the annual rainfall (1200 - 2000 mm).   The rainy season starts from May to November.

Acid sulphate soils cover about 45% of the area of the Mekong Delta. Most of these are actual ASS which were formed by oxidation of the Holocene sulphidic sediments deposited by the Mekong River during the last 10,000 years.   The majority of ASS concentrates in low back swamps, far from the main rivers.   Some potential ASS can be found in the saline mangrove area along the southeast coast of the delta and under permanent water bodies in the lowest depressions.

Because of it flat topography, the hydrology of the delta is completely governed by the upstream discharge, tidal fluctuations and rainfall.   Corresponding to the dry and the rainy season, there is also a low flow season and a flood season.   Maximum inundation depths, which occur in October, can reach 3 m in the upstream portions of the delta.   In the depressed areas, inundation periods can extend until January or February because of inadequate drainage facilities.   Fig. 1 illustrates typical dynamics of rainfall and water depth in an ASS area.

Rainfall (mm)   Water level (m)

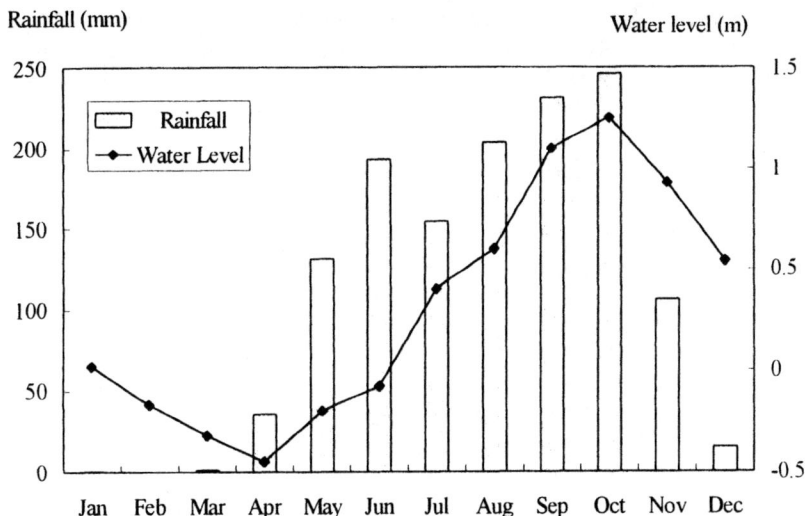

Fig. 1. Monthly rainfall and water level (above the mean ground level) in the Plain of Reeds, Mekong Delta, Vietnam.

Originally the ASS areas had only small natural streams and were covered by thick mats of vegetation. The most dominant natural vegetative indicators of ASS include Chinese water chesnut (*Eleocharis dulcis*), reed (*Phragnmites karka*) and *Melaleuca* (*melaleuca cajuputi*). To facilitate irrigation, drainage and transportation, a network of canals was constructed in the area. Thought some main canals were dug during the 1930's, 1940's, it was only after the end of the war in 1975 that the canal network has been expanded at an unprecedented rate. Apart from draining water during the flood season, these canals also act as irrigation canals in the dry season, facilitating the flow of water from the Mekong river to areas which are remote from the river. Previously, fresh water could not reach these areas by natural tidal action during the dry season.

3 Generation of acidity: Oxidation of sulphidic materials

3.1 The process
When sulphide sediments (pyrite) are exposed to oxygen, such as in periods of prolonged drought, or after draining, they are oxidized to sulphuric acid. Oxidation processes have been described in details elsewhere [4, 5]. In short, the processes involve the conversion of solid pyrite to dissolved iron and sulphate through a series of steps.

When completely oxidized, each mole of pyrite produces two moles of acid sulphuric. The acid soil solution in the oxidized sulphidic sediments reacts with clay minerals to release silica, and metal irons, principally aluminum, iron, potassium and manganese. Acidic toxicities, especially aluminum, are particular hazardous for fish and aquatic organism, since their threshold concentrations are far less than those for plant roots [4].

3.2 Oxidation of the Mekong Delta ASS: effects of human intervention

Oxidation of pyritic materials can occur in the dry season when the water table is lower than the layer in which pyrite starts to occur. The depth of the oxidized layer is determined by the depth of the lowest water table, which occurs at the end of the dry season.

There is a general belief that the recent excavation of canals lowers the water table and thus increases the oxidation of the ASS of the Mekong Delta. Validated modeling and field data however showed that changes in the canal network increased the dry season flow from the Mekong river to the ASS areas and raised the water level in the canal network [6]. Tuong [2] showed that at the end of the dry season, groundwater tables in Mekong Delta soils are lower than the water level in the canals. An increased water flow into the ASS area and more intensive canal network may increase the recharge from the canals to the groundwater.

Most of oxidation of ASS may have occurred before the period of intensive agriculture in the ASS areas. Natural vegetation, such as *Melaleuca*, with their deep root can extract groundwater to maintain high rate of evapotranspiration rate during the dry season and lower the water table considerably in years with severe drought. White [5] also indicated that oxidation of the suphidic sediments in backswamp areas of Australia occurred before engineered drainage.

Deepening existing canals and the digging new canals may easily lead to excavation of the pyrite-containing layer. Spoils of pyritic materials, being exposed directly to atmospheric oxygen, acidify much more rapidly than pyritic materials in situ. By this manner, canal excavation produced a tremendous amount of acidity. Construction of raised beds (ridges formed by piling up soil materials excavated from adjacent lateral ditches) for upland crop cultivation also produces addition oxidation.

4 Contamination of surface water

4.1 Agro-hydrological processes

Major water flow and transport processes in ASS in areas with tropical monsoon climate are summarized in Fig. 2. During dry season, water and soluble substances are transported by the capillary rises and accumulate in the topsoil layers. As the rainy season arrives, the soluble substances are dissolved by rainwater. They are leached (vertical removal from the root zone) or flushed (horizontal removal from the surface layers). The vertical movement of water produces leaching, while flushing is by surface drainage or surface runoff.

4.2 Pollution: the source

All or part of the leached or flushed substances will finally go into the drainage systems. In that sense, they become the pollutants to the surface water. Leaching processes in ASS are dominated by the bypass flow through the macro pores. Soil management and land uses, which affect the water-conduction pore system will have marked influence on leaching of ASS [7]. Leachate from raised beds have lower pH (from 2.9 to 3.4) and

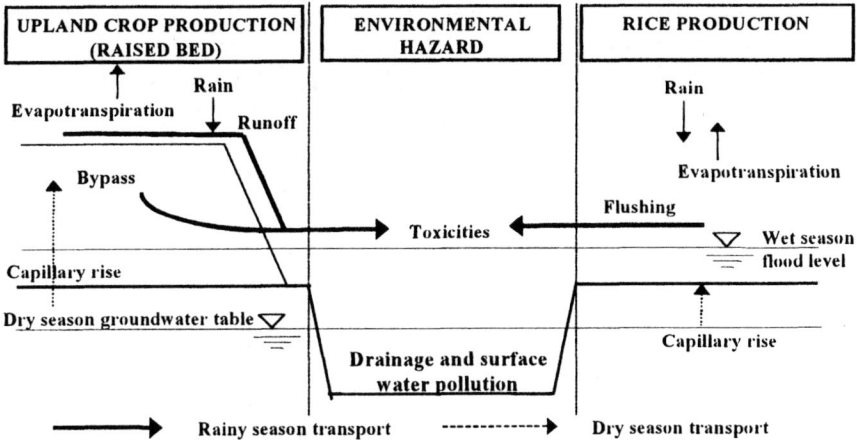

Fig. 2. Main solute transport processes in acid sulphate soils reclaimed for agricultural production.

higher Al^{3+} concentration (from 7 to 15 mmol(+) l^{-1}) than that from the rice fields (pH = 3.5 to 4.2; aluminum 4-5 mmol(+) l^{-1}).

Monthly total amount of aluminum released to the surroundings depends also on the volume of drainage water, which is also higher in upland crop lands than in rice fields. Rice fields with good water management facilities release less than 2 kmol(+) ha^{-1} of aluminum even on severely acid soils, while fields provided with raised beds for upland crop cultivation pollute the surface water much more severely, up to 16 kmol(+) ha^{-1} of aluminum (Fig. 3). Consolidation and crust forming in raised beds reduced the concentration and amount of aluminum released with respect to the age of the raised beds [7].

4.3 Spreading of pollution
The acid solutes are transported to the interconnected canal networks and contaminate much larger areas than from which they originate. Pollution from ASS leaching is most hazardous to the environment at the beginning of the rainy season due to a combination of highest total acidity released to the canal network and low river discharge. After heavy rainfalls in June, water is built up in the backswamp areas and drains toward the Mekong river [6].

Recent canal deepening, widening and new construction have drastically increased the flow capacity of the canal network and has increased the rate of transmission to the surrounding surface water of the oxidation product. White ([5] also showed that the engineered drainage has decreased the time scale for drainage of floodplains in Australia from 100 days to 5 days.

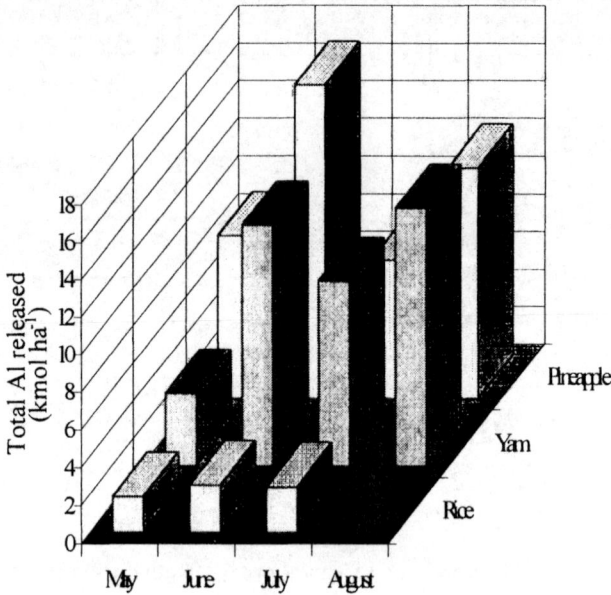

Fig. 3.   Monthly mean amount of aluminum released to the surrounding canal network from pineapple and yam raised beds and rice fields. In the same months, means (represented by column height) with the same letters are not significantly different at the 5% level by DMRT [7].

5 Management options to reduce acidic pollution

5.1 Land use selection and crop management

For potential ASS, i.e. ASS before pyritie oxidation and acidification, it is best to keep soil in reduced condition so that acidity will not be produced. In the coastal area of Vietnam, farmers rely on a crop of rice during the rainy season, then raise shrimp in the dry season [8]. The field is always kept flooded, reducing the oxidation hazards.

On actual ASS, growing *Melaleuca* requires the least soil disturbance and produce the least amount of acidity to the surrounding. Unfortunately this requires long term investment and is economically not as attractive as rice or other agricultural crops [9]. Forming raised beds to grow acid-resisting crops such as yams, sugar cane or pineapples are the heaviest surface water polluters. For the farmer interest, however, they are also the most efficient ways to improve severe ASS, especially for land-hungry people with no other possibilities. Growing rice in these areas seems more desirable from the environmental point of view but there may not be enough irrigation water to support rice in the dry season.

Practices like straw mulching, surface tillage can significantly lower the dry season toxicity accumulation in the topsoil and reduce pollution hazards from the raised beds and rice fields [10].

5.2 Adjusting cropping season

Traditionally, rice is cultivated in the rainy season and preceded by leaching in May, June. Minh et al. [11] showed that the environmental hazards could be reduced when the leaching treatment is carried out at the flood recession followed by an irrigated rice crop, which is the period from December to March. Acidity of the drainage water is more than 2 times lower than that from soils which are leached at the beginning of the rainy season. This is because the soil becomes less acidic after being submerged throughout the flood season. The discharge of the Mekong River and canal networks at the end of the flood season is two to three times that of the discharge at the beginning of the rainy season. A halved acid concentration in the drain water and a two to three times greater discharge at the end of the flood season would potentially result in a four to six times lower concentration in the surrounding water than leaching the beginning of the rainy season. This new cropping pattern is now widely practice in the Mekong Delta.

5.3 Reducing pollution from canal embankment

The oxidation of the embankment of the newly dug canals probably produce acidity comparable to the raised beds. Leachates from canal banks however flows directly to the canal water, and thus probably imposes higher pollution hazards to the surface water networks than that from the raised beds.

A different method of bank construction may help reduce the rate of oxidation and the concentration of the leachates. Covering the bank with non-acid materials and soil compaction may reduce the rate of oxygen diffusion and therefore the rate of the oxidation of the sulphidic materials. Compaction will also reduce the amount of bypass flow through the oxidized materials.

5.4 Rehabilitation of natural wetland by reflooding

In most of moderately acid soils (pH 4-6), water logging causes an increase in pH to a value between 6 and 7 after a few weeks of flooding. An attractive and economically viable treatment strategy for preventing further oxidation and acidic export is to rehabilitate the natural wetland by impounding fresh water with embankment and other hydraulic structures.

Several "forest reservoirs" in the Mekong Delta were established along the above principle. For example, in Tram Chim National Reserve, more than 5,000 hectares were surrounded by a system of dikes and gates which aim at retaining flood water for a longer period than the surroundings [12]. After a few years of implementation, water quality of the surface water in the reserve was improved and became superior to that in areas without water control. Fish yield, species richness and biomass of plankton and zoobenthos also increased significantly [13]. The reserves can serve as refuges for fauna.

Initial results are encouraging. However, most of studies on the reserves have been carried out in isolation of the surroundings. There is a need to quantify the effects of the adjacent intensive rice cultures on these natural wetlands and vice versa. It is important, for example, to determine the minimum total area of the natural wetlands, their optimum spatial distribution within the reclaimed ASS areas so that they can perform their ecological functions as refuge and for water quality improvement of the whole reclaimed areas.

It should be noted that reflooding of drained, partially oxidized floodplains with freshwater might not always be a panacea. This is because the large volume of acid stored in the soil causes a lack of metabolizable organic matter in the sediments needed to reduce sulphate and also because of irreversible changes to the soil due to oxidation [4,5]. Tidal flooding with brackish water seems more promising because of its greater buffering capacity and the daily water exchanges due to strong tidal action [5].

5.5 Trapping pollution with *Melaleuca* forest
The outflows from the agricultural lands can be routed to an adjacent Melaleuca forest, instead of flowing to the surrounding canal networks. The acidic discharge would not be harmful to the forest because Melaleuca has very high resistance to acidity. Ni et al. [14] showed that the water improved its quality as if flowed through the Melaleuca forest. This seems to support the hypothesis that Melaleuca forests can improve the surface water quality biologically and chemically by converting pollutants to less harmful forms or storing them within the soil. Water stored in the forest can be recycled for irrigation of the surrounding agricultural lands.

6 Concluding remarks

Improving agricultural production and reducing pollution hazards are the two sides of the same coin when reclaimed ASS. Pollution hazards are however not easily recognized by the farmers who cultivate the land for their daily basic food. This is because the environmental impact is at some point downstream from his field. The government has to come up with policies, which are conducive to balancing agricultural production and environment protection at a regional scale. This needs an ability to quantify the amount of acidity released and the spatial and temporal dynamics of the acidic concentration in surface water on a regional scale. There is also a need for economic assessment of the negative impact on fisheries, aquaculture and human health of the acid pollution from reclaiming acid sulphate soils.

7 References

1.   Breemen, N. van and Pons, L. J. (1978) Acid sulfate soils and rice, in *Soil and rice*, The International Rice Research Institute, Los Baños, Laguna, Philippines. pp. 739-761.
2.   Tuong, T. P. (1993) An overview of water management of acid sulphate soils, in *Selected papers on the Ho Chi Minh city symposium on acid sulphate soils*, (eds. D.L. Dent and M.E.F. van Mensvoort), International Institute for Land Reclamation and Improvement Publication No. 53, pp. 281-287.
3.   Willet, I.R., Melville, M.D. and White, I. (1993) Acid drainwaters from potential acid sulphate soils and their impact on esturine ecosystem, in *Selected papers on the Ho Chi Minh city symposium on acid sulphate soils*, (eds. D.L. Dent and M.E.F. van Mensvoort), International Institute for Land Reclamation and Improvement Publication No. 53, pp. 419-425.
4.   Breemen, N. van. (1993) Environmental aspects of acid sulphate soils, in *Selected*

papers on the Ho Chi Minh city symposium on acid sulphate soils, (eds. D.L. Dent and M.E.F. van Mensvoorst), International Institute for Land Reclamation and Improvement Publication No. 53, pp. 391-402.

5. White, I., Melville, M.D., Wilson, B.P. and Sammut, J. (1997) Reducing acid discharges from coastal wetlands in eastern Australia. *Wetlands ecology and management*, Vol. 5, No. 1. pp. 55-72.

6. Khue, N.N. (1993) *Hydraulic regime in the Plain of Reeds in the non-flooding period*, Sub-Institute of Water Resources Planning, Ho Chi Minh, Vietnam.

7. Minh, L.Q., Tuong, T.P., Mensvoort, M.E.F. van and Bouma, J. (1997) Contamination of surface water as affected by land use in acid sulphate soils in the Mekong delta, Vietnam. *Agriculture, Ecosystems & Environment*, Vol. 61, No. 1. pp. 19-27.

8. Xuan, V.T. (1993) Recent advances in integrated land uses on acid sulphate soils, in *Selected papers on the Ho Chi Minh City symposium on acid sulphate soils*, (eds. D.L. Dent and M.E.F. van Mensvoort), International Institute for Land Reclamation and Improvement Publication No. 53, pp. 129-135.

9. Thien, N.H. (1996) Winning support for conservation from local communities around Tram Chim, in *Towards sustainable management of Tram Chim National Reserve, Vietnam* , (eds. R.J. Safford, D.V. Ni, E. Malby, and V.T. Xuan), London, Royal Holloway Institute for Environmental Research, pp. 17-46.

10. Minh, L.Q., Tuong, T. P., Mensvoort, M. E. F. van, and Bouma, J. (1998) Soil and water table management effects on aluminum dynamics of an acid sulphate soil. *Agriculture, Ecosystems & Environment* (In press)

11. Minh, L.Q., Tuong, T.P., Mensvoort, F. van and Bouma, J. (1997) Tillage and water management for increasing riceland productivity in acid sulphate soils of the Mekong delta, Vietnam. *J. Soil and Tillage*, Vol. 42. pp. 1-14.

12. Safford, R.J., Ni, D.V., Malby, E. and Xuan, V.T. (1997) *Towards sustainable management of Tram Chim National Reserve, Vietnam.*, London: Royal Holloway Institute for Environmental Research.

13. Xuan, T.T. (1997) Management, conservation and utilization of aquatic resources at Tram Chim, in *Towards sustainable management of Tram Chim National Reserve, Vietnam*, (eds. R.J. Safford, D.V. Ni, E. Malby, and V.T. Xuan), London, Royal Holloway Institute for Environmental Research, pp. 103-108.

14. Ni, D.V., Maltby, E., Tuong, T.P., Safford, R.J. and V.T. Xuan (1997) The role of Melaleuca in tropical wetlands. Paper presented at the V Symposium on the Biogeochemistry of wetlands Royal Holloway University of London, 16-19 September 1997.

CONTROL OF FURROW EROSION AND INFILTRATION ON A MEDITERRANEAN SOIL USING POLYACRYLAMIDE

F.L. SANTOS and R.P. SERRALHEIRO
Department of Rural Engineering, University of Évora, Évora, Portugal
F.S. MELHORADO
Department of Ecology, University of Évora, Évora, Portugal
M.R. OLIVEIRA
Department of Crop Science, University of Évora, Évora, Portugal

Abstract
The effects of irrigation water-added polyacrylamide (PAM) on erosion control and infiltration were studied on a highly erodible Mediterranean soil organised on contour and on traditionally slopped furrows. Water added PAM significantly reduced sediment concentration and increased infiltration on the slopping ones. Obtained reductions on the contour site were less dramatic as cut back of initially high inflow rates and furrows implanted on a terrace of very mild slope helped to alleviate the differences between PAM treatment and the control. As for the infiltration, water-added PAM in the slopped furrows avoided the slacking and seal formation observed in the wetted perimeter of the non treated ones, and helped to maintain infiltration rates higher for longer periods of the irrigation set. Cumulative infiltration was also improved, showing high percentage increases for all events. In spite of the visible improvement on aggregate stability of the soil, on contour furrows the infiltration results were not significantly different from control. Regular use of PAM in the irrigation water also helped to reduce fine particle movement in the furrows, preventing tailwater pollution with fine sediments.
Keywords: Erosion, infiltration, polyacrylamide, sediment loss, soil particle movement, water quality.

1 Introduction

Irrigated crop production is critical to global agricultural output. The total irrigated cropland accounts for only 18% of the total Earth's cropland and surface irrigation, mostly furrow irrigation, accounts for 60% of this area [1].

The irrigated agriculture's high productivity on the estimated 18% of Earth's cropland makes up for over 50% of the total crop world production. Yet, with highly erodible soils supporting most irrigated agriculture and high reported furrow outflow

Water and the Environment: Innovative Issues in Irrigation and Drainage. Edited by Luis S. Pereira and John W. Gowing. Published in 1998 by E & FN Spon. ISBN 0 419 23710 0

soil losses [2] [3] [4], the irrigated agriculture is seriously under risk. Several conservation practices for furrow irrigation have been developed to eliminate runoff-carried sediment [5]. However, according to Sojka [1] few of these practices have been widely adopted, largely because they are regarded as inconvenient or intrusive by furrow irrigators, or they require additional or unfamiliar field operations that occur during otherwise busy periods in the farming schedule. Recently, high molecular weight anionic water-soluble polyacrylamide (PAM) has been reported to increase soil aggregate stability and flocculate suspended sediments, thereby reducing sediment detachment and transport in irrigation furrows [6] [7] [8] [9] [10].

In Portugal, of the 720 000 irrigated hectares 85% is under surface irrigation, and primarily furrow-irrigated (74% of the total irrigated area) [11]. In the south it is currently under construction a large irrigation scheme (Alqueva´s project) which will add 110 000 more irrigated hectares to the already existing 720 000 ha by the end of year 2025. This area, today under dry-farming agriculture, is mainly in sensitive, highly erodible and eroded Mediterranean soils. Its conversion to irrigation will expose it too to gradual irrigation-induced erosion if no lessening measures are taken in time.

The Mediterranean soils are very typical of the irrigation districts in southern Portugal, where they represent 40% of all soils. The particular characteristic of the profile of Mediterranean soils with a rapidly permeable A-horizon overlaying a B-horizon of very low permeability determines that infiltration rate of the irrigation water becomes rapidly limited to the low permeability of the B-horizon [12]. Consequently, late in an irrigation the loss of potential gradient coupled with the irrigation wetting front encountering this zone of low permeability significantly reduce infiltration. The main consequence of this soil infiltration behaviour is an almost constant advance velocity along the furrow [12] and, consequently, a high potential for the advance stream to induce soil erosion, sediment losses, and pollution. Best management irrigation practices for these soils require short irrigation, cut back of initially high inflow rates and conservation practices to control erosion. We hypothesised that irrigation water-added PAM could increase aggregate stability of the soil, flocculate sediments and help to maintain infiltration rates high.

The objectives of this study, under the field conditions provided by contour and slopping furrows on a highly erodible Mediterranean soil, were to:

- Determine the effect of irrigation water-added PAM on furrow erosion control.
- Determine the effect of water-added PAM on furrow infiltration.
- Determine the effect of water-added PAM on sediment movement in the field.

2 Water and erosion management with PAM

Over the last several years it has been recognised that some irrigated areas have developed water quality and other environmental problems largely through practices which have caused irrigation induced erosion and other related problems. A priority to resolve these problems stimulated research and methods to define and address the need for better water and land management in irrigated agriculture. Out of these efforts has resulted the use of polyacrylamide treatment of the irrigation water to control erosion and enhance infiltration.

According to Sojka [1] polyacrylamide treatment of irrigation water may be the fastest growing conservation technology in irrigated agriculture in the USA. PAM-use has proven highly effective for erosion control by reducing erosion losses below soil-loss-tolerance limits on furrow slopes ranging from 0.5-3.5%, and is well received by furrow irrigators [5]. Work published by several authors [13] [14] [15] concluded also that PAM is an excellent cost effective and safe technology to reduce both sediment and chemical loading in agricultural runoff. In addition to soil erosion control benefits, the use of polyacrylamide in furrow irrigation has been shown to highly improve total furrow infiltrated volumes [7] [8] [9] [10] [14] [15] by reducing sediment deposition and seal formation. Lentz [8] reported net infiltration increase of about 15% in field scale tests when furrows advance water were treated with up to 20 ppm PAM, and Trout [8] reported infiltration increase of 30% on the same soil with tests using a recirculating infiltrometer. They both claim that PAM-use prevented pore blockage along the wetted perimeter, promoted greater lateral flow and removed sediment in the flowing water.

3 Materials and methods

3.1 Field layout
The study described in this article was designed to determine whether low concentration of anionic polymer in irrigation water would appreciably reduce irrigation erosion and enhance infiltration on contour and slopping furrows in a typical Mediterranean soil.

Studies were conducted from June to September of 1997 on two fields of a co-operating farmer located south of Portugal, at the Herdade do Cabido near Arraiolos: one on site 1 with furrows on organised contour terraces following closely contour lines of 0.2% slope, and the other on site 2 with traditional slopped furrows averaging 1.2%. Surface soil texture in these two sites was similar and sandy loam, overlaying a clay loam B-horizon of very low permeability. Such a characteristic has imposed on site 1 the organisation on contour lines, to reduce slopes and erosion. On both fields seedbeds were disked, then roller-harrowed, and planted to corn (Zea Mays) on row spacing of 0.75 m. Furrows were shaped with a furrow-forming tool and irrigation water was applied from adjustable spigots on plastic gated pipes. On the contour furrow field water was delivered by a cablegation system used to continuously cut-back the inflow during irrigation. The system delivered flow rates of 2.7-0.5 L/s into furrow lengths varying from 160-220 m. For slopping furrows the delivery flow rate was set to 1.7 L/s on lengths of 140 m.

3.2 Polyacrylamide application
A high molecular weight anionic PAM (20% hydrolysis) manufactured and marketed under the trade name Superflock A836 by Cytec Industries was applied to the irrigation water. The white granular material was used to prepare a 2400 ppm aqueous stock solution concentration which was metered into monitored stream flows at each furrow head to achieve a concentration of 10 ppm (g/m^3) in the advancing water flow [1] [13]. Six contour and three slopping furrows were randomly selected for PAM delivery, and regularly monitored. Equally on both fields control furrows, with no PAM treatment, were selected and monitored.

Five irrigations were applied to the contour furrows and three to the slopping ones. Polymer treatment was employed during all irrigations. Irrigation and runoff times were noted and runoff volumes were periodically monitored using calibrated V-notch flumes [16]. Net furrow infiltration was determined from the differences between inflow and runoff volumes.

One-litre runoff samples were collected from free-flowing flume discharges at each flume reading (at the inflow and tail end of the monitored furrows), and from the middle furrow length. Samples were collected every 20 min during each irrigation event, and the settled volume per litter of sediment in Imhoff cones was determined every 30 min. The weight of sediment in the cone was determined by calibrating the volume of settled sediment and sediment weight per unit volume of runoff according to the procedure described in Sojka et al. [17]. ($R^2 = 0.983$). Sediment reduction and infiltration increase was computed as suggested by Lentz and Sojka [13]. Analysis of variance and the Scheffé multiple comparison procedure were employed to test for significance of treatment effects and to examine mean separations.

4 Results and discussion

4.1 Erosion control
PAM applications produced highly visual results in the two sites, with flowing and runoff water having a transparent or clear appearance. PAM treatments reduced mean sediment concentrations in all irrigations, showing the greatest benefits on the slopping furrows and after the first two irrigation events.

4.1.1 Contour furrows
Average sediment concentration data for the five monitored irrigations at experimental site 1 are presented in Table 1.

Table 1. Average sediment concentration at the three monitored location in the contour furrows

Monitored location	Sediment concentration (kg/ha) - PAM applied furrows irrigations				
	1st	2nd	3rd	4th	5th
inflow	2513.8	490.1	129.6	55.0	0.0
middle	108.4	57.9	-	-	14.2
tail end	15.4	30.0	7.4	7.4	11.8
Monitored location	Sediment loss (kg/ha) – Control furrows applied irrigations				
	1st	2nd	3rd	4th	5th
inflow	1652.5	703.0	298.9	297.8	181.1
middle	231.3	-	100.9	160.5	285.1
tail end	99.0	255.0	151.0	80.5	112.6

- no data available.

PAM application reduced sediment concentration for all five irrigations and locations, except for the inflow location at 1/3 of the furrow length. This location showed a 52% PAM increase in sediment with the first irrigation, fact that is explained by the rapid flocculation and deposition of soil sediment lifted and transported by the high initial inflow rate. The 2nd and subsequent PAM irrigations reduced sediment by 30.3, 56.6 and 100%, respectively. Concerning the furrow tail end percent sediment reduction were consistently greater for the PAM treatment, with values ranging from 84.4% for the first irrigation up to 90% for the last one. In spite of the observed percentage differences in sediment concentration and visual improvement in the aggregate stability of the furrows, the perceived reductions were not statistically significant. For the field conditions of this site PAM small differences are difficult to differentiate statistically as they are obscured by the contour furrows, an erosion conservation practice on its own right, and by the provided gradual cut-back of inflow rates during irrigation.

4.1.2 Slopping furrows

Average sediment concentration data for the monitored irrigations at site 2 are presented in Table 2. PAM application reduced sediment concentration for all three irrigations and monitored locations. At 1/3 furrow length location near the inflow head sediment concentration from 1st to 3rd irrigation averaged reductions with PAM application ranging from 87.3-98.9% while at the tail end reductions were higher, and in the order of 96.0-98%.

Table 2. Average sediment concentration at the three monitored location in the slopping furrows

Monitored location	Sediment concentration (kg/ha) – PAM applied furrows irrigations		
	1st	2nd	3rd
inflow	2820.9	112.3	51.0
tail end	693.9	504.6	60.6
Monitored location	Sediment concentration (kg/ha) – Control furrows irrigations		
	1st	2nd	3rd
inflow	22279.7	4373.9	4049.6
tail end	17532.3	8653.4	2607.3

Contrastingly with contour furrows here sediment concentrations were dramatically more important, showing the control treatment concentrations up to 79 times higher than PAM treated furrows. Significant differences between PAM and control were observed at the tail end location.

4.2 Sediment movement

Analysis of data presented in Tables 1-2 show that there was a continuos shifting of sediments in the furrows with each irrigation. Due to the number of irrigations this fact is more conspicuous in the control treatment of Table 1. There, the sudden increase in sediment concentration at the middle location of the control furrow after the 3rd irrigation and at the tail end after the 4th irrigation could be resulting from sediment being carried in one place and deposited on another.

Samples collected at the monitored locations in the contour furrows and analysed for particle sizes (Table 3) tend to reveal this continuos shifting of fine deposits. No data were collected for the slopping furrows.

Table 3 Average percentage of sand, silt and clay in each contour furrow monitored locations

Monitored location	PAM applied furrow			Control furrow		
	Sand (%)	Silt (%)	Clay (%)	Sand (%)	Silt (%)	Clay (%)
inflow	77.6	7.7	15.9	85.2	5.8	8.9
middle	73.3	12.3	15.9	81.7	6.4	11.3
tail end	66.7	14.3	21.8	82.0	6.3	11.2

For clay particles PAM-applied furrows present on average a relative increase of 44.2, 28.7 and 48.4% at the inflow, middle and end tail when compared to the control. Taking that all furrows presumably had the same surface texture prior to the application of PAM in the water, treated furrows show a relative increase of 41.3% for clay particles, fact which can lead to conclude that clay particles were retained in PAM treated furrows and carried away in the control ones. Similar fact takes place for silt, which also showed a relative increase of 45.6% in the PAM treated furrows, when compared to control. Contrary to the observed for silt and clay, in the control furrows a relative increase in sand was observed (8.9% at the inflow, 10.3 % at the middle and 18.7% at the tail end) which also leads to conclude that clay and silt particles are transported away from them, and into the tail end water runoff, increasing the relative size of the final sand fraction.

This removal, transport and deposition of sediments in the flowing water, as they show to affect surface soil structure and particle size distribution, had a significant impact on water intake rates and net infiltration.

4.3 Infiltration

As referred above, PAM applications produced highly visual results in the two sites, showing flowing and runoff water with a transparent or clear appearance. Besides stabilising the treated furrows from slacking, field observations revealed that irrigation water-added PAM prevented seal formation, promoted a greater lateral flow, and increased the high initial infiltration rate for a longer proportion of the irrigation set. The greatest benefits were observed for the slopping furrows.

4.3.1 Contour furrows

Average infiltration rates and cumulative infiltration for the five monitored irrigations at experimental site 1 are presented in Table 4. High variability among tests within the same treatment was observed. For all PAM treated furrows intake rates and cumulative infiltration were higher than the control, and PAM benefits were noticed after the second irrigation, where the percentage increases were more relevant. These facts can be explained by the higher stability of the wetted perimeter and reductions in seal formation induced by PAM-use. They were highly perceptible in the field.

The percent increase in cumulative infiltration was higher than the reported increases of 15-30% described in some similar studies [7] [8]. However no significant differences were observed between treatments and irrigation events.

Table 4. Average intake rates and cumulative infiltration for all contour furrow irrigations

Monitored irrigation	Infiltration					
	Rate			Cumulative		
	Control (L/m min)	PAM applied (L/m min)	Increase %	Control (L/m)	PAM applied (L/m)	Increase %
1st	0.51	0.59	13.3	46.0	78.8	41.6
2nd	0.27	0.34	20.5	31.9	43.8	27.2
3rd	0.16	0.33	52.7	36.3	43.8	16.9
4th	0.11	0.24	52.7	12.8	31.8	59.7
5th	0.10	0.21	49.5	22.1	45.6	51.5

4.3.2 Slopping furrows

Average infiltration rates and cumulative infiltration for the three monitored irrigations at experimental site 2 are presented in Table 5. Significant differences were observed between treatments. High variability among tests within the same treatment still persisted, with tests showing no significant differences between irrigation events. For all PAM treated furrows intake rates and cumulative infiltration were higher than the control, with percentage increases of 58-85%. PAM infiltration benefits were noticed since the first irrigation, probably due to furrow shape stability throughout irrigations, greater lateral flow and less seal formation, visibly perceptible in the field. Seal formation occurred rapidly in control furrows and PAM surface seal prevention helped increase advance-phase infiltration rates on treated furrows.

Table 5. Average intake rates and cumulative infiltration for all slopping furrow irrigations

Monitored irrigation	Infiltration					
	Rate			Cumulative		
	Control (L/m min)	PAM applied (L/m min)	Increase %	Control (L/m)	PAM applied (L/m)	Increase %
1st	0.05	0.17	70.6	4.0	20.4	80.3
2nd	0.13	0.31	58.0	7.4	50.0	85.2
3rd	0.05	0.24	79.2	3.4	11.3	70.0

5 Conclusions

The effects of irrigation water-added PAM was studied on a Mediterranean soil. Furrows were implanted on contour terraces and on maximum field slope. Irrigation water-added PAM was applied to selected furrows, which were monitored at locations for sediment concentration, infiltration volumes and particle size distribution. PAM applications significantly reduced sediment concentration and increased infiltration on slopping furrows but had no significant impact on the contour furrows. Furrows on contour terrace and cut back inflow could be responsible for the small sediment concentration differences observed vis à vis the control furrows.

Sediment movement was observed to occur, with all particle fractions being transported and carried away at the tail end (outflow) of control contour furrows. In spite of no particle size data for slopping furrows, the association between the obtained sediment concentrations shown in Table 2 and the contour furrow results allows one to predict greater sediment movement in the slopping furrows. PAM used to decrease erosion should also reduce this particle movements and help to control tailwater pollution.

As for infiltration, PAM-use in slopping furrows reduced the gradual slacking and seal formation in the wetted perimeter observed in the control treatment and, for each irrigation set, it helped to maintain high infiltration rates for longer periods of time. Results on the contour furrows were less dramatic as cut back of initially high inflow rates and furrows implanted on terraces of mild slope helped to alleviate the differences between PAM treatment and the control.

6 References

1. Sojka, R.E and Lentz, R.D. (1996) A PAM primer: A brief history of PAM and PAM-related issues, in *Managing irrigation-induced erosion and infiltration with polyacrylamide*, (ed. R.E. Sojka and R.D. Lentz), Proc. Twin Falls, ID 6-8 May, University of Idaho/USDA-ARS & NRCS, Kimberly, ID.
2. Berg, R.D and Carter, D.L. (1980) Furrow erosion and sediment losses in irrigated cropland. *J. Soil and Water Conserv.* 35:367-370.
3. Kemper, W.D., Trout, T.J., Brown, M.J., and Rosenau, R.C. (1985) Furrow erosion and water and soil management. *Trans. ASAE* 28:1564-1572.
4. Trout, T.J. (1996) Furrow irrigation erosion and sedimentation: on field distribution. *Trans. ASAE* : 715-723
5. Sojka, R.E. (1997) Research contribution to the understanding and management of irrigation-induced erosion. 50th anniversary symposium on Soil Conservation Research, *J. Soil and Water Conservation special publication* (ed. F.J. Pierce).
6. Lentz, R.D., Shainberg, I., Sojka, R.E. and Carter, D.L. (1992). Preventing irrigation furrow erosion with small applications of polymers. *Soil Sci. Soc. Am. J.* 56:1926-1932.
7. Lentz, R.D. and Sojka, R.E. (1994a) Field results using polyacrylamide to manage furrow erosion and infiltration. *Soil Sci.* 158:274-282.
8. Trout, T.J., Sojka, R.E. and Lentz, R.D. (1995) Polyacrylamide effect on furrow erosion and infiltration. *Trans. ASAE* 38:761-765.
9. Zhang, X.C. and Miller, W.P. (1996a) Physical and chemical processes affecting runoff and erosion in furrows. *SSSA J.* 60(3):860-865.
10. ___1996b. Polyacrylamide effect on infiltration and erosion in furroww. *SSSA J.* 60(3):866-872.
11. Raposo, J.R. *A rega: dos primitivos regadios às modernas técnicas de rega.* Fundação C. Gulbenkian, Lisboa, Portugal.
12. Serralheiro, R.P. (1995) Furrow irrigation advance and infiltration equations for a mediterranean soil. *J. agric. Engng. Res.* 62, 117-126.
13. Lentz, R.D. and Sojka, R.E. (1996) Five year research summary using PAM in furrow irrigation , in *Managing irrigation-induced erosion and infiltration with polyacrylamide*, (ed. R.E. Sojka and R.D. Lentz), Proc. Twin Falls, ID 6-8 May, University of Idaho/USDA-ARS & NRCS, Kimberly, ID.
14. Ben-Hur, M. (1994) Runoff, erosion and polymer application in moving-sprinkler irrigation. *Soil Sci.* 158 (4): 283-290.
15. Sojka, R.E., Lentz, R.D., Ross, C.W. and Trout, T.J. (1996) Net and tension infiltration effects of PAM in furrow irrigation in *Managing irrigation-induced erosion and infiltration with polyacrylamide*, (ed. R.E. Sojka and R.D. Lentz), Proc. Twin Falls, ID 6-8 May, University of Idaho/USDA-ARS & NRCS, Kimberly, ID.
16. Replogle, J.A. and Bos, M.G. (1982) Flow measurement flumes: applications to irrigation water management, in *Advances in Irrigation*, (ed. D. Hillel), Academic Press, New York vol. 1, 148-217.
17. Sojka, R.E., Carter, D.L., and Brown, M.J. (1992) Imhoff cones determination of sediment in irrigation runoff. *Soil Sci. Soc. Am. J.* 56: 884-890.

ASSESSMENT OF GROUNDWATER VULNERABILITY TO POLLUTION USING THE DRASTIC METHOD. APPLICATION TO THE ALQUEVA AREA

J.P. LOBO-FERREIRA and M. M. OLIVEIRA
Laboratório Nacional de Engenharia Civil, Lisboa, Portugal

Abstract

This paper contains an assessment of groundwater vulnerability to pollution of the aquifers of the Alqueva Project area, in Portugal, based on the DRASTIC method, evaluated both for fertilisers (standard DRASTIC) and for pesticides (pesticide DRASTIC). A general description of the concept of groundwater vulnerability to pollution is presented. This concept was applied aiming the assessment of groundwater vulnerability to pollution of the 110,000 ha potential area to be irrigated by the Alqueva Project, in southern Portugal, using the DRASTIC method, developed by Aller *et al.* [1] for the U.S. Environmental Protection Agency. In the paper the results evaluated both for fertilisers (standard DRASTIC) and for pesticides (pesticide DRASTIC) are presented.
Keywords: Groundwater contamination, pesticides, vulnerability assessment.

1 Introduction

The term vulnerability has been defined and used before in the area of water resources, but within the context of system performance evaluation, e.g. the definition given by Hashimoto *et al.* [2]. These authors present an analysis of system performance which focuses on system failure. They define three concepts that provide useful measures of system performance: (1) how likely the system is to fail is measured by its *reliability*, (2) how quickly the system returns to a satisfactory state once a failure has occurred is expressed by its *resiliency*, and (3) how severe the likely consequences of failure may be is measured by its *vulnerability*.

The concept of vulnerability defined in the context of system performance may also

Water and the Environment: Innovative Issues in Irrigation and Drainage. Edited by Luis S. Pereira and John W. Gowing. Published in 1998 by E & FN Spon. ISBN 0 419 23710 0

be used in the context of groundwater pollution if we replace *"system failure"* by *"pollutant loading"*. We believe that the most useful definition of vulnerability is the one that refers to the intrinsic characteristics of the media, which are relatively static and beyond human control. Lobo-Ferreira and Cabral [3] proposed groundwater vulnerability to pollution to be defined in Portugal, for watershed planning and management, in agreement with the conclusions and recommendations of the international conference on "Vulnerability of Soil and Groundwater to Pollutants", held in 1987 [4], as *"the sensitivity of groundwater quality to an imposed contaminant load, which is determined by the intrinsic characteristics of the aquifer"*.

Thus defined, vulnerability is distinct from pollution risk. Pollution risk depends not only on vulnerability but also on the existence of significant pollutant loading entering the subsurface environment. It is possible to have high aquifer vulnerability but no risk of pollution, if there is no significant pollutant loading; and to have high pollution risk in spite of low vulnerability, if the pollutant loading is exceptional. It is important to make clear the distinction between vulnerability and risk because risk of pollution is determined not only by the intrinsic characteristics of the aquifer, which are relatively static and hardly changeable, but also on the existence of potentially polluting activities, which are dynamic factors that can in principle be changed and controlled.

2 The "classic" vulnerability map of Portuguese southern regions of Alentejo and Algarve

The "classic" vulnerability map of Portuguese southern regions of Alentejo and Algarve, by Lobo-Ferreira and Calado [5], considers four vulnerability categories, *very high*, *high*, *variable*, and *low*. The factors are mostly hydrogeologic in nature and were estimated from the "Lithological Map of Portugal for Hydrogeologic Use" by Rodrigues *et al.* [6].

The key factors considered were:

1. Permeability of the unsaturated zone and of the aquifer
2. Aquifer recharge rate
3. Attenuation capacity of the unsaturated zone
4. Velocity of pollutant propagation in the unsaturated zone

According to these factors, the vulnerability categories were made correspondent to the following hydrogeological groups:

- *Very high* vulnerability:
  - Karst formations of medium to high permeability. Velocity of pollutant propagation in the order of 10 m/d.
  - Fluvial and dune alluvial formations of medium to high permeability. Velocity of pollutant propagation in the order of 1 to 10 m/d.
- *High* vulnerability:
  - Gabbro and diorite rock formations. Velocity of pollutant propagation in the order of 1 m/d.
  - Detritic cover formations of medium to low permeability.

- Mesozoic and Cenozoic porous and fractured sedimentary formations of medium to low permeability.
- Clay, sand and marl formations (e.g. *Grés de Silves*).
- Highly *variable* vulnerability:
  - Fractured igneous formations, mainly granites.Variable pollutant propagation velocity.
  - Fractured metamorphic formations, mainly schists and grauvaques.

Fig. 1 shows the "classic" vulnerability assessment of southern Portugal, developed by Lobo-Ferreira and Calado [5] for Alentejo and Algarve, and extended to the rest of the country in Rodrigues *et al.* [6]. In this assessment the solely consideration of the above mentioned factors, heavily dependent on hydrogeological factors, was not considered to be the optimal one, but a first assessment of the hydrogeological data available in Portugal.

3 Suggested system of index vulnerability evaluation and ranking for Portugal

3.1 General comments
Lobo-Ferreira and Cabral [3] suggested that a vulnerability index be used in the vulnerability ranking performed for Portugal. Such a standardised index has been adopted in the U.S., Canada and South Africa, and is currently used in those countries: the index DRASTIC, developed by Aller *et al.* [1] for the U.S. EPA. This index has the characteristics of simplicity and usefulness that we think to be necessary. The index DRASTIC is briefly reviewed below.

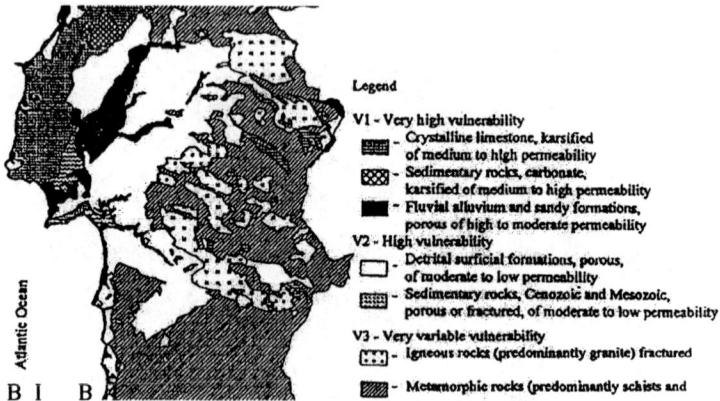

Fig. 1. "Classic" vulnerability assessment of southern Portugal (Alentejo and Algarve regions)

3.2 The index of vulnerability DRASTIC
The index of vulnerability DRASTIC (cf. [1]), corresponds to the weighted average of 7 values corresponding to 7 hydrogeologic parameters:

1. **D**epth to water (D)
2. Net **R**echarge (R)
3. **A**quifer media (A)
4. **S**oil media (S)
5. **T**opography (T)
6. **I**mpact of the vadose zone media (I)
7. Hydraulic **C**onductivity of the aquifer (C)

We attribute a value between 1 and 10 to each parameter, depending on local conditions. High values correspond to high vulnerability. The attributed values are obtained from tables, which give the correspondence between local hydrogeologic characteristics and the parameter value. Next, the local index of vulnerability is computed through multiplication of the value attributed to each parameter by its relative weight, and adding up all seven products. Thus, each parameter has a predetermined, fixed, relative weight that reflects its relative importance to vulnerability. The most significant factors have weights of 5; the least significant a weight of 1. A second weight has been assigned to reflect the agricultural usage of pesticides. In the following table the factors are presented together with the weights respectively for normal DRASTIC applications and for DRASTIC pesticide applications:

	normal DRASTIC	pesticide DRASTIC
1. **D**epth to water	5	5
2. Net **R**echarge	4	4
3. **A**quifer media	3	3
4. **S**oil media	2	5
5. **T**opography	1	3
6. **I**mpact of the vadose zone	5	4
7. Hydraulic **C**onductivity of the aquifer	3	2

The minimum value of standard DRASTIC index is 23 (26 for pesticide DRASTIC) and the maximum value is 226 (256 for pesticide DRASTIC). Such extreme values are very rare, the most common values being within the range 50 to 200.

4 Assessment of groundwater vulnerability to pollution of the Alqueva Project using the DRASTIC method

4.1 General comments
Several maps of the aquifer systems, hydrogeological parameters, aquifer's recharge and the final map of DRASTIC aquifer's vulnerability of Portugal, all in scale 1:500,000 were developed in ARC/INFO (the maps are presented in a 1:1,500,000 scale in Lobo-Ferreira *et al.* [7]).

In http://www-dh.lnec.pt/gias/novidades/DRASTIC_e.html, a one page map of the DRASTIC index vulnerability assessment of Portuguese groundwater, developed by Lobo-Ferreira and Oliveira [8], is presented.

Several other studies, that included DRASTIC groundwater vulnerability assessment, were developed in Portugal following this methodology. Among those study we highlight the *"Study for evaluation of the vulnerability of the reception capacity of coastal zone water resources in Portugal. The receiving water bodies: groundwater systems"*, by Lobo-Ferreira *et al.* [9]. In this study, the aquifers of the coastal areas of Portugal and the situation concerning groundwater exploitation were characterised. The main pollution problems affecting those areas, with a special attention to the salt water intrusion phenomenon, were also described. Finally, an application of the DRASTIC method for evaluation of the vulnerability to pollution of the aquifer formations of the coastal areas was presented and the vulnerability mapping at the scale 1:100,000 was developed.

In http://www-dh.lnec.pt/gias/novidades/nato97_vulner_internet.html a vulnerability map at the scale 1:100,000 is presented for the Peniche area, in Portugal's central coastal zone, based on the studies developed in [9].

Finally, Lobo-Ferreira and Oliveira [10] describe the assessment of groundwater vulnerability to pollution of the aquifers of the Alqueva Project area, according to the DRASTIC method, evaluated both for fertilisers (standard DRASTIC) and for pesticides (pesticide DRASTIC), in a 1:250,000 scale. A general description of the hydrogeology and hydrogeochemistry of the area's groundwater resources is also presented in [10].

4.2 Standard DRASTIC vulnerability assessment
The seven DRASTIC parameters were assessed for the Alqueva Project area. The index evaluated ranges from 109 to 182, as may be seen in Fig. 2. The highest assessed value is far from the potential maximum of 226. Therefore, the mean regional Alqueva area DRASTIC vulnerability index was considered to be *"medium vulnerability"*.

4.3 Pesticide DRASTIC vulnerability assessment
For the case of pesticide DRASTIC vulnerability assessment, as may be seen in Fig. 3, the index ranges from 129 to 210, for a potential maximum of 256. Therefore, the mean regional pesticide DRASTIC vulnerability index of the Alqueva area was also considered to be *"medium vulnerability"*.

5 Conclusions

It is our opinion that the vulnerability evaluation procedure should correspond to a well-defined computation of an index, in order to reduce subjectivity involved in the ranking.

Based on the experience gathered in Portugal, briefly described in the previously presented sections of this paper, we recommend that the index DRASTIC, be selected for the elaboration groundwater vulnerability maps, including those for the assessment of irrigation areas. The index DRASTIC presents, in our opinion the basic desired characteristics of simplicity and soundness.

An important advantage offered by DRASTIC is the amount of existing experience on its application, in the U.S., Canada, Portugal and other countries.

Research based on the aquifer vulnerability concept and the corresponding data acquisition process, exemplified in this paper for Portuguese conditions, allows a sounder application of mathematical groundwater flow and mass transport models.

Periodical groundwater quality analysis, especially for nitrates, nitrites and pesticides, should be made in the wells used for water supply in irrigation areas, such as the area of the Alqueva project.

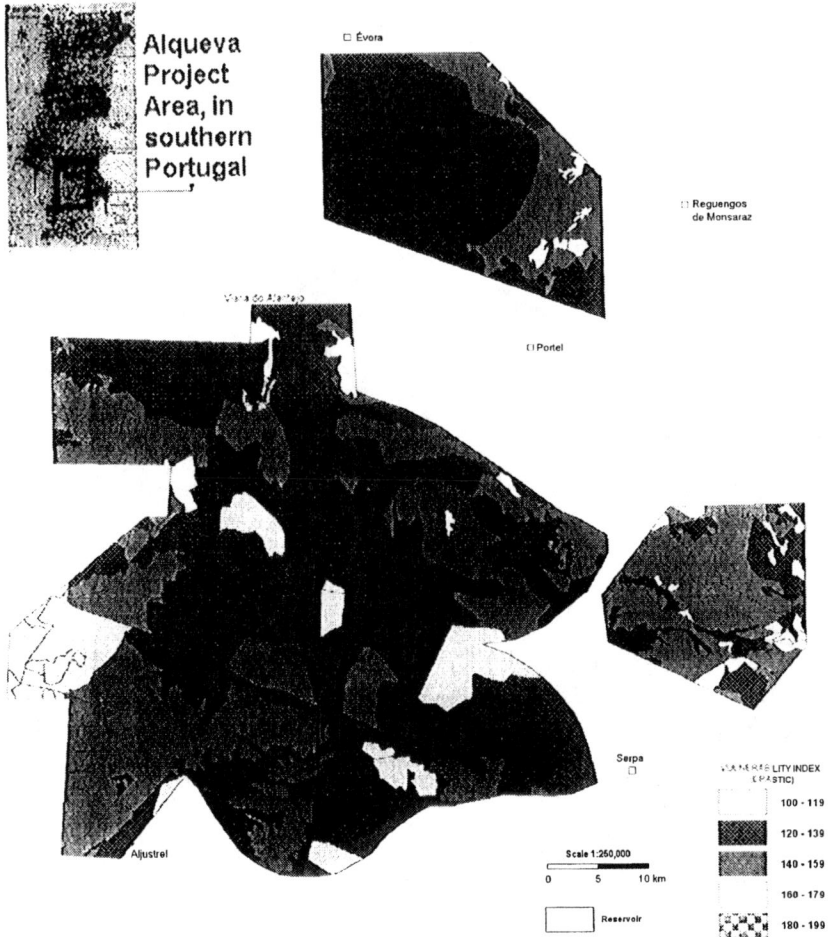

Fig. 2. DRASTIC (standard) vulnerability assessment of the Alqueva Project area

The water quality samples should be collected directly from the well. In those cases where no water treatment is available, samples should also be collected in the water supply reservoirs. In Portugal, the periodicity should be the one defined in Portuguese legislation (cf. Decreto Lei 74/90, article 14, number 2 and Annexes VII and VIII), i.e. four times a year for nitrites, two times a year for other N compounds and once a year for pesticides.

Over the aquifer's depth, a control of groundwater pollution transport, i.e. the monitoring of the movement of pollutants caused by agricultural practices, should also be performed. The areas to be monitored with greater detail should be those considered more vulnerable, such as the ones defined and mapped in Figs. 2 and 3 for the Alqueva Project area.

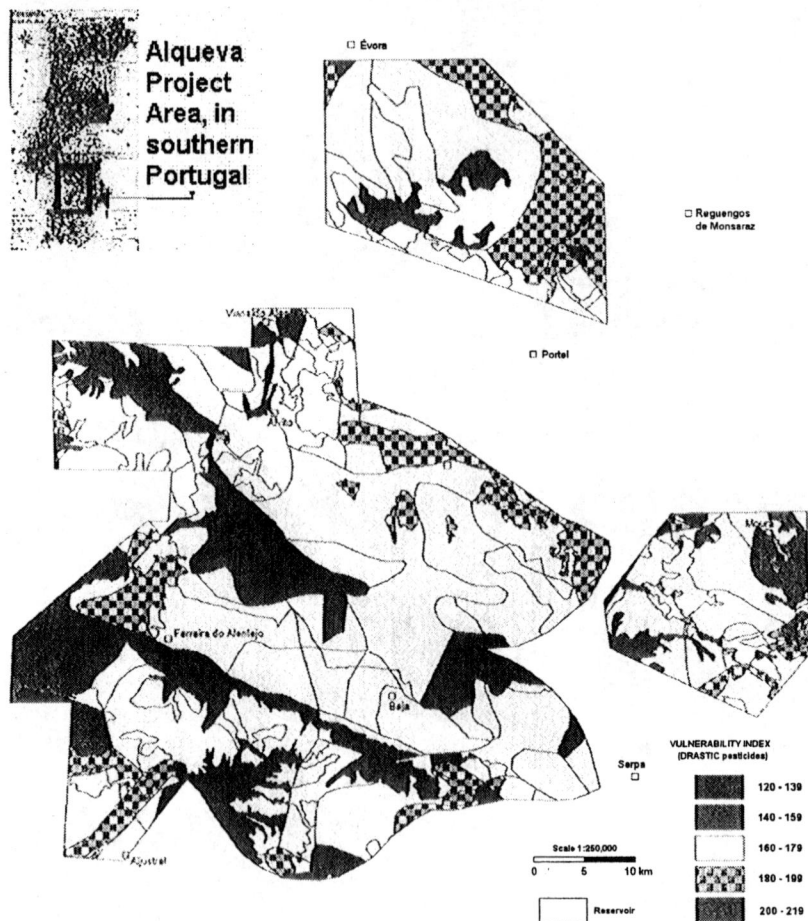

Fig. 3. DRASTIC pesticide vulnerability assessment of the Alqueva Project area

6 References

1. Aller, L., Bennet, T., Lehr, J.H. and Petty, R.J. (1987) *DRASTIC: a standardized system for evaluating groundwater pollution potential using hydrogeologic settings*, U.S. EPA Report 600/2-85/018.
2. Hashimoto, T., Stedinger, J.R. and Loucks, D.P. (1982) *Reliability, Resiliency, and Vulnerability Criteria for Water Resource System Performance Evaluation*, Water Resources Research, 18(1), p14-20.
3. Lobo-Ferreira, J.P. and Cabral, M. (1991) *Proposal for an Operational Definition of Vulnerability for the European Community's Atlas of Groundwater Resources*, in Meeting of the European Institute for Water, Groundwater Work Group Brussels, Feb. 1991.
4. Duijvenbooden, W. van and Waegeningh, H.G. van (Ed.) (1987) *Vulnerability of Soil and Groundwater to Pollutants*, Proceedings and Information No. 38 of the International Conference held in the Netherlands, in 1987, TNO Committee on Hydrological Research, Delft, The Netherlands.
5. Lobo-Ferreira, J.P. and Calado, F. (1989) *Avaliação da Vulnerabilidade à Poluição e Qualidade das Águas Subterrâneas de Portugal*. Lisboa, Laboratório Nacional de Engenharia Civil.
6. Rodrigues, J.D., Lobo-Ferreira, J.P., Santos, J.B, and Miguéns, N. (1989) *Caracterização sumária do recursos hídricos subterrâneos de Portugal*. Memória No. 735, Laboratório Nacional de Engenharia Civil, Lisboa and in Groundwater in Western Europe, Natural Resources Water Series, No. 27, 1992, United Nations, New York.
7. Lobo-Ferreira, J.P., Oliveira, M. Mendes and Ciabatti, P.C. (1995) *Desenvolvimento de um Inventário da Águas Subterrâneas de Portugal. Volume 1*. Laboratório Nacional de Engenharia Civil, Lisboa.
8. Lobo-Ferreira, J.P. and Oliveira, M.M. (1993) *Desenvolvimento de um Inventário das Águas Subterrâneas de Portugal, Caracterização dos Recursos Hídricos Subterrâneos e Mapeamento DRASTIC da Vulnerabilidade dos Aquíferos de Portugal*. Lisboa, Laboratório Nacional de Engenharia Civil. Relatório Final 179/93-GIAS.
9. Lobo-Ferreira, J.P., Oliveira, M.M., Moinante, M.J., Theves, T. and Diamantino, C. (1995) *Mapeamento das Águas Subterrâneas da Faixa Costeira Litoral e da Vulnerabilidade dos seus Aquíferos à Poluição*. Relatório Específico R3.3, LNEC, Relatório 237/95 - GIAS, 585 pp., Laboratório Nacional de Engenharia Civil, Lisboa.
10. Lobo-Ferreira, J.P. and Oliveira, M.M. (1994) *EIA Integrado para o Empreendimento do Alqueva. Análise dos Aquíferos da Zona do Perímetro de Rega, em Particular da sua Produtividade e Vulnerabilidade à Poluição*. Lisboa, Laboratório Nacional de Engenharia Civil, Relatório 300/94 - GIAS.

MANAGING THE EFFECTS OF AGRICULTURAL PRACTICES ON GROUNDWATER QUALITY

M.C. CAPUTO, G. PASSARELLA and M. VURRO
Istituto di Ricerca Sulle Acque, CNR, Bari, Italy
G. GIULIANO
Istituto di Ricerca Sulle Acque, CNR, Roma, Italy

Abstract
The development of methodologies able to study and to protect groundwater from contamination is an important task. The tools should have the character of managing instruments and be innovative with respect to the conventional modeling of nitrogen transport in groundwater. The methodology, proposed in this paper, is based on a mathematical model describing both water flow and nitrogen migration in saturated and unsaturated part of an underground domain. The mathematical model is governed by two equations describing the two-dimensional water flow through a porous matrix in a vertical plane and the transport and dispersion phenomena in the unsaturated and saturated zones. The equations are solved numerically using a finite difference approach. The model is applied to a study area, which comprises a part of the Po river basin, south of Modena, bounded by the Secchia and Panaro rivers, both tributaries of Po. This area is intensively used for civil, industrial and agricultural purposes and intensive cattle and pig breeding takes place on the south section of the study area. The influence of the agriculture is focused on the groundwater quality resulting in high nitrate concentrations that reach unacceptable values.
Keywords: Agricultural practices, groundwater, saturated and unsaturated flow, mathematical models, water quality, nitrogen migration.

1 Introduction

The study area is a part of the Po river basin, located south of Modena. The river Secchia and the river Panaro, both tributaries of river PO, form the respectively western and eastern boundary of the study area. The quality of groundwater and

Water and the Environment: Innovative Issues in Irrigation and Drainage. Edited by Luis S. Pereira and John W. Gowing. Published in 1998 by E & FN Spon. ISBN 0 419 23710 0

surface water in the Po river basin has deteriorated during the last decades due to the intense utilization of land for civil, industrial and agricultural purpose [1]. This study focuses on the influence of the agriculture on the groundwater quality degradation. Intensive cattle and pig breeding takes place on the high plain sector, south of the river Po resulting in high manure load and high nitrate concentrations in groundwater [2]. Several management strategies can be applied to mitigate this adverse effect. A computer-based model has been used to evaluate land management strategies with respect to nitrogen loads [3,4]. The study is divided in three phases. In the first phase a model calibration is made based on measured values of rainfall, groundwater levels and hydrogeologic parameters. In the second phase a model simulation is made for the study area. A third managerial phase should be based on the results of the second one.

A computer program, SATINSAT [5], is available to describe the nitrogen leaching to the groundwater in agricultural regions. This model calculates the groundwater level and the soil moisture profiles relate to given values of evapotranspiration, drainage and recharge. The model is also able to calculate nitrate concentration in the soil profile and the loads to the groundwater related to soil features and to agricultural practices. A simplified setting of the study area has been prepared for the model calibration. Different vertical profiles have been identified averaging over the depth the texture and the hydrogeological parameters. Two crop types have been distinguished, referring to the different period of manure and fertilizer loads. A calibration has been carried out but further calibrations on the basis of additional data should be performed in order to confirm the preliminary results.

2 Mathematical model

The unsaturated-saturated approach is the name adopted for the set of equations which treats the water flow through a porous matrix as the result of the movement of a two fluids (water-air) mixture, even though they neglect the direct description of the gaseous phases (i.e. air, water vapor). In view of this limitation the name of *limited unsaturated-saturated approach* seems more appropriate. The effects of the interaction between the liquid phase and air are taken into account only by means of empirical correlation between hydraulic head and liquid volumetric fraction (water retention model) and between these parameters and the effective permeability (hydraulic conductivity model). Relationships of this type have to be considered an essential part of the definition of the considered system as well as of the drying and/or wetting cycles, which are involved. Considering an idealized, two-dimensional, vertical infiltration domain, the governing equations of the flow field can be written as [6]:

$$\left(\beta S_s + (1-\beta)\frac{\partial \theta_w}{\partial \psi}\right)\frac{\partial \psi}{\partial t} = \frac{\partial}{\partial x}\left(K\frac{\partial \psi}{\partial x}\right) + \frac{\partial}{\partial z}\left(K\left(\frac{\partial \psi}{\partial z}+1\right)\right) + Q \tag{1}$$

where β is *1* or *0* inside saturated or unsaturated zone respectively, ψ is $(p/\rho g)$ or-$(p_c/\rho g)$ inside saturated or unsaturated zone respectively, g is the gravitational field strength, K denotes the hydraulic conductivity, p is the pressure, p_c is the capillary pressure, t the time, u and w, respectively, the x and z component of the specific discharge, ρ is the density of the water, φ is the piezometric head, θ_w is the water content (volumetric fraction of liquid), S_s the specific storage, and the relationships

$$\theta_w = \theta_w(\psi)$$
$$K = K(\psi)$$

(2)

are given functions.

Concerning the transport of contaminants from the soil surface to the saturated zone and assuming that the concentration is low, so that the density distribution of the total liquid phase cannot be changed and the exchanges between solid and liquid phases are described by a first order kinetic isotherm, the equation can be written as [6]:

$$\frac{\partial}{\partial t}(\theta_w RC) = -\frac{\partial}{\partial x}(uC) - \frac{\partial}{\partial z}(wC) + \frac{\partial}{\partial x}\left(\theta_w D_{xx}\frac{\partial C}{\partial x}\right) + \frac{\partial}{\partial z}\left(\theta_w D_{zz}\frac{\partial C}{\partial z}\right) +$$
$$+ Q\hat{C} + (-1)^\gamma \theta_w k_d C$$

(3)

where

$$R = \left(1 + \frac{\rho_b k_d}{\theta_W}\right) \quad \gamma = \{1, massaccretion; 2, massdecreasing$$

(4)

and C is the concentration in the liquid phase, $\hat{C}$ is C_{in} if $Q>0$ or C if $Q\leq 0$, C_{in} is the concentration assigned on the boundary, ρ_b is the bulk density, k_d the distribution coefficient and D_{xx} and D_{zz} are the dispersion coefficients. After the initial and boundary conditions have been defined, the solution of flow and transport equations can be obtained using the computer code SATINSAT [5] based on a finite difference (FD) procedure. An Eulerian mesh of rectangular cells of variable size covers the integration domain. The fluxes are stored at the cell faces whereas the pressure, moisture contents, hydraulic conductivities and other scalars are stored in the center of each cell or in the nodes of the main grid.

3 Study area

The study area is situated in Italy, in the province of Modena, covering an area of about 24000 ha. The western and eastern boundaries are given by the river Secchia and Panaro, the southern boundary is the Apennines margin and the northern boundaries a line crossing the town of Modena. The altitude ranges from 30 m above sea level near Modena to 150 m above the sea level at the southern border. The location of the area is shown in figure 1.

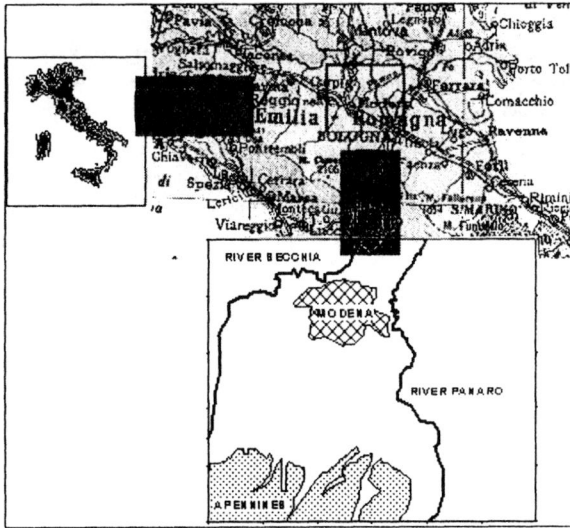

Fig. 1. Study area.

From a geologic standpoint, the high and average plain, included between the Apennine on the south and the sedimentary basin of river Po on the north, is basically characterized by a system of alluvial fans. These fans have been produced by sediments from the main rivers (Secchia and Panaro) and from the small torrents of the secondary network [7]. The Apennine is mainly made by clay sequences that date back to the Pliocenic-Calabrian marine cycle. The sediments of the low alluvial plain, which extends to the river Po, are made by sand with an important fraction of mud and clay. In the apical part, gravel and sand alternate with pelitic bodies whose thickness gradually increases when approaching the distal section prevalently makes up the fans.

The total thickness of the deposits of the Modena plain is very variable and generally increases from 100 m near the Apennine to the 400 m towards the north. This pattern is due to the deep tectonic structures, which caused a differential subsidence across the plain [8]. The water table decreases from the southwestern area to the northeast, near Modena, and is normally higher in spring than in autumn. The flow pattern is strongly affected by recharge from rivers Secchia and Panaro and groundwater extraction in the Modena area. A change in hydraulic conditions from free to confined or semi-confined aquifer, in the central sector of the plain, was pointed out in a former study [8]. A land use map derived from aerial photographs was provided by *Servizio Cartografico della Provincia di Modena* [9]. Agricultural use extends over 80% of the area. The principal crops are wheat, other cereals (maize), fodder, vegetables, vineyard and orchard [10].

All crops are irrigated in the dry season, except for wheat, which is scarcely sprinkled. In the study area, fertilizers are applied to crops in addition to the manure. Data about the yearly use of fertilizers in the Modena province were provided by the *Assessorato Agricoltura della Regione Emilia Romagna* [10] The yearly amount of nitrogen applied to the different crops was deduced from literature [11].

Meteorological data such as temperature, precipitation and evapotraspiration refer to the station of Modena. As it was pointed out in a former hydrogeologic study, precipitation increases from northeast to southwest, due to the topographic gradient. The study area can be divided into two zones: the northern part, where the precipitation is assumed as that in the station of Modena, and the southern part where the precipitation is 1.2 times the values measured in Modena. Temperature and evapotranspiration are assumed constant in the area. The precipitation in Modena is about 700 mm/year, the average yearly temperature varies between 10 and 14°C, the potential evapotranspiration is about 720 mm/year and the actual evapotranspiration is about 540 mm/year. Concerning the presence of undesirable parameters in groundwater, ammonia concentrations are extremely low in the rivers fans area. On the contrary, large nitrate concentrations can be found in the aquifer. The presence of nitrates is prevalently due to agricultural soil exploitation and produces a low qualitative grade in most of the study area.

4 Model application

The mathematical model has been calibrated by using data related to three different sites, named San Cesario, Modena and Sassuolo. These three sites have been chosen because their mean hydrogeological characteristics can be considered representative of the whole study area. The first step of applying mathematical models to saturated–unsaturated systems is the evaluation of the parameters of the water retention and hydraulic conductivity models. Two different ways can be followed to estimate these parameters: laboratory tests or indirect methods based on empirical relationship [12]. The second way has been chosen and, starting from the grain size distribution of the sites [8], the values of the parameters have been evaluated. The values obtained have been calibrated by using groundwater levels, irrigation and rainfall rates, of a particular period extracted from a long data set collected during last years.

The site of San Cesario has been chosen to understand the nitrate behavior by simulating its movement from the ground surface, through the vertical cross-section. The considered system has been schematized as a two-dimensional vertical domain 100 m long and 31 m depth. The water table has been initially positioned at 21 m under the ground surface, the eastern boundary has been considered as a symmetry axis, the southern one as an impervious layer and the western one as a free flow boundary. On the top of the system a variable inflow rate (rainfall and irrigation) has been imposed over the whole boundary but only over a 1m long cell a mass flow rate of nitrate has been assumed. This cell is assumed as the top of an observation column. Four cases have been simulated in order to reproduce different conditions for recharge and nitrate loads. Rainfall values of the driest (about 450 mm) and the wettest (about 950 mm) year of the 80s have been chosen. Two crop types, wheat and maize, have been supposed over the considered area, referring to the different rates of irrigation and period of manure and fertilizer loads. The four simulations can be summarized as follows: (1) wheat during a dry year, (2) wheat during a wet year, (3) maize during a dry year, (4) maize during a wet year. In details simulations (1) and (2) considered no inflow contributions due to irrigation and two periods of nitrate loads (300 and 150 kgN/ha/year for maize and wheat respectively), from February to April and in October.

The only difference between these two simulations was due to different rainfall rate. Simulations (3) and (4) considered the contribution of irrigation during June and July and a nitrate load in March, May and June (about 1800 m³/ha). Again the difference between these two simulations was due to the different rainfall rate.

5 Results and final remarks

The four simulations considered the whole year and boundary conditions were updated every ten days. Results in the observation column, over each simulated year, are reported in figures 2-3. Figure 2 shows the inflow rates due to rainfall and irrigation and the related simulated water table changes.

As the figures clearly show, during the wet year, the water table is a few centimeters higher then during the dry year. Irrigation produces the elevation of few more centimeters of the water table both during the dry year and the wet year. However, these changes cannot be considered of particular interest and confirm that the water table is located around 21 m below the ground surface during the full year. The results have been only considered in the observation column to understand the movement of the nitrogen front from the ground surface toward the aquifer.

Figure 3 shows the depth reached by nitrate, considering a concentration front of 10^{-3} mg/l and the nitrate load (g/d/m²) imposed on the top of the observation column.

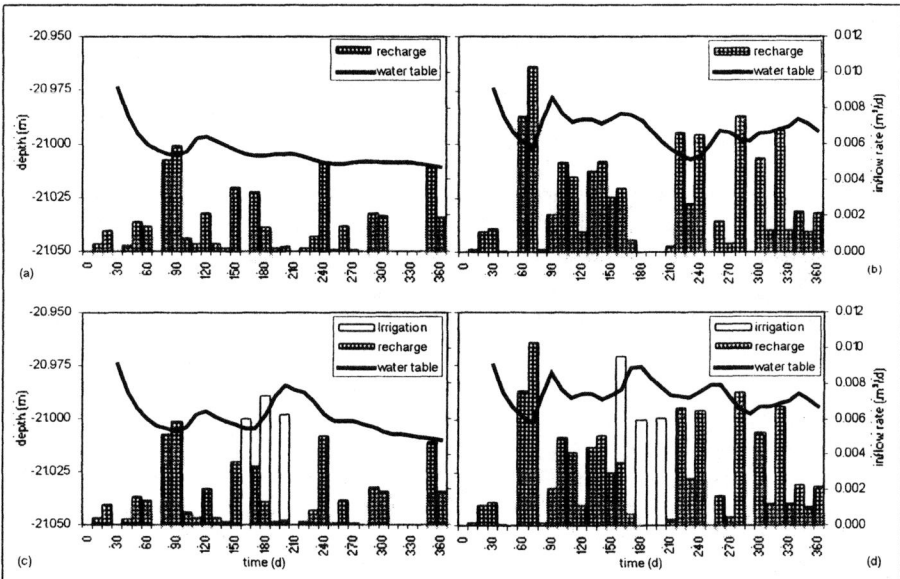

Fig. 2. Inflow boundary conditions and water table depth: (a) minimum recharge without irrigation; (b) maximum recharge without irrigation; (c) minimum recharge with irrigation; (d) maximum recharge with irrigation.

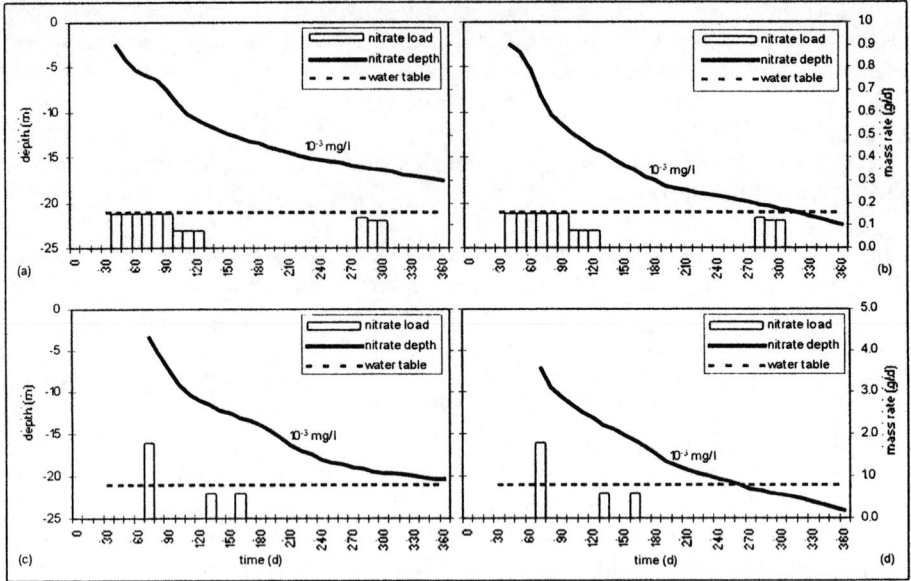

Fig. 3. Nitrate load boundary conditions and simulated nitrate depth: (a) minimum recharge without irrigation; (b) maximum recharge without irrigation; (c) minimum recharge with irrigation; (d) maximum recharge with irrigation.

The figures show that in the dry years the considered nitrate front does not reach the water table (simulation (1) and (3)). In the wet years it reaches the water table after about 250 and 300 days, respectively, depending on the irrigation is considered or not (simulations (4) and (2)). The time necessary because fronts of higher concentrations (i.e. 1 or 10 mg/l) reach the water table is notably larger.

A simulation tool to analyze problems of nitrate contamination of groundwater due to agricultural practices was developed. The model evaluates the response of water table to precipitation and irrigation inputs and the nitrate movement in the underground originating from the application of loads on the soil surface. The study is performed on representation site of the underground of the area which are schematized in a two dimensional vertical domain. At the present stage, the model is able to analyze the risk of groundwater contamination in terms of percolation of nitrate at a given concentration. The analysis, performed for different water infiltration rates cropping patterns, fertilizer application rates, can supply information on the potential effects of agricultural practices on the groundwater quality and on strategies for contaminant mitigation. The preliminary application of the model suggests, however, the necessity for further simulations in order to evaluate extreme precipitation events, higher concentration fronts and different crop types.

6 Acknowledgement

This research has been funded in the framework of CNR - Strategic Project on "Critical Aspects of Water Availability for Drinking Use". Thanks are due to Dr. A Zavatti for his kindly collaboration in furnishing data used in this work.

7 References

1. Gelmini, R., Paltrinieri, N. and Pellegrini, M. (1990) Vulnerabilità all'inquinamento delle acque sotterranee, in *Studi sulla Vulnerabilità degli Acquiferi, Vol.12. Quaderni di Tecniche di Protezione Ambientale - Protezione delle Acque Sotterranee*, Pitagora, Bologna, pp. 73-80.

2. Vicari, L. and Zavatti, A. (1990) Idrochimica, in *Studi sulla Vulnerabilità degli Acquiferi, Vol.12. Quaderni di Tecniche di Protezione Ambientale - Protezione delle Acque Sotterranee*, Pitagora, Bologna, pp. 27-72.

3. Zavatti, A. (1993) Esempio applicativo alle falde della Pianura Modenese, *Inquinamento*, vol. 12, pp. 11-12.

4. Passarella, G., Caputo, M.C., Giuliano, G. and Vurro, M. (1997) Groundwater Management in an Area of PO Valley Using Criteria Based on Water Quality, in *Water Management, Salinity and Pollution Control Towards Sustainable Irrigation in the Mediterranean Region*, vol II, pp. 151-168.

5. Masciopinto, C., Passarella, G., Vurro, M. and Castellano, L. (1994) Numerical Simulation for the Evaluation of the Free Surface History in a Porous Media. Comparison between Two Different Approaches, *Adv. in Engin. Soft.*, vol. 21, pp. 149-157.

6. Bear, J. (1979) *Hydraulics of Groundwater*, McGraw-Hill, New York.

7. Barelli, G., Marino, L. and Pagotto, A. (1990) Caratterizzazione idraulica degli acquiferi. *Studi sulla vulnerabilità degli acquiferi, Vol.12. Quaderni di Tecniche di Protezione Ambientale - Protezione delle Acque Sotterranee*, Pitagora, Bologna, pp. 15-26.

8. Paltrinieri, N. and Pellegrini, M. (1990) Comportamento idrodinamico dell'acquifero. *Studi sulla Vulnerabilità degli Acquiferi, Vol.12. Quaderni di Tecniche di Protezione Ambientale - Protezione delle Acque Sotterranee*, Pitagora, Bologna, pp. 9-14.

9. Regione Emilia Romagna (1993) *Carta Pedologica della Provincia di Modena*, Bologna.

10. ISTAT (1992) *Censimento Generale dell'Agricoltura*, Poligrafico dello Stato, Roma.

11. AA.VV. (1988) *Agricoltura e Ambiente*, Pitagora, Bologna.

12. Van Genuchten, M.Th. (1980) A close-form equation for predicting the hydraulic conductivity of unsaturated soils, *Soil Sci. Soc. Am. J.*, vol. 44, pp. 892-900.

SECTION II

MANAGING DEMAND

ENVIRONMENTAL CRITERIA IN A DECISION SUPPORT SYSTEM FOR SURFACE IRRIGATION DESIGN

J. M. GONÇALVES
Escola Superior Agrária, Bencanta, Coimbra, Portugal
L. S. PEREIRA and P. L. SOUSA
Department of Agricultural Engineering, Institute of Agronomy, Technical University of Lisbon, Portugal

Abstract
To assist planners and managers in the process of design and selection of on-farm surface irrigation systems, an interactive DSS software has been developed. Simulation models are utilised to determine the alternative attributes characterizing main impacts of surface irrigation systems, covering financial, economic, yield, and soil and water quality aspects. A multiple criteria decision-making methodology is applied. The DSS was tested with data collected from field experiments, and its usefulness was evaluated through the application to a sector of the Mondego Valley Project, Coimbra, Portugal. Selected results concerning the environmental sensitivity of the model relative to decision-maker priorities are presented.
Keywords: Economic impacts, environmental impacts, multicriteria analysis, surface irrigation.

1 Introduction

Surface irrigation systems play a very important role in Portuguese agriculture, being practised in 80% of the total irrigated area. Unfortunately, the majority of these systems do not produce the most desirable economic and environmental results concerning the use of water, soil and labour resources. A sustainable irrigated agriculture may be supported by modernised surface irrigation systems when these are properly designed and managed, providing for farm economic feasibility and natural resources conservation.

Main *economic impacts* of farm surface irrigation systems concern the costs relative to the investment, the operation and the maintenance of the system and, directly depending on the benefits, crop yields, and the social advantages to the farmer and his

Water and the Environment: Innovative Issues in Irrigation and Drainage. Edited by Luis S. Pereira and John W. Gowing. Published in 1998 by E & FN Spon. ISBN 0 419 23710 0

family. *Environmental impacts* include [1]:

- *Water quality impacts*, mainly ground water contamination by nitrate and pesticides, which are associated with excess use of agro-chemicals and excess water application inducing for solute transport; and surface water contamination with agro-chemicals transported in surface return flows. When low quality water is applied, other substances may be added to the fresh water resources.
- *Water quantity impacts*, related to uncontrolled demand and/or supply and excessive water application due to inappropriate irrigation practices and non-uniform water distribution over the field.
- *Soil quality impacts*, referring to soil erosion due to surface flow, soil degradation when inappropriate land grading causes detrimental modifications of the soil profile, soil structure and soil hydraulics properties deterioration, and changes in the chemical characteristics of the soil. The last two aspects are mainly related to soil salinization, when saline and low quality water is applied, and when excess water application induces the rising of a saline water table.

Economic results may be maximised and negative environmental impacts can be controlled when appropriate irrigation techniques are utilised. This implies the combination of the water application techniques with irrigation scheduling methods. To assist designers and managers to select the systems which may provide for the best combination of economic and environmental consequences, a decision support system (DSS) using the multicriteria analysis [2] was developed.

2 The DSS model SADREGA

The DSS conceptual structure is represented in Fig. 1. The design component applies database information and produces a set of alternative designs. These alternatives are then utilized in the selection component, where they are compared and ranked. The decision-maker participates in the decision process through appropriate interface dialog structures, both for design and selection of alternatives. The interface system-user has a crucial role in the learning process, improving the feedback cycles. The DSS provides for "learning by doing", thus increasing the perception of the problem by the decision-maker.

The DSS SADREGA is developed for the MS-DOS environment, on 486 PC's. The modules were coded using Borland C++ v.3.1 programming language. It has a friendly interface, with a menu driven structure and graphical outputs which allows easy data handling and models inter-communicability. The interface pull-down menus, mouse support and dialog boxes were programmed with Borland Turbo Vision application framework.

The modular components of DSS are represented in Fig. 2. The data base refers to input data required to the design models, mainly field characteristics, soil hydraulic properties, crop data and respective yield functions, irrigation scheduling data produced by the ISAREG model [3], and system costs. The surface irrigation systems are designed using the SIRMOD simulation model [4], a land levelling calculation algorithm using an iterative optimisation to determine the design slopes that minimize

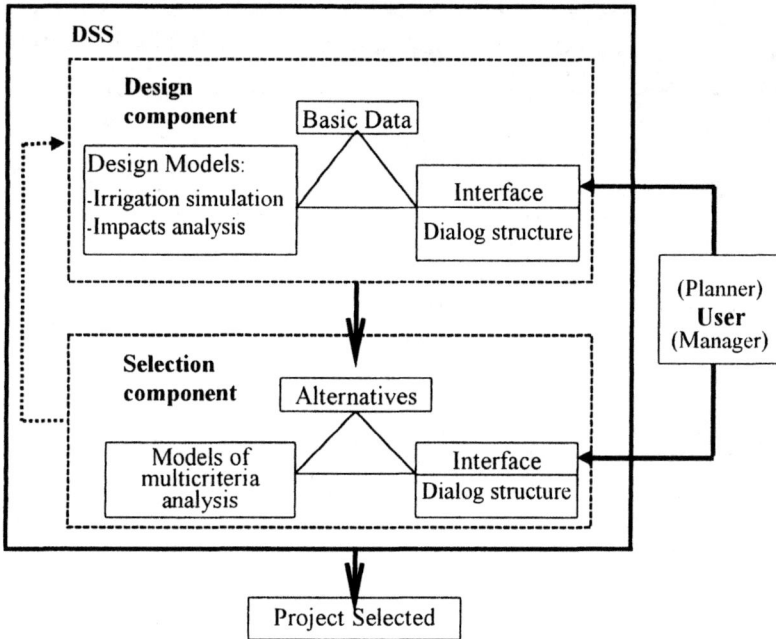

Fig. 1. SADREGA conceptual structure

the soil cut volumes, and several models or algorithms used to design on-farm water supply structures. The last include the models CABLE [5] and BORDER [6] for cablegation applied to furrows and borders or basins, respectively. The impact analysis module is used to compute the costs and benefits relative to every design alternative, and to determine corresponding environmental impact attributes.

Table 1. Design variables

Type	Implementation flexibility	Variable Designation	
Structural	Rigid	- field length	
	(constant for the irrigation season)	- field width	
		- transversal slope	
		- longitudinal slope	
		- water supply system	
		- reuse system	
	Flexible	- tail end conditions:	- open
	(dependent from other structural variables,		- diked
	with seasonal options)		- with reuse
Operational	Flexible	- unit inflow rate	
	(variable in each irrigation within some	- cutoff time	
	limits)	- applied water depth	

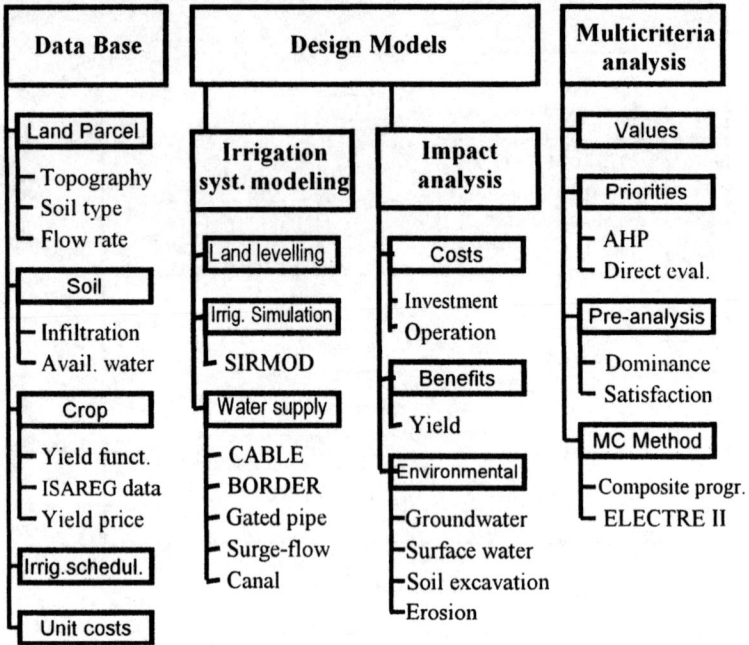

Fig. 2. SADREGA modular components

Table 2 – Design options

Irrigation method	Inflow rate	Water supply systems	Tail end
Level basins (LB) (flat or furrowed)	Constant	· layflat tubing · gated pipe · cablegation · canal (with siphons or orifices)	· diked
Graded borders (GB)	Constant	· layflat tubing · gated pipe · cablegation · canal (with siphons or orifices)	· diked · open · reuse system
Graded furrows (GF)	Constant	· layflat tubing · gated pipe · canal (with siphons)	· diked · open · reuse system
	Surge-flow	· gated pipe + surge valve	· diked · open · reuse system
	Variable	· cablegation	· diked · open · reuse system

The DSS is applied to a farm field, assumed with a rectangular shape and uniform intake characteristics. The field supply outlet is supposed to have constant flow rate and hydraulic head.

The design variables are presented in Table 1. They are classified as structural or operational and according to the flexibility for selecting their values. The design options to be taken in consideration are shown in Table 2 and refer to level basin, graded borders, level and graded furrows adopting several techniques of water supply and tail water management [7] [8], including reuse.

After the design alternatives are computed using the irrigation system models (Fig. 2), the impact analysis models are applied to determine the attributes characterizing these alternatives. Objectives and attributes are presented in Table 3. The objectives relative to economic impacts, which are defined in the perspective of farm economic feasibility, are evaluated through attributes 1 and 2 characterizing the system costs (land levelling, supply structures and labour costs), and the attributes 3 and 4, representing the economic impacts of irrigation performances in terms of water and yield losses. The objectives relative to the control of environmental impacts are associated with attributes relative to seasonal water losses by deep percolation and tail water runoff (calculated from the irrigation simulation), and to soil degradation due to the land levelling operations and erosion.

Table 3. Objectives and attributes for design selection

Objectives	Alternative Attributes	Units
Minimize project cost	1. investment costs	PTE/year
	2. operation and maintenance costs	PTE/year
Minimize economic impacts due to	3. water losses (cost)	PTE/year
poor irrigation performances	4. yield losses (cost)	PTE/year
Minimize environmental impacts due	5. water lost by deep percolation	m^3/year
to water losses	6. water lost by runoff	m^3/year
Maximize soil conservation	7. depth of land levelling cuts	cm
	8. erosion	dimensionless index

When each design alternative is characterized by the respective attributes listed in Table 3, the selection of the best alternatives becomes a multiple objective problem which solution requires a multicriteria analysis [2]. The multicriteria methodology integrates the different types of attributes on a trade-off analysis, allowing the comparison between environmental and economic criteria. This methodology helps a better understanding of the irrigation impacts and allows achieving a satisfactory compromise between adversative decision-maker objectives.

To make all decision criteria commensurable, the attributes are scaled to a measure of utility or value, which expresses the decision-maker preferences. Logistic and linear type value functions are applied for the environmental and economic criteria, respectively. To aggregate the values and ranking the design alternatives, the distance-based composite programming [9] and the outranking ELECTRE II [10] methods are applied. To express the decision-maker priorities, relative importance coefficients (weights) are assigned to the criteria. They can be directly evaluated or calculated by the AHP method [11].

3 Application

The DSS SADREGA was applied to the sector Tentúgal of the Mondego Valley irrigation project, Portugal. At present, this area is irrigated by basins and cultivated with corn and rice. The application was performed to test the capabilities of the model and to evaluate the adequacy of several irrigation and water supply systems to the area. Main results are presented in Table 4. The best irrigation systems for different field sizes and soil types were identified using the composite programming method. Results obtained when selecting the farm water supply systems did not show significant differences among those listed in Table 2.

Table 4. Best alternatives for different field sizes and soil types, sector Tentugal, Mondego.

Field size (m x m)	Sandy soil	Sandy loam soil	Loam soil	Silty loam soil
100x75	-GF, S_y=0.20, diked	-GF, S_y=0.10, diked	-GF, S_y=0.05, diked	-LB (furrowed)
	-GF, S_y=0.20, reuse	-GF, S_y=0.05, diked	-GB, S_y=0.05, diked	-GF, S_y=0.05, diked
	-GF, S_y=0.10, diked	-LB (furrowed)	-LB (furrowed)	-GF, S_y=0.05, reuse
150x75	-GF, S_y=0.20, diked	-GF, S_y=0.10, diked	-GF, S_y=0.10, diked	-GF, S_y=0.05, diked
	-GF, S_y=0.20, open	-GF, S_y=0.20, diked	-GF, S_y=0.05, diked	-LB (furrowed)
	-GF, S_y=0.30, diked	-GF, S_y=0.05, diked	-GF, S_y=0.05, reuse	-GB, S_y=0.05, diked
200x125	- alternatives not accepted	-GF, S_y=0.20, diked	-GF, S_y=0.10, diked	-GF, S_y=0.05, diked
		-GF, S_y=0.05, diked	-GF, S_y=0.05, diked	-LB (furrowed)
		-GF, S_y=0.10, diked	-GF, S_y=0.05, reuse	-GB, S_y=0.05, diked
265x75	-alternatives not accepted	-GF, S_y=0.10, diked	-GF, S_y=0.05, diked	-GF, S_y=0.05, diked
		-GF, S_y=0.05, diked	-GF, S_y=0.10, diked	-GF, S_y=0.05, reuse
		-GF, S_y=0.05, open	-GF, S_y=0.05, reuse	-GB, S_y=0.05, reuse
200x340	-alternatives not accepted	-GF, S_y=0.20, reuse	-GF, S_y=0.05, diked	-GF, S_y=0.05, diked
		-GF, S_y=0.20, open	-GF, S_y=0.05, open	-LB (furrowed)
		-GF, S_y=0.20,diked	-GF, S_y=0.05, reuse	-GB, S_y=0.05, diked

GF = graded furrows; LB = level basins; GB = graded borders; S_y = slope (%)

A sensitivity analysis was performed to determine if the selection of alternatives would be influenced by the weights assigned to the criteria. The different weighing schemes representing four priority scenarios defined by the decision-maker are listed in Table 5. Results of the sensitivity analysis for a 200x125m field with a silty loam soil

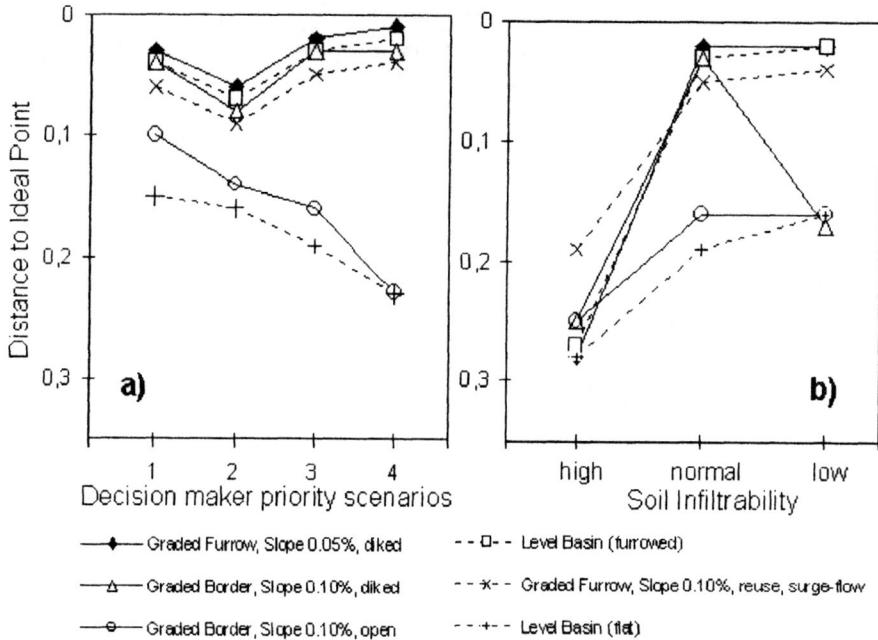

Fig. 3. Sensitivity analysis: a) relative to decision-maker priority scenarios defined in Table 5; b) relative to soil infiltrability (application to a field with size 200 x 125 m with a silty loam soil)

are given in Fig. 3a). One can observe that the ranking does not change significantly with the priority scenarios. However, the environmental criteria play a great role in discriminating the alternatives.

Table 5. Weighing system for decision-maker priority scenarios

Criteria	Priority scenarios*			
	1	2	3	4
Investment costs	0.2	0.05	0.125	0.05
Operation and maintenance costs	0.2	0.65	0.125	0.05
Costs due to water losses	0.2	0.05	0.125	0.05
Yield losses due to poor performances	0.2	0.05	0.125	0.05
Water losses by deep percolation	0.05	0.05	0.125	0.2
Losses by runoff	0.05	0.05	0.125	0.2
Erosion	0.05	0.05	0.125	0.2
Depth of land levelling cuts	0.05	0.05	0.125	0.2

* Scenario 1: priority to minimize total costs; Scenario 2: priority to minimize labour (operation and maintenance cost); Scenario 3: balance between economic and environmental impacts; Scenario 4: priority to minimize environmental impacts

Figure 3b) shows the results of the sensitivity analysis for soil infiltrability in the same field. Because this soil property highly influences the performances of irrigation, the ranking changes when soil infiltrability varies. This indicates that selection is sensitive to the quality of input data, namely soil data. Since performances are the mostly influenced, because environmental attributes are mainly related with performances, environmental criteria are also affected by the quality of soil data.

4 Conclusions

This study shows that the use of a DSS for design of surface irrigation systems is a powerful tool for selecting the best alternative systems for fields with different sizes and soil types. The DSS is able to appropriately combine economic and environmental criteria. The sensitivity analysis performed has shown that when higher weights are assigned to the environmental criteria there is a better discrimination of alternatives. Results also show that the model is sensitive to the data influencing the systems performance, thus requiring good quality data.

5 References

1. Pereira, L. S. (1996) Inter-relationships between irrigation scheduling methods and on-farm irrigation systems, in *Irrigation Scheduling: From Theory to Practice* (ed. M. Smith, L. S. Pereira, J. Berengena, B. Itier, J. Goussard, R. Ragab, L. Tollefson and P. van Hofwegen), Water Reports 8, FAO, Rome, pp. 91-104.
2. Janssen, R. (1992) *Multiobjective Decision Support for Environmental Management*. Kluwer Academic Publisher, Dordrecht.
3. Teixeira, J. L. and Pereira, L. S. (1992) ISAREG, an irrigation scheduling simulation model, in *Crop-water models* (ed. L. S. Pereira, A. Perrier, M. A. Kadi and P. Kabat), Special issue of ICID Bulletin, Vol. 41 (2): 29-48.
4. ISED. (1989) *SIRMOD, Surface Irrigation Simulation Software. User's Guide.* Irrigation Software Engng. Div., Dep. Agric. and Irrig. Engng., Utah St. Univ., Logan.
5. Kincaid, D. S. and Stevens, J. L. (1992) *CABLE - A design program for the Cablegation automatic surface irrigation system*, USDA-ARS, Kimberly.
6. Trout T. (1987) *Documentation for BORDER cablegation design program.* USDA-ARS, Kimberly.
7. Walker, W. R. and Skogerboe, G. (1987) *Surface Irrigation: Theory and Practice*. Prentice-Hall, Inc., Englewood Cliffs, New Jersey.
8. Pereira, L. S. (1996) Surface Irrigation Systems, in *Sustainability of Irrigated Agriculture* (ed. L.S. Pereira, R.A.Feddes, J.R.Gilley and B.Lesaffre). Kluwer Academic Publ., Dorchecht, pp. 268-289.
9. Bogardi, I. and Bardossy, A. (1983) Application of MCDM to Geological Exploration, in: *Essays and Surveys on Multiple Criteria Decision Making* (ed. Hansen, P). Springer-Verlag, New York.

10. Roy, B. and Bouyssou, D. (1993) *Aide Multicritère: Méthodes et Cas.* Economica, Paris.
11. Saaty, T. L. (1990) How to make a decision: the AHP. *Europ. J. Operat. Res.*, **48** (1): 9-26.

SURFACE IRRIGATION EFFICIENCY IN CRACKING SOILS AS INFLUENCED BY WATER RESTRICTIONS

A. ZAIRI and A. SLATNI
Institut National de Recherches en Génie Rural, Eaux et Forêts – Tunis, Tunisie
J.C. MAILHOL
Cemagref – Division Irrigation - Groupement de Montpellier, France
H. ACHOUR
Ecole Supérieure des Ingénieurs de l'Equipement Rural - Medjez El Bab, Tunisie

Abstract
An experimental study has been carried out in order to improve surface irrigation practices in cracking soils of the Lower Medjerda Valley. The experiments deal with soil depletion impact on the irrigation efficiencies. A comparison between border and furrow practices was conducted on wheat usually irrigated by the border technique in Tunisia. The advance infiltration process was analysed using the calibration of the infiltration parameters on the advance trajectory (closed-end furrow and closed-end border practices). This calibration was conducted by means of two modelling approaches allowing hydraulic efficiencies to be estimated. The two main conclusions which can be derived from this study are: (1) water losses due to border practice under cracking soils can be significantly reduced by adopting the furrow practice without yield decreases; (2) according to the furrow practice, the irrigation strategies optimizing the efficiencies (hydraulic and agronomic) consist of giving two water applications to the crop. The first one insures crop emergence whereas the second has to be given when the soil water depletion reaches 90% of available water for the central region (low rainfall occurrence) and 30% for the northern region of Tunisia. The interest of coupling a water balance estimation with an operative furrow model in order to identify the best irrigation strategy on the whole irrigation campaign is here underlined.
Keywords: surface irrigation, water restrictions, soil moisture deficiency, cracked soils, wheat production.

1 Introduction

In Tunisia, 80% of the irrigated area depends upon surface methods. Although this dominance tends to decrease, surface irrigation will still continue to be much more used

Water and the Environment: Innovative Issues in Irrigation and Drainage. Edited by Luis S. Pereira and John W. Gowing. Published in 1998 by E & FN Spon. ISBN 0 419 23710 0

than other irrigation techniques. Because it is not modernized yet, its application efficiency is often as low as 30% [1] [2].

Irrigation is often applied under heavy water stress conditions inducing a substantial contribution of the soil water reserve. In a such soil water depletion context, the watering is conducted in heavily cracked soil (even for soils having low clay content) except for the sowing irrigation. In such conditions the irrigation is not easy to manage even when the farmer can use a substantial daily water volume. Excessive water losses are generally observed, some times attaining 1000 to 2000 m^3/ha/irrigation [3].

Loss reduction requires an adequate choice of parameters such as slope, inlet discharge and plot length. Investigations in Tunisia are often focussed on the discharge-length relationships and their impact on the water application depth owing to the high cost of levelling [4]. Generally these studies were conducted in moderately dry soil conditions with relatively low water application depths (near 50 mm). Recent field experiments conducted on the same soil type but under cracked conditions have given very low efficiencies [5].

The field experiments conducted during the wheat campaign of 1996/97 deal with the impact of cracking and the spatial variability of water application depths. In order to improve the advance process, a comparison between border and furrow irrigation has been conducted on a wheat crop, usually irrigated using the border technique. The impact on the yields and water saving has been analyzed. Irrigation strategies based on a soil water depletion level were tested in order to identify those which both optimize yields and water saving.

2 Material and method

2.1 Experimental site
The field experiment were conducted at the experimental site of Cherfech managed by the INRGREF institute, located in the Lower Medjerda Valley, 20 km north of Tunis. The Lower Medjerda Valley climate is Mediterranean (semi-arid upper step) characterized by an average rainfall near 450 mm/year. This annual rainfall is not well temporally spread with maximal values in winter and spring. The soil can be characterized by :

- a loamy-clay soil having 60 to 80 cm of thickness (clay from 20 to 45%)
- a clay-loam to clay soil at a deep level.

The hydro-physical characterization of the plot give on a 1 m depth a bulk density of 1.53 and moisture content (in weight) of 27.3% at pF2.7 and 16.5% at pF2.4. That gives a maximal available soil water (MASW) of 160 mm/m.

The surface irrigation system consists of a canal supplying the gated pipe system which delivers the water to the upstream part of border and closed-end furrows. The irrigation is stopped when the water front attains the few last meters in plot end. The slope from upstream to downstream is 0.2% whereas traversal slope equals zero.

2.2 Experimental procedure

The soil water status impact on the irrigation performance is analyzed by choosing different soil water depletion coefficient values of MASW (p= 0.3, 0.6, and 0.9) to decide to irrigate. That corresponds to different soil moisture deficit (SMD) presented in Table 1.

The irrigation technique impact is analyzed by comparing border and furrow irrigation (for a 1.5 m inter furrow spacing). This furrow spacing is used by the Tunisian farmers for potatoes to make easy the machinery traffic during the treatments and for harvesting.

The whole plot represents an area of 1.5 ha divided in 3 blocks having 90 m length (with 3 repetitions). Each block is divided according to the length in border and furrow having the same width (10 m). Two plots of 700 m² are used as non-irrigated treatment.

According to previous studies [6] dealing with surface irrigation improvement, an available inflow rate of 15 L/s was chosen and remained steady for the whole of experiments. As a result, an inflow value of 1.5 L/s is delivered by unit plot width.

Table 1. Adopted treatments for the study of the soil moisture deficit and irrigation technique impact

Irrigation technique	soil water depletion coefficient (p)	SMD (mm/m)	Soil water content before irrigation (mm/m)
Border	0.3	50	370
Furrow	0.3	50	370
Border	0.6	100	320
Furrow	0.6	100	320
Border	0.9	150	270
Furrow	0.9	150	270

The measurements consisted of advance and recession monitoring. In the case of the furrow, 7 furrows were monitored at a time in order to derive the advance process of an averaged furrow. The depth of water infiltrated and stored in the root zone (root depth = 100 cm) was measured by determining the soil water profile before and 48 h after irrigation using the gravimetric method. The soil water profiles were established until 1.5 m depth for 6 stations, from upstream to downstream.

Ten local samples were taken in each treatment for the evaluation of yield components: number of plants at the emergence, number of ears/m², number of grains/ear and weight of 1000 grains (WTG).

2.3 Performance parameters of irrigation

In order to evaluate the irrigation performance, the parameters described by Merriam and Keller [7] were adopted :

• Application efficiency (E_a): is the ratio of the average depth of the irrigation water infiltrated and stored in the root zone to the average depth of irrigation water applied.
• Distribution uniformity (DU) : is the ratio of the average low-quarter depth of irrigation water infiltrated (or caught) to the average depth of irrigation water infiltrated (or caught).

- Storage ratio (SR): is the ratio of the average depth of water stored in the root zone to the average depth storable.

The determination of these parameters was conducted using soil moisture measurements before and after irrigation and also using two simulation models: SIRMOD [8] and RAIEOPT [9] [10].

The SIRMOD model enables the different irrigation phases to be simulated by combining St Venant (advance and recession of surface flow) and Kostiakov-Lewis (cumulative infiltration) equations. According to the closed-end furrow or border conditions, the zero inertia option of SIRMOD was adopted.

The RAIEOPT model was initially developed for the furrow case then extended to the border case. Different modelling options, such as those allowing closed-end furrow simulations, were added to initial version of the model [11].

Kostiakov-Lewis equation used in SIRMOD is derived from the numerical solution of the water balance equation using the program IDENTABC [12]. The good agreement between observed and simulated advance with SIRMOD is verified. Sensitive modifications may be done eventually to improve the quality of the advance simulation.

3 Results and discussions

The 1996/97 experimental campaign was characterized by a substantial rainfall deficit. During the crop cycle (from mid November to mid June), 200 mm of rainfall were observed representing 60% of the mean rainfall. The soil moisture measurements at the sowing date attest that the soil water content is around 10% of MASW. As a result, the water applications were as follows:

- $p = 0.3$: 1 irrigation at the sowing date and 4 irrigations between earing and grain filling.
- $p = 0.6$ and 0.9: 1 irrigation at sowing and 2 irrigations between earing and grain filling.

3.1 Evaluation of irrigation performance using measurements

The experiment show significant differences between the first irrigation and those after earing. The application at sowing is conducted in tilled soil whereas those after earing are conducted under cracking soil conditions. In the field, the degree of cracking can be easily observed between the $p = 0.3$ treatment and that of 0.6 or 0.9. Table 2 shows the average application depth (AWD) for the irrigation of sowing and those after earing for a same p value ($p = 0.9$).

Table 2. Applied water depths comparison between tilled and cracking soil

Irrigation Technique	Sowing irrigation		Irrigation after earing	
	AWD (mm)	E_a (%)	AWD (mm)	E_a (%)
Furrow	75	75	156	80
Border	140	75	287	37

A substantial deviation is observed between the AWD during the first irrigation and those following. This deviation is estimated at 81 mm for the furrow practice and at 147 mm for the border. Owing to the substantiale cracking after earing AWD is much greater than for the first irrigation conducted in tilled soil conditions (after sowing). The contrary phenomenon is generally observed in surface irrigation for crops requiring high irrigation frequencies and cultivated on line (on the inter furrows) such as corn for instance. In our context, the efficiency half decreases between first and second irrigation for the border. Figure 1 underlines the difference in the advance trajectory between a tilled soil and a cracking soil.

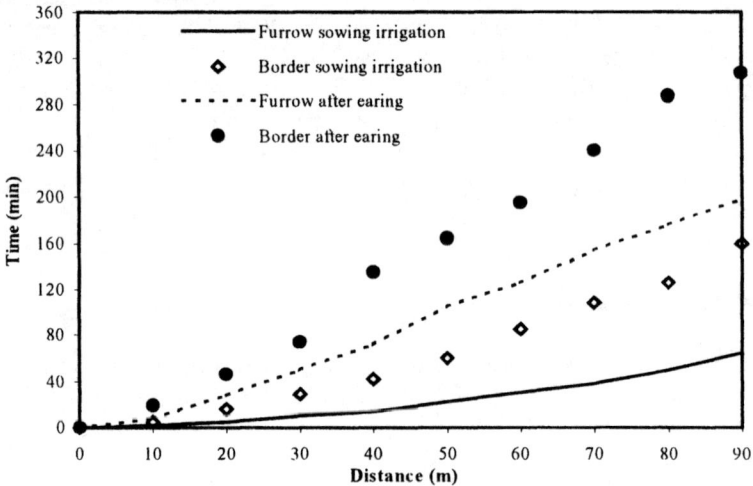

Figure 1. Comparison of the advance process between tilled soil (sowing irrigation) and cracking soil (after earing) for the same soil water depletion conditions (p = 0.9).

Whatever the irrigation order, border practice requires the highest applied water depths (compared with the furrow practice). After earing the applied water depths are approximately twice soil moisture deficit. This fact encourage to analyze particularly the furrow practice in order to propose irrigation strategies to improve water efficiency. To argue this purpose, Table 3 presents a comparison (between border and furrow) of the performance of the irrigation conducted after earing.

Table 3. Comparison of hydraulic performance of an irrigation conducted after earing

Irrigation technique	p	AWD (mm)	SMD (mm)	E_a (%)	DU (%)	SR (%)
Furrow	0.9	156	125	80	98	83
Border	0.9	287	105	37	72	70
Furrow	0.6	150	80	53	80	83
Border	0.6	264	85	32	80	88
Furrow	0.3	71	50	70	96	100
Border	0.3	145	50	35	94	100

As mentioned above, border irrigation induces the highest AWD compared with the furrow whatever the initial soil moisture status. Although these AWD decrease with p, E_a does not exceed 40%. For the border, these AWD are much greater than the SMD. The excess is estimated at 95, 164 and 137 mm, respectively for p = 0.3, 0.6 and 0.9. One still have to be careful regarding the measurement accuracy of the soil water content after irrigation due to a high probability of lateral losses in heavy cracking soil context. These eventual losses could explain that AWD is always greater than measured SMD.

In furrow irrigation, the AWD are contrasted when p increases from 0.3 to 0.6, but there is little change from p = 0.6 to p = 0.9. Nevertheless, for these two p values, the water storage in the root zone seems to be proportional to the SMD. That can explain the relatively low efficiency observed when p = 0.6. It would seem that cracking does not change significantly from p= 0.6 to p = 0.9 on this soil type. This fact still have to be confirmed by other field experiments. In the case of furrow, the experiments show that hydraulic performance criteria are almost comparable and satisfactory for p= 0.3 and p = 0.9 except for SR according that only 83% of the storable water is insured for p = 0.9.

3.2 Performance evaluation using modelling approach

The confirmation of the previous results by the modelling approach seems to be relevant. The AWD estimation was derived from the calibration of infiltration parameters conducted on the advance process for the irrigation after earing (i.e in heavy cracking soil context). Results are presented in Table 4.

Table 4. Irrigation evaluations criteria of hydraulic performance

Irrigation technique	p	SIRMOD model		RAIEOPT Model		Measured value	
		E_a (%)	SR (%)	E_a (%)	SR (%)	E_a (%)	SR (%)
Furrow	0.9	88	94	96	98	80	83
Furrow	0.6	62	97	65	97	53	83
Furrow	0.3	70	100	70	100	70	100
Border	0.9	52	100	54	100	37	70
Border	0.6	38	100	42	100	32	88
Border	0.3	33	100	38	100	35	100

One can observe that comparable results are proposed by the two models. Both overestimate the criteria derived from measurements probably due to the reasons previously evoked (lateral losses). According to the operative aspect of RAIEOPT and its adequacy to the physical context of cracking soil it appears relevant to select this model in order to estimate AWD using the advance process monitoring.

3.3 Yield evaluations

It is now necessary to evaluate the impact of the furrow choice against that of border on the agronomic efficiency. Table 5 shows the temporal evolution of emergence and the ratio of final plant number for a sowing made with 350 grains/m^2. The first irrigation has immediately followed the sowing according to soil dryness status. Since the soil water depletion level was the same for all the treatments (p = 0.9), the deviation of emergence is not significant between the treatments irrigated with the same irrigation technique. It can be noticed that the emergence losses are from 10 to 50% respectively for the border and the furrow. The substantiality of the losses can be attributed to the perturbation of the sowing bed and/or due to a too low lateral water diffusion (cut-off time too low despite of substantial capillarity forces during the first irrigation) in the case of furrow.

Table 5. Emergence evolution (plants/m²) and ratio (%) of final emergence

Irrigation technique	Days after sowing irrigation			% final emergence
	15	20	30	
Border	241	267	310	93
Furrow	83	130	182	52

The earing density underlines the difference between border and furrow in Table 6. Nevertheless, this deviation is relatively lower than which observed at the emergence according to the compensation effect due to tillering phenomenon.

The impact of irrigation after earing is highly evident on the grain filling and does not seem to be affected by the irrigation technique. So, the deviation between the weight of 1000 grains (WTG) is not significant for a same p value. Regarding the more efficient treatments (those of furrow), WTG seems to depend on the SMD for irrigations after earing with the values of 250, 200 and 160 mm, respectively for the p values :0.9, 0.3 and 0.6, (Table 3). Table 6 presents the observed yields (Y_a), the variation coefficient of which are around 5%, and the total applied water depth (TAWD) during the crop cycle. The irrigation effect is perceived using the additional yield (Ya-Y0) obtained in comparison with the non irrigated treatment where Y0= 2.3 T/ha. The ratio (E_i) between additional yield and TAWD is a criterion of comparison of the irrigation efficiency between the different treatments.

Table 6. Yield comparison for the different treatments

Irrigation technique	p	Number of ears/m²	WTG (g)	Y_a (T/ha)	Y_a-Y0 (T/ha)	TAWD (mm)	E_i (Kg/m³)
Border	0.9	350	40.3	5.1	2.8	714	0.39
Furrow	0.9	307	41.7	4.6	2.3	387	0.59
Border	0.6	351	30.0	3.8	1.5	668	0.22
Furrow	0.6	336	31.8	3.8	1.5	375	0.40
Border	0.3	366	36.1	4.9	2.6	720	0.36
Furrow	0.3	340	36.0	4.6	2.3	359	0.64

The lowest E_i values are observed for the border although this technique seems to insure, in our field conditions, a higher yield. These results militate in favour of the choice of the furrow technique in limited available water context. These results still have to be associated to the hydraulic efficiencies of the different treatments. As previously shown, the more efficient treatment (E_a = 80%) corresponded to the furrow technique with p = 0.9.

Thus results were applied to a climatic scenario (probability for exceeding of net irrigation requirement for wheat = 30%) in order to test irrigation strategies optimizing yields and efficiencies. This application uses the results of a study based on the simulation of irrigation strategies for wheat in Tunisia [13] [14] using ISAREG model [15]. The authors have established probability distributions of irrigation water requirements, using the water balance simulation on 40 years (1940-41 to 1979-80) accounting for climatic differences between northern part and central part of Tunisia. According to this analysis, the net irrigation water requirements (NIR) for a 30% probability are the following:

- NIR = 244 mm for the central region (annual rainfall: 330 mm).
- NIR = 122 mm for the northern region (annual rainfall: 450 mm).

Regarding the net requirements, the analysis shows a filling ratio of MASW at sowing around 30 to 40%, which signifies that an irrigation is necessary to insure the crop emergence.

The relative yield (Y_a/Y_m) is derived from Stewart model as described by Doorenbos and Kassam [16] using a yield response factor for winter wheat Ky = 1.05. The choice of an optimal treatment has to account for this ratio and for application efficiency.

Finally, the relative yield and the application efficiency E_a both are not sufficient to characterize the efficiency of the irrigations since the amount of water stored is not always utilised by the crop. In case of over run of NIR, a part of stored water remains in the soil. We account for that in our analysis the results of which are presented in Table 7 and 8.

For central region the best results are obtained with p = 0.3. Using a sowing irrigation (75 mm) and 3 irrigations (3 x 70 mm) spread from earing to grain filling, it is possible to increase the application efficiency until 72% with solely a 10% yield loss compared with maximum yield (Y_m). The maximization of yields requires an additional irrigation and does not affect the application efficiency. Whatever the yield objective, this irrigation strategy (p = 0.3) implies substantial equipment and man power requirements with management difficulties.

An interesting value of E_a is obtained with two irrigations for p = 0.9, one at sowing (75 mm) and the other at flowering (156 mm). This alternative assumes that a yield decrease of 18% is accepted but still appears better in limited water conditions because the amount applied does not exceed 230 mm. Moreover, the risk of plant lodging, which sometimes induces yield losses, is limited. For both the same p and E_a values, yield maximization requires a greater water supply (387 mm) which only 60% are profitable to the crop (63 mm of rest). The irrigation according to p = 0.6 would imply either a substantial reduction of yields (30%) or very high applied water (375 to 525 mm) inducing losses around 50%.

For the northern region, the optimal solution consists of adopting p= 0.3 using two irrigations, the first at sowing and the other at flowering. A third irrigation does not seem to be necessary according to the correct yield level with only two water applications. In this region, when irrigation is made later (p = 0.6 to 0.9), comparable yields can be obtained but a supplement of water around 80 mm will be lost.

Table 7. Simulation Results for several irrigation scheduling strategies using furrow irrigation experimental data
Central Tunisia   P= 30%, NIR = 244 mm

p	Sowing irrigation		Irrigation after earing			Total AWD (mm)	Total SMD (mm)	E_a (%)	Rest (mm)	ET_a/ET_m (%)	Y_a/Y_m (%)
	AWD (mm)	SMD (mm)	Number of irrigations	AWD (mm)	SMD (mm)						
0.3	75	57	3	213	150	288	207	72	-37	90	89
0.3	75	57	4	284	200	359	257	71	13	100	100
0.6	75	57	1	150	80	225	137	61	-107	72	70
0.6	75	57	2	300	160	375	217	58	-27	93	92
0.6	75	57	3	450	240	525	297	56	53	100	100
0.9	75	57	1	156	125	231	182	79	-62	84	82
0.9	75	57	2	312	250	387	307	79	63	100	100

Table 8. Simulation Results for several irrigation scheduling strategies using furrow irrigation experimental data
Northern Tunisia   P= 30%, NIR = 122 mm

p	Sowing irrigation		Irrigation after earing			Total AWD (mm)	Total SMD (mm)	E_a (%)	Rest (mm)	ET_a/ET_m (%)	Y_a/Y_m (%)
	AWD (mm)	SMD (mm)	Number of irrigations	AWD (mm)	SMD (mm)						
0.3	75	57	1	71	50	146	107	73	-15	96	96
0.3	75	57	2	142	100	217	157	72	35	100	100
0.6	75	57	1	150	80	225	137	61	15	100	100
0.9	75	57	1	156	125	231	182	79	60	100	100

ET_a = actual evapotranspiration Y_a = actual yield
ET_m = maximum evapotranspiration Y_m = maximum yield

4 Conclusion

For a wheat crop grown on a cracking soil we have shown that furrow irrigation practice (inter furrow spacing = 1.5 m) is more efficient from a hydraulic and an agronomic point of view than border irrigation. From the hydraulic point of view this conclusion is based both on measured water application in the root depth and results derived from modelling approaches of the advance-infiltration process on closed-end border and furrow. Under comparable initial soil water content, a water saving around 130 mm/irrigation after earing was obtained against the border with no significant yield decrease. In our field context, an adequate strategy of irrigation scheduling optimizing the agronomic efficiency has been proposed. It consists of supplying water to plots when the soil water depletion attains the value of 90%.

This strategy for the furrow irrigation practice was selected and tested on a climatic scenario corresponding to crop water requirements corresponding to 3 years return period for the central and northern region of Tunisia. The result of this test show that the strategy which optimizes the efficiencies consists of delivering 2 applications to wheat. The first insures the emergence, the second has to be applied after earing when the soil water depletion reach 90% for central region and 30% for northern region.

5 References

1. Rebai, M. (1994) Analyse - diagnostic du périmètre irrigué de Chaouat. Rapport technique, Projet de formation et de développement de l'économie de l'eau en irrigation, PNUD/FAO/TUN/91/002, pp. 59.
2. Fatnassi, A. (1995) Analyse - diagnostic du périmètre irrigué de Borj El Amri. Rapport technique, Projet de formation et de développement de l'économie de l'eau en irrigation, PNUD/FAO/TUN/91/002, pp. 80.
3. Zairi, A., Achour, H., Nasr, Z. et Ben Mechlia, N. (1994) Impact de certains paramètres sur la qualité de l'irrigation par calant dans un sol fendillant. Rapport technique,. Rapport de recherche, Projet PNUD/RAB/90/005, Programme de Tunisie 1994, pp. 32-43.
4. Research Center for the Utilisation of Saline Water in Irrigation, (1970) Research and training on irrigation with saline water. Tech. Rep. 1962-1969. UNDP-UNESCO/CRUESI. TUN 5. Paris. 241 p + Annexes.
5. Zairi, A., Achour, H., Nasr, Z. et Ben Mechlia, N. (1993) Irrigation par planche - interaction débit/longueur d'un calant et effet sur la dose. Rapport de recherche, Projet PNUD/RAB/90/005, Programme de Tunisie 1993, pp. 21-23.
6. Zairi, A., Achour, H. (1996) Comparison de la qualité de l'irrigation par calant et de l'irrigation à la raie sur un sol fendillant. Rapport de recherche, Projet SERST/PNM94 "Amélioration de l'irrigation de surface en conditions hydriques restrictives", pp. 30.
7. Merriam, J.L and Keller, J. (1978) *Farm Irrigation System Evaluation : A Guide for Management.* Utah State University, Logan, 285 p.
8. Walker W.R. (1993) *SIRMOD, Surface Irrigation Simulation Software.* Utah State University. Logn, Utah, USA. pp. 25.

9.   Mailhol, J.C. (1992) Un modèle pour Améliorer la conduite de l'irrigation à la raie *ICID Bulletin*, Vol. 41 No. 1, pp. 43-60.

10. Mailhol, J.C. et Gonzalez, J.M. (1993) Furrow Irrigation Model for Real-Time Applications on Cracking Soils. *J. Irrig. And Drain. Engrg.*, 119(5) : 768-783p.

11. Mailhol , J.C Baqri, M. et Lachhab M. (1997) Operative irrigation modelling for real-time applications on closed-end furrows. *Irrigation and Drainage Systems* No. 11, Netherlands, 347-366p.

12. Mailhol , J.C. , Gitrousse J.C. et Rakotoarisoa A. (1988) Essai de prédetermination de la qualité hydraulique d'un arrosage : Application aux raies courtes. *C.I.G.R. Int. Sympos.* Sept. 5-10 Ilorin, Nigeria, pp. 403-410.

13. Pereira, L.S. (1991) Rapport technique sur l'irrigation. Projet d'appui au programme national d'irrigation d'appoint des céréales PNUD/FAO/TUN/86/014, pp. 53.

14. Teixeira, J.L., Fernando, R.M., and Pereira, L.S. (1995) Irrigation scheduling alternatives for limited water supply and drought. *ICID J.* , Vol 44, No.2, pp.73-88.

15. Teixeira, J.L., and Pereira, L.S. (1990). ISAREG, an irrigation scheduling simulation model. *ICID Bulletin*, Vol 41, No. 2, pp. 29-48.

16. Doorenbos, J., and Kassam, A.H. (1979) *Yield response to water.* Irrigation and Drainage Bulletin, n 33, FAO, Rome.

FURROW SURGE IRRIGATION AS A WATER SAVING TECHNIQUE

I. VARLEV
Research Institute of Irrigation, Drainage and Hydraulic Engineering, Sofia, Bulgaria
Z. POPOVA
Institute of Soil Science and Agroecology 'N. Pushkarov', Sofia, Bulgaria
I. GOSPODINOV
Experimental Station, Stara Zagora, Bulgaria

Abstract
Surge irrigation is defined as on and off cycles of stream delivered at the head of the furrows. Usually on - and off - time are constant during surge irrigation event. Surge irrigation have been tested in Bulgaria for different furrow length (from 100 to 700 m), soil and terrain conditions since 1965. Water saving is studied separately over first and second phase of stream advance and registered surge effect is compared for the studied situations. During the first phase streams advance traditionally non - uniformly over the surface of a furrow set which usually includes more then 15 - 20 simultaneously irrigated furrows. This non-uniformity is substantially reduced by applying water in surges that saves 20 - 30% of the irrigation water. The post - advance phase aims at increasing the opportunity time at the furrows tail. Runoff water losses reach 30 - 40% of the delivered water when realizing continuos furrow irrigation of high uniformity performances. Surge irrigation in post - advance phase could produce equivalent uniformity and reduce runoff to 10 - 15% of the total delivered water. Overlapping surges of short on/off - time during post-advance phase turn out continuos opportunity time at the furrow tail that improves uniformity of water distribution over the furrow length. Automatic surge have been manufactured. The price of the latter is unreasonably high for the farming system in developing countries. Surge irrigation could be realized in such countries, including Bulgaria, by means of additional labor requirement instead of installment of automatic valves. Several surge irrigation technology have been tested and their description is provided in the full text of the report. On basis of the reported analytical and experimental research it is concluded that surge irrigation could save 30 - 45% of irrigation water required for continuous irrigation.
Keywords: Advance uniformity, furrow irrigation, runoff, surge.- flow, water saving.

Water and the Environment: Innovative Issues in Irrigation and Drainage. Edited by Luis S. Pereira and John W. Gowing. Published in 1998 by E & FN Spon. ISBN 0 419 23710 0

1 Introduction and background

Furrow irrigation is widespread throughout the world. In the USA, Russia, Bulgaria and many other countries it is used in over 60% of the irrigated areas. The furrow surge irrigation as a new technique gives an opportunity to achieve high technical and economical results.

With the simultaneous irrigation of a set of furrows (for example a set more than 10 - 15 furrows) the streams advance over the surface nonuniformly [1]. This phenomenon cannot be characterized precisely by dividing the furrows only to 'wheeled' and 'nonwheeled' [2]. Complete description of furrows' nonuniformity of stream advance in the set is obtained by the coefficient K_{nun} [3]:

$$K_{nun} = 2(L_{max} - L_{min})/(L_{max} + L_{min}) \qquad (1)$$

where L_{max} and L_{min} are respectively the maximum and the minimum lengths of the streams obtained by averaging test results with a straight line (Fig. 1).

The coefficient K_{nun} should be obtained when the 'average' stream reaches the end of the furrows - L_f. For example, $K_{nun} = 1,0$ is obtained at $L_{max} = 3 L_{min}$. The most common values of K_{nun} with simultaneous irrigation of 30-100 furrows range from 0,3 to 1,0.

The nonuniformity of irrigation has been subject to hundreds of studies. Along with the famous coefficient of Christiansen [4] Cu, scores of other relationships have also been proposed giving only statistical description of the nonuniformity.

Following dozens of his own and other experiments the author has shown [5] that in over 80% of the studied cases the relationship 'yield - water infiltrated in the soil during irrigation' is precisely described by a quadratic equation. In these cases the lost yield due to the nonuniform irrigation, Δy, with regard to the maximum possible yield, Y_{max} is described by the relationship:

$$\Delta y = (1 - y_0)m^2 F_{nun} \quad [\%] \qquad (2)$$

where: $y_0 = Y_0/Y_{max}$ is the relative yield under rainfed conditions; Y_0 - yield without

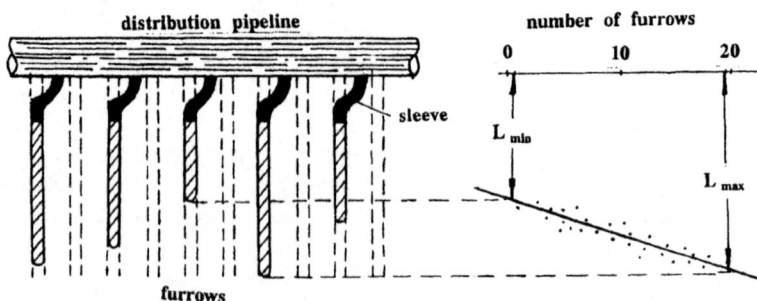

Fig. 1 Non - uniformity of stream advance of a furrow set

irrigation, [kg/ha]; Y_{max} - maximum yield [kg/ha], obtained with irrigation depth M_{max} [m³/ha]; $m = M/M_{max}$ - average relative irrigation depth for the area considered; M - average irrigation depth [mm]; F_{nun} is the coefficient of nonuniformity of water distribution [6] and is equal to:

$$F_{nun} = -\frac{100}{n} (M_{\omega} / M - 1)^2 = 100 \ V^2 \ [\%] \tag{3}$$

where: M_{ω} - total (individual) infiltrated irrigation depth [mm] delivered to the different points of the area considered; n - number of points; V - coefficient of variation.

It may be useful to note that between the coefficient F_{nun} and the coefficient of Christiansen exists an approximation:

$$F_{nun} \approx 1.63(1 - Cu)^2 \ [\%] \tag{4}$$

In arid and semiarid conditions the relative yield without irrigation is practically zero, $y_0 \approx 0$. The frequently delivered irrigation depths are close to M_{max}, i.e. m=1. Therefore in such cases from equation (2) it follows that: $\Delta y' = F_{nun}$, [%], i.e. the yield lost due to the nonuniformity of irrigation is equal to the coefficient F_{nun} [%]. This means that the coefficient of nonuniformity F_{nun} has both statistical (equation 3) and economic importance (equation 2), which marks it out among the other coefficients.

After the streams reach the end of the furrows, irrigation water is still delivered so that the soil in the lower furrow end can be wetted (second phase of the irrigation). To describe the duration of this phase, the parameter I -relative extension of application time - is used:

$$1 = t_{ad}/t_l \tag{5}$$

where t_{ad} is the time for additional water application after the stream has reached the furrow end for the time t_l.

Experiments with furrow surge irrigation during the second phase were carried out in Bulgaria from 1965. Their results were published in Proceedings of the Research Institute for Irrigation and Drainage, Sofia, [7] in Bulgarian language with short summaries in Russian and English. Although these bulletins were send out to a number of European libraries, the furrow surge irrigation did not become popular outside Bulgaria at the time. Similar were the cases with other studies which were published in small countries and in languages which are not widely used by the international scientific community.

After 1978, in the USA and in other countries [8] extensive experimental and theoretical studies were conducted with furrow surge irrigation. A number of mathematical models were designed and automatic valves launched in production which were operated through ordinary or solar batteries. These valves provide continual on and off streams according to a pre - determined programme and make automatic almost the entire irrigation process.

2 Objectives and methods of research

The research objectives are:

1. To establish, at all other conditions being equal, the real savings of water with furrow surge irrigation as compared with the traditional method;
2. To describe in brief technologies through which furrow surge irrigation can be applied without using automatic valves.

The experiments with furrow surge irrigation with furrows of 100 to 700 m in length are carried out in Bulgaria under various soil and terrain conditions. The velocities of stream head advance and the runoff of the water accumulated in the furrows, the volume of runoff and other indicators are measured (Fig. 2).

The nonuniformity with which stream heads advance along the dry furrow bed (K_{nun}) is measured either by the lengths achieved at a given moment (T_i = const) (Fig. 1, equation 1) or by the time for reaching furrows' end $(t_1)_i$.

3 Nonuniformity of stream advance in the furrows' set

At simultaneous irrigation of 23 furrows during 2 consecutive days in 1991, the measured average K_{nun} with furrow surge irrigation was 0.41. With continuous irrigation it was 0.91 (Fig. 3). Figure 3 also shows the average results for K_{nun} for 1992 and 1993. These results were achieved with pre-irrigation soil moisture of 70 to 92% FC. They show that with furrow surge irrigation the stream heads advance along the furrows considerably more uniformly. This can be explained with the 'preliminary' wetting of the furrow bed by the previous surges. During the interval between two surges the humidity spreads out, the soil swells and part of the micro and macropores get closed. In this way the simultaneously irrigated furrows under surge irrigation become more 'homogeneous'. The soil moisture before the irrigation has a significant effect on the coefficient K_{nun}. It can be seen from Fig. 3 that in general the

Fig. 2 Runoff measurements by volumetric method at the end of the furrow

decrease of the pre - irrigation soil moisture leads to a decrease of the coefficient K_{nun}.

It should be noted that the coefficient K_{nun} must be determined by observations in 20-25 furrows. Then its values are representative for the simultaneous irrigation of 40-100 furrows. With less observed furrows (7 - 9 in 1992 and 1993) (Fig. 3) the values of K_{nun} are 30 - 40% lower.

In real field conditions as well as in some of the experiments the delivery of water takes place in daytime only due to a number of reasons. When this time is insufficient to wet the furrows (for example, furrows 250 - 400 m length) the irrigation has to continue on the following day. The break during the night (for some 10 - 14 hours) has an effect similar to the intervals with furrow surge irrigation. This explains the very low values of K_{nun} of 0,06 to 0,13 in the experiments from August 12 - 13th and 17 - 19th, 1992 with the pause during the night. On August 20th, 1992, the experiment with traditional method has coefficient of nonuniformity $K_{nun} = 0,5$.

The nonuniformity of stream advance (K_{nun} - equation 1) has a direct effect on the nonuniformity of water distribution over the irrigated area (F_{nun} - equation 3) and on the yield (equation 2).

The relationship between K_{nun} and F_{nun} can be seen on Fig. 4. To achieve one and the same uniformity of irrigation (for example, $F_{nun} = 10\%$), at $K_{nun} = 0,5$ and $K_{nun} = 1,0$, the parameter I should be 0,3 and 0,7, respectively (Fig. 4). This means that in the second case ($K_{nun} = 1,0$) the total duration of the irrigation (1+I) and the consumed water will be 30% greater than that of the case with $K_{nun} = 0,5$ (1,7:1,3). It may be noted that nonuniformity characterised by $K_{nun} = 0,5$ is achieved easily with surge irrigation.

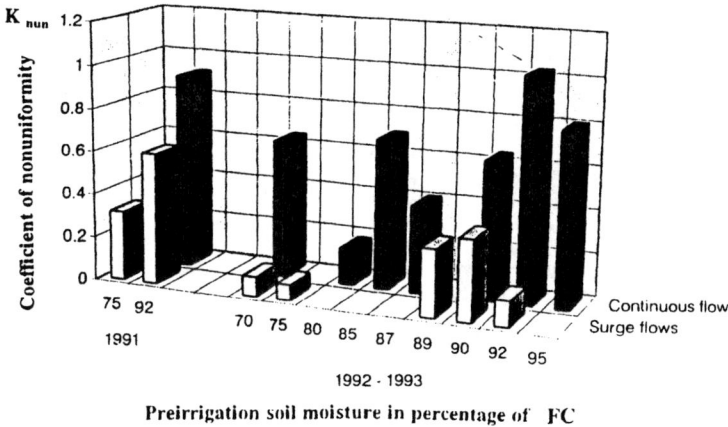

Fig. 3 Coefficient K_{nun} in different pre-irrigated soil moisture condition with two irrigation techniques

The economically optimal values of F_{nun} coefficient for the field crops grown in Bulgaria vary from 6 to 16%. For the prevalent climatic conditions in the country, the mean multiyear values of y_0 vary from 0,4 to 0,6. When m = 1,0 it follows that the economically reasonable yield losses (equation 2) are in the range of 3 - 9%. It can be seen (Fig. 3) that on average with surge irrigation, the coefficient K_{nun} is twice lower compared with continuos flow. Given these data and using the chart presented in Fig. 4, it can be found that the water savings with surge irrigation will be up to 20 - 30% in comparison with the traditional method of continuous flow (for Bulgarian conditions).

In countries with drier climate (lower values of the relative yield y_0), in order to stay in the same range of yield losses (3 - 9%) it is necessary to achieve higher uniformity, i.e. lower values of F_{nun}. In such conditions the water savings resulting from surge irrigation (Fig. 4) will be higher than the ones in Bulgaria (20 - 30%).

4 Estimation of runoff losses

When streams reach the furrows' end, runoff takes place (the second phase). With the traditional method of continuous streams, the runoff grows constantly (Fig. 5 - the 'dot-dash' curve). After some time, when the intake rate gets closer to the coefficient of filtration, the runoff becomes constant. Depending on the necessary uniformity (F_{nun}), the runoff losses with the traditional method amount to 50 - 60% of the delivered water during the second phase.

With surge irrigation the runoff losses are lower due to the intermittent streams (Fig. 5). At this point it is necessary to raise the problem of the time for contact between water and soil along furrows' length (opportunity time). In previous studies [5] it has been established that the velocity of the last part of water accumulated in the furrows v_{rec} is lower than the velocity of stream head advance v_{adv} (Fig. 6). As a result of many experiments it was determined that at a suitable duration of the surges (t_{on}), the stream of the next impulse ($i+1$) reaches the water from the previous impulse (i) (Fig. 6).

On Fig. 7 is shown the variation of the opportunity time along the furrow lenght in 'usual' and 'overlapped' surges. This happens on a distance from the furrow head $l_{overlapped}$ which is practically constant in subsequent surges. Thus the lower part of

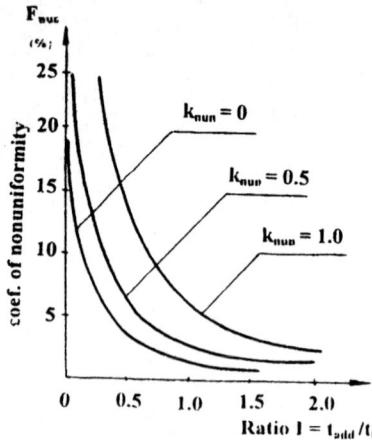

Fig. 4 Coefficients of non-uniformity of infiltrated water over the irrigated area for different K_{nun} and I

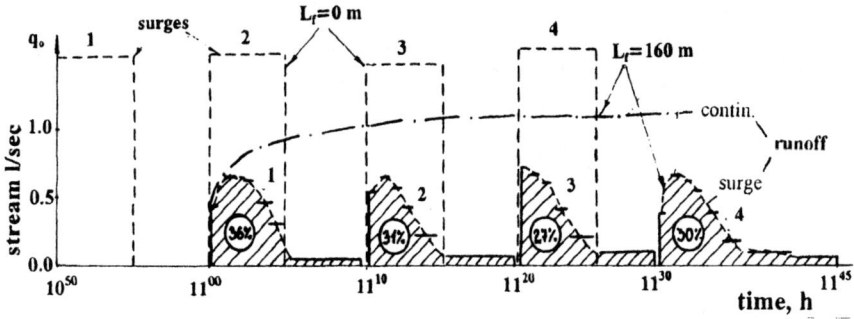

Fig. 5 Streams diagrams at the head and runoff from the tail of the furrow in surge irrigation. The runoff in continuos irrigation is shown with sign "dot-dash"

the furrows with length L_f - $l_{overlapped}$ is constantly covered by water. During the same period the opportunity time at the head of the furrows is twice shorter. The uniformity of distribution during the second phase is improved by the relative increase the opportunity time in the lower part of the furrows.

Particularly interesting is the diagram of the runoff from the furrows' end in the case of overlapping surges (Fig. 5). The diagrams of inflow and runoff streams, at 4 consecutive surges and on and off time of 5 minutes, are presented (Fig. 5). The phase of "intensive" runoff lasts 5 minutes after which come another 5 minutes of minimal runoff. The average runoff quantity for the four surges is 30% of the water delivered at the head of the furrow. If the duration of the second phase is equal to the first one, i.e. I = 1,0, the runoff losses are reduced to 15% of the water delivered at the furrow head. In comparison to the irrigation with continuous streams, water savings of 10 - 15% are achieved. If these savings are combined with the savings resulting from the more uniform stream advance during the first phase, the savings with surge irrigation amount to 30 - 45% in comparison to the irrigation with continuous streams.

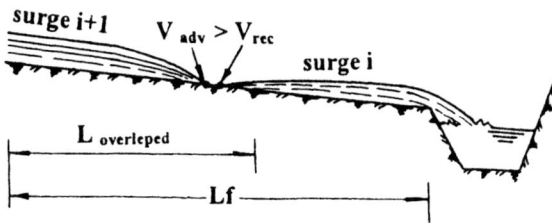

Fig. 6 Scheme. The stream head of the surge (i+1) reach the stored water from surge (i)

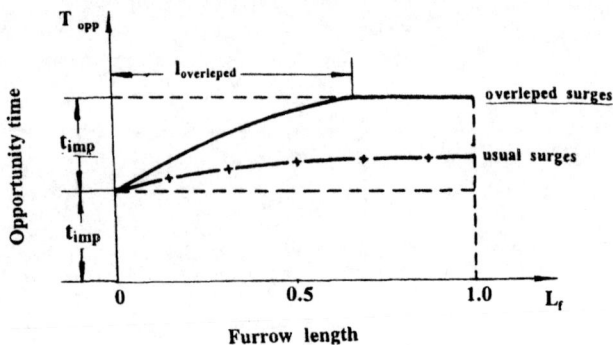

Fig. 7 Opportunity time along the furrow length in 'usual' and 'overlapped' surges

5 Surge irrigation technologies without automatic valves

It is appropriate to note again the perfection and the automation of the irrigation process achieved by using the valves for automatic switch on and off streams designed in the USA.

Unfortunately, the price of these valves (some 1500 USD) is too high for countries with insufficiently developed economies (including Bulgaria). In these countries the price of labor is very low (100 - 150 USD/month), which makes the use of automatic valves economically unviable.

This part of the paper presents in brief some well known and not well known technologies through which it is possible to apply furrow surge irrigation without automatic valves. In such cases, to turn on and off the streams labor requirement up to 20 - 25% of the labor needed for the remaining irrigation operations.

If the water is delivered during the daytime only (as practiced in Bulgaria), a two fold decrease of the coefficient K_{nun} is obtained when during the first day the furrows are wetted to 60 - 70% of the length. On the second day the wetted length reaches 90-100% of the total length. On the third day enough water is delivered to wet the lower end of the furrows. The irrigator may irrigate in this manner 2 - 4 sets to be busy full day.

Another way to apply surge irrigation is in the case when the water is delivered by portable pipelines distributing the water by flexible sleeves. In this case the surge regime is realize by moving the sleeves between two adjacent furrows (Fig. 1).

If the pipeline outlets have regulators the surges are carried out by their successive opening and closing.

The automatic valves may be replaced by manually operated 'butterfly' valves which deliver water to the left and right in two adjacent furrow sets.

6 Conclusions

The stream advance along the dry furrows beds takes place with certain nonuniformity, which is characterised well enough by the coefficient K_{nun} (equation 1). This coefficient is about twice lower with surge irrigation as compared to traditional furrow irrigation (Fig. 3). If equal uniformity is achieved, water savings with surge irrigation will be 20 - 30%.

The runoff from furrows' end with surge irrigation is reduced with 10 - 20% in comparison to the traditional irrigation with continuous streams. With 'overlapped' surges the lower part of the furrows is constantly covered by water which improves the uniformity of irrigation during the second phase. On the whole the water savings with surge irrigation reach 25 - 40%.

Due to the high price of the valves which provide automation of the irrigation process, the surge irrigation can be applied by manual switch on and off streams. These manual technology solutions are applicable in countries where labor cost is comparatively low. A brief description of the tested technology in Bulgaria is given in this paper (section 5).

7 References

1. Garbunova, E. H., (1963) *Organization and Surface Irrigation Practice for Cotton in the Conditions of Tadjic Soviet Union State*, PhD Thesis, Moscow, (in Russian).
2. Mangs, H. L., M. L. Hooker, (1984) *Field Comparison of Continuous, Surge and Cutback Irrigation*, paper N 84-2093, ÀSAE.
3. Varlev, I., (1983) *Optimization of Irrigation Uniformity*, Academy of Agriculture, Sofia, p. 148 (in Bulgarian)
4. Cristiansen, F. E., (1941) *The Uniformity of Application of Water by Sprinkler Systems*, Agricultural Engineering, Mar.
5. Varlev, I., N. Kolev, I. Kirkova (1994) *'Yield - Water' Relationships and their Changes during Individual Climatic Years*, 17th European Regional Conference on Irrigation and Drainage, Varna, Bulgaria
6. Varlev, I., (1965) *Necessary Uniformity in Distribution of Irrigation Rate*, Hydrothechnics and Melioration, No 9 (in Bulgarian)
7. Varlev, I., (1971) *Improvement in the Uniformity along the Furrows and Beds with Surge Irrigation*, Scientific Works of Institute of Land Improvement, Sofia, Vol. XII, pp. 5-21 (in Bulgarian)
8. Walker, W. R. G. Skogeboe (1987) *Surface Irrigation*, Prentic haal Inc. New Jersey 07632.

EVALUATION OF PERFORMANCE OF SURFACE IRRIGATION SYSTEMS: OPTIMIZATION OF INFILTRATION AND ROUGHNESS PARAMETERS

M. J. CALEJO
Hidráulica Agrícola, Hidrotécnica Portuguesa, Lisboa, Portugal
P. L. SOUSA and L. S. PEREIRA
Department of Agricultural Engineering, Institute of Agronomy, Technical University of Lisbon, Portugal

Abstract
Field evaluation of the performances of irrigation systems is of great importance to improve water use in surface irrigation. Using field data in surface irrigation models allows to estimate the infiltration and roughness parameters by the inverse solution of the irrigation model. A programme based on the flexible tolerance algorithm was developed to optimize these parameters, which works interactively with the surface irrigation model SRFR. Two objective functions are utilized, one relative to the advance phase, the second to both advance and recession. The analysis, relative to furrow irrigation, shows that computations of the advance and recession times, infiltration profiles, distribution uniformity and application efficiencies are appropriate when the second objective function is utilized.
Keywords: Advance and recession, parameters optimization, furrow irrigation, infiltration, modelling, performance, roughness.

1 Introduction

Performance of surface irrigation systems is commonly evaluated by the distribution uniformity DU (%) and the application efficiency e_a (%) [1] [2]. DU is defined by the ratio between the average infiltrated depth in the low quarter of the field (Z_{lq}, mm) and the average infiltrated depth over the field (Z_{avg}, mm), i.e. $DU = 100\, Z_{lq} / Z_{avg}$. The application efficiency represents the ratio between the average depth added to the root zone storage (Z_r, mm) and the average depth applied to the field (D, mm), i.e. $e_a = 100\, Z_r / D$.

DU is an indicator of the system performance and mostly depends on the parameters characterising the irrigation event. e_a is a management performance indicator depending

Water and the Environment: Innovative Issues in Irrigation and Drainage. Edited by Luis S. Pereira and John W. Gowing. Published in 1998 by E & FN Spon. ISBN 0 419 23710 0

upon both the system characteristics and the irrigation scheduling [2] [3]. Both parameters are good indicators of the environmental appropriateness of the systems design and management. High uniformities provide for improved use of water by the field crops, avoid overirrigation in a part of the field, and contribute to control leaching of agro-chemicals. High efficiencies are associated with appropriate conditions to control water wastes and water contamination, and to use low quality water and irrigating saline soils. DU e e_a are better estimated when using field data from field evaluations and, often, with help of simulation models [4].

Surface irrigation evaluation [1] includes observation of: infiltration characteristics; the slope/topography of the field being irrigated; furrow geometry and spacing; soil moisture before and after irrigation; inflow-outflow balance; irrigation phases, particularity advance and recession; depth of water; irrigation time; and management rules currently utilized by the irrigator. Field evaluations provide accurate estimation of irrigation performances and play a fundamental role in improving surface irrigation. On the one hand, they offer the most appropriate conditions for advising irrigators and to support real time irrigation management decisions; on the other hand, they provide the information required for design, model validation and updating, and optimization programming.

The use of surface irrigation models is, worldwide, far behind potentialities [5]. Some of the reasons are related to insufficient infiltration and roughness input data required by the models. Several infiltration models may be utilized to describe infiltration under surface irrigation [4] [5]. Empirical and semi-empirical equations are usually adopted but deterministic approaches can be utilized [6]. Field techniques [4] can be utilized to estimate the parameters of infiltration equations. Several studies have been developed to define methodologies to estimate the infiltration parameters using volume balance approaches to the furrow advance [4] [7] [8] [9] [10].

Infiltration and the roughness parameters can be estimated through the inverse solution of surface irrigation. Using field data on advance, recession, inflow rates, furrow form, length and slope, those parameters are searched by iterative manipulation of a simulation model, and minimizing the differences between simulated and observed advance and recession [11] [12].

Further advances are reported from using optimization techniques coupled with the the inverse solution of the zero-inertia model [13] [14] [15]. The study reported herein follows this line and concerns the application of an optimization model coupled with a surface irrigation model to obtain the infiltration and roughness parameters for furrow irrigation.

2 Material and methods

The SRFR model [16] is selected to simulate the furrow irrigation events. This model adapts particularly well to describe level furrows and the impacts of geometry of furrows on irrigation performances.

Infiltration is described in the SRFR model by the modified Kostiakov equation:

$$I = ak\tau^{a-1} + I_0 \tag{1}$$

where I (mm h^{-1}) is the infiltration rate, a and k (mm h^{-a}) are empirical parameters, I_0 (mm h^{-1}) is the empirical steady state infiltration rate and τ (h) is the time of opportunity for infiltration.

This equation is used in the model with a variable charges represented by the flow depth, and a variable intake area by unit length of the furrow, represented by the wetted perimeter.

The roughness in the flow equations is expressed by the Manning coefficient n:

$$ n = \frac{AR^{2/3}S_f^{1/2}}{Q} \tag{2} $$

where A (m^2) is the wetted area, R (m) is the hydraulics radius, S_f is the furrow slope and Q (m^3 s^{-1}) is the discharge rate at the section. In SRFR, it is assumed that the n remains constant during the irrigation.

The flexible tolerance algorithm and the flexible polyhedron search of Nelder and Mead well described by Himmelblau [17], is used to implement the optimization technique. A successful application of this optimization procedure is referred in [14]. A computer programme SRFRINV [18] was developed to apply this optimization procedure to the estimation of the infiltration parameters a, k and I_0 (Eq.1) and the roughness n (Eq. 2). The programme runs with the model SRFR producing the data required for the SRFR to compute the advance and recession times and using the respective output data for the optimization. Model SRFRINV, summarized in Fig.1, aims at minimizing one of the following objective functions:

$$ OF_1 = \sum_{i=x}^{L} \left(t_{av,o} - t_{av,s} \right)_i^2 \tag{3} $$

$$ OF_2 = \sum_{i=x}^{L} \left(t_{av,o} - t_{av,s} \right)_i^2 + \sum_{i=x}^{L} \left(t_{rec,o} - t_{rec,s} \right)_i^2 \tag{4} $$

where $t_{av,o}$ and $t_{av,s}$ are, respectively, the observed and simulated advance times (min) at the distance x (m) from the upstream end, $t_{rec,o}$ and $t_{rec,s}$ are the observed and simulated recession times at the same distance x, and L is the distance to the point of maximum advance.

Observations have been performed in the graded furrows with 200 m length, average slope 0.2%, and spaced 0.75%. Furrows were installed in a silty-loam soil, in the Experimental Station António Teixeira, Coruche. Soils are similar to those described in [6]. Field observations utilized in this study correspond to two first irrigations events in 1994, where the inflow rate was $q_{in} = 1.0$ l s^{-1} and the time of cutoff was $t_{co} = 85.2$ min, and two other first irrigation events in 1995, with $q_{in} = 1.47$ l s^{-1} and $t_{co} = 81$ min. Observations of advance and recession times were performed each 20 m from the upstream end. The initial values for the infiltration parameters (Table 1) were obtained by furrow infiltrometer measurements. The parameter n was estimated from observations of the flow depth with help of Eq. 2.

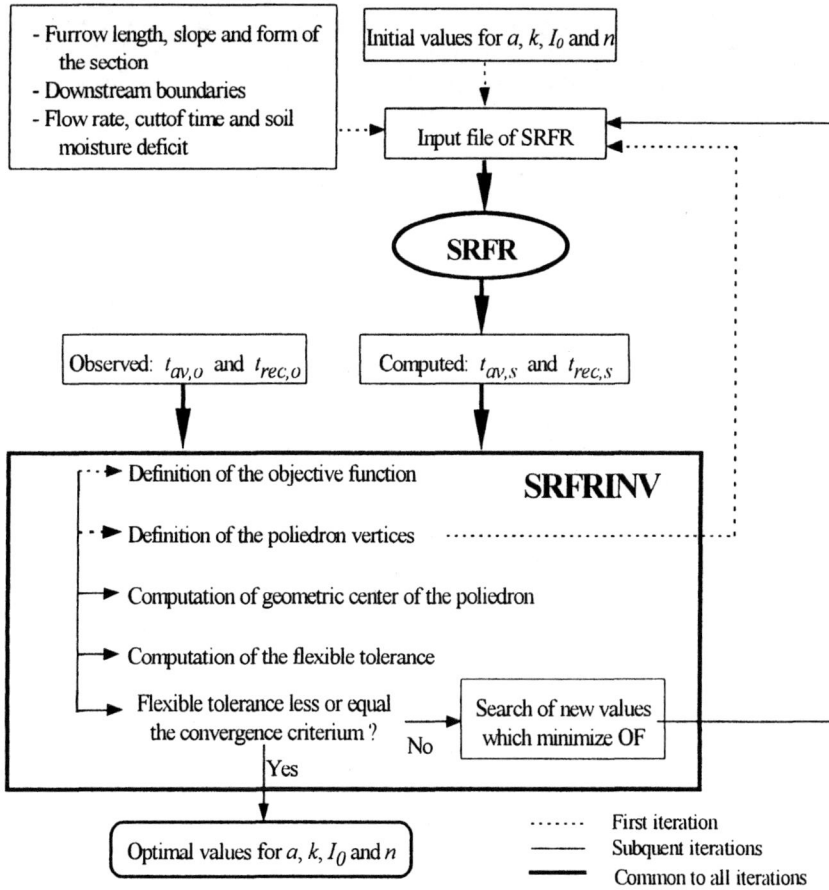

Fig. 1. Simplified representation of programme SRFRINV adopting the flexible tolerance algorithm and running in conjunction with the model SRFR.

For both objective functions (Eq. 3 and 4) the following alternatives were considered: search of the 2 parameters n, and k, assuming a and I_0 are known; search of the 3 parameters n, k, and a, assuming I_0 known; and search of the 4 parameters n, k, a, and I_0.

Table 1. Initial values of infiltration and roughness parameters

	Irrigation events			
	MC-1/94	MI-1/94	MI-1/95	MM-1/95
k (mm h^{-a})	49.84	39.67	74.65	74.65
a	0.28	0.24	0.23	0.23
I_0 (mm h^{-1})	11.65	11.38	13.67	13.67
n (m$^{-1/3}$ s)	0.02	0.02	0.03	0.04

The parameters are optimized in their dimensionless form, in order to reduce the problems concerning the different scale of the parameters. Constraints were applied to each parameter to confine it to its normal expected range of values for alluvial soils. The criterion of convergence was fixed at 1.0×10^{-5}, which refers to the distance of the optimal solution to the center of the polyhedron.

3 Results

Infiltration and roughness parameters computed through the optmization procedure using the objective functions OF_1 and OF_2 (Eq. 3 and 4) when searching 2, 3 or 4 optimal values for the parameters, i.e. $X = (n, k)$, $X = (n, k, a)$ and $X = (n, k, a, I_0)$ have been utilized with the model SRFR to simulate the irrigation events under analysis. From these simulations, selected results were compared with those observed in the field. Results of these comparisons by means of linear regression through origin are given in the Table 2. The goodness fit is evaluated through the regression coefficient b, the coefficient of determination r^2, the average absolute error of estimates AAE (min) and the average relative error of estimates ARE (%).

The advance time (t_{av}) is well simulated for every case. However, results are better when the objective function OF_1 is utilized since parameters are optimized for the advance time only. When function OF_2 is utilized, there is a slight trend to overestimate t_{av} but the maximum AAE is only 3.9 min. Most of ARE values are below 10 %.

The recession time is poorly estimated when the function OF_1 is used, with errors of estimates increasing when the number of optimized parameters also increase. For the 1994 events, the errors are unacceptable. When the objective function OF_2 is utilized, errors for t_{rec} are much smaller, with all values for AAE bellow 11 min and ARE ≤ 7.3 %.

In Fig. 2 is illustrated how $t_{av,s}$ and $t_{rec,s}$ are modified when using the function OF_1 or OF_2 in searching $X = (n, k)$.

Results in Table 2 and Fig. 2 show the superiority of the optimization using the objective function OF_2. This is also illustrated in Fig. 3 where infiltration profiles are compared. When OF_1 is used, the errors in the recession time, and consequently in the opportunity time for infiltration have a first order influence on the infiltration profile. When the function OF_2 is utilized the infiltration profiles are much more coherent and close to observation, and there is no evidence of differences from searching the optimal values for 2, 3 and 4 parameters.

The distribution uniformity DU and the application efficiency e_a (definitions given in the Introduction section) were computed by the model using the different sets of optimized infiltration and roughness parameters. Results in Table 3 show that indicators computed when using parameters optimized through the objective function OF_2 have smaller variation than those computed from objective function OF_1. This is explained by the fact that parameters computed from OF_2 produce a smaller variation in the infiltration profiles, as represented in Fig. 3.

Considering all results analysed before, it may be accepted that performance indicators computed with parameters optimized usin the equation OF_2 are closer to the reality than the other ones. Results in Table 3 also make evident that the discharge rate utilized for the 1995 irrigation events is execessive, comming downstream losses which originate the lower values for the performance in these two events.

Table 2. Statistical parameters of goodness fit of simulated vs. observed data

Irrigation event, objective function and parameters searched	Advance time				Recession time			
	b	r^2	AAE (min)	ARE (%)	b	r^2	AAE (min)	ARE (%)
MC-1/94								
OF_1								
$X = (n, k)$	0.986	0.99	1.98	7.56	1.381	0.25	43.9	34.8
$X = (n, k, a)$	0.993	1.00	1.26	3.93	1.672	0.19	77.2	61.1
$X = (n, k, a, I_0)$	0.995	1.00	0.94	2.81	6.798	0.08	658.	514.
OF_2								
$X = (n, k)$	1.045	0.98	2.47	8.32	0.984	0.70	4.69	3.81
$X = (n, k, a)$	0.990	0.98	3.38	13.6	0.997	0.70	4.81	3.92
$X = (n, k, a, I_0)$	1.000	0.98	3.27	12.4	0.994	0.68	4.40	3.61
MI-1/94								
OF_1								
$X = (n, k)$	0.996	1.00	1.31	4.55	1.025	0.40	23.9	14.8
$X = (n, k, a)$	0.999	1.00	0.82	3.13	1.169	0.22	52.4	31.8
$X = (n, k, a, I_0)$	0.968	1.00	1.42	4.44	3.980	0.10	414.	247.
OF_2								
$X = (n, k)$	1.121	0.99	3.90	7.58	0.978	0.63	9.80	6.59
$X = (n, k, a)$	1.079	0.99	2.73	5.62	0.966	0.64	11.0	7.35
$X = (n, k, a, I_0)$	1.083	0.99	2.93	6.24	0.975	0.62	9.99	6.72
MI-1/95								
OF_1								
$X = (n, k)$	0.998	1.00	0.51	6.29	1.040	0.67	8.82	7.68
$X = (n, k, a)$	0.997	1.00	0.53	6.94	1.184	0.61	19.1	16.6
$X = (n, k, a, I_0)$	0.998	1.00	0.46	5.32	1.589	0.22	58.4	48.6
OF_2								
$X = (n, k)$	1.056	0.99	1.27	8.38	0.972	0.84	4.53	4.17
$X = (n, k, a)$	0.975	0.99	1.03	11.6	1.002	0.90	1.80	1.71
$X = (n, k, a, I_0)$	1.029	1.00	0.76	6.71	0.982	0.90	3.10	2.85
MM-1/95								
$\bar{O}F_1$								
$X = (n, k)$	1.000	1.00	0.30	2.65	1.093	0.36	7.45	7.64
$X = (n, k, a)$	1.000	1.00	0.50	3.12	1.427	0.14	39.8	40.7
$X = (n, k, a, I_0)$	0.999	1.00	0.29	2.77	1.438	0.08	40.2	40.5
OF_2								
$X = (n, k)$	1.104	1.00	1.90	8.13	1.047	0.90	4.46	4.68
$X = (n, k, a)$	1.067	1.00	1.26	5.90	1.064	0.96	6.16	6.43
$X = (n, k, a, I_0)$	0.965	0.99	0.86	8.86	1.025	0.55	2.41	2.57

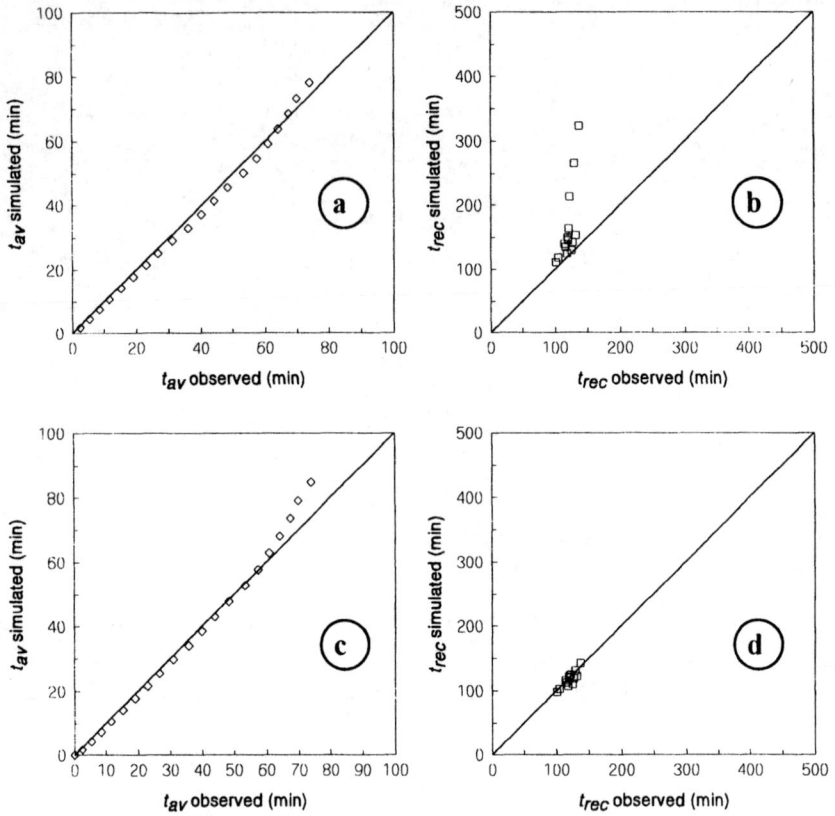

Fig. 2. Simulated vs. observed times for advance (t_{av}) and recession (t_{rec}) when simulation are performed with k and n optimized using the objective functions F_1 (a and b) and F_2 (c and d) for the irrigation event MC-1/94. (goodness fit parameters in Table 2).

Table 3. Computed values of UD and e_a when infiltration and roughness parameters are estimated through different optimization approaches

Irrigation	Performance parameters	OF_1			OF_2		
		$X=(n,k)$	$X=(n,k,a)$	$X=(n,k,a,I_0)$	$X=(n,k)$	$X=(n,k,a)$	$X=(n,k,a,I_0)$
MC-1/94	UD	69.5	60.7	46.2	82.6	81.9	82.3
	e_a	69.4	60.7	82.4	82.4	81.7	82.1
MI-1/94	UD	72.2	64.6	52.9	81.0	79.9	80.2
	e_a	72.1	64.5	52.8	80.8	79.7	80.0
MI-1/95	UD	55.2	51.4	41.9	59.8	62.3	60.7
	e_a	55.1	51.4	41.9	59.7	62.3	60.6
MM-1/95	UD	57.3	39.5	44.3	62.9	62.8	75.7
	e_a	57.2	39.4	44.2	62.8	62.7	75.5

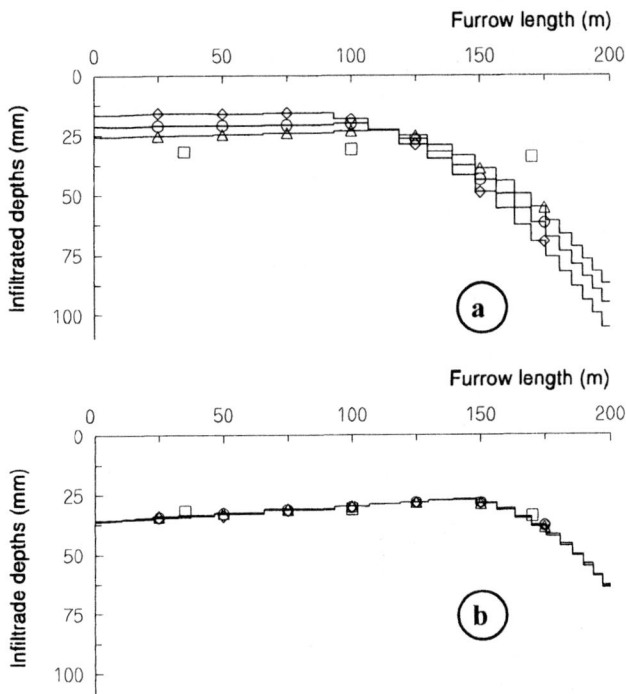

Fig. 3. Infiltrated profiles generated from infiltration and roughness parameters optimized using the objective functions OF$_1$ (a) and OF$_2$ (b) for the irrigation MC-1/94 (Δ) for X = (*n*, *k*); (O) for X = (*n*, *k*, *a*); (◊) for X = (*n*, *k,a*, *I$_0$*); (□) for observations.

5 Conclusions

When just the advance phase is used to optimize the infiltration and roughness parameters the errors in the estimation of recession are very high. Errors increase when more than two parameters are searched. The errors in predicting the recession are greatly reduced when the objective function relates to both the advance and recession phases. Furthermore, it was observed that the infiltration profiles computed with parameters optimized with this objective function present very little differences when 2, 3 or 4 parameters are searched. Similarly, DU and e_a have very small variation when computed with parameters searched using the same OF$_2$ function. This indicates that optimization of infiltration and roughness shall be done when the objective function concerns both the advance and recession. Results also indicate that searching the parameters *k* and *n* only could be the must appropriate in view of decreasing the computation time. It was verified that the performance parameters DU and e_a are sensitive to the quality of the input infiltration and roughness parameters, which justifies the use of appropriate methodologies to estimate these parameters.

6   References

1.   Merriam, J.L. and Keller, J. (1978) *Farm Irrigation System Evaluation. A Guide for Management*. Dep. of Agri. and Irrig. Engrg., Utah Univ., Logan.

2.   Pereira, L.S. and Trout, T.J. (1998). Irrigation Methods. Chapter I.5.4 of the *Handbook of Agricultural Engineering. I. Land and Water*. GIGR and ASAE (in publication).

3.   Pereira, L.S. (1997) Inter-relationships between the on-farm irrigation systems and irrigation scheduling methods. in *Irrigation Scheduling. From Theory to Practice* (ed. M., Smith *et al.*), Water Report 8, FAO, Rome, pp. 91-104.

4.   Walker, W.R. and Skogerboe, G.V. (1987) *Surface Irrigation. Theory and Practice*. Prentice-Hall, Englewood Cliffs.

5.   Pereira, L.S. (1996) Surface irrigation systems. in *Sustainability of Irrigated Agriculture* (ed. L.S., Pereira *et al.*), Kluwer, Dordrecht, pp. 269-289.

6.   Tabuada, M.A., Rego, Z.J.C., Vachaud, C. and Pereira, L.S. (1995) Two-dimensional infiltration under furrow irrigation: modelling, its validation and applications. *Agri. Water Manag*. 27: 105-123.

7.   Elliott, R.L. and Walker, W.R. (1982) Field evaluation of furrow infiltration and advance functions. *Transactions of the ASAE* 25: 369-400.

8.   Smerdon, E.T., Blair, A.W. and Reddel, D.L. (1988) Infiltration from furrow advance data. *J. Irrig. Drain. Engrg*. 114: 4-17.

9.   De Tar, W.R. (1989). Infiltration function from furrow stream advance. *J. Irrig. Drain. Engrg*. 115: 722-730

10. Shepard, J.S., Wallender, W.W. and Hopmans, J.W., 1993. One point method for estimating furrow infiltration. *Transactions of the ASAE* 36: 395-404.

11. Katapodes, N.D., Tang, J. and Clemmens, A.J. (1990) Estimation of surface irrigation parameters. *J. Irrig. Drain. Engrg*. 116: 676-695.

12. Katapodes, N.D. (1990) Observability of surface irrigation advance. *J. Irrig. Drain. Engrg*. 116: 656-675.

13. Walker, W.R. and Busman, J.D. (1990). Real time estimation of furrow infiltration. *J. Irrig. Drain. Engrg*. 116: 299-318.

14. Turral, H. (1993) The evaluation of surge-flow in border irrigation with particular reference to cracking soils. Ph.D. Dissertation, Univ. of Melbourne, Parkville.

15. Malano, H.M., Turral, H.N. and Wood M.L. (1997). Surface irrigation management in real time in South Eastern Australia: Irrigation scheduling and field application. in *Irrigation Scheduling. from Theory to Practice* (ed. M., Smith *et al.*), Water Report 8, FAO, Rome, pp. 105-118

16. Strelkoff, T. (1993) *SRFR 20.5: a computer programme for simulating flow in surface irrigation*. USDA.-ARS, Water Conservation Lab., Phoenix.

17. Himmelblau, D.M. (1972). *Applied Nonlinear Programming*. McGraw-Hill Book Company, New York.

18. Calejo M.J. (1996). Optimization of infiltration and roughness parameters in surface irrigation. M.Sc. dissertation, Instituto Superior de Agronomia, Univ. Téc. de Lisboa, (in Portuguese).

REAL TIME MANAGEMENT OF FURROW IRRIGATION WITH A CABLEGATION SYSTEM

S. SHAHIDIAN, R. P. SERRALHEIRO, L. L. SILVA
Department of Agricultural Engineering, University of Évora, Portugal

Abstract
An electronic controller for Cablegation surface irrigation systems was developed, that uses real time irrigation data to optimize water application efficiency, and minimize run-off in irregularly shaped fields with varying slopes. The system adjusts the admission time of each furrow to its length and infiltration parameters determined in real time at the beginning of each irrigation event. The computerized system receives advance data from two points in a selected furrow with which it calculates the parameters of the Kostiakov equation, using the two point method of Elliot and Walker. The results of 19 irrigation events carried out on a 3.5 ha corn field are very promising and indicate that the system can calculate in the ideal pace of irrigation adapted to existing infiltrability and soil conditions of each irrigation event. It was also shown that adequate gate opening diameter is fundamental for obtaining high application efficiencies.
Keywords: Automation, cablegation, furrow irrigation, gated pipe, infiltration, modeling advance.

1 Introduction

Furrow irrigation has the potential to efficiently use water and energy resources while maintaining high levels of crop production. In order to realize this potential, the irrigation systems must be properly designed and operated.

Cablegation is a recent method using gated pipe to irrigate long furrows. A piston moves inside the pipe at predetermined speeds, allowing water to flow from the gates for irrigating the furrows.

On the sloped lands of Alentejo, as on many other Mediterranean regions, the farmer

Water and the Environment: Innovative Issues in Irrigation and Drainage. Edited by Luis S. Pereira and John W. Gowing. Published in 1998 by E & FN Spon. ISBN 0 419 23710 0

sometimes plants up and down the hillsides. In order to overcome the soil surface irregularities and ensure water flow, the furrows are often along steep slopes, which lead to erosion and result in poor irrigation performances.

In order to install surface irrigation and prevent erosion of these soils it is highly recommended to use contour terraces. The land strips between terraces are smooth and gently sloped lengthwise. The resulting furrows are of varying length, needing thus different admission times, which makes surface irrigation difficult.

This problem has been a great limitation to the use of Cablegation systems on this type of fields, as it needs a constant re-adjustment of the pace of the Cablegation system. The advance in electronics makes it interesting to develop electronic Cablegation controlling and operating systems that can operate with furrows of varying length.

2 Cablegation system

Cablegation uses a commercially available gated pipe to irrigate long furrows. The "gates" or outlets are near the top side and are left open. The pipe is laid on a precise grade and a piston moves slowly through the pipe causing water to flow, in sequence, to the furrows [1].

Water flows in the pipe below the level of the outlets until it reaches the piston. This obstruction causes the water to fill the pipe and run out of the outlets near the plug. A light cable or line from a reel at the standpipe is attached to the upstream end of the piston. The cable is reeled out according to the desired rate at which the irrigation is to progress across the field.

In the proposed method a small electric motor is used along with a reduction mechanism to reel out the piston at desired rates. The motor is controlled by a computerized controlling system using a relay.

A simple electronic controller was developed (CaboGest) using a computer to run a specially developed program which controls irrigation in real time.

Two water sensors located half way and at the end of one of the initial furrows transmit to the computer the position of the advance front.

In 1997, a 3 ha field was organized in six contour blocks, each 30 m wide. The furrows were between 80 and 280 m in length, with 0.2% slope, and spaced 0.75 m.

3 The principle of CaboGest method

Due to the fact that in Cablegation systems water is supplied simultaneously to a number of furrows, it is necessary to have a waiting period in the first set of furrows in order for these to be watered [2]. During this period the piston is stationary. CaboGest method uses this period of advance in the furrows with a constant inflow to determine the infiltration equation.

In order to do this, CaboGest receives in real time the advance times to the middle and end of a specific furrow and with these data determines the infiltration equation using the two point method of Elliot and Walker [3].

Based on the determined infiltration equation and using a volume balance model specifically developed here, CaboGest calculates the advance time for the furrow. The modeled advance time is compared with the observed one and is adjusted through the determination of Manning's roughness coefficient.

Based on these data, the system simulates advance in the remaining furrows, taking into consideration the gradual cut-back characteristic of Cablegation systems, and the slope and length of the furrow; thus determining the pace of the piston so that the desired depth of water is infiltrated. This information is stored and the field watered accordingly.

3.1 The volume balance equation

According to the volume balance method, the total volume applied is equal to the sum of the surface stored water and the infiltrated water. Supposing an exponential advance equation

$$x = p \ t^r \tag{1}$$

then

$$Q_0 \ t = \sigma_y \ A_0 x + \sigma_z k \ t^a x + \frac{f_0 \ t \ x}{1+r} \tag{2}$$

where Q_0 is inlet discharge, m^3/min; σ_z subsuperface shape factor, t elapsed time, min; σ_y surface storage shape factor, which is defined as a constant (0,7 to 0,8); A_0 cross-sectional area of flow at the inlet, m2 f_0 basic infiltration rate, p and r are coefficients of the advance equation, determined by the two point method, x is distance, m [4].

The parameters of the Kostiakov equation,

$$z = k \ t^a \tag{3}$$

k and a, are calculated using the two point method,

$$a = \frac{\ln\left(V_l / V_m\right)}{\ln\left(t_l / t_m\right)} \qquad \text{and} \qquad k = \frac{V_l}{t_l^a} \tag{4 \& 5}$$

where V_m and V_l are the volumes infiltrated during advance to the middle and end of the furrow, and t_m and t_l are the advance time to the middle and end of the furrow. The basic infiltration rate, f_0, is considered as zero.

3.2 CaboGest flow simulation model

The proposed model uses a time cycle, and for each time interval t, determines the space covered by the advance front. In its turn, the furrow is divided into n cells, each corresponding to the space covered by the advance front in each of the successive time intervals. For each one minute time increment, the space covered is numerically equal to speed, and is determined by the following expression [5]:

$$V_{el} = a^{2/(5u-2)} Q \left(Q \frac{n}{\sqrt{S}} \right)^{-3u/(5u-2)} \tag{6}$$

Where Vel is the advance speed, in m/min.; Q flow rate in the increment in m^3/min; S is the friction slope (m/m); a and u are the coefficients of the furrow geometry equation which relates the wetted perimeter Wp, with cross-sectional area of the furrow, A:

$$A = a\, Wp^u \tag{7}$$

In each cell, surface storage is determined using a similar equation [6]

$$A = a^{-2/(5u-2)} \left(Q \frac{n}{\sqrt{S}} \right)^{3u\ /(5u-2)} \tag{8}$$

The volume infiltrated in each cell is determined multiplying the instantaneous infiltration by the length of the cell.

At the end of each time increment, the volume of water available to the next time increment t_{+1} is calculated by a volume balance of all the cells of the furrow covered until time t. In this way the cycle is repeated until the end of the furrow is reached, or there is an interruption in the water supply.

An advantage of the proposed advance model is that it does not need any input parameters besides those already used for determining the infiltration equation.

4 Results and discussion

A total of 19 irrigation events were carried out. The infiltration equations and roughness were calculated in a satisfactory manner by CaboGest. Table 1 presents the infiltration equations and Manning's roughness coefficient for the first ten irrigation events. It is possible to observe a reduction in the permeability and a decrease in the advance times. This can result in excessive run-off if the inflow rates are not decreased. Thus in order to maximize application efficiency, the opening of the gates was reduced from 51 mm (diameter) to 23 mm in the 7th irrigation event.

In the remaining irrigation events infiltrability had basically stabilized although the roughness coefficient increased again at the end of the season.

Irrigation events 7 and 8 were carried out on consecutive days in order to study the effect of higher initial soil humidity on advance and the infiltration equation. The results indicate that higher initial soil moisture reduce infiltration and increase the speed of advance.

Table 1. Advance times, infiltration equations, Manning's roughness coefficient and other data relative to the first ten irrigation events.

Irrig. Event	Date	Advance times		Infiltration Equation (mm)	Manning Nm	Gate opening
		t1	t2			$\varnothing$ (mm)
1	08/ 07	22	66	$z= 3.68 \ t^{0.598}$	0.065	51
2	18/ 07	18	48	$z= 2.51 \ t^{0.540}$	0.047	42
3	25/ 07	22	59	$z= 2.20 \ t^{0.530}$	0.045	35.5
4	28/ 07	24	66	$z= 2.09 \ t^{0.537}$	0.050	32
5	31/ 07	22	57	$z= 1.97 \ t^{0.510}$	0.045	32
6	04/ 08			same as Irrig. 5		32
7	06/ 08	37.5	105	$z= 2.10 \ t^{0.449}$	0.055	23
8	07/ 08	20	50	$z= 0.60 \ t^{0.629}$	0.033	23
9	11/ 08	32	79	$z= 3.25 \ t^{0.344}$	0.050	23
10	13/ 08	23	61	$z= 0.92 \ t^{0.592}$	0.040	23

Fig. 1 compares the advance phase simulated in the field by CaboGest for furrows M1-M4, and the actual times observed (irrigation event 7). The observed advance times were within ± 15 min. of the simulated. Generally speaking, the model used underestimates advance rate at the beginning of the furrows where the flow depth is high, and overestimates the advance rate at the lower end of the furrows where the flow depth is shallow. It can be speculated that this fact is associated with a particular characteristic of Manning's roughness coefficient, n, used in equation 6 to calculate

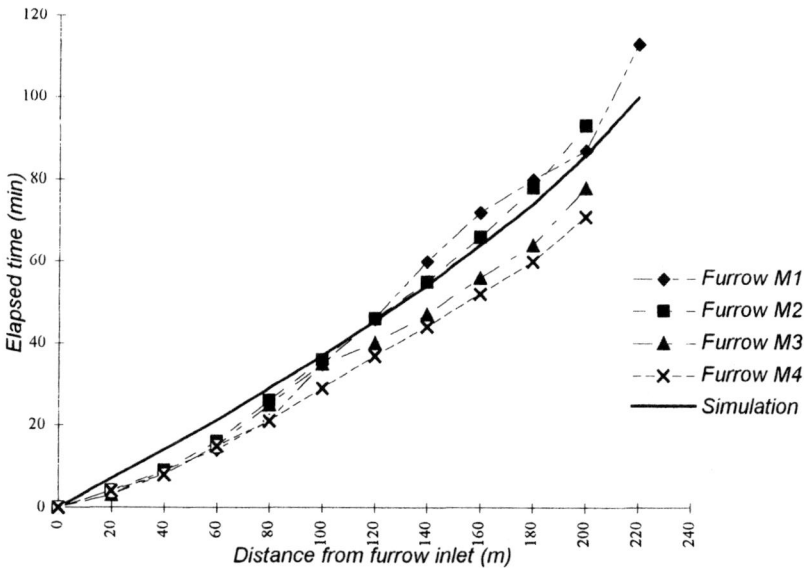

Fig. 1. Comparison of simulated and observed advance times for furrows M1-M4, (Irrigation event 7).

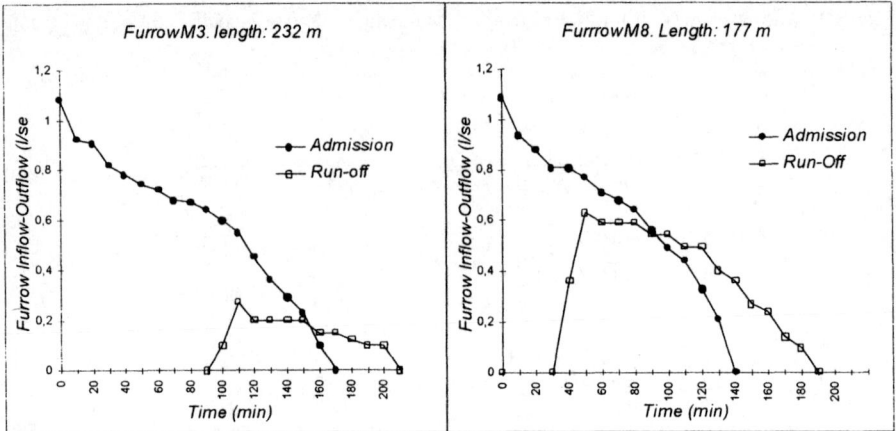

Fig. 2. Comparison of inflow-runoff hydrograph of furrows of different length, irrigation 7

advance rates. As already pointed out by [7], the value of n depends on flow depth. That is, as the flow depth decreases, the furrow's surface roughness assumes greater influence on advance rates.

As the furrows become smaller, CaboGest calculates shorter admission times, that is a faster cut-back. This can be seen in Fig. 2, which compares inflow and outflow of furrows of different length. It is interesting to note that although the inflow times were reduced by CaboGest in an attempt to maintain the same intake opportunity time in both furrows, the outflow was significantly greater in the shorter furrow. This is due to the shorter length of furrow M8, which implies a smaller infiltration surface, resulting in a 35% application efficiency compared to 81% for furrow M3. This is a clear indication that in a field with different furrow lengths it is not sufficient to adjust the speed of the piston, but also the opening of the gates have to be adjusted to the length of the furrows, that is, noticeably shorter furrows must have smaller gate openings.

Fig. 3 demonstrates the same principle for the whole field. Total run-off increased not only due to greater total inflow but also due to greater run-off from the shorter furrows at the lower side of the field.

It was also observed that infiltration increases with flow rate, due to greater infiltrating surface, and thus an infiltration equation measured in a furrow is acurate only for a given flow rate.

5 Conclusions

This economical system, costing less than 900 USD per hectare (150 m long furrows), is capable of automating gated pipe surface irrigation, with minimum human supervision.

Real-time determination of infiltration parameters are important for optimizing long-furrow irrigation, as furrow hydraulic characteristics evolve along the irrigation season, and depend on soil water deficit.

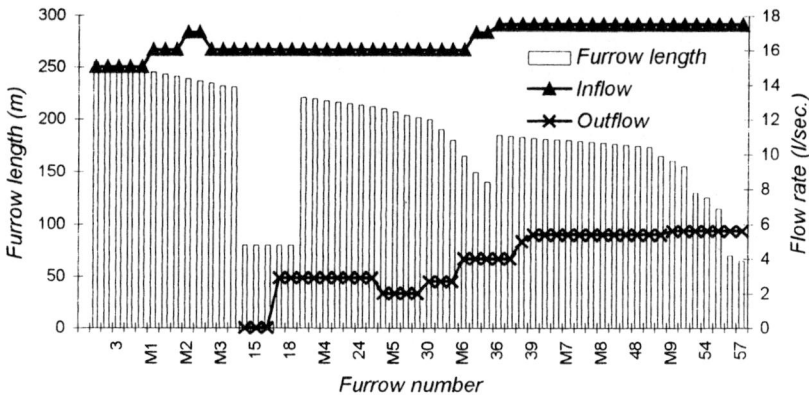

Fig. 3. Inflow-runoff hydrograph of the entire field (number of simultaneously irrigated furrows: 17). The outflow was measured at the end of a run-off collection canal (Irrigation event 15)

The opening of the gates should be adjusted to the infiltration rate along the season and also to the different furrow lengths within each irrigation.

The advance simulation model, CaboGest appears to be insensitive to changes in flow depth. This is associated to the fact that the influence of surface roughness on flow varies with the flow depth.

Furrow transversal geometry has a great influence on advance rate in the furrow.

6 References

1. Kemper W.D, D.C. Kincaid D.C, R.V.Worstell, W.H.Heinemann, T.J.Trout, J.E.Chapman, F.W.Kemper e M.Wilson. (1985) *Cablegation type irrigation systems: Description, design installation and performance* (Draft copy) Kimberly, Idaho, USDA.

2. Kincaid, D.C., e W.D.Kemper (1984) Cablegation: IV. The bypass method and cutoff outlets to improve water distribution. *Transactions of the ASAE* 27(3): 752-768.

3. Elliot R.L, W.R.Walker (1982) Field Evaluation of Furrow Infiltration and advance functions. *Trans. of the ASAE*, 25(2):396-400

4. Walker, W.R. e G.V. Skogerboe (1987) *Surface Irrigation. Theory and practice.* Prentice-Hall, Englewood Cliffs, NJ.

5. Trout, T.J. (1988) *Evaluation of cablegation system performance.* Kimberly, ID, USDA.

6. Shahidian, S. (1996) Desenvolvimento por automatização de um sistema de CaboRega, Msc thesis, Universidade de Évora.

7. Basset, D.L., D.D. Fangmeier e T.Strelkoff (1983) Hydraulics of surface irrigation, in *Design and Operation of Farm Irrigation Systems*, (ed. Jensen, M. E.) St.Joseph, MI.ASAE.

PREDICTING RUNOFF UNDER SPRINKLER IRRIGATION USING EUROSEM

B. GHORBANI
Agricultural College, Shahre-Kord University, Shahre-Kord, Iran

Abstract

Although sprinkler systems are usually designed so that the water application rate is less than the soil infiltration rate to avoid runoff, there are a growing number of sprinkler systems where runoff has become a serious problem. Moving sprinkler irrigation systems such as centre pivots and rainguns can have excessively high water application rates and are prone to runoff problems. This is exacerbated by the trend to low pressure operation, for the reason of saving energy, which adds to the problem.
Runoff problems are traditionally dealt with in the field either by adjusting the irrigation system or the cultivation practices. This paper describes a more fundamentally based approach to predict runoff from sprinkler irrigation systems under different conditions (stationary and moving conditions). The European Soil Erosion Model, EUROSEM was used for this purpose. Statistical and graphical techniques were used to compare the simulated data with the observed values in the laboratory and in the field. The comparison showed that the model simulates observed data successfully.
Keywords: Modeling, runoff, sprinkler irrigation

1 Introduction

Sprinkler irrigation is growing in popularity as a method of applying water to crops, primarily because it is seen as one way of increasing the efficiency of water use in areas of scarcity. But there are pressures which tend to reduce its effectiveness. The need to conserve energy, for example, has encouraged manufacturers to reduce operating pressures, which in turn has increased water application rates and the potential for water loss from runoff. Moving sprinkler irrigation systems such as

Water and the Environment: Innovative Issues in Irrigation and Drainage. Edited by Luis S. Pereira and John W. Gowing. Published in 1998 by E & FN Spon. ISBN 0 419 23710 0

centre-pivots and rainguns, where instantaneous application rates can exceed 200mm/hr are particularly prone to runoff problems   Pair [1] reported that the maximum application rate under centre pivot machines often exceeds the infiltration capacity of soil. The higher application rates associated with low pressure systems also exacerbate the problem.

In some cases runoff problems can be overcome by modifying irrigation equipment and its use, but another approach is to examine ways of holding water on the soil surface so that it can infiltrate into the root zone [2]. The implication of this is that the selection of appropriate cultivation practices to reduce or eliminate runoff becomes as much a part of the sprinkler design process as the hydraulic selection of appropriate pipes, pumps and sprinklers.

In order to 'design' the cultivation it is essential to be able to predict runoff rather than deal with it, as is normally done, on a trial and error basis, once the system is installed.   To do this, a well established   mathematical model, the European Soil Erosion Model (EUROSEM) normally used to predict runoff and sediment movement on large agricultural catchments, was evaluated to see if it could also be used to predict runoff from a ridge and furrow cultivation under sprinkler systems.

2 The model

EUROSEM was developed by a joint effort of European scientists [3]. It is based on KINEROS, a Kinematic Runoff and Erosion Model [4], which is a distributed single event model developed to predict surface runoff as rainfall excess and route runoff and sediment over land [5].

EUROSEM was developed for relatively large hydrological catchments and is based on the kinematic wave theory which uses the equations of continuity and momentum. The continuity equation is

$$\partial q/\partial t + \partial h/\partial x = Q \tag{1}$$

in which q = discharge per unit width and h = equivalent normal water depth.   The momentum equation can be expressed as a general unit discharge-depth relationship

$$q = \alpha h^{m} \tag{2}$$

where m = an exponent whose value depends on flow regime and α = a constant depending on the slope and surface roughness.   Depending on the flow regime the Darcy-Weisbach or Manning's equation is used to find the runoff flow per unit width.

To use this model for predicting runoff from ridge and furrow cultivation the catchment was defined as three interconnected planes from which runoff can develop during the constant application of irrigation water (Figure 1).   Runoff first collects in the bottom of the furrow and then flows down the furrow.

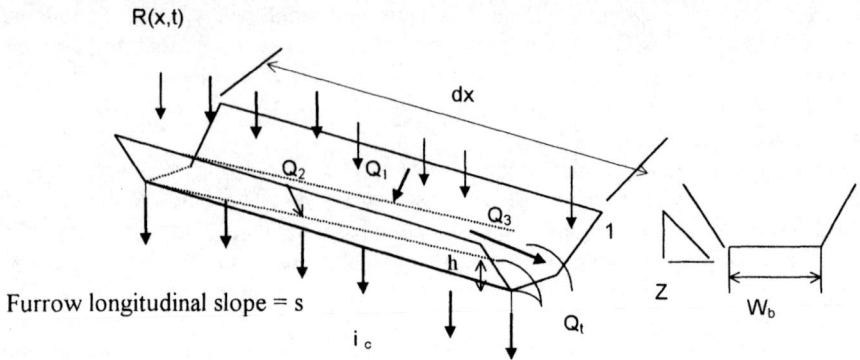

Fig. 1 Runoff from a ridge and furrow cultivation

Water application and infiltration occur on all three planes. When the application rate (R) exceeds the infiltration capacity (i_c), then runoff (Q) will occur from both furrow side slopes and from the bed, i.e.

$$Q = R - i_c \tag{3}$$

EUROSEM uses the following relationship for infiltration capacity (i_c) developed by Smith and Parlange [6]:

$$i_c = \frac{K_{sat}\ \exp(\dfrac{d_i}{B})}{(\exp(\dfrac{d_i}{B}) - 1)} \tag{4}$$

where K_{sat} = the saturated hydraulic conductivity of soil, d_i = the amount of water already absorbed by soil and B = the product of the capillary drive (G) and the saturated deficit of soil.

$$B = G(\theta_s - \theta_i) \tag{5}$$

in which θ_i and θ_s are the initial and saturated soil water content respectively. Capillary drive (G) is a soil characteristic and has the unit of length and is one of the key parameters describing infiltration.

Depression storage is an important aspect of controlling runoff. This is described in two terms in the EUROSEM model [3]; roughness ratio along the slope (RFR) and roughness ratio across the slope (ASR). These ratios are the proportion of the difference between actual distance (X) and straight distance (Y) between two points either along the slope or across the slope as follows:

$$RFR(or(ASR)) = \frac{(X - Y)}{X} \times 100 \tag{6}$$

Roughness ratio was measured in the field using a metal chain. When *RFR* was less than 30% it had very little effect on the runoff hyrograph and so depression storage could ignored below this value. The sensitivity of runoff hydrograph to *ASR* is less than *RFR* [7].

Although EUROSEM is normally used for predicting runoff from a single stationary rainstorm, there seems to be no reason why it should not be used for a moving storm. This can be done by dividing the catchment into a number of elements so that each element receives water in turn as the storm passes over them. The model can allow for changes in rainfall intensity and soil infiltration capacity with time to be made with each element of the catchemnt if needed. This concept is very similar to the application of water from a moving sprinkler as it travels across the field. To use this model the catchment is divided into a number of elements having uniform soil and water application characteristics. Figure 2 shows how this can be done for a simple plane and a ridge and furrow catchment. The numbers and arrows show the arrangement of the elements and the direction of the runoff from one element to the next element respectively. Superimposed on the catchmnet is the moving circular wetting pattern from the mobile sprinkler system.

When a mobile irrigation machine moves along a catchment the water application pattern moves as well. So it takes a time for the pattern to pass successively over the beginning and end of a catchment. The length of this time depends on the speed of the irrigation machine. The greater the speed the less the water application depth and the lower the surface runoff from that element.

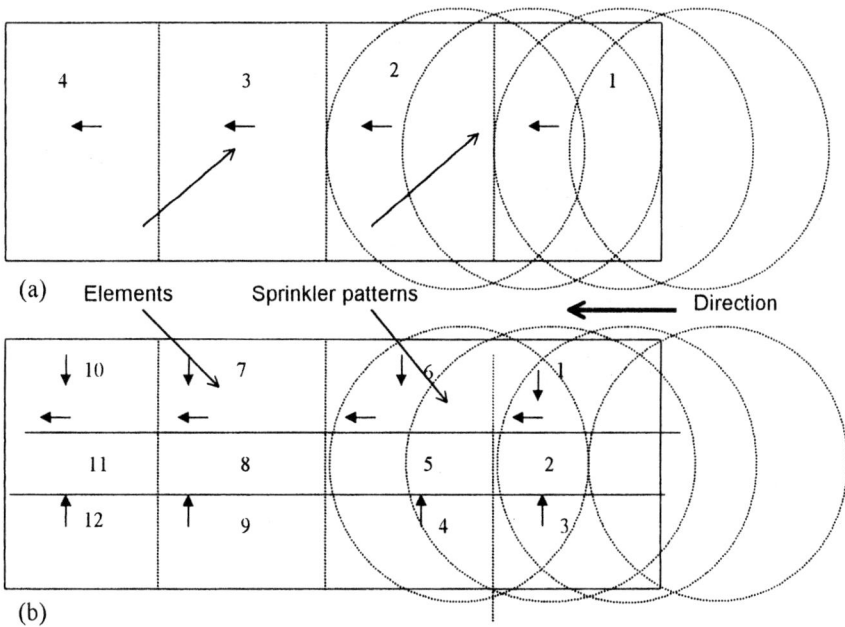

Fig. 2 Moving sprinkler irrigation application over (a) simple plane and (b) ridge and furrow cultivation

3 Model validation

The model was validated using both laboratory and field data. In the laboratory a soil tray , 2000mm ×1000mm and 200mm in depth was filled with a sandy loam soil and placed under an array of fixed spray nozzles to simulate stationary sprinkler irrigation. Runoff from the soil tray was collected and measured with respect to time. Two drain tubes were installed in the base of the tray to collect water in order to measure infiltration. Tests were conducted at different water application rates on ridge and furrows constructed in the soil for a range of furrow slopes. The range of experimental parameters are summarised in Table 1.

In the field, furrows 12 m long and 0.9 wide were established on a plot of sandy loam soil similar to that used in the laboratory for a stationary sprinkler (Table 1). Water was applied using commercially available rotary impact sprinklers and variations in application rate were achieved using the line source technique [8], [9]. The high application rates required to produce runoff where achieved by reducing the spacing between sprinklers.

A field experiment was undertaken to validate the EUROSEM model for moving condition. The soil was a sandy loam, similar to that used for the stationary sprinkler validation [10]. Water was applied using a Nelson 8401 Rain Train traveller fitted with a 6mm nozzle rotary impact sprinkler (Table 1).

Tests were conducted on two types of cultivation, a plane sloping surface and ridge and furrow cultivation with plots running down the main land slope (Table 1).

Table 1 Experimental parameters

Parameters	Laboratory tests	Field tests
Applied water		
Application rate, R^1 (mm/hr)	60-90, 200-225	15-44
Rainfall duration, Tr (min)	60	60-210
Soil characteristics		
Texture	sandy loam	sandy loam
Hydraulic conductivity, k (mm/hr)	20-45, 50-65	4-14, 18-30
Initial soil moisture content, θ_I(%)	24-33	7-16, 17-33
Capillary drive, G (mm)	17	11
Cultivation (Ridge and furrow)		
Ridge spacing (mm)	900	900
Ridge height (m)	200	200
Ridge slope (%)	55	55
Main slope (%)	3.5, 10, 23.5	3.5
Ridge length (m)	2.0	12.3
Nelson 8401 Rain Traveller		
Average discharge (l/s)		0.308
Nozzle diameter (mm)		6
Height of riser (mm)		1000
Average throw diameter (m)		24

1: Note also that R (Irrigation water application rate) is assumed to be constant for stationary and variable for moving condition.

3.1 Input parameters

To use the EUROSEM model, several input parameters are needed which require field measurement (Table 1). Water application rate was determined for each element by first measuring the amount of water collected in a catch can located in the centre of each element. The volume of water collected was then divided by the water application time (in the case of moving condition it was determined by the machine speed) and catching area to give an average value for each element. Generated surface runoff was measured at the end of each plane and furrows during each experiment using a volumetric method.

The values of saturated hydraulic conductivity of the soil were determined using soil infiltration capacity curves. Initial soil water content was measured in the laboratory using the gravimetric method.

The values of capillary drive (G) were calculated during the validation experiments by using a split sample test. G values for the sandy loam determined in this way ranged from 0 to 30mm with an average value of 17mm. This compares with values of 98 to 526mm from Woolhiser et al [4] which were local spot measurements and may not be representative of a whole catchment (Smith, R 1996, personal communication). Tests were carried out for a range of travel speed from 3-12 m/h. The faster the travel speed, the less amount water was applied for the same sprinkler.

3.2 Comparison between model and observed data

Figures 3 show examples of the results obtained from the laboratory and field tests together with those predicted by EUROSEM. There is clearly good visual agreement between experiments and model predictions for stationary condition. but additionally different statistical techniques such as root mean square error (RMSE), regression coefficient (R^2) and model efficiency (E) were used to compare the simulated values with the observed data [11]. Model efficiency values for time to peak, peak flow rate, time to runoff and volume of runoff were 0.70, 0.99, 0.54 and 0.99 and root mean square error values were 17.3, 9, 2.3 and 3.8 respectively.

Figures 4 show a comparison between simulated and observed runoff hydrographs for a moving sprinkler. based on 3m long elements. From Figure (4-a) can be seen that the travel speed is low (3.4m/hr) and average water application rate is high (42mm/hr), but runoff starts much later and peak flow rate is smaller compare to other runoff hydrograph (Figure 4-b). This is because the soil was quite dry (soil water content = 7%) for the first irrigation event and there was no crust on the soil surface so that the hydraulic conductivity was high. The soil water content has a positive effect and hydraulic conductivity has the negative effect on runoff. The sensitivity of peak flow rate and runoff volume to the soil water content and the hydraulic conductivity is also high. This is why the peak flow rate and runoff volume of the chart (a) is lower than that of chart (b). Visually there is good agreement between observed and predicted runoff hydrographs (charts a, b) and corresponding parameters. The statistical analysis of hydrograph paramete were also undertaken. Model efficiency values for time to peak, peak flow rate and time to runoff and volume of runoff were 0.99, 0.99, 0.90 and 0.7 and root mean square error values were 6, 2.6, 6 and 1.5 respectively.

The correlation analysis test also showed that the reliability of association between observed and predicted data at 95% confidence level was more than 90% for both stationary and moving conditions.

The T-test indicated that the discrepancy between predicted and observed data was not significant in each experiment. This was because, the slope of best fit line and the intercept values were not significantly different from unity and zero respectively.

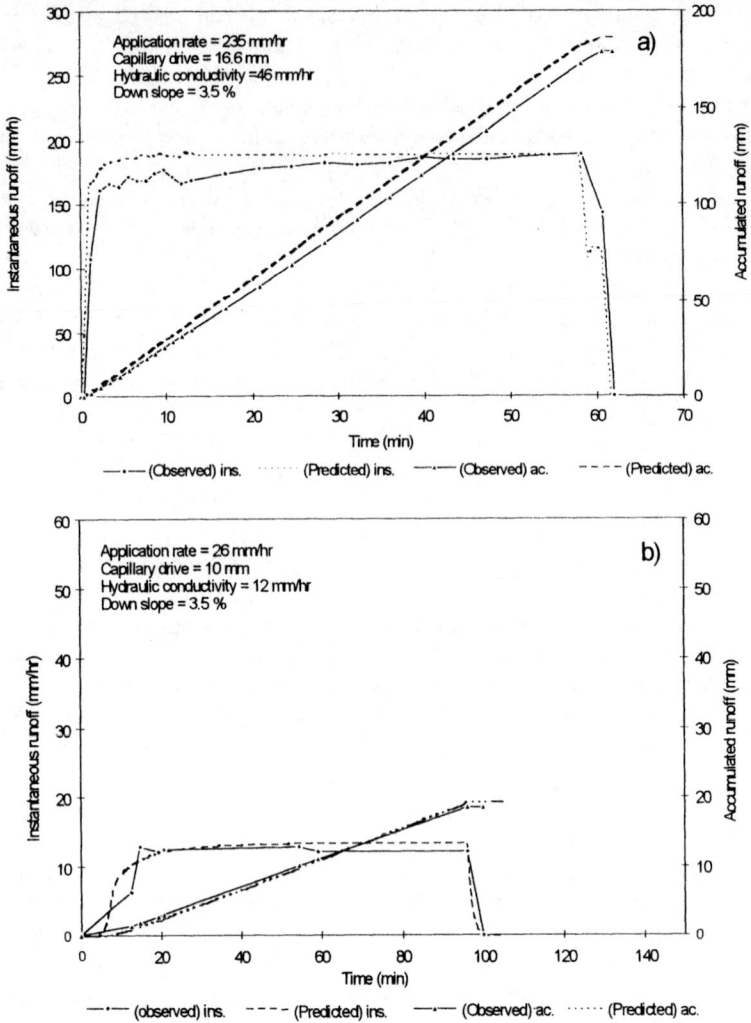

Fig. 3. Typical examples of comparison between observed and predicted runoff hydrographs for a ridge and furrow in the field for a stationary sprinkler (a - laboratory, b - field test)

4 Conclusion

The main objective of this application of EUROSEM was to predict the runoff hydrograph and corresponding parameters for sprinkler irrigation design and management. The prediction of runoff means that cultivation systems can now be designed to control runoff prior to installation and operation of sprinkler irrigation systems in the field.

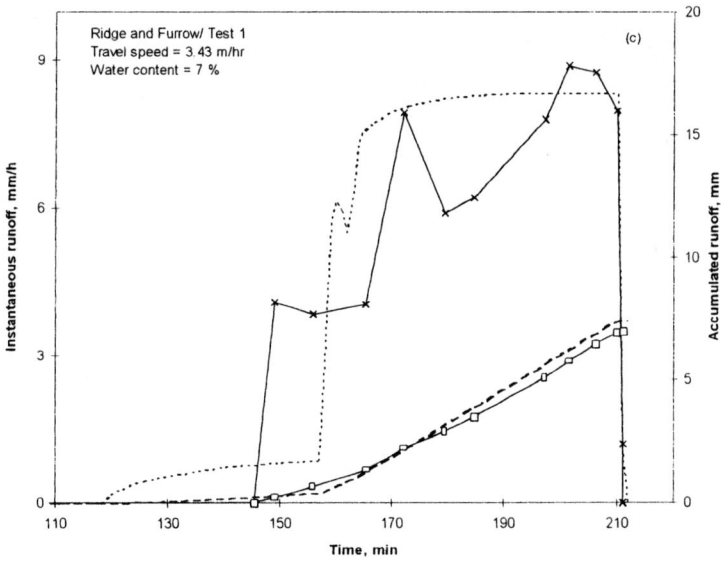

a- low water content

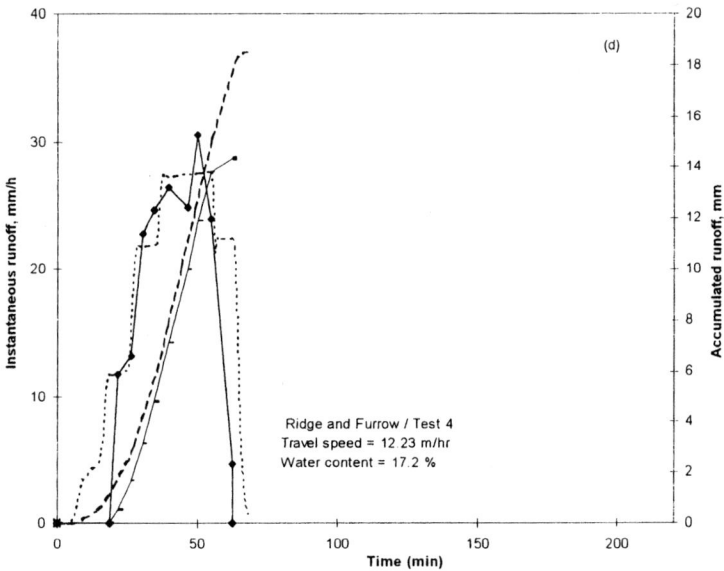

b-high water content

Fig. 4. Typical examples of comparison between observed and predicted runoff hydrographs for a ridge and furrow in the field for a moving sprinkler (a- low water content, b-high water content)

EUROSEM was developed for large hydrological watersheds but has been used to predict surface runoff from different cultivation under stationary and moving sprinkler irrigation systems. The model has been calibrated and validated using both laboratory and field tests and it simulates observed runoff hydrographs and corresponding parameters (time to runoff, time to peak, peak flow rate and volume of runoff) successfully.

5 Acknowledgement

I would like to express my thanks to the Ministry of Culture and Higher Education and the University of Gilan, the Islamic Republic of Iran for sponsoring this research at Silsoe College, Cranfield University. The enthusiastic support and constructive criticism of Mr Kay M, Professor R.P.C. Morgan, Dr J Quinton and other members of Silsoe College staff through this study is greatly appreciated.

6 References

1. Pair, C.H. (1968) Water distribution under sprinkler irrigation. *Transaction of American Society of Agricultural Engineering*, 11(5) 648-51.
2. Kay, M.G. and Abo-Ghobar, H. (1990) Design method to control runoff from high intensity sprinkler irrigation. *Irrigation and Drainage Systems*, 4:109-116
3. Morgan, R.P.C, Quinton, J.N.and Rickson, R.J., (1993) *EUROSEM: A user guide*. Silsoe College, Cranfield University, UK.
4. Woolhiser, D.A. Smith, R.E. and Goodrich, D. C. (1990) KINEROS, *A kinematic runoff and erosion model, Documentation and User Manual*. U.S. Department of Agriculture, Agricultural Research Service.
5. Morgan, R.P.C, Quinton, J.N. and Rickson, R.J. (1992) *EUROSEM: Documentation and manual*. Silsoe College, Cranfield University, UK.
6. Smith, R.E. and Parlange, J.Y. (1978) A parameter-efficient hydrologic infiltration model. *Water Resources Research*, 14, 533-538.
7. Ghorbani, B. (1997) A mathematical model to predict surface runoff under sprinkler irrigation conditions. PhD thesis, Silsoe College, Cranfield University, Bedford, UK.
8. Hanks, R.J.; Keller, J.; Ranssen, V.P. & Wilson, G.D. (1976) Line source sprinkler for continuity variable irrigation crop production studies. *J. ASSS*, volume, 40, 3:426-430.
9. Willardson, L.S., Ooslerhais, D.M. and Johnson, D.A. (1987) Sprinkler selection for line source irrigation system. *Irrigation Science*, 8, 56-76.
10. Ghorbani,, B. and Kay, M.G. (1997) Predicting runoff under stationary sprinkler Irrigation Using the European Soil Erosion Model (EUROSEM). Unpublished.
11. Green, I.R.A and Stephen, D. (1986) Criteria for comparison of single event models. *Hydrological Science Journal*, 31, 3: 395-411.

PERFORMANCE OF CENTRE PIVOT SYSTEMS IN FIELD PRACTICE

J. MONTERO, J.M. TARJUELO, F.T. HONRUBIA, J. ORTIZ, P. CARRIÓN, J.A. de JUAN and M. CALVO
Centro Regional de Estudios del Agua, Instituto de Desarrollo Regional, University of Castilla-La Mancha, Albacete, Spain

Abstract
In the order to study the problems of water distribution when using centre pivot systems, 84 assessments have been carried out in Albacete (Spain), a region where these systems have been used for fifteen years and cover a surface over 40000 ha. The average size of the pilot plot ranges 10 to 150 ha and several kinds of sprayer or sprinkler nozzles are present (operating pressure: 50-350 kPa). Results show that, in general, water distribution may be considered as acceptable (average CU: 86.6%). A major problem is reducing evaporation and drift losses during the water application process. In this way, the height of emitter over the ground and the type o emitter are two main factors. The effects of equipment size, type of nozzle, operating pressure and even wind action on water distribution uniformity are not significant. There are significant differences according to the existence of several shallow waterlogged formations (possible run-off in slope zones), the shorter is the wetted width, and the closer the nozzle is located above ground. The best balance between evaporation and drift losses and uniformity may be achieved locating sprayer or sprinkler nozzles two meters high, with a value of wetted width around 15 m, with a operating pressure requirement ranging 150 to 200 kPa.
Keywords: Centre pivot, evaporation losses, sprinkler irrigation, uniformity, wind drift losses.

1 Introduction

The goal of irrigation is maximise water application efficiency, that is, the amount of water discharged by the irrigation system being used to satisfy both requirements of crop

Water and the Environment: Innovative Issues in Irrigation and Drainage. Edited by Luis S. Pereira and John W. Gowing. Published in 1998 by E & FN Spon. ISBN 0 419 23710 0

and salt leaching. This task implies minimising evaporation and drift losses, run-off, deep seepage and other less important losses. To achieve this the irrigation system must be correctly designed, scheduled and maintained.

In order to know the process of water application in an irrigation system as well as to identify and solve possible problems related to performance and handling of irrigation plants, it is necessary to carry out evaluations of the irrigation systems.

The current irrigation farming in our region is characterised by using groundwater (84%); almost irrigation plants belong to the private sector (about 70%); the major irrigation system is sprinkler irrigation (i.e. centre pivot and buried solid set).

The objectives of this work are to study the current status of centre pivot irrigation in the province of Albacete and analyse several factors subject to influence water distribution uniformity with centre pivot irrigation.

2 Material and methods

To carry out evaluation of centre pivot systems the methodology by [1], [2] and [3] and [4] and [5] standards have been followed.

In tests, catch cans with a 16-cm opening diameter and a 15-cm height are used. Catch cans are located along a line extending radially from the pivot point at a 2-m spacing. This line is far away from the pipeline for allowing the system to achieve working conditions before arriving to the test site. The discharge of the system is measured with an ultrasound probe located at the beginning of the pipeline away from the centre pivot elbow. Bourdon gauges are used to measure pressure at several points along the pipeline as well as the pumping. Environmental conditions during testing (i.e. air temperature and relative humidity, wind speed and direction) are registered with portable anemometers and thermohygrometers.

When analysing results it is necessary to take into account that uniformity values obtained from calculations refer to concrete set times performed under specific conditions. With a irrigation frequency of 2-4 days both uniformity and efficiency for the entire irrigation season are considerably higher. In addition, if the effect of soil water redistribution is considered, water uniformity and Water Use Efficiency (WUE) by crops are even higher if an adequate irrigation scheduling is designed. High yields can be attained both with high uniformity and rather lower uniformity. The difference lies in the amount of water required in every case, which is lower the higher the uniformity. In addition, with high uniformity areas either with water deficit or with seepage are avoided.

The values observed in catch cans must be weighted since the amount of water corresponding to every point represents a larger surface far away from the centre pivot. This weighting can be done by multiplying the amount of water observed in every catch by either the distance from the centre pivot or its number of location.

A set of parameters are calculated from data collected during the test: Efficiency of discharge (Ed), Distribution Uniformity (DU), Coefficient of Uniformity by Heermann and Hein (CU_h) [6], Coefficient of Uniformity of Variation (CU_v) [7], Mean Collected Depth (MCD), Mean Observed Depth (MOD)

The definition of the uniformity parameters is given below:

- Mean Observed Depth (MOD in mm)

$$MOD(mm) = \frac{\sum P_i D_i}{\sum D_i} \tag{1}$$

where P_i is the depth water collected in catch cans (in mm) and D_i is the distance of every catch can from the centre pivot (in m). This parameter represents the mean weighted value of the water collected by the whole catch cans.

- Efficiency of discharge (Ed as %)

$$Ed = \frac{AMR}{AMA} * 100 \tag{2}$$

- Mean Collected Depth (MCD in mm)

$$MCD = \frac{\sum P_i D_i}{\sum D_i} \tag{3}$$

where P_i is the depth water collected in catch cans (in mm) and D_i is the distance of every catch can from the centre pivot (in m).

- Heermann and Hein's Coefficient of Uniformity (CU_h)

$$CU_h(\%) = \left[1 - \frac{\sum D_i \left[P_i - \frac{\sum P_i D_i}{\sum D_i} \right]}{\sum P_i D_i} \right] \cdot 100 \tag{4}$$

- Coefficient of Uniformity of variation (CU_v)

$$CU_v(\%) = \left[1 - \frac{1}{\frac{\sum P_i D_i}{\sum D_i}} \cdot \sqrt{\frac{\sum \left(P_i - \frac{\sum P_i D_i}{D_i} \right)^2 D_i}{\sum D_i}} \right] \cdot 100 \tag{5}$$

This coefficient is more susceptible to extreme variations of rainfall collected by catch cans than CU_h.

3 Results and discussion

During 1996 and 1997, 58 centre pivot catch can tests have been conducted in several farms in the province of Albacete, in order to identify possible factors influencing both uniformity and efficiency.

There is a wide range of climate conditions existing during tests. We can highlight the following average-maximum-minimum values: air temperature: 20.3-4.9-30.9 °C; air relative humidity: 45.5-21.3-81.5 %; and wind speed: 2.23-0.0-7.2 m/s. Working pressure at the end of the pipeline lateral ranged from 55 to 375 kPa, depending on using impact sprinklers or spray heads. The area irrigated by centre pivots ranged from 6 to 166 ha (average surface: 45 ha).

Table 1 summarises the analysis of uniformity parameters as a function of factors such as working pressure (at the end of the pipeline), wind speed, sprinkler height and type and pivot size (surface wetted).

In addition the correlation analysis among the uniformity parameters and the quantitative variables considered are shown in table 2. There is a negative correlation between wind speed and working pressure according to DU, although correlation coefficients are very low. Nevertheless there is no relationship among these variables and CU_h and CU_v.

There is a positive relationship between DU and the sprinkler height. The highest uniformities are achieved at 4 m above the ground.

The size of the machine has no influence on irrigation uniformity.

Concerning the sprinkler type, slightly higher uniformity is achieved with impact sprinklers than with sprayer heads, although differences are no significant.

Table 1. Computed uniformity indicators for centre pivots.

	N° test	DU (%)			CU_h (%)			CU_v (%)		
		Avg.	Max.	Min.	Avg.	Max.	Min.	Avg.	Max.	Min.
Pressure*										
55 - 150 kPa	20	81.21	88.37	68.82	87.16	92.58	71.75	82.65	88.90	63.36
150 - 250 kPa	18	79.70	88.48	67.91	86.52	90.26	82.70	81.27	86.28	75.67
250 - 375 kPa	5	73.87	84.60	63.31	84.25	91.01	78.24	78.68	87.67	70.31
Wind speed										
0 - 2 m/s	28	80.68	88.48	67.03	86.86	92.58	71.75	82.30	86.88	63.36
2 - 4 m/s	24	75.01	88.37	49.09	84.31	91.37	68.34	78.57	88.90	58.94
4 - 7.23m/s	6	76.00	82.97	67.91	85.40	89.53	80.62	80.85	86.73	75.67
Sprinkler height										
1.5 m	4	65.46	72.70	49.09	78.06	81.95	68.34	71.49	76.70	58.94
2.5 m	8	73.54	81.14	64.07	83.58	87.96	77.55	78.09	83.36	69.34
4.0 m	46	79.68	88.48	61.31	86.68	92.58	71.75	81.84	88.90	63.36
Sprinkler type										
Impact sprinkler	20	79.18	86.42	63.31	86.84	92.58	78.24	82.01	87.67	70.31
Spray heads	38	77.15	88.48	49.09	85.03	91.37	68.34	79.86	88.90	58.94
Area irrigated										
6 - 30 ha	20	77.39	88.37	63.31	85.41	91.37	77.55	80.51	88.90	69.34
30 - 70 ha	31	77.58	88.48	49.09	85.54	92.58	68.34	80.35	86.88	58.94
70 - 166 ha	7	80.35	88.02	69.30	86.89	90.45	81.52	82.01	86.40	74.98

Only 43 data are available

In order to remark the inter-relation among the uniformity parameters we offer the following equations (parameters expressed as %):

$$CU_h = 43,89 + 0,53 \, DU \qquad\qquad R^2: 0.81 \qquad\qquad (6)$$

$$CU_h = 24.28 + 0.76\ CU_v \qquad R^2: 0.95 \qquad (7)$$

$$DU = -15.96 + 1.16\ CU_v \qquad R^2: 0.80 \qquad (8)$$

Figure 1 shows DU, CU_h and CU_v values achieved in every test in DU-increasing order. Both sprinklers and spray heads located at 4-m in height have the same high range of uniformity with the exception of isolated tests with particular problems (i.e. emitter clogging, lateral pipeline leaks and incorrect maintenance of the emitter package). However, uniformity at 2.5 m has a wider range of variability; eventually uniformity is strongly lower at 1.5 m due to the design of the emitter itself as is shown below.

To analyse evaporation and drift losses and uniformity as a function of the emitter height, three different heights were studied for the same emitter (spray head) 4 m, 2.5 m and 1 m, in a centre pivot of 400 m (50.26 ha) with seven towers (Fig. 2). Tests were performed during both day and night-time (when there is no wind at all and evaporation is negligible). Results show that in night irrigation the highest uniformity is achieved in locating the emitter at 4-m height (CU_h: 91%) (Fig. 2). If spray heads are located at 2.5 m

Table 2. Correlation analysis among uniformity parameters according to the quantitative variables analysed (top value: correlation coefficient; bottom value: significance level).

	Wind speed	Working pressure	Surface	Sprinkler height
DU	-0.402	-0.351	0.027	0.385
	0.008	0.021	0.863	0.011
CU_h	-0.212	-0.142	-0.011	0.288
	0.171	0.362	0.946	0.061
CU_v	-0.243	-0.188	-0.039	0.237
	0.116	0.228	0.802	0.127

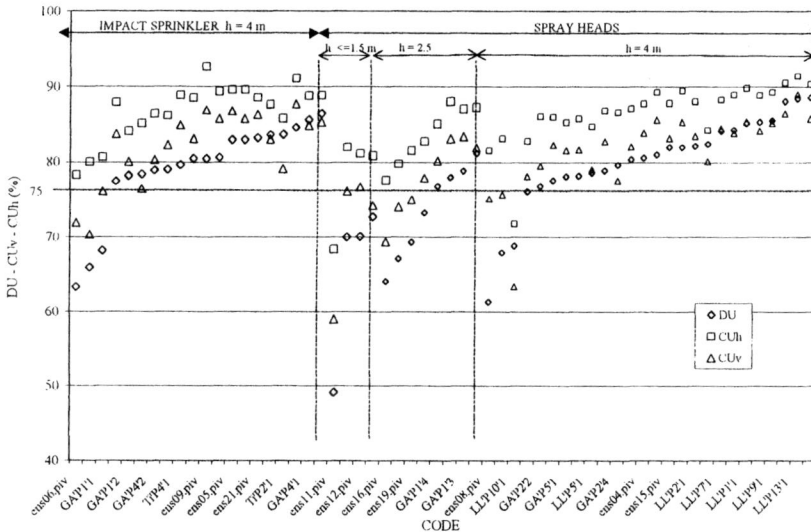

Fig. 1. Uniformity values (DU, CU_h, CU_v) as a function of sprinkler type and sprinkler height above the ground (h), ordered for each group on the increasing order of DU.

uniformity slightly decreases (CU_h: 87%), but strongly decreases if they are located at 1 m (78%). Uniformity decreases since the spray head design itself does not include any rotary device, which helps to the complete jet disruption with view to achieve a regular rainfall, when the emitter height is not sufficient to complete the total jet disruption. In consequence when locating the spray head at 1 m above the soil, there are soil points that receive a large amount of water, whereas water distribution in other points is very scarce. The behaviour of Ed may be also explained for that reason (Fig. 2).

However, in day-time irrigation, both uniformity and efficiency of discharge values strongly vary. Uniformity of nozzle heads located at 1 m increase because of the effect of wind, which plays a major role in the jet breaking-up process and results in a more regular rainfall (Fig. 2). On the other hand uniformity keeps almost the same values in the case of 2.5 and 4 m. The greater dispersion of CU values at 2.5 m are due to that, as a consequence of the experimental design and the dominant direction of wind from the pivot point to the end point, a proportion of the water discharged by the set of spray heads located at 4 m was drifted by wind towards this tower resulting in an increase of rainfall at the beginning of the tower. This is also the reason for the high values of efficiency of discharge in this tower. Efficiency at 4 m strongly decreases since in this case both wind and evaporation action lower the eventual collected rainfall. Efficiency in the 1-m tower is high since it is little influenced by wind and evaporation due to closeness of emitters to soil (Fig. 2).

Production analysis was done to evaluate the effect of the three different sprinkler heights on the yield of a wheat crop. Three replications were performed fore every sprinkler height. No significant differences were found between 1 m and 2.5 m. Yields averaged 7463 and 7573 kg/ha, respectively However significant differences were noticed between them and the yield attained in the tower with emitters located at 4 m, which averaged 6420 kg/ha.

4 Conclusions

- Water distribution can be globally considered as good (average CU_h: 85.6%), although some isolated problems have been detected relating to pipeline leaks, inadequate (either excessive or insufficient) pressure at the centre pivot, faulty emitters, etc.
- The major variability in uniformity has been detected at the outer end of the pipeline. Moreover it is necessary to note that this is the most susceptible zone to wind action. In addition, it corresponds to the biggest surface wetted as a function of the lateral length. This can be overcome by locating a medium-size raingun at the outer end of the lateral located at 2 m above the ground when designing the emitter chart.
- No clear differences in water distribution have been observed when analysing factors such as equipment size, emitters, working pressure or wind speed and direction. The major factor is undoubtedly the correct design, scheduling and maintenance of the emitter chart.

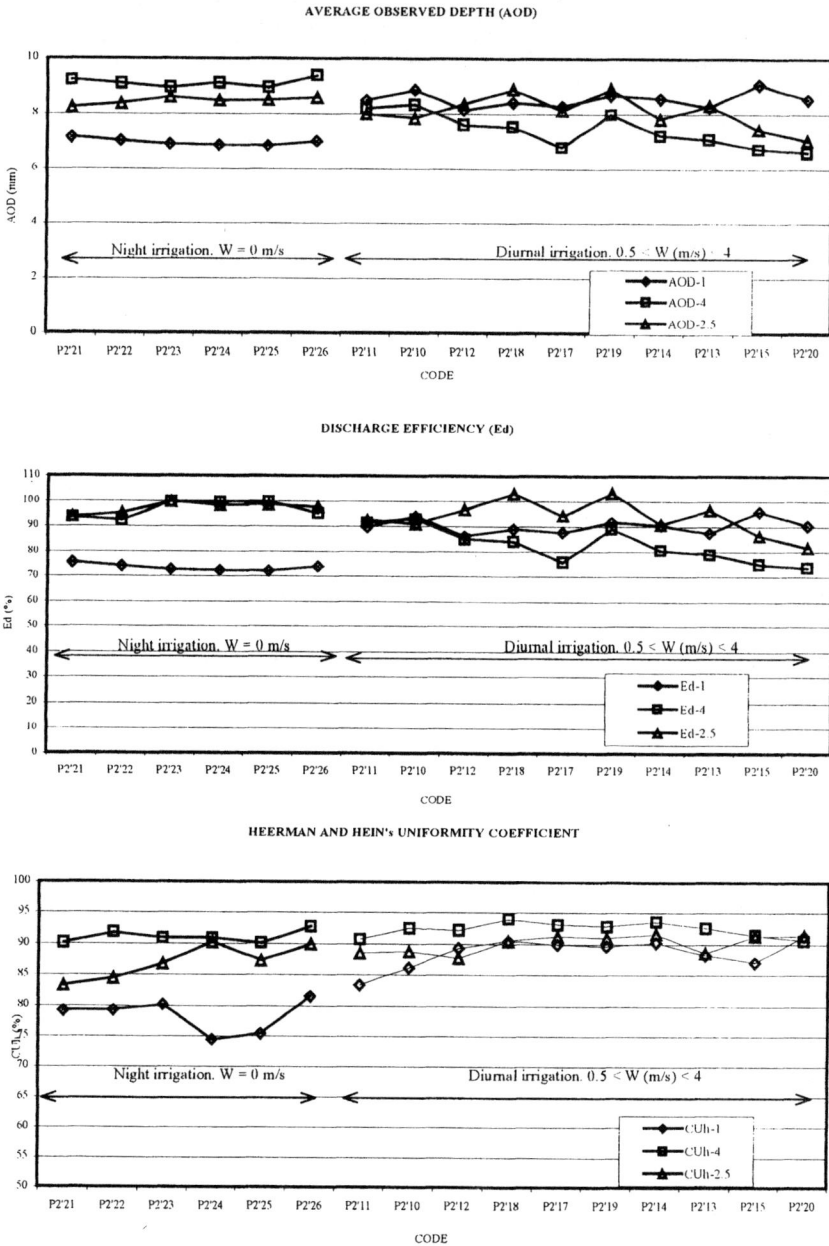

Fig. 2. Average Observed Depth AOD [mm], discharge Efficiency Ed [%] and Heermann and Hein's Coefficient of Uniformity CU_h [%] for night and diurnal irrigation conditions, when sprinklers are placed at 1.0, 2.5 and 4.0 m above the ground, ordered in the increasing order of wind speed.

- Evaporation and drift losses lower if emitters are located closer to the soil, but this fact reduces the wetted radius, implies shallower flooding (with possible runoff in slope areas) and reduces water distribution uniformity.
- The most suitable emitter layout in order to balance evaporation and drift losses and uniformity is to locate emitters about 2 m in height. The wetted radius is then about 15 m, which requires 150-200 kPa. In this case the spacing of emitters ranges 2.5 to 3 m.

5 Acknowledgements

This research was partly funded by the FAIR CT 950088 European Project (NIWASAVE) and the HID96-1373 and PTR94-0107 CICYT Projects. The writers are indebted to Katryn Walsh, Department of Modern Philology, College of Agricultural Engineers, University of Castilla-La Mancha, for providing translation and review services.

6 References

1. Merrian, J.L. and Keller, J. 1978. *Farm irrigation system evaluation: a guide for management*. Utah State University. Logan. Utah. USA.
2. Merrian, J.L. Shearer, M.N. and Burt C.M. 1980. Evaluating irrigation systems and practices, in *Design and operation of farm irrigation systems* (ed. M.E Jensen) ASAE monograph n° 3 pp. 721-760.
3. Tarjuelo, J.M. 1995. *El riego por aspersión y su tecnología*. Madrid: Mundi-Prensa S.A. pp 271-348.
4. ISO-11545, 1994. Agricultural irrigation equipment. Centre pivot and moving lateral irrigation machines with sprayer or sprinkler nozzles. Determination of uniformity of water distribution.
5. ANSI/ASAE standards S436. 1995. Test procedure for determining the uniformity of water distribution of centre pivot, corner pivot, and moving lateral irrigation machines equipped with spray or sprinkler nozzles.
6. Heermann, D.F., and Hein P.R. 1968. Performance characteristics of self-propped centre pivot sprinkler irrigation system. *Transactions of the ASAE* 11(1): 11-15.
7. Bremond, B and Molle, B. 1995. Characterisation of rainfall under centre pivot: influence of measuring procedure. *J. of Irrig. and Drain. Eng.* 121(5): 347-353

OPTIMIZATION OF WATER APPLICATION FOR SPRINKLER IRRIGATION

V. P. S. PAZ
Department of Rural Engineering, ESALQ/USP-FAPESP, Piracicaba, SP, Brazil.
J. A. FRIZZONE, T. A. BOTREL and M. V. FOLEGATTI
Department of Rural Engineering, ESALQ/USP, Piracicaba, SP, Brazil.

Abstract
Water availability has ever more becoming a limiting factor to the agricultural development. Efficient water use can be achieved by appropriate water management. This leads to higher yield, reduced production costs and, ultimately, maximum investment returns. Furthermore, the use efficient water management and application techniques are essential for sustainable agricultural development, based on the best use of the natural resources. This paper has as objectives to study the reduction of the net revenue due to deficient or excessive water application, determine the effects of uniformity of water distribution on the irrigation depth that maximizes economical profits, and provide information to optimize operation of sprinkler irrigation systems for a bean crop in Piracicaba, Brazil. The results showed that minimization of revenue losses is related to the uniformity of water distribution of the irrigation system, the commodity price, and the irrigation costs. The best operation of irrigation systems, determined from optimization of strategies to maximize net profits, showed satisfactory to improve water application efficiency, water storage efficiency, and appropriate irrigated area. When maximization of net revenue is the objective function, the uniformity of water distribution, the cost of the applied water and the commodity price are decisive factors for the optimization of irrigation systems.
Keywords: Optimization, sprinkler irrigation, uniformity.

1 Introduction

Reliable yield estimation as a function of available resources has been object of several recent studies. Among the production factors, water and nutrients are the ones most limiting to crop production. Thus, the improvement of irrigation and soil fertility management constitutes an essential approach to succeed in agricultural enterprises

Water and the Environment: Innovative Issues in Irrigation and Drainage. Edited by Luis S. Pereira and John W. Gowing. Published in 1998 by E & FN Spon. ISBN 0 419 23710 0

[4]. Both in arid or semiarid regions, where water is scarce, or even in humid regions, higher yields relies on rational use of water.

Fresh water has been ever more recognized as of vital importance for future generation. On the other hand, the use of irrigation has showed unpredictable impacts to the environment. Therefore, is fundamental to establish standards for evaluation of the resulting impact of irrigation to the natural resources, in order to predict its use in a larger area without the problems occurred in existing projects. The planing and designing of irrigation projects is the best stage for diagnosis of the possible environmental impacts and for making necessary adjustments to mitigate future damage to the natural resources [5].

Distribution of water application in irrigation systems is not perfectly uniform. High application uniformity can be reached at higher operational costs. It is common to describe water distribution uniformity by coefficients that express the variability of the applied water over the area [6]. After Christiansen [7], who introduced the uniformity coefficient (CUC), many other methods were proposed to express the uniformity of irrigation application. However, any of them were not sufficiently complete to substitute, with advantages, Christiansen's CUC [8]. A further aspect to be considered is the soil moisture uniformity in the soil. The effects of surface water distribution on the moisture redistribution in the soil profile was analyzed by [9,10]. Higher values of the uniformity coefficients were achieved over time for all depths.

Assuming a distribution model for application of irrigation water, there exists an optimal amount of water that should infiltrate to minimize the reduction of the expected net revenue, caused by excessive and deficient soil moisture over the area. This amount is named as optimal applied depth. It expresses the average depth that should infiltrate to obtain the maximum net revenue, by minimizing reduction due to excess and deficit irrigation [1,2].

The reduction of net revenue due to the deficient and excessive water application considering the costs of water, the price of the commodity, and the uniformity water distribution was studied by [11], following the methodology proposed by [1,2,3]. They concluded that the irrigation systems with better uniformity provided higher economic returns for any combination of agricultural product price and water cost.

This research has as objectives to study the reduction of the net revenue due to deficient or excessive water application, determine the effects of uniformity of water distribution on the irrigation depth that maximizes economical profits, and provide information to optimize operation of sprinkler irrigation systems.

2 Material and methods

The coefficient of uniformity proposed by Christiansen (CUC) was used to represent the distribution of the applied water over the area by the irrigation system. The economical analysis of a bean crop (*Phaseolus vulgaris*, L), was performed following the approach of English [3]. The response crop production function (Y in kg/ha) to the seasonal applied irrigation (W in mm) was described as following [12]:

$$Y(W) = 63,9656 + 8,9745 \, W - 0,008503 \, W^2 \qquad (1)$$

According to Equation 1, the application of 528 mm during the crop cycle leads to the maximum yield, which is equal to 2432 kg/ha.

The net revenue is given by Equation 2:

$$IL(W) = Y(W)P_i - C(W) \tag{2}$$

where Pi is the commodity price and C(W) is the production cost of the irrigated crop, which was considered as a lineal function [13]:

$$C(W) = C_0 + C_W W \tag{3}$$

where Co is the production cost not related to the application of irrigation, Cw is the production cost per unit of applied water.

The value of Co was taken as US 411/ha [14], and Cw varied from US 0.32/mm.ha to US 0.75/mm.ha [14] and Pi from US 0.44/kg to US 1.62/kg [15].

Assuming land as a limiting factor, there exists an optimal irrigation depth (W_1*) that provides the maximum net revenue [3], which is given by the following expression:

$$W_1^* = \frac{C_W - P_i B}{2P_i C} \tag{4}$$

The following assumptions were made for the estimation and analysis of the economic loss due to deficit and excess applied water:

- The irrigation system applies the seasonal depth of water in each point over the area. Since each point represents a portion of the total, this results in a distribution of the applied water over the irrigated area;
- Net revenue can be calculated for each point of the area from the total applied depth in each point;
- The optimal net revenue can be obtained assuming W_1* as the seasonal depth applied uniformly over the area;
- The economic loss (DEC) in each point i of the irrigated area can be estimated by the difference between the optimal net revenue (IL*) and the net revenue calculated for each point of the area (ILi), according to Equation 5:

$$\Delta EC_i = IL^* - IL_i \tag{5}$$

- Revenue loss in each point can occur due deficit or excess when Wi is less or greater than W*, respectively;
- Applied water follows a normal distribution. Therefore, the reduction of net revenue due to excess (α) and deficit (β) water application was calculated using the methodology by [16].

In the analysis of the irrigation system operation, efficiency parameters expressing the quality of the irrigation were used [17].

3 Results and discussion

The optimal applied depth (W*) increased with increase in product price and decrease in water cost. However, there was a little effect of water cost on W* when product price was high (Fig. 1).

The value of W* approached the seasonal depth required for maximum yield when the product price was high and the cost of water was low. To achieve the optimal economic revenue W* can be as much as 18.6% less than the depth required for maximum yield. This indicates that a significant amount of water can be saved.

The effects of uniformity of water distribution were given by the ratio between reduction of net revenue due to deficit (β) and excess (α) applied water. In this analysis net revenue loss was higher due to deficit than excess ($\beta/\alpha > 1$). The ratio β/α decreased with increase in cost of water. Therefore, for a given uniformity, β/α increased as water cost decreased, regardless product price (Fig. 2). For any combination of product price and water cost, β/α was higher for systems with lower uniformity of application (Fig. 2b). The results in Fig. 2 show also that higher uniformity is required when product price is higher, even for condition of inexpensive water. This would induce to saving in water application, since there will be an equilibrium between excess and deficit in the other fraction of the area.

Taking into account an optimal economic level for crop production and considering a scenario including different water costs and product prices a further analysis was performed. In that, a relative optimal irrigation depth was defined as the ratio between a given required depth (X_r) and the average irrigation depth (X_m*) that has to be applied to attend X_r, in order to minimize economic losses due to deficit and excess water.

The results of the analysis are illustrated in Fig. 3. Systems with higher application uniformity are more appropriate to supply crop water requirements, resulting in lower values for the optimal applied depth. As an example, assuming a required depth X_r equal to 12 mm and a system with β/α equal to 3.93 and CUC equal 90%, the average irrigation depth X_m has to be 13.3 mm, thus 10.8% higher than the required depth (Fig. 3a). This was similar to the results of [2] who found X_m 10% higher than X_r for β/α equal to 4. As a second example, for the same situation but with CUC equal to 80%, the system has to apply an average depth equal to 14.9 mm to supply the required

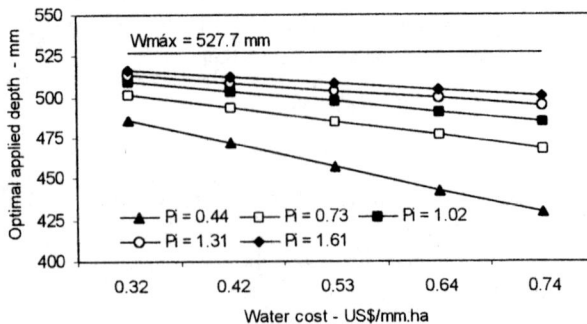

Fig. 1. Optimal applied depth as a function of water cost and product price (Pi).

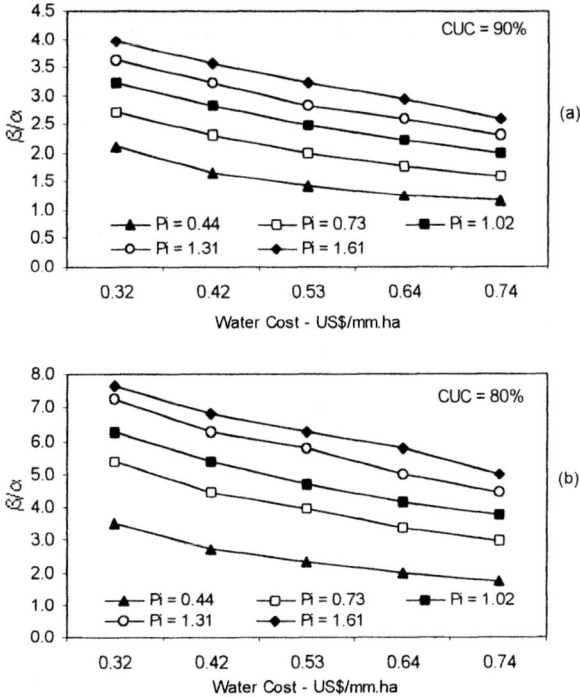

Fig. 2. Relative effect of deficit to excess of applied water (β/α), for different water costs and product prices (Pi), for irrigation systems with uniformity of distribution (CUC) equal to (a) 90% and (b) 80%.

areas with deficit and excess, increasing β/α. Consequently, a higher optimal irrigation depth is required to attend the demanded application.

The results of analysis on the operation of the irrigation system as a function of the distribution uniformity are shown in Fig. 4. The quality of irrigation, evaluated by the efficiency parameters, was reduced as distribution uniformity decreased, for any optimal economic scenario of crop production. Systems with capability of application close to the required depth, as given by the low relative optimal average depth (Xm^*/Xr), had higher application efficiency. However, the quality of irrigation was reduced, as given by lower storage efficiency, and consequent, there was a reduction in area irrigated appropriately. This was more evident for systems with low uniformity of distribution (Fig. 4b). For a scenario in which the relative effect of deficit over excess water is known, and for a given uniformity of application of the irrigation system, it is possible to optimize system operation to minimize the effects of deficit and excess applied water, and then, to reduce resulting water loss.

4 Conclusions

The reduction of net revenue losses is related to the uniformity of water application of the irrigation system, the product price and irrigation costs.

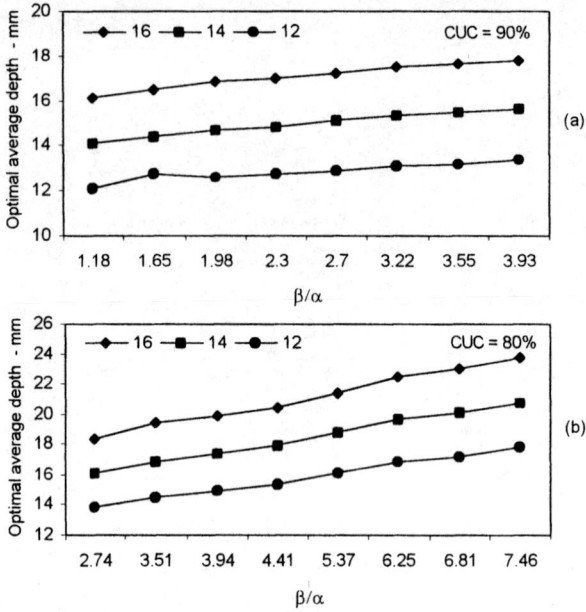

Fig. 3. Optimal average irrigation depth as a function of β/α, for different required depths and uniformity of distribution (CUC) equal to (a) 90% and (b) 80%.

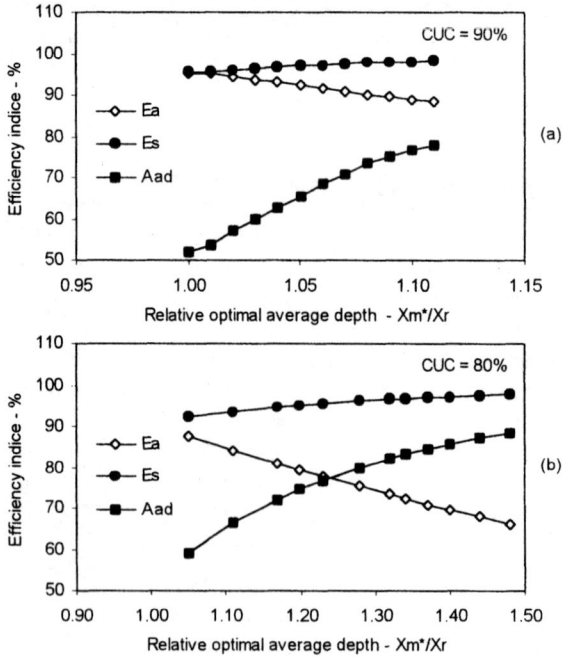

Fig. 4. Efficiency indices for optimal irrigation system operation, as a function of relative optimal average depth (Xm^*/Xr) and uniformity of distribution (CUC) equal to (a) 90% and (b) 80%.

The best operation of irrigation systems, determined from optimization of strategies to maximize net profits, showed satisfactory to improve water application efficiency, water storage efficiency, and appropriate irrigated area.

When maximization of net revenue is the objective function, system application uniformity, cost of water and product price are determinant factors to optimize irrigation systems.

5 References

1. Norum, D.I.; Peri, G.; Hart, W.E. (1979) Application of systems optimal depth concert. *Journal of the Irrigation and Drainage Division*, v.105, n.4, p.357-367.
2. Peri, G.; Hart, W.E.; Norum, D.I. (1979) Optimal irrigation depths - A method of analysis. *Journal of the Irrigation and Drainage Division*, v.105, n.4, p.341-354.
3. English, M. J. (1990) Deficit irrigation. I: Analytical framework. *Journal of the Irrigation and Drainage Division*, v.116, n.3, p.399-412.
4. Frizzone, J.A. (1993) *Funções de resposta das culturas à irrigação*. Piracicaba: ESALQ/DER, 42p.
5. Bernardo, S. (1997) Impacto ambiental da irrigação no Brasil, in *Recursos Hídricos e Desenvolvimento Sustentável da Agricultura*. Viçosa, pp.79-88.
6. Solomon, K. H. (1984) Yield related interpretations of irrigation uniformity and efficiency measurements. *Irrigation Science*, v.5, n.3, p.161-72.
7. Christiansen, J. E. (1942) *Irrigation by Sprinkler*. Berkeley, California Agricultural Station. 124p. (Bul.670).
8. Elliot, R.L.; Nelson, J.D.; Loftis, J.C.; Hart, W.E. 1980) Comparison of sprinkler uniformity models. *Journal of Irrigation and Drainage Division*, v.106, n.IR4, p.321-30.
9. Paiva, J.B.D. de. (1980) *Uniformidade de aplicação de água, abaixo da superfície do solo, utilizando irrigação por aspersão*. São Carlos, EESC/USP, 333p.
10. Rezende, R.; Frizzone, J.A.; Botrel, T.A. (1993) Desempenho de um sistema de irrigação pivô central quanto à uniformidade de distribuição de água abaixo e acima da superfície do solo. *Engenharia na Agricultura*, v.1, n.1, p.1-7, 1993.
11. Paz, V.P.S.; Frizzone, J.A.; Botrel, T.A.; Folegatti, M.V. (1997) Redução da receita líquida por déficit ou excesso de água na cultura do feijoeiro. Pesquisa Agropecuária Brasileira, v.32,n.9,p.869-75.
12. Frizzone, J.A. (1986) Funções de resposta do feijoeiro (*Phaseolus vulgaris*, L.) ao uso de nitrogênio e lâminas de irrigação. Piracicaba, ESALQ/USP, 133p.
13. Hart, W.E.; Norum, D.I.; Peri, G. (1980) Optimal seasonal irrigation application analysis. *Journal of the Irrigation and Drainage Division*, v.106, n.3, p.221-235.
14. IPT (1994) Relatórios técnicos. São Paulo, 27p.
15. Preços agrícolas (1994), Piracicaba, n.8, p.10-52.
16. Paz, V. P. S. (1995) *Condições ótimas de operação de sistemas de irrigação por aspersão*. Piracicaba, ESALQ/USP, 125p.
17. Walker, W. R. (1979) Explicit sprinkler irrigation uniformity: efficiency model. *Journal of the Irrigation and Drainage Division*, v.109, n.3, p.317-32.

NEW NON-WATER-STRESSED BASELINES FOR IRRIGATION SCHEDULING WITH INFRARED THERMOMETERS

I. ALVES and L. S. PEREIRA
Department of Agricultural Engineering, Institute of Agronomy, Technical University of Lisbon, Portugal

Abstract
Surface temperature measured with infrared thermometers is an important tool for irrigation scheduling that has been in practice for some decades. Several indices have been developed to time irrigation events. The most useful is the Crop Water Stress Index (CWSI). Its use however relies in a non-water-stressed baseline that up to now, although having a theoretical basis, is to be determined experimentally given the uncertainties on the surface resistance of the crop. Drawbacks of this procedure, besides the non-transferability of the lines from local to local, are that the surface temperature measurements are to be made always at the same time of the day and under similar weather conditions. A new definition of a non-water stressed baseline theoretically based and driven by weather variables that can be easily measured and/or estimated is proposed that allows measurements at any time of the day, thus simplifying the task of the irrigator.
Keywords: Crop water stress index, energy balance, evapotranspiration, infrared surface temperature, irrigation scheduling, vapour pressure deficit.

1 Introduction

One of the main decisions an irrigator is faced with is the timing of irrigations. Several methods exist either plant, soil or atmosphere based, that may be more or less time consuming or that may rely on expensive equipment.

The arrival of commercial portable infrared thermometers for surface temperature measurements represented a great advance in this area. They are used mainly to detect crop water stress, and several indices have been proposed as an aid to irrigation

Water and the Environment: Innovative Issues in Irrigation and Drainage. Edited by Luis S. Pereira and John W. Gowing. Published in 1998 by E & FN Spon. ISBN 0 419 23710 0

scheduling, like Stress Degree Day (SDD), Critical Temperature Variability (CTV), Temperature Stress Day (TSD), and, specially, Crop Water Stress Index (CWSI) [1].

The CWSI method relies on two baselines: the non-water-stressed baseline, that represents a full watered crop, and was first introduced by Idso *et al.* [2], and the stressed baseline, that corresponds to a non-transpiring crop (stomata fully closed). Both lines are drawn on a T_s -T_a *vs.* *VPD* plot (Fig. 1), where T_s and T_a are, respectively, the surface and the air temperature at the reference level (°C) and VPD is vapour pressure deficit (Pa).

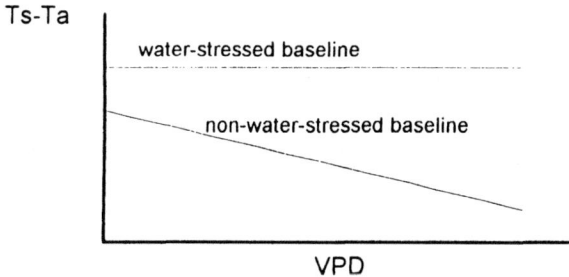

Fig. 1. Baselines used for the computation of CWSI (schematic representation).

Though the CWSI evolved from experimental findings, it can be given a theoretical basis. From the Penman-Monteith equation:

$$\lambda E = \frac{\Delta(R_n - G) + \rho_a c_p VPD/r_a}{\Delta + \gamma(1 + r_s/r_a)} \tag{1}$$

where λE is latent heat flux, R_n is net radiation and G the soil heat flux (all with units of W/m^2), Δ is the slope of the saturated vapour pressure *vs.* temperature curve (Pa/°C), γ is the psychrometric constant (Pa/°C), ρ_a is air density (kg/m^3), c_p is specific heat at constant pressure (J kg^{-1} °C^{-1}), VPD is vapour pressure deficit at the reference level (Pa), r_a is the aerodynamic resistance to heat flow between the surface and the reference level (s/m) and r_s is the surface resistance (s/m), and the sensible heat flux (H) (W/m^2) equation:

$$H = \rho c_p \frac{T_o - T_a}{r_a} \tag{2}$$

where T_o is the temperature at the surface level (°C) and T_a is air temperature at the reference level (°C), it can be derived:

$$T_o - T_a = \frac{r_a}{\rho_a c_p} \frac{(R_n - G)\gamma(1 + r_s/r_a)}{\Delta + \gamma(1 + r_s/r_a)} - \frac{VPD}{\Delta + \gamma(1 + r_s/r_a)} \tag{3}$$

From this equation it can be deduced that for the same level of net radiation and similar temperature and windspeed values:

$$T_o - T_a = a - b \, VPD \tag{4}$$

that, when applied to a full watered crop, represents the non-water-stressed baseline, as in Fig. 1. Also, if stomata are closed, $r_s \rightarrow \infty$ and Eq. (3) becomes:

$$T_o - T_a = \frac{r_a}{\rho_a c_p} (R_n - G) = a' \tag{5}$$

that is the equation of the horizontal line that represents the non transpiring crop in Fig. 1.

Given this theoretical basis, one would expect that the non-water-stressed baseline could be easily computed. Net radiation can be measured or estimated, using the methodology proposed by Allen *et al.* [3]. Since infrared thermometers only sense the top layer of the canopy, aerodynamic resistance can be advantageously computed from the top of the canopy (h) up to the reference level (z) as ([4]):

$$r_a = \frac{\{\ln[(z-d)/z_{oM}] - \psi_M\}\{\ln[(z-d)/(h-d)] - \psi_H\}}{k^2 u_z} \tag{6}$$

where d is zero plane displacement height (m), z_{oM} is the roughness length (m) for momentum, k is the von Karman constant (0.41), u_z is the wind speed (m/s) at the reference height z (m) and Ψ_M and Ψ_H are the integrated stability functions for describing the effects of the buoyancy or stability on momentum transfer and heat flux between the surface and the reference level. Parameters d and z_o can be estimated with no great error, for complete cover crops, from crop height (h) ([5][6]).

But it is not possible to proceed further as

- the knowledge on r_s and the values it can assume for a given climatic situation and throughout the day is still scarce;
- T_o is an "aerodynamic" temperature that is different from the surface radiometric temperature T_s.

So, to be applied in practice, the baseline still has to be determined experimentally. This however has two main drawbacks: baselines can not be transposed to other places, since they will be site specific; and measurements have to be done always at the same time of the day and with the same (clear) sky conditions to assure similar R_n, temperature and r_a values from day to day. This clearly complicates the task of the irrigator, that has to have a strict schedule to perform the surface temperature readings.

Meanwhile, an original view of surface temperature has been presented by Wanjura *et al.* [7] and Wanjura and Upchurch [8]. They state that the wet bulb temperature (T_w) is the lower temperature limit for an evaporating surface. Wanjura *et al.* [7] estimated that a canopy is likely to only cool to about 2ºC above the ambient wet bulb temperature. They attributed this difference to the different geometry of plant leaves when compared to that of an aspirated wet bulb thermometer. However, these authors failed to realise that the air at the surface level, being the source of vapour flux, is more humid than the air at the reference height which actually leads to a lower wet bulb depression.

Also, as H is mostly positive during the day, which implies from Eq. (2) that $T_o > T_a$, a higher T_w at the surface is effectively to be expected ($T_{w \; surf.} > T_{w \; air}$).

This study was then performed to evaluate the infrared surface temperature, T_s, as a "wet bulb" temperature ($T_s = T_{w \; surf.}$).

2 Materials and methods

2.1 Site and crop characteristics
The field trial is essentially the one that has been already described in Alves *et al.* [9], where further details can be found. It was conducted at an Experimental Station belonging to INIA (National Institute for Agricultural Research) located at Coruche (lat. 38° 57' N, long. 8° 32' W, alt. 30 m), some 80 km north-east from Lisbon, Portugal, during the summer of 1992. An iceberg lettuce crop (*Lactuca sativa* var. *capitata* cv Saladin) was planted on the 28 May with a density of 8 plants/m^2 on a 0.5 ha field (50 m × 100 m) of a sandy soil. The crop was drip irrigated almost every day, mostly during the night or early morning, so maintaining the humidity in the root zone permanently near soil capacity. Measurements were made between 25 June and 30 July when the crop completely covered the soil.

2.2 Instrumentation and data acquisition
The equipment consisted of:

* 2 anemometers, Young (model 12102D), at the heights of 0.85 and 1.46 m;
* 1 wind direction sensor, Vector Instruments, model W200P, at 1.65 m height;
* 2 psychrometers, made from ventilated, double-shielded copper-constantan thermocouples (at the same heights as the anemometers), with an accuracy of ± 0.02°C;
* 1 net radiometer, Schenk, at 1.5 m height and south oriented;
* 1 infrared thermometer, AGEMA model Thermopoint 80 Scope, with an accuracy of ±1% and a field of view of about 2°; the instrument was south oriented, perpendicularly to the row, and positioned at an angle of 60° below the horizontal in order to view the top leaves of the plants at some 0.40 m distance.

These instruments were installed in a measurement tower that was placed near the edge of the plot, in order to benefit from approximately 80 m of fetch. The infrared thermometer was installed inside a white-painted wood shelter.

An emittance of 0.98 was used when setting the infrared thermometer. This model allows the introduction of the ambient temperature to correct the readings for reflected energy. We used a temperature of 20°C, considered to be an average for the diurnal period.

All the meteorological instruments were calibrated/checked before the trial ([9]). The infrared thermometer was checked against a portable calibration source from Everest, model 1000, the values being within ±0.5°C.

The instruments were connected to a Campbell Scientific 21X datalogger that scanned the sensors every second and stored the average values at 10-minute intervals.

2.3 Data handling

Only the values recorded during the periods when the fetch was adequate and $R_n > 0$ were kept.

Aerodynamic resistance was calculated according to Eq. (6), with $d/h = 0.67$, $z_0/h = 0.126$, and $z = 0.85$ m, the height of the lowest instruments. Stability conditions of the atmosphere were evaluated using the Richardson number (Ri); only values obtained for $|Ri| < 0.2$ were kept, allowing to use no stability corrections.

The latent and sensible heat fluxes were computed with the Bowen ratio (β) method. As the crop was drip irrigated almost every day, no water stress was expected to develop, which could influence the infrared readings. In fact, computed Bowen ratio values (β) were between -0.2 and 0.2, which are representative of a fully watered crop ([10]).

Soil heat flux (G) was estimated to be 10% of the measured net radiation, following the studies of Clothier et al. [11] on closed canopies. All additional necessary parameters were calculated with the algorithms proposed by Allen et al. [3].

3 The radiometric surface temperature as a "wet bulb" temperature

From the saturation pressure curve, it can be derived ([12]):

$$e \approx e_s(T) - (\Delta + \gamma)(T - T_w) \tag{7}$$

where T is air temperature (dry bulb thermometer) (ºC), T_w is wet bulb temperature (ºC), e is actual vapour pressure (Pa) and $e_s(T)$ is saturated vapour pressure at temperature T (Pa). Then, the change of actual vapour pressure between two heights, and given the definition of Δ, will be:

$$e_1 - e_2 = (\Delta + \gamma)(T_{w_1} - T_{w_2}) - \gamma(T_1 - T_2) \tag{8}$$

Applying Eq. (8) to determine the vapour pressure difference between the surface level (subscript o), with T_s and T_o, respectively, as the wet bulb temperature and the air temperature at this level, and the reference height (subscript z), where measurements of T_w and T_a are made, we obtain:

$$e_o - e_z = (\Delta + \gamma)(T_s - T_w) - \gamma(T_o - T_a) \tag{9}$$

Then, if we combine Eq. (9) with the latent heat flux equation

$$\lambda E = \frac{\rho_a c_p}{\gamma} \frac{e_o - e_z}{r_a} \tag{10}$$

and the sensible heat flux Eq. (2), from the energy balance equation

$$R_n - G = \lambda E + H \tag{11}$$

we obtain:

$$T_s - T_w = \frac{\gamma}{\Delta + \gamma} \frac{r_a}{\rho_a c_p} (R_n - G) \qquad (12)$$

which may be considered a new definition of the non-water-stressed baseline.

4 Results and discussion

The range of values of R_n, r_a and T_a recorded during the trial and belonging to the set of data that was kept can be found in Fig. 2. Although the values of R_n do not vary greatly from day to day, aerodynamic resistance can differ greatly.

Given the theoretical basis provided by Eq. (3), it is to be expected that the relationship between surface temperature and VPD will change throughout the day, mainly due to the evolution of R_n, as it is clearly illustrated in Fig. 3, where four possible baselines are drawn. This implies that if only one curve is to be used measurements are to be made at the same time of the day and with the same climatic conditions, which restricts the freedom of the irrigator. The use of a multitude of curves, however, is not a practical alternative.

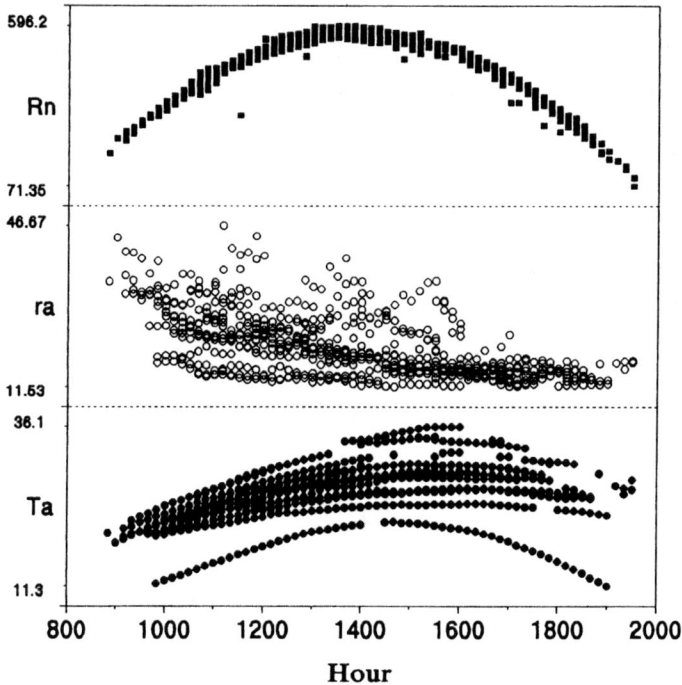

Fig. 2. Values of net radiation (R_n), aerodynamic resistance (r_a) and air temperature (T_a) recorded during the trial and belonging to the set of data kept for analysis.

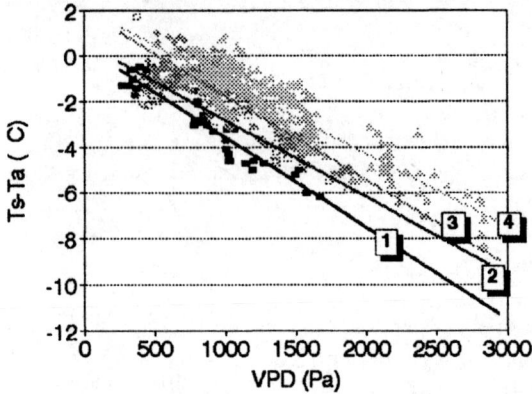

Fig. 3. Four possible baselines, depending of the level of net radiation (R_n). 1) R_n < 300 W/m^2; 2) 300 < R_n < 400 W/m^2; 3) 400 < R_n < 500 W/m^2; R_n > 500 W/m^2.

The proposed approach to the non-water-stressed baseline would be easier to apply. To test this formulation, the measured values of surface temperature were compared with those calculated with Eq. (12) and plotted in Fig. 4. It can be seen that the agreement is good, supporting the validity of the approach. There are however small differences that Fig. 5 shows to depend on the time of the day. These differences can be attributed either to the fact that G was not measured (giving an error in the R_n-G term) either to errors in T_s itself, as the sun height and azimuth of the measurements, which are known to influence the readings ([13][14]), change throughout the day.

We can therefore conclude that infrared surface temperature of fully transpiring crops can be regarded as a wet bulb temperature and that Eq. (12) can be used as a non-water-stressed baseline. This formulation has the advantage over Eq. (3) that it does not require any knowledge on r_s and it can easily be computed with measured or estimated climatic parameters.

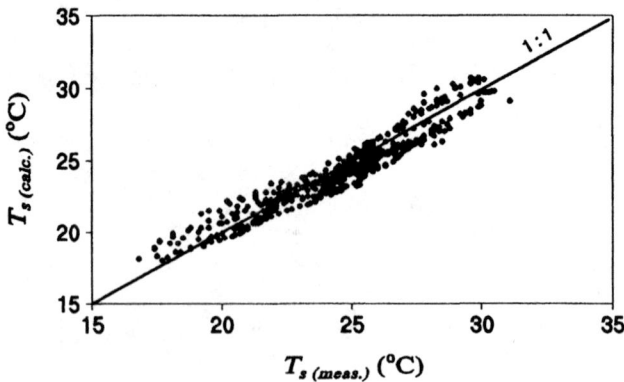

Fig. 4. Comparison of calculated surface temperature, $T_{s(calc)}$, using Eq. (12) and measured surface temperature, $T_{s(meas.)}$.

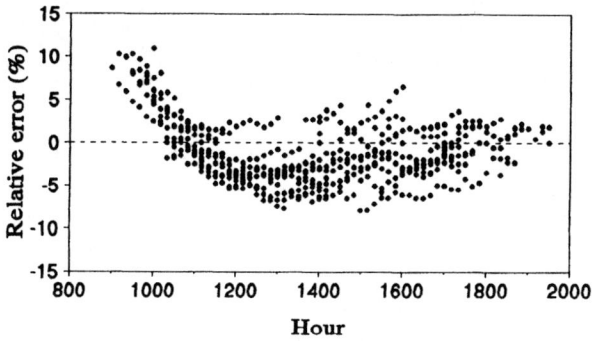

Fig. 5. Relative error of calculated surface temperature, $(T_{s(calc.)} - T_{s(meas.)}) / T_{s(meas.)}$, and its evolution throughout the day time.

Values of calculated T_s-T_w are plotted in Fig. 6. It shows that it may be highly variable throughout the day and that the value of 2ºC considered by Wanjura *et al.* [7] can only be used as a rough "rule of thumb". For more accurate applications (namely overcast days) Eq. (12) is to be used to determine the temperature of the fully watered crop.

5 Conclusions

The non-water-stressed baseline is a useful concept that can effectively guide the irrigator when maximum yields are to be obtained. The Idso *et al.* [2] baseline has to be determined experimentally, which precludes its transfer to different climatic conditions (other places, other time of the day). In this study, we conclude that surface temperature of fully irrigated crops can be regarded as a wet bulb temperature which allows a theoretical derivation of its value when R_n, r_a and air temperature are known. It may be possible then to monitor the crop at whatever time of the day and even in overcast days, which greatly eases the task of the irrigator.

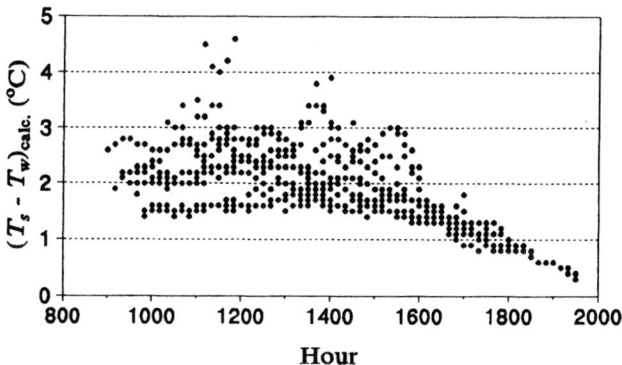

Fig. 6. Calculated difference between surface temperature and wet bulb temperature $(T_s - T_w)$ and its evolution throughout the day.

6 References

1.   Jackson, R.D. (1982) - Canopy temperature and crop water stress. *Advances in Irrigation* 1: 43-85.
2.   Idso, S.B.; Jackson, R.D.; Pinter, P.J.; Reginato, R.J.; Hatfield, J.L. (1981) - Normalizing the stress-degree-day parameter for environmental variability. *Agricultural Meteorology* 24: 45-55.
3.   Allen, R.G.; Smith, M.; Pereira, L.S.; Perrier, A. (1994) - An update of the calculation of reference evapotranspiration. *ICID Bulletin* 43 (2): 35-92.
4.   Perrier, A. (1975) - Etude physique de l'évapotranspiration dans les conditions naturelles. III. Evapotranspiration réelle et potentielle des couverts végétaux. *Annales Agronomiques* 26: 229-243.
5.   Brutsaert, W. (1982) - *Evaporation into the atmosphere.* R. Deidel Pub. Co., Dordrecht, Holland.
6.   Allen, R.G., Pruitt, W.O., Businger, J.A., Fritschen, L.J., Jensen, M.E., Quinn, F.H. (1996) - Evaporation and Transpiration. In: Wootton *et al.* (eds.), *ASCE Handbook of Hydrology*, pp. 125-252 ASCE, New York.
7.   Wanjura, D.F.; Upchurch, D.R.; Sassenrath-Cole, G.; DeTar, W.R. (1995) - Calculating time-thresholds for irrigation scheduling. *Proceedings Beltwide Cotton Conferences (Jan 4-7, San Antonio, Texas)*, pp. 449-452.
8.   Wanjura, D.F.; Upchurch, D.R. (1996) - Time thresholds for canopy temperature-based irrigation. In: Camp, C.R.; Sadler, E.J.; Yoder, R.E. (eds) - *Evapotranspiration and Irrigation Scheduling.* Proceedings of the International Conf. (San Antonio, Texas, 3-6 Nov), ASAE/IA/ICID, pp. 295-303.
9.   Alves, I.; Perrier, A.; Pereira, L.S. (1998) - Surface resistance of complete cover crops: How good is the "big leaf"? *Transactions of the ASAE* 41 (in press).
10. Angus, D.E.; Watts, P.J. (1984) - Evapotranspiration - How good is the Bowen ratio method? *Agricultural Water Management* 8: 133-150.
11. Clothier, B.E.; Clawson, K.L.; Pinter, P.J. Jr.; Moran, M.S.; Reginato, R.J.; Jackson, R.D. (1986) - Estimation of soil heat flux from net radiation during the growth of alfalfa. *Agricultural and Forest Meteorology* 37: 319-329.
12. Monteith, J.L.; Unsworth, M.H. (1990) - *Principles of Environmental Physics.* 2nd edition, Edward Arnold, London.
13. Choudhury, B.J.; Reginato, R.J.; Idso, S.B. (1986) - An analysis of infrared temperature observations over wheat and calculation of latent heat flux. *Agricultural and Forest Meteorology* 37: 75-88.
14. Huband, N.D.S.; Monteith, J.L. (1986) - Radiative surface temperature and energy balance of a wheat canopy. I: comparison of radiative and aerodynamic temperature. *Boundary-layer Meteorology* 36: 1-17.

CROP WATER STRESS INDEX FOR BEANS OBTAINED FROM TEMPERATURE DIFFERENCE BETWEEN CANOPY AND AIR

P.E.P. ALBUQUERQUE and R.L. GOMIDE
EMBRAPA, Sete Lagoas - MG, Brazil
A.E. KLAR
Dept. Rural Engineering, State University of S. Paulo, Botucatu, Brazil

Abstract

Canopy temperature (Tc), besides serving as a parameter to estimate crop evapotranspiration (ET_c), can also be used to define water stress indexes that have the advantage to serve as a reference for the rational scheduling for irrigation for a crop. One of these indexes, which has been well disseminated, is the crop water stress index (CWSI), based on the ratio between crop resistance and aerodynamic resistance (r_c/r_a), whose value can be estimated by the difference between the cropy canopy and the air (Tc - Ta). A study was carried out with the objective of obtaining CWSI values for beans (*Phaseolus vulgaris* L.), under field conditions, in which the crop was under different irrigation scheduling: 2, 4, 8, 12 and 16 day intervals. The experiment was conducted at Sete Lagoas, Minas Gerais State, Brazil, in the period from July through October, 1995. Tc values were measured by means of infrared thermometry at two times of the day: between 10:00 and 11:00 a.m., and between 01:00 and 02:00 p.m. The obtained results pointed to a limit of 0.15 as a reference for the beginning of water stress to occur, which is consistent with other work and other index obtained under similar conditions.

Keywords: Crop water stress index, evapotranspiration, irrigation scheduling.

1 Introduction

Irrigation science advances will necessarily imply in the gradual acquisition and utilization of knowledge which will conduct toward an accurate and precise control about the quality and quantity of applied water. This control has the purpose of optimizing the irrigation and other crop production practices.

Water and the Environment: Innovative Issues in Irrigation and Drainage. Edited by Luis S. Pereira and John W. Gowing. Published in 1998 by E & FN Spon. ISBN 0 419 23710 0

A way of rationalizing the use of irrigation water is to define the irrigation scheduling by utilizing the crop water stress index (CWSI) which has its bases on actual or real crop evapotranspiration (ET_c). The CWSI may serve as a parameter to schedule irrigation and to reduce the amount of applied water and improve the crop water use efficiency, within well established levels, without compromising desirable commercial and economical yield. The CWSI can be defined as

$$CWSI = 1 - \frac{ET_c}{ET_m} \tag{1}$$

where ET_m is the maximum or potential crop evapotranspiration. Both, ET_c and ET_m are expressed in mm/h, mm/day or mm/period. They can be express also as the latent heat flux per unit area (λE_p in W/m^2). The CWSI is an adimensional index. ET_m can be defined as the maximum water flow rate which is directly evaporated from the soil and plants (foliage intercepted water) and evaporated from the leaves stomatas (transpiration), with the plants presenting no water or health limitation. On the other hand, water can be a limited parameter in the ET_c situation. Thus, the CWSI values will change from 0 to 1. The CWSI will be 0 when ET_c reaches ET_m (no water stress).

The canopy temperature (Tc) is used today as a parameter to predict ET_c as well as to define CWSI ([1], [2], [3], [4], [5]). The relationship between plant water status and Tc parameters are quite complex. Several other variables, which are not constant during the plant growth period, should be taken into account in the relation between Tc and ET_c. However, Tc has been easily measured and used as a key parameter for a great number of stress indexes. According to Clawson et al. [3], Idso et al. [7] were the first to use the temperature difference between the crop canopy and the air (Tc - Ta) as a way of establishing a crop water stress index.

The Tc values can be obtained from remote sensing equipments which are able to detect the surface temperature by means of infrared radiation. These instruments are known as infrared thermometer (IRT). Inoue [6] pointed out advantages of IRT and also verified IRT performance to detect (Tc - Ta) and plant physiology depression in parameters such as photosynthesis rate, transpiration and stomata conductance in corn and wheat.

One of their first parameters utilized as a way to establish the crop water stress index was the well known stress degree day (SDD) which presented good results in arid regions. Later, the SDD was normalized for humid regions conditions [3].

Idso et al. [7] also presented an empirical method for dealing with the CWSI problem. However, the theoretical basis of the CWSI was studied by Jackson [1] who used Tc measurements obtained from IRT to predict ET_c, from a vegetated surface energy balance equations.

Several authors ([2], [4], [5], [8], [9], [10]) studied the CWSI developed by Jackson [1]. Grimes et al. [9], Silva [10] and Kobayashi [5] obtained CWSI values of about 0.15, 0.3 and 0.20 for alfalfa, cotton and beans, respectively, as limits above which crop water stress will begin. These limits could be used as reference for irrigation scheduling. Pitts et al. [4] and Folegatti [2] pointed out the occurrence of clouds as an important limitation for the use of the CWSI.

The objective of the present work was to obtain CWSI values for beans (*Phaseolus vulgaris* L.) submitted to five different irrigation frequency intervals (2, 4, 8, 12, and 16 days) under field condition, utilizing the IRT technique approach (Jackson [1]).

2 Material and methods

The field experiment was conducted at EMBRAPA, in Sete Lagoas, Minas Gerais State, Brazil, during the months of July through October of 1995.

The crop was planted in 26/07/95 with a crop density equivalent to a stand of about 220,000 plant per hectare. Completely randomized plots were used as statistical design, with 5 treatments and 4 replications. The treatments were formed with different irrigation intervals (2, 4, 8, 12, and 16 days) which started being applied 29 days after crop seedling (DAS), when the plants presented the third leaf emission.

The water volume to be added in each plot during irrigation was obtained by means of the soil water content determination (tensiometer and gravimetric) in the day of irrigation, in such a way as to bring the soil water level up to the "field capacity". The soil water holding capacity curve was used to carry out this computation.

Samples of the plants (3 for each plot) were collected weekly for determination of plant height (hc) and leaf area index (LAI). Whenever clear sky conditions occurred, at selected crop growth stages, readings of abaxial stomata resistance (r_s) were obtained from some well developed and lightened leaves (from 4 plants in each plot) between 10:00 and 12:00 hours, with a diffusion porometer. In the last sample, crop yield and yield components were evaluated.

A portable electronic weather station, installed 2 m above soil surface, provided the following local daily climatic data: average, maximum and minimum air temperature, rainfall, and global solar radiation (R_s). Also, a totalizing anemometer was installed 2 m above soil surface, close to the field trial. Continuous data registration were obtained from an analogical electronic diagram of net radiation (R_n), soil heat flux (G), and albedo ($\propto$). In order to carried out the R_n and $\propto$ measurements, a net radiometer and albedometer sensors were installed 1.5 m above soil surface in the treatment irrigated each 2 days (T2). Soil heat flux plates sensors were placed 2 cm below soil surface in the treatments irrigated each 2, 8, and 16 days (T2, T8, and T16 respectively) to obtain G readings. Other climatic parameters, such as Class A pan water evaporation, sunshine hours, atmospheric pressure, and air relative humidity were obtained from EMBRAPA's main standard Climatological Weather Station, located about 1500 m away of the field trial.

The canopy temperature (Tc) was taken daily, within two periods: from 10:00 to 11:00 hours and from 13:00 to 14:00 hours, always with under clear sky conditions. The following IRT was used: 1 ° C field of view, temperature range -30 to 300 ° C, precision of ± 1.5 ° C, sensibility of 0.1 ° C, response time of about 1 s, spectral band of 8 to 10 μm, and emissivity of 0.98. Readings were taken by keeping the IRT field of view at about 45 ° with the canopy surface and 2 m of distance from the target (leaves). Ten to twelve Tc readings were used in the average Tc calculation, and the variation coefficients were maintained within 2 to 3 % limits.

The aerodynamic resistance (r_a) was estimated through the equation presented by Allen *et al.* [11] and Jensen *et al.* [12]. Daily CWSI were obtained for morning and afternoon periods with the following equation ([13] and [1]):

$$CWSI = \frac{\gamma \cdot \left(1 + {}^{r_c}\!/\!_{r_a}\right) - \gamma^*}{\Delta + \gamma \cdot \left(1 + {}^{r_c}\!/\!_{r_a}\right)} \tag{2}$$

in which the symbol Δ represents the slope of the saturated vapor pressure-temperature relation (Pa / °C), and was estimated according to Jackson *et al.* [14] mathematical model. The term γ (in units of Pa / °C) is the psychrometric constant and was obtained as described in Brunt's procedure, according to Smith *et al.* [15]. The ratio r_c/r_a was determined according to Jackson [1], in terms of R_n, (Tc-Ta), vapor pressure deficit (VPD), and r_a. Afterwards, the r_c values were obtained from (Tc-Ta) and r_a daily values.

The crop situation in the maximum evapotranspiration condition (ET_m) was provided by Treatment 2, as well as the canopy resistance at ET_m (r_{cp}) and the modified psychrometric constant (γ^* in units of Pa / °C), according to the formula

$$y^* = y\left(1 + r_{cp}/r_a\right) \tag{3}$$

3 Results and discussion

All treatments received a total of 118.8 mm of applied water from seedling to 28 days after seedling (DAS), with 4-day average irrigation interval. Just after the 28 DAS, treatment differentiation started, with the application of pre-determined depths of water, according to the irrigation intervals. From seedling to harvest, the total amounts of applied water, including 66 mm of rainfall, were 482.0, 439.6, 385.6, 326.5, and 290.4 mm for T2, T4, T8, T12 and T16 treatments, respectively.

The crop water stress index results considered for analysis were from the growth period between the 29 and 80 DAS. Table 1 lists the crop yield and other yield components parameters. It was not observed statistical significance among the treatments for the number of pods per plant. The same behavior was observed for harvest index (HI) and grain yield, with no response for irrigation frequencies of 2, 4, and 8 days. Also no responses were verified when comparing 8 and 12 days, and 12 and 16 days irrigation intervals. Satisfactory results were obtained for HI, grain yield, number of seeds per pod, and the weight of 1000 seeds in the treatment T8. Results show that irrigation frequency beyond 8-day intervals decreased the magnitude of the studied parameters, causing injury to the crop.

Table 1 also shows strong water stress levels for T12 and T16, resulting in reduction of grain yield and yield components parameters. Grain yield was the most affected one.

Table 1. Number of pods per plant, number of seeds per pod, 1000-seed weigh, harvest index, and grain yield responses to different irrigation frequencies on beans (*Phaseolus vulgaris* L.)

Irrigation frequency (days)	Pod Number per plant	Seed number per pod	1000 seed weight (g)	Harvest index[1]	Grain yield[2] (Kg.ha^{-1})
2	11.4 a	4.8 a	239.6 a b	0.55 a	2755 a
4	11.1 a	4.7 a b	245.5 a	0.57 a	2750 a
8	9.5 a	4.0 b	241.6 a b	0.53 a b	2294 a b
12	7.4 a	3.3 c	259.4 a	0.46 b c	1583 b c
16	6.8 a	3.2 c	225.0 b	0.42 c	1057 c
Average	9.2	4.0	242.2	0.51	2088
VC	25.16	7.33	3.73	5.65	15.44

[1] Dry matter of the grain / Total dry matter (at the harvest)
[2] Values expressed for 13 % humidity
Means followed by the same letters do not present statistical significance by the Tukey test at 5 % probability level.
VC = Variation Coefficient

Assuming the grain yield of treatment T2 as the potential yield (no water stress condition), it was verified a reduction of 0.181, 16.7, 42.5, and 61.6 % in the grain weight caused water shortage in treatments T4, T8, T12, and T16, respectively.

Figures 1 and 2 show the crop water stress index variation with the crop growth period (expressed as days after seedling) at four irrigation frequencies (T4, T8, T12, and T16) and two measuring periods (morning and afternoon), respectively. Treatment T2 was assumed to have a crop with adequate water supply and maximum or potential evapotranspiration rate in the CWSI determination. The results show that T4 presented less oscillation in the CWSI values which were maintained close to the zero level. Therefore, four-day irrigation frequency did not cause water stress damage to the crop. This result can also be verified in the Table 1 where the crop yield and yield components behavior of T2 and T4 treatments are quite similar. By analyzing the two measuring periods (morning and afternoon), it can be noticed a similar tendency in the CWSI variation. But the highest CWSI values were obtained during afternoon readings. At first, this suggests that the best time for canopy temperature (Tc) reading is between 13:00 and 14:00 hours for CWSI studies.

The treatment that caused most water stress was T16, where the CWSI reached values of 0.26 and 0.40 for morning and afternoon periods, respectively. Table 1 results emphasize the damage caused to the crop.

Overall, the results show that a CWSI value of about 0.15 may be used as a limit for irrigation water management strategies to diferentiate the irrigated crop from a non-stressed to a stressed condition, in order to avoid signifficant yield loss.

This limit of 0.15 derives from the observation that T4 and T8 treatments presented maximum CWSI values of roughly 0.10 and 0.20, respectively. At T4, results indicate that the crop was also adequately supplied with water (no-stress, similar to T2). On the other hand, at T8 the crop may have experienced a certain degree of water stress. These results agree with the results from Kobayashi [5] for the same crop.

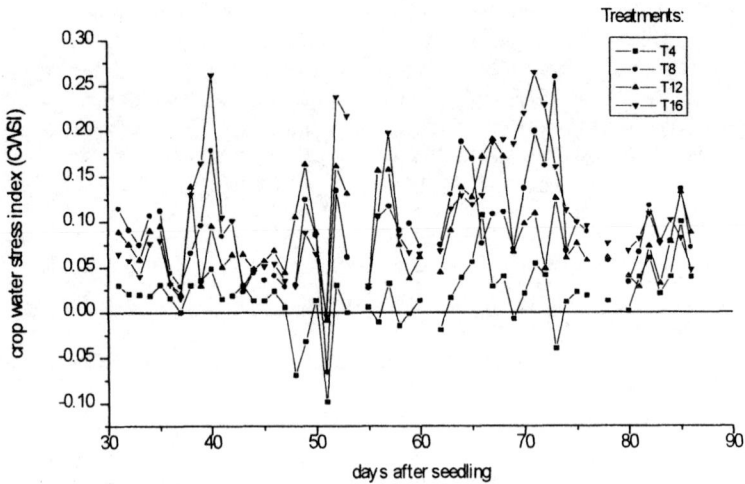

Fig. 1. Crop water stress index versus crop growth period (expressed as days after seedling) at four irrigation frequencies (T4, T8, T12, and T16) for measurements in the morning period (EMBRAPA, Sete Lagoas, MG, Brazil, 1995).

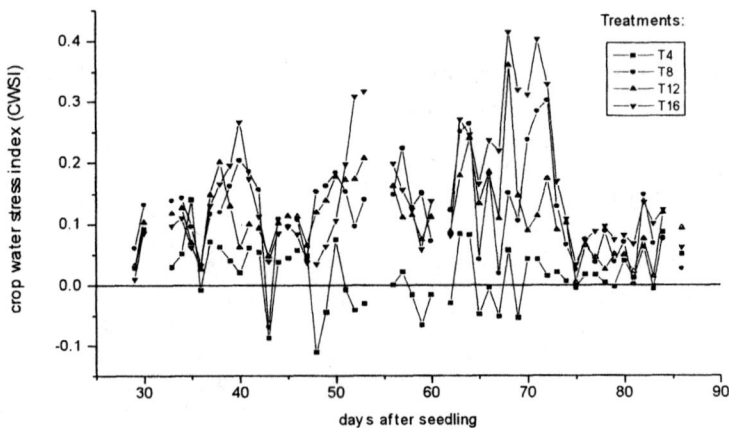

Fig. 2. Crop water stress index versus crop growth period (expressed as days after seedling) at four irrigation frequencies (T4, T8, T12, and T16) for measurements in the afternoon period (EMBRAPA, Sete Lagoas, MG, Brazil, 1995).

4 Conclusions

The potential grain yield (2,755 Kg.ha^{-1}) was obtained with the irrigation frequency of two days, under no water stress. A reduction of 0.18, 16.7, 42.5, and 61.6 % was verified in the grain weight due to treatments T4, T8, T12, and T16, respectively (tretments with water shortage). There was no statistical significance for grain yield among the 2-, 4-, and 8-day irrigation intervals.

The total depth of applied water, including 66 mm of rainfall, were of 482.0, 439.6, 385.6, 326.5, and 290.4 mm for 2-, 4-, 8-, 12- and 16-day of irrigation frequency, respectively.

It was verified a similar tendency of the CWSI variation in the periods of measurement (morning and afternoon). The highest CWSI values were obtained during afternoon readings, suggesting that as the best period for canopy temperature (Tc) measurements. The highest water stress was obtained with 16-day irrigation frequency, when the CWSI reached values of 0.26 and 0.40, for morning and afternoon periods, respectively.

A crop water stress index value of about 0.15 may be used as a limit for irrigation water management to differentiate the irrigated crop from a non-stressed to a stressed condition, as a criteria to avoid signifficant yield loss.

5 References

1.   1.Jackson, R.D. (1982) Canopy temperature and crop water stress. *Advances in irrigation.* New York, Vol.1, pp.43-85.
2.   Folegatti, M.V. (1988) Avaliação do desempenho de um "scheduler" na detecção do estresse hídrico em cultura do feijoeiro (Phaseolus vulgaris L.) irrigada com diferentes lâminas. Piracicaba: ESALQ, 1988. 188p. Thesis (Doctor in Agronomy) - Escola Superior de Agricultura "Luiz de Queiroz", Universidade de São Paulo, Brazil.
3.   Clawson, K.L., Jackson, R.D. and Pinter Jr., P.J. (1989) Evaluating plant water stress with canopy temperature differences. *Agronomy Journal.* Vol.81, pp.858-63.
4.   Pitts, D.J., Wright, R.E., Kimbrough, J.A. and Johnson, D.R. (1990) Furrow irrigated cotton on clayed soil in the Lower Mississipi River Valley. *Applied Engineering in Agriculture*, Vol.6, pp.446-52.
5.   Kobayashi, M.K. (1996) Determinação do índice de estresse hídrico da cultura do feijoeiro (Phaseolus vulgaris L.) por meio da termometria a infravermelho, e do fator de disponibilidade de água no solo, em minilisímetro de pesagem. Viçosa: UFV, 1996. 90p. Thesis (M.Sc. in Agricultural Engineering) - Universidade Federal de Viçosa, Brazil.
6.   Inoue, Y. (1991) Remote and visualized sensing of physiological depression in crop plants with infrared thermal imagery. *Japan Agricultural Research Quarterly*, Vol.25, pp.1-5.
7.   Idso, S.B., Jackson, R.D., Pinter Jr., P.J., Reginato, R.G. and Hatfield, J.L. (1981) Normalizing the stress-degree-day parameter for environmental variability. *Agricultural Meteorology*, Vol.24, pp.45-55.

8.   Andrews, P.K., Chalmers, D.J. and Moremong, M. (1992) Canopy-air temperature differences and soil water as predictors of water stress of apple trees grown in a humid, temperate climate. *Journal American Society Horticultural Science*, Vol.117, pp.453-8.

9.   Grimes, D.W., Wiley, P.L. and Sheesley, W.R. (1992) Alfalfa yield and plant water relations with variable irrigation. *Crop Science.*, Vol.32, pp.1381-7.

10. Silva, B.B. (1994) Estresse hídrico em algodoeiro herbáceo irrigado evidenciado pela termometria infravermelha. Campina Grande: UFPB, 1994. 139p. Thesis (Doctor in Civil Engineering) - Centro de Ciências e Tecnologia, Universidade Federal da Paraíba, Brazil.

11. Allen, R.G., Jensen, M.E., Wright, J.L. and Burman, R.D. (1989) Operational estimates of reference evapotranspiration. *Agronomy Journal*, Vol.81, pp.650-62.

12. Jensen, M.E., Burman, R.D. and Allen, R.G. (1990) *Evapotranspiration and irrigation water requirements.* New York: ASCE, 332p.

13. Jackson, R.D., Idso, S.B., Reginato, R.J. and Pinter Jr., P.J. (1981) Canopy temperature as a crop water stress indicator. *Water Resource Research*, Vol.17, pp.1133-8.

14. Jackson, R.D., Kustas, W.P. and Choudhury, B.J. (1988) A reexamination of the crop water stress index. *Irrigation Science*, Vol.9, pp.309-17.

15. Smith, M., Segeren, A., Pereira, L.S., Perrier, A. and Allen, R. (1991) *Report on the expert consultation on procedures for revision of FAO guidelines for prediction of crop water requirements.* Rome: FAO, 45p.

REAL-TIME IRRIGATION SCHEDULING MODEL FOR COTTON

Y.H. LI and B. DONG
College of Water Resources, Wuhan University of Hydraulic and Electric Engineering,
Wuhan, 430072, P.R.China

Abstract
This paper presents the principle and methods of computerizing irrigation scheduling of cotton fields in the Jianghan Plain of China where the groundwater table is shallow. According to the changing patterns of daily reference evapotranspiration (ET_0) estimated from the modified Penman Equation, models for forecasting daily ET_0 are introduced. Based on the experimental data from the Tianmen Irrigation and Drainage Experiment Station (TIDES) for many years, the methods for calculating crop coefficient and water stress coefficient are discussed. Another feature of the model concerns the accounting for the contribution from groundwater to the root zone of cotton. An empirical equation for estimating the capillary rise from groundwater table is suggested in line with the measured data. The computer program is compiled in Lotus 123 worksheets and is available as a menu-driven package. The application shows that the principle and mathematic models introduced in the paper are feasible and simple to use.
Keywords: Cotton, evapotranspiration, groundwater, irrigation scheduling, modeling, real-time.

1 Introduction

More and more attention is being paid to irrigation water management of cotton fields as cotton is one of the most important cash crop in China. The Jianghan Plain in the Yangtze Delta of China is a big base for planting cotton. Traditionally, cotton is grown without irrigation in this area because the rainfall is plenty and the groundwater table is shallow. However, the temperature is very high during the growing season of cotton, and the distribution of precipitation both in seasons and years is quite uneven. To raise the cotton yields, irrigation has been considered desirable, and the real-time irrigation

Water and the Environment: Innovative Issues in Irrigation and Drainage. Edited by Luis S. Pereira and John W. Gowing. Published in 1998 by E & FN Spon. ISBN 0 419 23710 0

scheduling for cotton fields is an effective way to reduce drain water disposal needs and improve the water use efficiency.

The real-time forecasting of cotton evapotranspiration (ET) is a key procedure for the irrigation scheduling of cotton fields. Usually, daily reference evapotranspiration (ET0) and crop coefficient (Kc) which form the basis of estimating daily ET are not predicted with real-time information [1], [2], [3], [4]. This will likely result in that the values of predicted ET are more or less than actual ET. Some previous studies [5], [6], [7], have indicated that ET0 could be forecasted according to the weather type and julian day number. Li and Cui [7] used the percentage of green leaf cover (CC) to estimate Kc of rice. But there is no available model for forecasting ET of cotton in the area.

Shallow groundwater uptake (Gu) is an important component of water balance in cotton fields [8]; however, many irrigation scheduling models do not take into account groundwater contributions to crop water use [9], [10], [1], [2]. Li [11]; Teixeira et al. [4] and Danuso et al. [12] combined Gu into soil water balance simulation models for irrigation scheduling. But the effect of either ET or groundwater depth on Gu was not considered in their approach. This paper introduces mathematic models for forecasting daily ET of cotton and Gu based on experimental data and presents a real-time irrigation scheduling model for cotton.

2 Materials and methods

2.1 Irrigation experiments
Cotton irrigation experiments have been done since 1963 at the TIDES, Hubei Province. 16 non-weighing type concrete lysimeters (2m wide by 3.33m long, 2.7m high) were constructed to accommodate 4 groundwater depths; i.e., 50-75cm depth, 75-100cm depth, 100-140cm depth and without free water surface of groundwater. 4 lysimeters with the same sizes but without bottom have the same groundwater depths as the cotton fields around the station. The groundwater depths are measured in the underground corridor. CC is measured by an illuminometer.

2.2 The software and hardware
The program was written using the macro programming language of Lotus 123. The main menus are Input, Requirements, Schedule, Graph, Output and Exit. Each main menu has some options forming the menu tree. Once the first letter of an option is chosen, the relevant subroutine is invoked. The real-time irrigation scheduling program for cotton has the following characteristics:

1. it allows up to 300 blocks of cotton to be modelled at the same time;
2. it is menu-driven with available options and their meaning on the screen;
3. all factors related to the daily water balance for each field and all irrigation events are kept in the worksheets over a whole irrigation season;
4. multi worksheets are used for the macro program, data tables, calculation forms and output forms for ease of checking;
5. there is appropriate use of graphics to check and monitor results;
6. parameters which vary between different districts are not fixed, so the program could have multiple uses.

To use the program friendly, a version of Lotus 123 Release 4 for Windows 3.1, or later, in Chinese must be installed on the computer. The computer should have a minimum of 1 MB of RAM, consisting of a combination of 640 KB conventional memory, a hard disc and a colour monitor.

3 Forecasting of cotton evapotranspiration

The relationship between ET and its influencing factors can be expressed as Equation 1 under either available water supply or water deficit conditions [5], [13]:

$$ET = f_1(S) \times f_2(P) \times f_3(A) \tag{1}$$

where ET is actual cotton evapotranspiration (mm d-1), $f_1(S)$ is soil factor function, $f_2(P)$ is plant factor function and $f_3(A)$ is meteorological factor function.

3.1 Forecasting of reference evapotranspiration

For the effects of meteorological factors $f_3(A)$, ET0 calculated by the Modified Penman Equation is the most suitable parameter [5], [13], [14]. Daily ET_0 is estimated from Equation 2 (Smith et al., 1992):

$$ET_{oi} = \frac{0.408 \Delta_i(R_{ni} - G_i) + 900\gamma U_{2i}(e_{ai} - e_{di})/(T_i + 273)}{\Delta_i + \gamma(1 + 0.34 U_{2i})} \tag{2}$$

where ET_{0i} is reference crop evapotranspiration on day i (mm d^{-1}), i is julian day number, Δ_i is slop vapour pressure curve on day i (KPa °C^{-1}), Rni is net radiation at crop surface on day i (MJ $m^{-2} d^{-1}$), Gi is soil heat flux on day i (MJ $m^{-2} d^{-1}$), γ is psychrometric constant (KPa °C^{-1}), U_{2i} is windspeed measured at 2m height on day i (m s^{-1}), e_{ai} is saturation vapour pressure at temperature T_i (KPa), e_{di} is actual vapour on day i (KPa), T_i is average temperature on day i (°C).

However, the Equation 2 can not be used to forecast daily ET_{0i} since some required meteorological factors can not always be predicted. Based on the changing patterns of ET_0, daily ET_{0i} can be predicted from [6]:

$$f_3(A) = \phi_i ET_{o\max} \exp(-((i - i_{\max})/A_0)^2) \tag{3}$$

where ϕ_i is a weather type coefficient on day i obtained from the ratios of actual ET_0 under clear, cloudy, overcast and rainy weather conditions to the long term mean values on the same day. Table 1 gives values of ϕ_i in Tianmen cotton planting district, Hubei Province, $ET_{0\max}$ is a mean value of the maximum ET_0 over the years and equal to 5.12 mm d^{-1} in the district, $i_{\max}$ is the julian day number on which ET_0 reaches $ET_{0\max}$ related to latitude. $i_{\max}$ can be set equal to 197 in most regions and to 180 in high latitude regions in China [6], A_0 is an empirical parameter related to latitude and the julian day number. Mao et al. [6] and Li and Cui [7] reported that A_0 could be set equal to 97, 112, 130 and 150 in high latitude regions, northern, central and southern China, respectively.

As A_o suggests the gradient of ET_{0i} with the day number i, A^o changes along with i and can be obtained from:

$$A_0 = \frac{|i - i_{max}|}{\sqrt{\ln ET_{omax} - \ln ET_{oavr}}}$$

(4)

Table 1. Values of weather type cofficient ϕ in Tianmen, China

Weather	Term[a]	Apr.	May	Jun.	Jul.	Aug.	Sept.	Oct.
Clear	1		1.50	1.47	1.48	1.37	1.29	1.35	1.56
	2		1.47	1.43	1.46	1.32	1.29	1.36	1.53
	3		1.44	1.43	1.40	1.23	1.34	1.35	1.41
Cloudy	1	1.22	1.25	1.21	1.16	1.04	1.12	1.19	
	2		1.18	1.33	1.18	1.06	1.08	1.29	1.26
	3		1.08	1.19	1.10	1.01	1.09	1.03	1.01
Overcast	1		0.93	0.87	0.88	0.91	0.84	0.75	0.87
	2		0.83	0.80	0.84	0.86	0.85	0.72	0.91
	3		0.86	0.84	0.87	0.88	0.83	0.77	0.83
Rain	1		0.76	0.75	0.75	0.61	0.69	0.64	0.71
	2		0.73	0.69	0.76	0.62	0.78	0.64	0.77
	3		0.77	0.75	0.70	0.74	0.68	0.59	0.64

[a] 1,2 and 3 refer to the first, second and third 10-day terms in a month, respectively.

where ET_{0avr} is the long term mean value of ET_0 over years on day i (mm [d-1]).

Ao, imax, ET_{0max} and ϕ_i can be obtained according to measured meteorological data. Then ET_{0i} can be forecasted once the julian day number i is determined and the weather type on that day predicted.

3.2 Crop coefficient

Usually, the crop coefficient (Kc) represents the contributions of plant factors to ET or $f_2(P)$. For real-time forecasting of ET, the leaf area index (Mao, 1994) or percentage of crop cover (Li and Parkes, 1993) was used to calculate Kc. According to the measured data, Kc of cotton has a good relationship with CC (Figure 1). Therefore, $f_2(P)$ can be expressed as:

$$f_2(P) = Kc_i = a + bx\, CC_i^n$$

(5)

where Kc_i is the crop coefficient of cotton on day i, a, b, n are empirical values related to crop type and can be set equal to 0.507, 2.32×10^{-5} and 2.27, respectively, for cotton on the basis of simulating analysis of the experimental data and CC_i is the percentage of cotton green leaf cover on day i (%), estimated from:

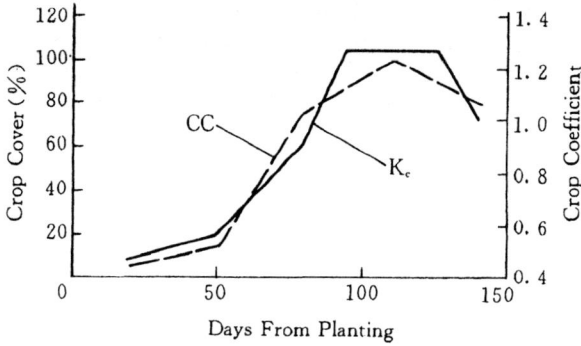

Fig. 1. Measured CC and calculated Kc of cotton at TIDES

$$CC_i = CC_o + (CC_t - CC_o)i_r /T \tag{6}$$

where CC_o is the percentage of green leaf cover on the initial day and sent to the computer operator by field observers as real-time information (%), CC_t is the percentage of green leaf cover on day T (%) and predicted in line with the variation of cotton cover with time and modified continually according to CC_o, i_r is the number of days from the initial day to day i and T is the number of days from the initial day to the day on which cotton cover percentage equals to CC_t.

3.3 Moisture stress coefficient

The effects of soil factors or $f_1(S)$ is considered according to the moisture status within cotton root zone [13]:

$$f_1(S) = 1 \qquad\qquad \text{if } \Theta_i \geq \Theta_f \tag{7}$$

$$\text{or } f_1(S) = \ln(1 + \Theta_i)/\ln 101 \qquad \text{if } \Theta_c \leq \Theta_i < \Theta_f \tag{8}$$

$$\text{or } f_1(S) = \alpha \exp((\Theta_i - \Theta_c)/\Theta_c) \quad \text{if } \Theta_i < \Theta_c \tag{9}$$

where Θ_i is the actual moisture content on day i as a percentage of field capacity (%), Θ_f is the field capacity in volumetric percentage (vol.%), Θ_c is a critical moisture content for soil stress and set equal to 60% for cotton, α is an empirical coefficient and set equal to 0.89 for cotton.

4 Shallow groundwater contribution

Table 2 gives measured values of shallow groundwater uptake (Gu) which indicates that Gu is strongly influenced by groundwater depth and ET. With same groundwater depth, Gu increases as ET does. In the same year, Gu decreases when groundwater table falls. Li [11] and Mao et al. [6] claimed that Gu was also affected by soil texture since the potential capillary rise from groundwater is different. Based on the simulation of measured data, Gu can be estimated from:

$$Gu_i = ET_i \exp(- h_i) \tag{10}$$

where Gu_i is the shallow groundwater uptake on day i (mm d^{-1}), β is an empirical value related to soil texture and can be set equal to 2.1, 2.0 and 1.9, respectively, for sand, loam and clay, h_i is the groundwater depth on day i (m).

5 Application

After validated by experimental data, the program was used to model the moisture contents and forecast the irrigation water requirements of cotton fields in the Tianmen cotton planting district, Jianghan Plain, China. Before irrigation scheduling, the field file was built up. The field file includes location, topography, soil texture, planting date, area of all representative blocks and other parameters related to water balance. The real-time information such as forecasted weather types, rainfall, CC_o, irrigation depth and groundwater depth was reported by meteorological and field observers on the morning of first day of each 10 day period. After the moisture contents are revised using measured meteorological data, crop cover, rainfall, groundwater depth and actual irrigation depth, water balance is performed day by day from initial day and block by block. Once heavy rainfall or a full irrigation appears, the modelled moisture contents are modified. Figure 2 shows the predicted and measured moisture contents in a typical block near by the TIDES.

Table 2. Values of Gu (mm) under different conditions at TIDES, China

Groundwater depth	50 - 75 cm			75 -100 cm			100 - 140 cm		
Year[b]	1991	1995	1981	1991	1995	1981	1991	1995	1981
May	20.3	39.4	51.7	18.0	22.2	45.1	13.4	18.8	31.8
Jun	54.4	46.1	56.0	48.3	42.8	48.9	35.7	44.7	34.5
Jul.	69.6	95.9	121.1	61.8	87.0	105.7	45.8	63.7	74.5
Aug.	62.8	77.7	94.3	55.7	66.1	82.3	41.2	46.0	58.1
Sep.	47.0	28.9	38.3	41.8	33.2	33.5	30.9	28.2	23.6
Whole season	254.1	288.0	361.4	225.6	251.3	315.5	167.0	201.4	222.5

[b] in 1991, 1995 and 1981, it was wet, moderate and dry, and ET of cotton in whole season was 558.8 mm, 609.5 mm and 650.9mm respectively.

6 Conclusions

Not only does the program which was originally designed for real-time irrigation scheduling satisfy the demand to improve the efficiency of irrigation, but it is also beneficial for drainage of cotton fields because moisture contents are modelled. The mathematic models and the program introduced in this paper focus on improving the reliability and predictability of water use, and required inputs are simple and may easily be collected. Before irrigation scheduling, empirical parameters in models should be obtained on the basis of simulating analysis of measured data over years. Under shallow groundwater conditions, the factors influencing Gu are very complicated, but Gu can be estimated according to ET, groundwater depth and the soil texture. The case study shows that the principle and methods proposed are feasible and rigorous and the computer program is easy to apply.

7 Acknowledgements

This research is part of project Upland Crops Irrigation funded by Hubei Provincial Scientific and Technical Committee. Sincere appreciation is due to the help and collaboration of the TIDES.

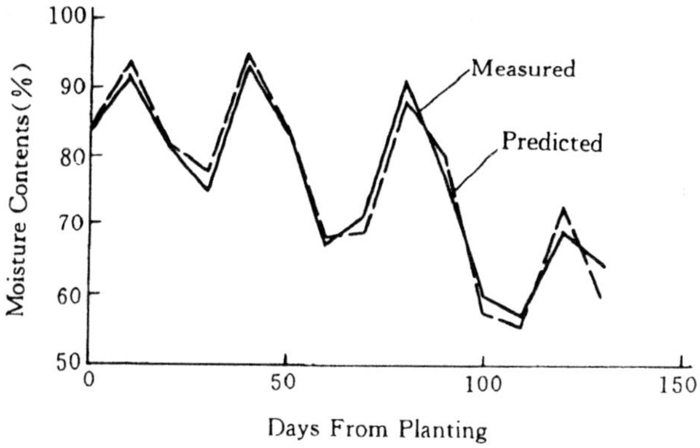

Fig. 2 Measured and predicted Θ_i in a typical field near by TIDES

8 References

1. Spark, P. and Makin, I.W. (1992) A practical guide to irrigation water scheduling. Rep. OD/TN 61, Hydraulics Research, Wallingford, UK, 28 pp. (Unpublished.)
2. Li, Y.H. and Parkes, M.E. (1993) *Revised irrigation scheduling program user guide.* Dep. Note 65, Scottish Centre of Agricultural Engineering, 24 pp.
3. Buchleiter, G.W. (1995) Improved irrigation management under center pivots with SCHED. In: L.S. Pereira, B.J. van den Broek, P. Kabat and R.G. Allen (Editors), *Crop-Water-Simulation Models in Practice.* Wageningen Pers, The Netherlands, pp. 27-47.
4. Teixeira, J.L., Fernando, R.M. and Pereira, L.S. (1995) RELREG: a model for real time irrigation scheduling. In: L.S. Pereira, B.J. van den Broek, P. Kabat and R.G. Allen (Editors), *Crop-Water-Simulation Models in Practice.* Wageningen Pers, The Netherlands, pp. 3-15.
5. Mao, Z. (1994) Forecast of crop evapotranspiration. *ICID Bulletin.*, 43(1): 23-36.
6. Mao, Z., Li, Y.H. and Li, H.C. (1995) Study of mathematic model for forecasting daily crop evapotranspiration. *J. Wuhan Univ. Hydraul. Electric Eng.*, 28(3): 253-

259. (In Chinese, with English abstract.)

7.   Li, Y.H. and Cui, Y.L. (1996) Real-time forecasting of irrigation water requirements of paddy fields. *Agric. Water Manage.*, 31(3): 185-193.

8.   Hutmacher, R.B., Ayars, J.E., Vail, S.S., Bravo, A.D., Dettinger, D. and Schoneman, R.A. (1996) Uptake of shallow groundwater by cotton: growth stage, groundwater salinity effects in column lysimeters. *Agric. Water Manage.*, 31(3): 205-223.

9.   Wallender, W.W., Grimes, D.W., Henderson, D.W. and Stromberg, L.K. (1979) Estimating the contribution of a perched water table to the seasonal evapotranspiration of cotton. *Agron. J.*, 71: 1056-1060.

10. Hutmacher, R.B. and Ayars, J.E. (1991) Managing shallow groundwater in arid irrigated areas. Paper No.912119, ASAE International meeting, Albuquerque, N.M., 17 pp.

11. Li, Y.H. (1994) Method and application of real-time irrigation forecast. *J. Hydraul. Eng.*, 2: 46-51. (In Chinese, with English abstract.)

12. Danuso, F., Gani, M. and Giovanardi, R. (1995) Field water balance: BIdriCo 2. In: L.S. Pereira, B.J. van den Broek, P. Kabat and R.G. Allen (Editors), *Crop-Water-Simulation Models in Practice*.Wageningen Pers, The Netherlands, pp. 49-73.

13. Li, Y.H., Zhang, M.Z., Yuan, W.C., Li, H.C. and Zhang, H.J. (1994) Patterns and calculation of evapotranspiration of upland crops under insufficient irrigation conditions. *J. Wuhan Univ. Hydraul. Electric Eng.*, 27(5): 506-512. (In Chinese, with English abstract.)

14. Smith, M., Allen, R.G., Monteith, J.L., Perrier, A., Pereira, L.S. and Segeren, A. (1992) *Report on the expert consultation on procedures for revision of FAO guidelines for predicting of crop water requirements*. UN-FAO, Rome, Italy, 54 pp.

USING SAP FLOW MEASUREMENTS TO QUANTIFY WATER CONSUMPTION IN THE OLIVE TREE

M.J. PALOMO, A. DÍAZ-ESPEJO, J.E. FERNÁNDEZ, I.F. GIRÓN and F. MORENO
Instituto de Recursos Naturales y Agrobiología, Apartado 1052, 41080 Seville, Spain

Abstract
We used the compensation heat-pulse technique to measure sap flows in 28-year-old olive trees, in order to determine their transpiration fluxes in real time. Measurements were made in irrigated and non-irrigated trees. Evaporation from the soil was estimated on days of high atmospheric demand by using an empirical equation obtained for the orchard conditions. Crop evapotranspiration was then determined from daily data of transpiration and evaporation (ET_c-2). Results were compared with the crop evapotranspiration (ET_c-1) calculated from the potential evapotranspiration (ET_o) in the area, corrected by reduction coefficients (K_c and K_r). Our aim was to carry out a preliminary study of a new method for calculating crop evapotranspiration, designed to avoid the errors derived from the scarce information on K_c and K_r, and for the lack of temporal resolution in the calculation of ET_o. For irrigated trees, ET_c-2 was 80% of ET_c-1. For non-irrigated trees, ET_c-2 was only 43% of ET_c-1, due to the lack of water in the soil. A good agreement between transpiration fluxes and environmental conditions was observed.
Keywords: Evapotranspiration, drip irrigation, olive tree, sap flow, water consumption.

1 Introduction

The olive tree is usually grown in areas where water for irrigation is scarce. This, together with the spectacular increment in irrigated area of this crop in the last few years, makes it crucial to improve the calculation of the crop water needs. The conventional method for calculating the crop evapotranspiration (ET_c) is based on the potential evapotranspiration in the area (ET_o), corrected by a crop coefficient (K_c) and a reduction coefficient (K_r) related to the degree of orchard floor plant cover [1] [2].

Water and the Environment: Innovative Issues in Irrigation and Drainage. Edited by Luis S. Pereira and John W. Gowing. Published in 1998 by E & FN Spon. ISBN 0 419 23710 0

We have denominated ET_c-1 the ET_c so calculated. This method, despite being widely accepted, has two main disadvantages. On one hand, the information on K_c and K_r is scarce. Consequently, the published values of both coefficients are often used in orchards with conditions different from those under which the coefficients were obtained. On the other, ET_o is usually calculated for a period of several days, normally 7 to 10, and the average daily value is used in determining the irrigation dose for the next period. Thus, daily variations are not considered, and there is a delay in the application of ET_o data. Therefore, any research oriented towards minimizing those errors, or towards developing new methods to improve the calculation of the ET_c, is welcome not only by the farmers, but also by social sectors concerned by the high amounts of water spent in agriculture.

Apart from the method described to calculate ET_c-1, there are other possibilities for determining water consumption in a fruit orchard. Soil water balance was used by Fernández [3] in an olive orchard, but, once again, the method proved to be time-consuming and of low temporal resolution, apart from the uncertainties in the estimation of drainage and surface runoff. Micrometeorological methods, such as the Bowen-ratio technique and eddy covariance can be used to determine ET_c with better temporal resolution, but these methods are complex, require expensive equipment and well trained people, and they can be used only in flat, uniform areas [4]. A promising approach is the use of sap flow measurements to obtain real-time data on water consumption. Cohen et al. [5], Smith and Allen [4] and Braun [6] have published comprehensive reviews on the topic. In the olive tree, sap flow measurements were first made by Moreno et al. [7] and Fernández et al. [8]. They used the compensation heat-pulse technique (CHP technique) [9] to measure sap flows both in the trunk and in main roots of mature olive trees under field conditions. These experiments proved the CHP technique a suitable tool to determine the water consumed by olive trees, which opened the door to calculating ET_c in olive orchards from independent measurements of transpiration (T) and evaporation (E) fluxes.

The objective of this work was to calculate the evapotranspiration of single mature olive trees under different water regimes, by measuring their T fluxes and the E fluxes from the corresponding soil surface. The CHP technique was used for determining T, and E was determined by using an empirical equation obtained for the orchard conditions. We denominated ET_c-2 the crop evapotranspiration so calculated. The agreement between ET_c-1 and ET_c-2 data was analysed.

2 Materials and Methods

The experiments were carried out at the experimental farm of the Instituto de Recursos Naturales y Agrobiología, in Seville, Spain (37° 17' N, 6° 3' W, elevation 30 m). The 0.5 ha experimental orchard was planted in 28-year-old olive trees (*Olea europaea* L., var. *Manzanillo*) at a spacing of 7 × 5 m. The soil is a sandy loam of about 1.5 m depth, with 73.5% coarse sand, 4.7% fine sand, 7.0% silt and 14.8% clay. The volumetric soil water content is 0.32 m^3 m^{-3} for a soil matric potential of 0 MPa, and 0.09 m^3 m^{-3} for 1.5 MPa. Two water regimes were imposed during the experiments: Treatment I was a daily drip irrigation to replace the crop water demand (ET_c, mm). We used a single pipe placed on the soil surface in each tree row, with five 3 L h^{-1}

emitters 1 m apart per tree. Water (418.3 mm) was applied throughout the dry season of 1997, from the middle of March to the beginning of October. The ET_c was calculated by the equation:

$$ET_c = K_c\, K_r\, ET_o \tag{1}$$

where K_c is the crop coefficient obtained by Pastor and Orgaz [1] increased by 0.05 (0.7 in March; 0.65 in April and October; 0.6 in May and September; 0.55 in June; and 0.5 in July and August), which has been proved adequate for the experimental orchard [10]. K_r is the coefficient relating the degree of orchard floor plant cover with the evapotranspiration [11]; in our case, $K_r = 0.7$, since the trees covered 35% of the orchard floor. ET_o (mm) is the potential evapotranspiration as calculated by the FAO-Penman equation, which Mantovani et al. [12] evaluated as the best for the area. We have denominated ET_c-1 the ET_c calculated by Eq. (1), as is mentioned in the Introduction. Treatment D was dry farming, with the rainfall as the only source of water supply (138.5 mm).

Transpiration fluxes were determined every half hour in one representative tree per treatment. To do so, we monitored sap flows by the CHP technique [9]. The technique was calibrated for the olive tree by Fernández et al. [13]. We installed three sets of probes in the trunk of each tree, equally distributed around the trunk, and at about 40-50 cm from the ground. Each probe was designed to monitor sap velocities at four different depths below the cambium: 5, 12, 22 and 35 mm. The probes were installed on July 16 and remained in the trees until October 10. The data of each tree were collected with a CR10X Campbell Scientific data logger, and processed with a computer programme developed by S.R. Green and B.E. Clothier, (Hort+Research, Palmerston North, New Zealand) who also made the probes and associated electronics. The system was powered with a 92 A battery.

In each tree, the single trunk divided in two main branches at about 1.2 m from the ground. The trunk diameter at the height of the probes was about 200 mm for both trees. We normalized the transpired amounts of each tree by dividing it by the leaf area of the tree and multiplying the result by the leaf area of an average tree. The leaf area of each tree was determined by counting the number of leaves of one main sector of the tree, determining the area of the leaves from leaf samples measured with a leaf area meter (SKYE Instruments, Ltd. UK), and then estimating how many sectors were on the tree. The leaf areas were 61.2 m^2 for the I tree and 64.4 m^2 for the D tree.

Additional transpiration measurements were made in October, after the dry season, to determine the effect of the changes in meteorological conditions on the transpiration of the olive tree. One tree of the orchard, called tree W, was instrumented with the CHP technique as described above, and transpiration measurements were carried out during several days of October, in which marked changes on meteorological conditions took place due to rainy events alternating with sunny, dry days. The tree was well irrigated during the measurements, to avoid the influence of lack of water in the soil.

To determine the evaporation fluxes from the soil surface of the experimental orchard, we obtained an equation from the data of two experiments. Experiment 1 was carried out in July 1995, on typically hot, dry summer days (average $ET_o = 8.0$ mm d^{-1}).

We inserted a set of microlysimeters randomly in the soil around one tree of the experimental orchard. Microlysimeters were made with 80 mm long pieces of an 80 mm diameter methacrylate tube of 3 mm wall. After installation of the instruments, the tree was abundantly irrigated by applying water in a 2.5 m radius pond, covering the area in which the microlysimeters were installed. Early in the morning of the next day, three microlysimeters were taken out, their bottom wrapped in plastic, weighted with an electronic weighing machine (10^{-2} g precision), and re-inserted in the soil. Twenty-four hours later, the same microlysimeters were re-weighed, and the water evaporated from their open surface was determined as the difference in weight. This was repeated for five consecutive days, using different microlysimeters each day. Experiment 2 was carried out by Moreno et al. [14]. They used the internal drainage method described by Hillel [15] to determine the hydraulic characteristics of the soil in the experimental orchard. From the data they collected on days of similar ET_0 to those in experiment 1, we obtained the values of E for the lowest range of volumetric soil water content (θ, cm^3 cm^{-3}) in the top 80 mm layer.

For our estimations of E, we measured the θ values of the top layer of the soil corresponding to the D and I trees by TDR with a cable texter (Tektronix 1502, Oregon, USA). In the drip-irrigated tree, we measured θ in locations affected and un-affected by the irrigation, and the wetted and dry areas of soil were taken into account in the calculations of E.

Soil matric potential head (h, MPa) and θ were monitored in the trees in which sap flow was measured. A set of five mercury tensiometers (at 0.30, 0.60, 0.85, 1.30 and 1.50 m depth) was installed in the soil of each tree, 1 m away from the trunk and close to the irrigation pipe. Tensiometric readings were taken every 3-4 days throughout the irrigation period. On each tree, four access tubes for a neutron probe (Troxler 3300, Research Triangle Park, North Carolina, USA) were also installed, at 0.5, 1.5, 2.5 and 3.5 m away from the trunk, parallel to the irrigation pipe and close to it. The volumetric soil water content was measured from 0.2 m down to 1.5 m every 0.1 m. For the top soil layer, θ was measured by gravimetry. Measurements were made every 7-10 days throughout the irrigation period. Soil moisture profiles were used to calculate a depth equivalent of water, expressed as the level of relative extractable water (REW, mm), defined by the equation [16]:

$$REW = (R - R_{min}) / (R_{max} - R_{min}) \qquad (2)$$

where R (mm) is the actual soil water content, R_{min} (mm) is the minimum soil water content measured during the experiments, and R_{max} (mm) is the soil water content at field capacity. In treatment I, REW values exceeded 0.9 for most of the irrigation period, and h values exceeded over -0.02 MPa at all the depths. In treatment D, the REW and the h values decreased throughout the dry season, REW being < 0.4 from June 20. Fernández et al. [10] found that a REW $\leq$ 0.4 can be considered a threshold for soil water deficit. Tensiometers were out of range in the treatment D by the end of June.

Weather variables were measured with the automatic weather station of the experimental farm. One-hour averages of temperature and relative humidity of the air, wind speed, rainfall, global solar radiation and photosynthetically active radiation were recorded and used to calculate ET_0 by the FAO-Penman equation [17], vapour pressure deficit of the air (D_a, kPa), and photon flux density (I_P, μmol m^{-2} s^{-1}).

3 Results and discussion

The CHP technique showed a satisfactory behaviour when working under field conditions for long periods of time. The system worked continuously for the three months the probes were inserted in the trees. The batteries needed to be recharged every 15 days, and with the same frequency, the information in the data logger was downloaded to a portable computer, though the memory capacity of the data logger allowed longer periods of time. The system, therefore, proved to be robust and to require little maintenance. Figure 1 shows an example of the transpiration fluxes recorded in each tree, on hot, sunny days. On such days, the transpiration of the D tree was much lower than the transpiration of the I tree, due to the lack of water in the soil. Figure 2 shows how the transpiration of the W tree changed with meteorological conditions. From DOY 291, the weather became increasingly unstable, with 13 and 7 mm of rainfall on DOY 293 and 294 respectively. Recorded daily transpiration fluxes followed the same pattern as that of atmospheric demand, shown in Figure 2 by the daily changes of D_a and I_P.

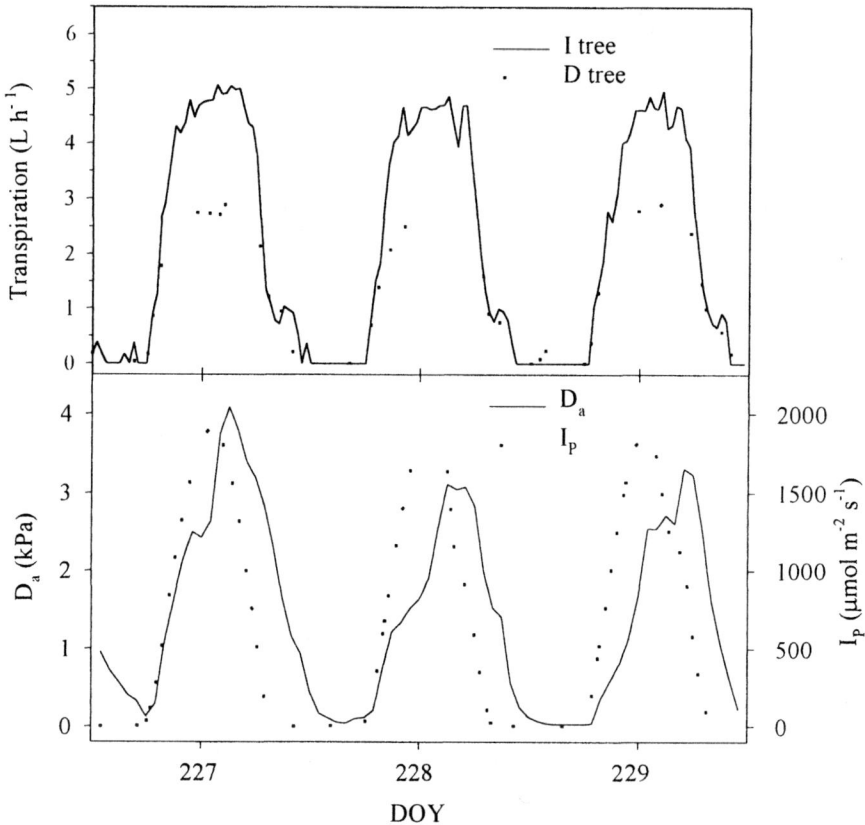

Fig. 1 Transpiration fluxes measured in august in the irrigated (I) and in the non-irrigated tree (D), on typical summer days (D_a = vapour pressure deficit of the air; I_p = photon flux density).

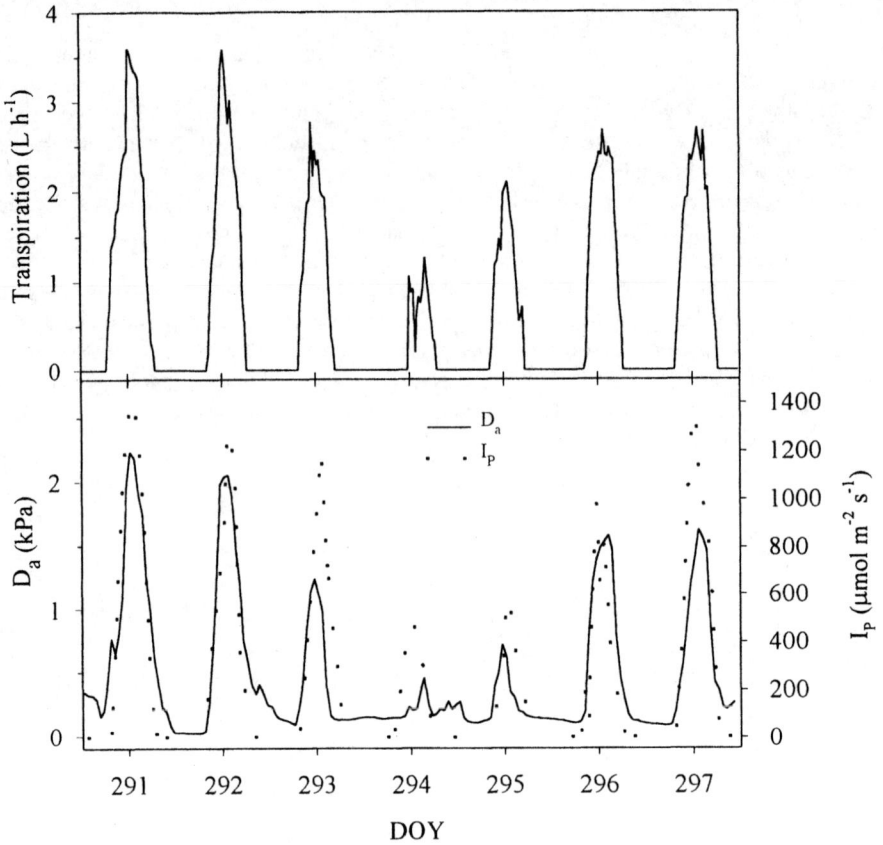

Fig. 2. Transpiration fluxes measured in October in the W tree (see Materials and Methods for details), on days with different meteorological conditions (D_a = vapour pressure deficit of the air; I_p = photon flux density)

Figure 3 shows the daily evaporation values recorded in the orchard on typically hot, dry summer days (average ET_o = 8 mm d^{-1}), for a wide range of humidity of the top 80 mm soil layer. The E values from experiments 1 and 2 fitted well (r^2 = 0.995) the exponential relationship E = 0.057 e$^{0.2\ \theta}$. We used this exponential relationship, together with measurements by gravimetry of θ in the top soil layer, to calculate the evaporation fluxes from the orchard soil on days when the atmospheric demand was similar to that of those on which the evaporation experiments were carried out. The E values were then used, together with the values of T measured on each tree, to calculate ET_c-2. Table 1 shows the total values of ET_c-1 and ET_c-2 measured on those days. For the I tree, ET_c-2 amounted to 80% of ET_c-1. For the D tree, ET_c-2 accounted for only 43% of ET_c-1, due to the lack of water in the soil. We cannot say, however, that ET_c-2 underestimated crop evapotranspiration, since the errors involved in the calculation of ET_c-1, mentioned in the Introduction, may be partly responsible for the difference between ET_c-1 and ET_c-2. In addition, headlosses caused that the

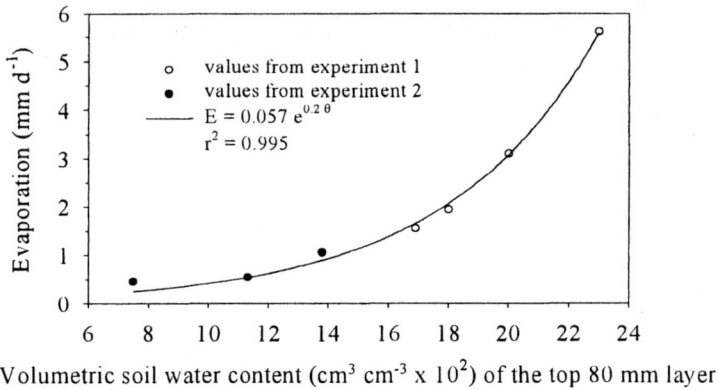

Fig. 3. Relationship between evaporation from the soil surface (E) and volumetric soil water content of the top layer (θ), determined from the evaporation values of experiments 1 and 2 (see Materials and Methods for details)

total amount of water applied was only 92% of ET_c-1. Part of this difference may also be due to difficulties in the integration process of the sap velocities determined at different depths below the cambium under conditions of water stress, as mentioned by Fernández et al. [13]. They calibrated the CHP technique for the olive tree. Before noon, they observed a good agreement between the transpiration fluxes estimated by the CHP technique and the actual transpiration fluxes. After noon, however, when the water stress of the tree was greater, estimated transpiration fluxes were lower than the actual ones. The authors suggested that some modification in the integration process might improve the results of the CHP technique when used in water-stressed trees.

Table 1. Comparison of the ET_c-1 with the ET_c-2 (see text for details) determined for both treatments on seven days in which soil evaporation was calculated ($L\ t^{-1}$ = Liters per tree)

Treatment	ET_c-1 ($L\ t^{-1}$)	ET_c-2 ($L\ t^{-1}$)	(ET_c-2 /ET_c-1) 100
I	607.5	485.2	80.0
D	561.0	242.1	43.2

4 References

1. Pastor, M. and Orgaz, F. (1994) Riego deficitario del olivar. *Agricultura*, No. 746. pp. 768-76.
2. Orgaz, F. and Fereres, E. (1997) Riego, in *El Cultivo del Olivo* (eds. D. Barranco, R. Fernández-Escobar and L. Rallo), Ediciones Mundi-Prensa y Junta de Andalucía, pp. 253-72.
3. Fernández J.E. (1989) Comportamiento del olivo (*Olea europaea* L., var. Manzanillo) sometido a distintos regímenes hídricos, con especial referencia a la dinámica del sistema radicular y de la transpiración. Ph. D. Thesis, Department of Agronomy, University of Córdoba, 71 pp.

4. Smith D.M. and Allen S.J. (1996) Measurement of sap flow in plant stems. *Journal of Experimental Botany*, No. 47. pp. 1833-44.

5. Cohen, Y., Takeuchi, S., Nozaka, J. and Yano, T. (1993) Accuracy of sap flow measurement using heat balance and heat pulse methods. *Agronomy Journal*, No. 85. pp. 1080-86.

6. Braun, P. (1997) Sap flow measurements in fruit trees - Advantages and shortfalls of currently used systems. *Acta Horticulturae*, Vol I, No. 449. pp. 267-72.

7. Moreno, F., Fernández, J.E., Clothier, B.E. and Green, S.R. (1996) Transpiration and root water uptake by olive trees. *Plant and Soil*. No. 184. pp. 85-96.

8. Fernández, J.E., Moreno, F., Clothier, B.E. and Green, S.R. (1996) Aplicación de la técnica de compensación de pulso de calor a la medida del flujo de savia en olivo. *Abstracts of the XIV Congreso Nacional de Riegos*, pp. 1-8. Aguadulce (Almería), Spain, 11-13 of June.

9. Green, S.R. and Clothier, B.E. (1988) Water use of kiwifruit vines and apple trees by the heat-pulse technique. *Journal of Experimental Botany*. No. 39. pp. 115-23.

10. Fernández, J.E., Moreno, F., Girón, I.F. and Blázquez, O.M. (1997) Stomatal control of water use in olive tree leaves. *Plant and Soil*, No. 190. pp. 179-92.

11. Fereres, E. and Castel, J.R. (1981) Drip irrigation management. Division of Agricultural Sciences, University of California. *Leaflet 21259*.

12. Mantovani, C.E., Berengena, J., Villalobos, F.J., Orgaz, F. and Fereres, E. (1991) Medidas y estimaciones de la evapotranspiración real del trigo de regadío en Córdoba. *Actas de las IX Jornadas Técnicas de Riegos*. Granada (Spain).

13. Fernández, J.E., Palomo, M.J., Díaz-Espejo, A. and Girón, I.F. (1997) Calibrating the compensation heat-pulse technique for measuring sap flow in olive. *Olea*. No. 24. pp. 24.

14. Moreno, F., Vachaud, G., and Martín-Aranda, J. (1983). Caracterización hidrodinámica de un suelo de olivar. Fundamento teórico y métodos experimentales. *Anales de Edafología y Agrobiología*, No. 5-6. pp. 695-721.

15. Hillel, D., Krentos, V.D. and Stilianou, Y. (1972). Procedure and test of an internal drainage method for measuring soil hydraulic characteristics *in-situ*. *Soil Science*, No. 114. 395-400.

16. Granier, A. (1987) Evaluation of transpiration in a Douglas-fir stand by means of sap flow measurements. *Tree Physiology*, No. 3. pp. 309-20.

17. Doorenbos, J. and Pruitt, W.O. (1977) Guidelines for predicting crop water requirements. *FAO Irrigation and Drainage*. Paper No. 24, 2nd ed. FAO Rome, Italy, 156 pp.

5 Acknowledgements

This work was supported with funds of the Spanish CICYT, project HID96-1342-CO4-01, and the *Junta de Andalucia* (Research Group AGR-151).

WATER-YIELD FUNCTION OF RICE AND ITS TEMPORAL AND SPATIAL VARIATION

Z. MAO and Y.L. CUI
Department of Irrigation and Drainage Engineering, Wuhan University of Hydraulic and Electric Engineering, Wuhan, China

Abstract
Based on the analysis of the experimental data, the models of water-production function (WPF) for rice, which can be applied to both northern and southern parts of China, are investigated. The regularity of the variation of water sensitivity parameters in various models has been discussed. It is suggested that Jensen model is the applicable model of WPF for rice in China. The regularity of the variation of water sensitivity index λ in Jensen model with the hydrology years and regions, especially, the relationship between λ and the reference evapotranspiration ET_0 has been presented. Thus, the principle and methods for widening the ranges of time and space for applying WPF of rice have been established.
Keywords: Rice, temporal and spatial variation, water-production function, water sensitivity index

1 Introduction

China is one of the most important countries of producing rice. The total yield and growing area of rice in this country are ranked first and second in the world respectively. In China, the deficit of water resources has been become serious year by year. In recent years, the water deficit-irrigation for rice have been applied widely and therefore, the importance of the research and application of the water-production function (WPF) of rice, which is one of the key factors for the rational application of water deficit-irrigation of rice has been enhanced. For this reason, the work for a research project of WPF for rice, which is funded by the National Natural Science Foundation of China has been conducted from 1988. The content of this paper is one of the main results of this project.

The experiments in the above mentioned project were conducted in Guilin Irrigation

Water and the Environment: Innovative Issues in Irrigation and Drainage. Edited by Luis S. Pereira and John W. Gowing. Published in 1998 by E & FN Spon. ISBN 0 419 23710 0

Experiment Station, South China and Tanghai Rice Experiment Station, North China during 1988~1993. The experiment in Guilin Station is for early and late season rice which is widespread in South China, and the experiment in Tanghai Station is for mid season rice because in North China mid rice is planted only. The conclusion from the experimental results of early and late rice in Guilin Station and mid rice in Tanghi Station are all similar, so the content of this paper includes mainly the results of late rice in Guilin Station.

2 General situation of experiment

Guilin Irrigation Experiment Station is located at 25°N and 110°E. The climate here is mild. The annual rainfall is 1400~1700 mm with uneven distribution: about 70% of the rainfall is concentrated in spring. The growing seasons (from transplanting to harvest) of early and late rice here are from the early in May to the late in July and the early in August to the late in October respectively. It is drought during the rice growing seasons especially the growing season of late rice. The soil texture is clay loam and the level of rice yield is 6.0~7.5 t/ha.

The experiment is conducted in 36 weighing lysimeters. The area of one lysimeter A is 0.30m², 12 nonweighing lysimeters (A=4.00m²) and 12 plots (A=66.67m²). The results from weighing lysimeters are compared and checked with the results from nonweighing lysimeters and plots. The growing season of rice is divided into 4 growing stages: I. revival of green and tillering, II. elongating and booting, III. heading and flowering, IV. milk ripening and yellow ripe. Based on the water condition in each growing stage, the following 12 treatments are arranged: (1)—normal irrigation (CK); (2),(4),(6) and (8)—light drought in stages I,II,III and IV; (3),(5),(7) and (9)—heavy drought in stages I,II,III and IV; (10),(11) and (12)—continual medium drought in stage I to stage II, stage II to stage III and stage III to stage IV. The standard of water condition in rice field for the normal irrigation is shallow water (0~40 mm) from revival of green to milk ripening and dry the field in the late part of tillering and yellow ripe. The standards of light, medium and heavy drought are that the lowest limits of soil moisture are 70%, 60% and 50% of saturation respectively. There are 3 replications in each treatment for weighing lysimeters. For each treatment and replication, the yield Y is measured, evapotranspiration ET is measured every day and the processes of growing and main physiological reaction indices of rice are measured every growing stage.

3 Model of water-production function for rice

A great deal of results from experiment, research and practice for crops have been indicated that the reduction in crop yield is caused by the timing and magnitude of water deficit or evapotranspiration deficit. The mathematical relationships between crop yield and evapotranspiration or other water conditions for crop with consideration of the time of water deficit during crop growing period are defined as crop water-production functions(WPF). The following 4 models of WPF are widely researched and applied in China. They are:

(1) Jensen model

$$\frac{Y}{Y_m} = \prod_{i=1}^{n} (\frac{ET_i}{ET_{mi}})^{\lambda_i} \qquad (1)$$

(2) Blank model

$$\frac{Y}{Y_m} = \sum_{i=1}^{n} A_i \cdot \frac{ET_i}{ET_{mi}} \qquad (2)$$

(3) Stewart model

$$\frac{Y}{Y_m} = 1 - \sum_{i=1}^{n} B_i \cdot (\frac{ET_{mi} - ET_i}{ET_{mi}}) \qquad (3)$$

(4) Singh model

$$\frac{Y}{Y_m} = \sum_{i=1}^{n} C_i \cdot [1 - (1 - \frac{ET_i}{ET_{mi}})^2] \qquad (4)$$

where Y is actual crop yield corresponding actual evapotranspiration ET, Y_m is maximum crop yield corresponding maximum evapotranspiration ET_m, λ and A,B,C are the sensitivity index and sensitivity coefficients of the crop to water deficit for respective models. i is an integer representing the number of crop growing stages. Π and Σ represent multiplication and summation respectively.

Based on the experimental data of lysimeters in 5 years (1988 and 1990~1993) in Guilin Station and using the mathematical statistics method, the values of λ and A, B, C of each growing stage for late rice are computed and given in Table 1.
Table 1 Shows:

(1) For Jensen model, the values of maximum λ in every year occur all in stage II and the processes of variation of λ from stage I to stage IV in every year are all similar. These appearances tally with the physiological reaction of rice. The values of regression coefficients R for this model in every year are more than 0.978 and higher than that for the other models. That the negative value of λ to be occurred is not reasonable, but it is occurred only two times in 5 years and the absolute values are quite close to zero. For these reasons, Jensen model is suitable for late rice in the South China.

(2) For Blank model, Stewart model and Singh model, the stages of occurring the maximum sensitivity coefficients A, B and C and the processes of variations of A,B and C from stage I to Stage IV between 5 years are different. The negative values of A, B and C are occurred many times and the absolute values of some negatives are very high. For these reasons, the above mentioned 3 models are not suitable for the late rice in South China.

Moreover, the results of early rice in Guilin Station and mid rice in Tanghai Station are all similar to the results of late rice in Guilin Station. Therefore Jensen model can

Table 1. Sensitivity index λ and coefficient A,B,C in 4 models in 5 years (late rice, Guilin)

Year	Growing Stage	Sensitivity index and coefficients in 4 models			
		(1) Jensen model λ	(2) Blank model A	(3) Stewart model B	(4) Singh model C
1988	I	0.1209	0.8492	0.1197	0.6905
	II	0.2376	-0.2864	0.2475	0.7759
	III	0.1300	0.7601	0.0451	-0.640
	IV	0.0213	-0.2661	-0.1076	0.2228
	Regression Coefficient R	0.980	0.960	0.973	0.483
1990	I	0.1940	0.3714	0.3620	0.7284
	II	0.5395	0.7879	0.9095	-0.2184
	III	0.2234	0.3722	0.5718	-2.7550
	IV	-0.0726	-0.4889	-0.8445	3.3016
	Regression Coefficient R	0.978	0.980	0.790	0.992
1991	I	0.1359	-0.0348	0.0948	0.6258
	II	0.3747	1.0055	0.4200	-0.0185
	III	0.1683	0.3161	0.1971	0.2214
	IV	-0.0612	-0.3049	0.0579	0.1276
	Regression Coefficient R	0.982	0.928	0.981	0.674
1992	I	0.2090	0.0525	0.1746	-0.4669
	II	0.7025	0.5575	0.6108	1.2223
	III	0.2199	0.2909	0.2680	0.6524
	IV	0.1523	0.0607	0.1609	-0.5750
	Regression Coefficient R	0.999	0.997	0.998	0.936
1993	I	0.3861	0.4524	0.3464	2.5342
	II	0.5898	0.3708	0.5970	-0.0500
	III	0.2889	-0.1628	0.2852	-0.4956
	IV	0.1204	-0.0268	0.1319	-1.1004
	Regression Coefficient R	0.999	0.999	0.989	0.999

be applied for both northern and southern parts of China. i. e. Jensen model is the suitable model of WPF for rice in China.

4 Variation of sensitivity index (λ) of Jensen model for rice in growing season

The type of hydrological year can be determined by the frequency P of the reference evapotranspiration ET_0 because ET_0 is one of the main indices of atmospheric dry and wet conditions. The values of λ in each stage of late rice and the frequency of ET_0—P in five years from Guilin Station are given in Table 2. The processes of variation of λ in growing season are shown in Fig 1.

The process of variation of λ in growing season can be analyzed with the average value of λ in the 5 years ($\bar{\lambda}$) because the drought, medium and wet year are all included in these years and the average of frequency of ET_0 in the 5 years is near 50%. Thus, WPF for late rice in South China can be expressed as Eqn.5 and the relationship between λ and time can be expressed as Eqn.6.

Where the numbers of 1,2,3 and 4 represent the growing stage I, II, III and IV

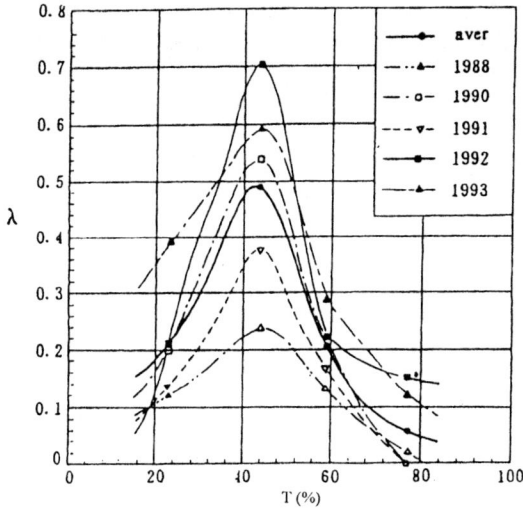

Fig.1. Process of variation of λ in growing season

Table 2. Values of λ in different hydrologic years

Year	Frequency of $ET_0,P(\%)$	Type of year	Value of λ in Jensen model			
			Stage I	Stage II	Stage III	Stage IV
1988	9.5	Wet	0.1209	0.2376	0.1300	0.0213
1990	71.4	Medium Drought	0.1940	0.5395	0.2234	-0.0726
1991	42.9	Medium Wet	0.1359	0.3747	0.1683	-0.0612
1992	85.7	Drought	0.2090	0.7025	0.2199	0.1523
1993	52.4	Medium	0.3861	0.5998	0.2889	0.1204
Average	52.38		0.2092	0.4888	0.2061	0.0588

Note: the negatives may be used instead of zero in the calculation of average

$$\frac{Y}{Y_m} = (\frac{ET_1}{ET_{m1}})^{0.21} \cdot (\frac{ET_2}{ET_{m2}})^{0.49} \cdot (\frac{ET_3}{ET_{m3}})^{0.21} \cdot (\frac{ET_4}{ET_{m4}})^{0.06} \tag{5}$$

$$\overline{\lambda} = \frac{1}{(10.08 - 17.09 \times 10^{-2}T - 5.14 \times 10^{-3}T^2 + 1.107 \times 10^{-4}T^3)} \tag{6}$$

respectively, T is the percentage of the days after transplanting with the days of growing season (from transplanting to harvest).

5 Variation of sensitivity index between hydrologic years and its application

It is shown in Table 2 that the absolute values of the differences between λ and $\overline{\lambda}$ in stage I, III and IV of every year except in stage I of 1993 are all smaller than 0.1, therefore the errors of calculating Y/Y_m are all less than 5% under the condition of that $\overline{\lambda}$ is substituted for λ in stages I, III and IV in every year. For this reason, $\overline{\lambda}$ can be used to substitute for λ except in the stage of occurring the maximum $\lambda - \lambda_m$. For the stage of occurring λ_m, the fluctuation of λ_m in different years is high and must be considered. λ_m in different years can be calculated with a modified

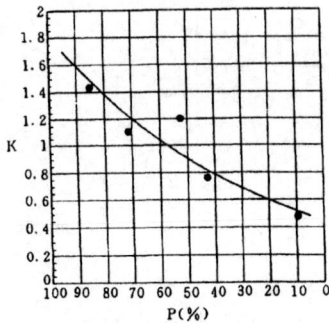

Fig. 2. Relation curve of modified coefficient K and frequency of ET_0,P

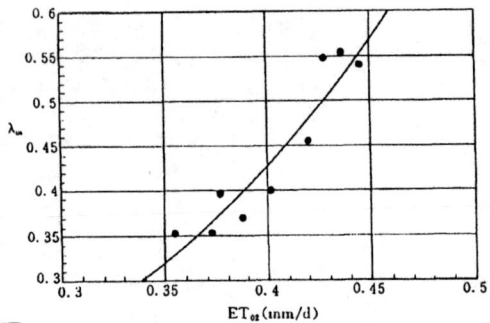

Fig. 3. Relation curve of maximum sensitivity index λ_m and reference evaportranspiration ET_{02}

coefficient K ($K = \lambda_m / \overline{\lambda_m}$) from the annual average value of $\lambda_m - \overline{\lambda}_m$. From Table 2, values of K in different years can be calculated and they are given in Table 3.The relationship between K and frequency of ET_0—P is shown in Fig.2, and the regression equation is given in Eqn. (7).

$$K = 0.4503 \times 1.014^P \qquad \text{(regression coefficient R=0.83)} \qquad (7)$$

where $K = \lambda_m / \overline{\lambda_m}$, P is the frequency of ET_0 in the growing season of late rice (%).

Table 3. Relationship between the modified coefficient K and frequency P

Year	Frequency of ET_0 P(%)	Maximum of λ λ_m	Modified coefficient K
1988	9.5	0.2376	0.486
1990	71.4	0.5395	1.104
1991	42.9	0.3747	0.767
1992	85.7	0.7025	1.437
1993	52.4	0.5998	1.207
Average	52.38	$\overline{\lambda}_m$=0.4888	1.000

For complementing and extending the annual series of the data of WPF of rice, the values of λ in different years can be determined by:
(1) In the stage of occurring maximum sensitivity index (λ_m),

$$\lambda_m = K \cdot \overline{\lambda_m} \qquad (8)$$

value of K can be fined in Table 4.
(2) In other stages, using the annual average values of $\lambda(\lambda)$ to substitute for λ.

Table 4. Modified coefficient K for λ_m in different hydrologic years

Type of year	Heavy drought	Drought	Medium drought	medium	Medium wet	Wet	Heavy wet
Frequency of ET_0, P(%)	95	85	75	50	25	15	5
$K(\lambda/\overline{\lambda})$	1.69	1.47	1.28	0.90	0.64	0.55	0.48

6 Spatial variation of the sensitivity index and its application

The experiments of evapotranspiration ET of late rice under 5 treatments were conducted in the 9 irrigation experiment stations which are scattered over the Guangxi Autonomous Region, China. The values of λ in Jensen model for late rice in 1990 from the 9 stations can be calculated by using the data from these experiments. The values of λ_m (occurred also in stage II) and corresponding average ET_0 from transplanting to the end of stage II are given in Table 5.

Table 5. Relationship between λ_m and ET_{02} of late rice in different stations

Station	λ_m (in stage II)	ET_{02}(mm/d)
1.Guilin	0.540	0.445
2.Baise	0.354	0.373
3.Hechi	0.353	0.355
4.Nanning	0.400	0.402
5.Luzhou	0.548	0.428
6.Wuzhou	0.370	0.388
7.Monshan	0.397	0.377
8.Lingshan	0.555	0.436
9.Hexian	0.456	0.420

From Table 5, the regression curve and equation are given in Fig.3 and Eqn.9 respectively.

$$\lambda_m = 0.04394 \times 295.7^{ET_{02}} \qquad \text{(regression coefficient R=0.9294)} \qquad (9)$$

where λ_m is the maximum value of λ in the given region, ET_{02} is the average value of reference evapotranspiration from transplanting to the end of stage II in the same region (mm/d). λ is the only factor which need to be determined in Jensen model. Values of λ_m in different regions can be calculated with Eqn.(9) by using the reference evapotanspiration in the corresponding regions, thus, the problems of the spatial variation and the replacement and extension in space for WPF of rice can be resolved by using the relationship between the values of λ_m and ET_{02}.

7 Conclusion

The following conclusions are reached based on the analysis of the rice experimental data:

(1) Jensen model is the applicable model of water-production function for rice in China. The concrete equation of Jensen model of late rice in Guilin Region, South China for multi-annual average value is:

$$\frac{Y}{Y_m} = (\frac{ET_1}{ET_{m1}})^{0.21} \cdot (\frac{ET_2}{ET_{m2}})^{0.49} \cdot (\frac{ET_3}{ET_{m3}})^{0.21} \cdot (\frac{ET_4}{ET_{m4}})^{0.06}$$

where numbers of 1, 2, 3 and 4 represent the growing stage of "revival of green and tillering", "elongation and booting", "heading and flowering" and "milk ripening and yellow ripe" respectively.

(2) The maximum values of sensitivity indices in Jensen model λ_m vary with the hydrologic years. The drier the year, the higher the λ_m. The values of λ_m in different hydrologic years can be calculated by:

$$\lambda_m = 0.5403 \times 1.014^P \overline{\lambda_m}$$

where $\overline{\lambda}_m$ is the average value of λ_m and P is the frequency of reference evapotranspiration in the given year(%).

(3) The values of λ_m in different regions vary with the reference evapotranspiration ET_0. The higher the ET_0, the higher the λ_m. Values of λ_m in different regions can be calculated by:

$$\lambda_m = 0.04394 \times 295.7^{ET_{02}}$$

where ET_{02} is the average value of ET_0 from transplanting to the end of stage of booting (mm/d) in the given region.

(4) The principle and methods for widening the ranges in time and space for applying the water-production function of rice have been presented based on the relationship between the sensitivity index in Jensen model and the reference evapotranspiration.

8 References

1. Vaux, H.J., Pruitt, W.O.(1983) Crop-water production functions. Advanced Irrigation, U.S.A. Vol.2.
2. Singh, P., Wolkewitz, H., Kumar, R.(1986) Comparative performance of different crop production function for wheat. Irrigation Science, U.S.A, Vol.8, PP.273-290.
3. Mao, Z., Cui, Y.L.(1994) Study of water production function for rice in South China. Journal of Hydraulic Engineering, Beijing, Vol.9, PP.21-31.
4. Cui, Y. L., Li, Y. H., Mao, Z.(1998) Models of crop water production function with consideration of influence of the reference evapotranspiration frequency. Journal of Hydraulic Engineering, Beijing, Vol.4.

SECTION III

MANAGING SUPPLY IN IRRIGATION SYSTEMS

PARTICIPATORY RESEARCH ON TECHNICAL INTERVENTIONS TO IMPROVE IRRIGATION PERFORMANCE: INFORMING PERCEPTIONS

B.A. LANKFORD
School of Development Studies, University of East Anglia, Norwich, UK
J. W. GOWING
Centre for Land Use and Water Resources Research, University of Newcastle upon Tyne, UK

Abstract
Interviews were carried out with irrigators on commercial sugarcane irrigation schemes in Swaziland to seek their opinions on what would promote 'good irrigation'. Ranking methods were used to define good irrigation and allow irrigators to rank five technical interventions; altering furrow layouts, ensuring a correct flow to an area, improving canal type and equipment, changing the type of turnout and fixing leaks and spills. The results showed that a) irrigation planning addressing complex technical questions could be conducted in a participatory way, b) irrigators clearly perceived a hierarchy of needs, c) the hierarchy of needs was affected by exposure to different technical options, and that therefore, d) on irrigation schemes, research and policy based on participatory methods should be aware of the gaps in information available to users, researchers and planners, and proceed accordingly.
Keywords: Canal irrigation, irrigation technology, participatory research, performance.

1 Introduction

The debate regarding management of natural resources appears to favour participation: "Sustainable agriculture cannot succeed without the full participation and collective action of rural people and land managers [1]. Among many definitions, participation is 'involving users and communities in all stages of the development process to achieve sustained benefits' [2]. Examples of some of the stages are research, policy-setting, delivering services and evaluation of services. Considerable problems need to be overcome in each of these stages, particularly in terms of using and interpreting participatory methodologies, and seeing them as tools, not as 'recipes for success' [3]. Clearly balances have to be struck between methodologies and expectations, as well as

Water and the Environment: Innovative Issues in Irrigation and Drainage. Edited by Luis S. Pereira and John W. Gowing. Published in 1998 by E & FN Spon. ISBN 0 419 23710 0

between peoples' needs and wants, between their needs and what can be provided, and between perceptions of the actors involved.

This paper addresses the last kind of balance; between the perceptions of local users and 'expert outsiders' and specifically in the area where **gaps** in knowledge and perceptions might exist. It does so by examining the research stage when experts attempt to discover user's needs as contextualised within their local knowledge. A case study is employed to demonstrate that participatory research certainly gives outcomes with meaning for those involved, but which may not capture 'unknown' technical interventions which may have a far greater impact on irrigation performance.

The case study involves interviews with irrigators of commercial irrigation schemes. Although it could be argued that this is not applicable to the world of farmers, several important parallels apply. Firstly such organisations are attempting to raise productivity and reduce costs by motivating employees and giving them greater responsibility and this requires a participatory process. Secondly employed irrigators have similar perceptions of surface irrigation schemes as do farmers and farmer groups. These perceptions are of water at the lower levels of the system (e.g. tertiary and field outlet) and are determined by a lack of involvement in the management of the system, and by a lack of comparative experiences of other parts of the system or other irrigation systems altogether.

2 The case study

2.1 Introduction
Four large-scale surface irrigation schemes in the Swaziland lowveld were visited during the period 1995 to 1997 as a part of a research programme investigating differing approaches to irrigation design [4,5]. During that programme, the research asked the question; does the design of the system affect the way it is managed? For the sake of anonymity, these schemes are called, A, B, C and D. All four are similar in respect of climate, crop (sugarcane), soil type, topography, cropping calendar (a 12 month cycle) and generally speaking estate management. In terms of irrigation operation, similarities also occur; lined tertiary canals, use of both syphon and spile pipes, furrow irrigation using 1.5 metre wide furrows, and during peak demand periods constant secondary canal flows and rotational flows at the tertiary level.

However, important differences exist; schemes A, B and C are termed "type 1", characterised by a wide variety of tertiary turnout designs, and (as flow gauging determined) variable flow rates from one tertiary canal to another. Also on these schemes, flows are rotated loosely within the secondary area to wherever fields are driest. Scheme D is "type 2", having fixed, measured flows at the tertiary level, standard modular turnouts and long-crested weirs. Tertiary flows on these schemes rotate strictly within groups of fields as dictated by the hydromodule (l/sec/ha).

2.2 Interviews, questionnaires and questions
Interviews were carried out on the four schemes with a total of 37 irrigators using semi-structured group interviews. A questionnaire was developed to gain an understanding of how the irrigators perceived their responsibilities and tasks. The interviews were usually conducted in the open, close to fields with an interviewer, translator and

approximately four to six irrigators. Irrigators were 'sampled' from several secondary systems within each scheme. The questionnaire consisted of open and closed questions, and time was allowed to follow new lines of inquiry. The final part of the questionnaire was devoted to a ranking exercise, called the "shopping list question", which is the focus of this paper.

The null hypothesis was that there is no difference between the irrigators' understanding of factors affecting performance dependent on the type of irrigation scheme. The alternative hypothesis is that there is a difference between the irrigators' understanding of factors affecting performance dependent on the type of irrigation scheme. In other words, the alternative hypothesis is that the ranking of factors by irrigators is contingent on the design and operation of the scheme.

The shopping list question was "what would you like to fix or keep so that you can irrigate well and get round in time before cane stress shows?" The question was asked in a number of ways and ended up with presenting the irrigators with five choices of interventions that might improve their irrigation. Irrigators were asked to discuss and rank them from 1 (most important) to 5 (least important). The five choices were:

1. *Layouts and furrow lengths*: improve layouts to reduce number of feeder furrows (which are small in-field feeder furrows leading off from the concrete side canal to irrigate small portions of fields), ensure high ridges of soil so breakouts of water from one furrow to another are minimised, and change furrow length where too long or too short by altering the width of the field (an imaginary situation).
2. *Flow to area groups*: ensure that the right flow can supply the right area, otherwise known as correct leadstream cycling groups.
3. *Canal type*: change the canal type which includes altering the diameter of the spile or syphon pipes, changing to spiles or syphons, improving head control in the tertiary canal by adding more checkboards or by altering the design of head control.
4. *Turnout type*: improve the design of turnout, or change to another design, either to reduce leaks, or to make it easier to operate or to add flow measurement
5. *Leaks, spills and flow rate*: alter the flow rate, either by changing the tertiary turnout, or increase the flow by reducing leaks and spills.

To test differences in irrigators beliefs, 2 x 5 contingency tables were drawn up so that Pearson's Chi-squared (χ^2) test could be used. Excel spreadsheets were used to enter ranked classes of answers, calculate expected frequencies and the chi-squared statistic, which was compared to a critical χ^2 statistic for a 95% level of confidence.

2.3 Results

Irrigators showed a high level of responsibility towards their jobs; irrigating as well as possible within the constraints set by the design and operation of the canal and field systems. They were concerned about cane stress and well-understood that summer was the critical time for achieving "good" irrigation. The ranking question took time to discuss but in the end the irrigators appeared sure about their responses.

Table 1 gives the results of the ranking of irrigators to improve irrigation on type 1 schemes and table 2 gives the results for the same exercise for the type 2 scheme (D). In both tables, five columns represent the five technical choices. The numbers in the columns are the frequency count of the rank of importance by the respondents. For example in table 1, on type 1 schemes, 13 out of 20 irrigators ranked furrow layouts as being the most important, and in table 2, 11 out of 17 irrigators ranked layouts as being the most important. At the bottom of each table, the mean response is calculated to show the order of ranking (in brackets) of each of the five interventions.

Table 1. Responses to questions about perceptions on means to improve irrigation for type 1 schemes

	Layouts	Flow to area groups	Canal type	Turnout type	Flow rate
Count of rank 1's	13	0	4	1	4
Count of rank 2's	4	4	0	4	7
Count of rank 3's	3	0	5	4	7
Count of rank 4's	0	7	10	1	2
Count of rank 5's	0	9	1	10	0
Mean (and order)	1.5 (1)	4.0 (5)	3.2 (3)	3.8 (4)	2.4 (2)
Total number	20	20	20	20	20

Table 2. Responses to questions about perceptions on means to improve irrigation for type 2 schemes

	Layouts	Flow to area groups	Canal type	Turnout type	Flow rate
Count of rank 1's	11	6	0	0	0
Count of rank 2's	6	11	0	0	0
Count of rank 3's	0	0	11	0	6
Count of rank 4's	0	0	6	0	11
Count of rank 5's	0	0	0	17	0
Mean (and order)	1.4 (1)	1.6 (2)	3.2 (3)	5.0 (5)	3.8 (4)
Total number	17	17	17	17	17

The frequency counts of each rank were used to test differences between the two types of schemes for the five factors. The outcome of each test is discussed in order:

1. *Layouts*: both types of irrigators, those on type 1 and type 2 schemes, ranked this intervention very highly, with 100% of irrigators putting it in the top three. The calculated chi-squared (χ2) statistic (3.345) was less than the critical value (9.49), and so there is no significant difference between the two types of irrigation system.
2. *Flow to area groups*: nearly 50% of the type 1 irrigators ranked this the least important (number 5), whereas 100% of the 17 type 2 irrigators put this in the top two most important slots. The statistical test, with a χ^2 of 25.189, greater than 9.49, indicates a significant difference at the 95% probability level.
3. *Canal type*: although the spread of answers is noticeable with the type 1 irrigators, with some ranking this most important, 50% of them ranked it number 4. When compared against the type 2 irrigators who placed this in either rank 3 or rank 4, this was found not to be a significant difference.
4. *Turnout type*: the range of answers was large for the type 1 irrigators, placing this across all five positions of order. This contrasts with the type 2 irrigators who placed these all within the lowest rank. It was found that was a significant difference at the 95% probability level (χ^2 value = 11.648).

5. *Leaks, spills and flow rate*: 90% of the type 1 irrigators systems ranked this in the top three, whereas the majority of type 2 irrigators placed this in rank 4. This difference was found to be significant ($\chi^2 = 17.177$).

With the exception of the need to improve layouts, an intervention which both sets of irrigators ranked highly, and the choice regarding canal types, an option both irrigators ranked in the middle, significant differences arose for three other interventions. Type 1 irrigators did not rank the correct flow to area highly, reflecting their lack of experience of this type of operation whereas type 2 irrigators who understand the operation of the leadstream cycling, ranked highly a correct flow for a group of fields. Their knowledge of this comes their experience of difficulties when the flow is too low for the area, or when the area is too large for the cycled flow.

Type 1 irrigators with experience of variable gates felt they would see some benefit if "better gates" were introduced, thought it is not clear what. This "turnout" intervention was dismissed by the type 2 irrigators with experience of modular gates.

Type 1 irrigators ranked more highly the need to fix leaks and ensure the right flow rate than type 2 irrigators. However, this may be more a reflection of the priorities of the type 2 irrigators rather than differences in leakage rate. Observations indicated that differences in leakage between the schemes were roughly similar and leaks were not great due to the use of lined canals placed within clay to clay loams soils.

In summary, type 1 irrigators ranked the interventions in the following order of priority; layouts; flow rate and leaks, canal type; turnout type and correct flow to area groups. The type 2 irrigators ordered them; layouts, correct flow to area; canal type, flow rate and leaks; and turnout type. It seems that on the evidence collected, the null hypothesis can be rejected and the alternative hypothesis can be accepted and say that on the whole there are significant differences in understanding of the factors affecting irrigation depending on the local technology and operation of the system.

3 Discussion on participatory research

Three main issues arise from this case study; inherent benefits in the participatory process; solving problems according to a hierarchy of needs; and the paradox of how to deal with technological options unknown to both researcher and user. Firstly, there can be no doubt that inherent 'first order' benefits occur via the participatory process; summarised by the words 'talking and listening' which includes expressing views, understanding another's view, accepting and encouraging, all of which help to 'generate alliances', an 'enduring value of PRA' (Jeanrenaud and Jeanrenaud, quoted in [3]). At the end of the interview sessions in Swaziland there was certainly a feeling of problems having been well aired, and of acknowledgement of those problems.

On the second issue; it is possible to respond to the perceived hierarchy of needs with interventions in the same order presented by the users. Responding is an important part of the process that begins with talking and listening, adding benefits besides solving the technical problems such as building trust, increasing motivation, and enriching future research cycles. Such benefits are termed here 'second order'.

The third issue is about known and unknown knowledge of both researcher and user. Unknown knowledge, or gaps, may be very important in terms of omitting significant interventions from a programme or influencing the order in which interventions are tackled. Although it may be argued that order of events is secondary to process, the omission of important factors affecting the performance of irrigation is a more serious matter. To discuss this further, four classes of experience or knowledge interchange between users and researchers are proposed, shown in table 3.

Table 3. Framework of knowledge interchange between users and researchers on a particular subject.

		Researcher's knowledge of the subject			
		Yes		No	
		Actor/process	Effect	Actor/process	Effect
User's knowledge of a subject	Yes	User (U) Researcher (R)	can agree can agree	User (U) Researcher (R)	gauges & teaches can learn
		Direction of learning (and class no.)	$U \Leftrightarrow R$ (1)	Direction of learning (and class no.)	$U \Rightarrow R$ (2)
	No	User (U) Researcher (R)	can learn can gauges & teaches	User (U) Researcher (R)	restricted learning restricted learning
		Direction of learning (and class no.)	$U \Leftarrow R$ (3)	Direction of learning (and class no.)	$U \neq R$ (4)

It should be understood that the teaching and learning as described in table 3 is additional to the first order benefits inherent to the participatory process. The knowledge interchange is of specific interventions intended to improve the performance of users and of irrigation. Referring to table 3, class 1 in the top left is where both researcher and user know about a proposed intervention. The second kind, top right, is where the user has knowledge of an intervention, which researcher does not know about. Here, the user passes knowledge to the researcher. The third class is where the researcher has knowledge of an intervention which the user does not, and where the flow of knowledge is from researcher to user. The fourth class is where both user and researcher do not have knowledge of a particular intervention.

The four kinds of perceptions are exemplified by reference to the case study. A class one interchange happened when the researcher who knew about leadstream cycling found agreement as to its importance with type 2 irrigators. A class two interchange might have happened if a researcher without experience of leadstream cycling was introduced to it by irrigators from type 2 irrigation schemes. A class three interchange happened when the researcher familiar with the idea of leadstream cycling introduced the irrigators on type 1 schemes to the idea of leadstream cycling (though this was not followed up). A class four interchange might have occurred if another researcher not acquainted with leadstream cycling had visited the type 1 irrigation schemes which the irrigators did not reveal through lack of experience.

In summary, this third issue acknowledges that errors in ranking might occur. This opens up the possibility of different orders of interventions which may be better reflections of what is required (measured by impact, rather than by perception). If these more effective interventions can be found then these might be called third order benefits to the participatory process.

4 Lessons for participatory research on irrigation technology and design

The discussion above examines the possibility that neither the user nor the researcher might know about interventions which may be relevant and significant to a particular context. Recent research on participation of farmers in the design process covers issues such as canal alignment through farmer's land or changes of turnout design from divisors to orifices [8]. Resolving unsuitable irrigation engineering need not be reason enough to demonstrate the worth of participatory process (beyond first order benefits of alliance building, and second order benefits of prioritised problem-removal) since these may be good examples of class 1 or class 2 interchanges of knowledge where the user or both parties readily understand the problem.

In the literature, contrasts are made between of top-down, non-participatory design by bureaucratic engineers and bottom-up farmer-participatory design by other types of engineers [7]. Although there is demand for participatory engineers, the implication is that farmer design is enlightened and bound to succeed. However, one has to question how participation will deliver answers to complex questions of water management on large-scale schemes, some of which are, by the nature of their complexity, are likely to fall within class 3 interchanges which may be seen as top-down and therefore unsuitable, or within class 4 type interchanges where neither party knows the answers. This paradox is more acute for irrigation when we think of how difficult it appears for experts to transfer best-practices amongst each other, and how farmers, who tend to be fixed to one locale, cannot through the market-place know other designs and operation. In particular, participatory design with farmers who have recently moved out of rainfed agriculture is open to question.

5 Conclusions

Research on four irrigation schemes in Swaziland revealed that irrigators' perceptions of irrigation technology were related to contexts that they were working in. Although the interviews provided results, the question must be asked what do these opinions mean for researchers who seek answers beyond the rewards of the participatory process. The following conclusions can be made:

- irrigation planning tackling quite complex technical questions can be conducted in a participatory way with users of irrigation schemes because users of irrigation schemes readily perceive a hierarchy of needs;
- those perceptions are informed by comparisons they may make within their geographical area or timeline of technology and operation;
- those perceptions are constrained by a lack of experience and comparison with other areas of technology and operational types;

- solving priority needs may reveal further needs, which in turn need to be solved, which is an acceptable manner of progressing;
- these perceptions may not occur in the same order as in their real impact on irrigation performance;
- research and policy based on participatory methods should be aware of the levels of information available to both users and planners. Perhaps the chief implication is to promote comparative experiences, as Meijers says; "attention should be paid to broadening the designers' frame of reference as well as the users' in order to come to a more effective interaction." [9]. If participation is to raise performance then it cannot not, as a primarily social science tool, ignore the effort required to address complex technical and water management issues. This is particularly so when the technical elements of large-scale network canal systems become so interwoven and hidden within the irrigation scheme's considerable social dimensions.

6 References

1.   Pretty, J. (1995) *Regenerating Agriculture. Policies and Practice for Sustainability and Self-Reliance.* Earthscan, London.
2.   Narayan, D. (1993) *Participatory Evaluation. Tools for managing change in water and sanitation. World Bank Technical Paper Number 207,* World Bank, USA.
3.   Biggs, S. and Smith, G. (1998) Beyond Methodologies: Coalition-Building for Participatory Technology Development. *World Development*, Vol. 26, pp. 239-248.
4.   Lankford, B. A. and Gowing, J. (1997) Providing a water delivery service through design management interactions and system management, in *Water: Economics, Management and Demand*, (eds. M. Kay, T. Franks, and L. Smith), E & FN Spon, London, pp. 238-246.
5.   Lankford, B.A. and Gowing, J. (1996) Understanding water supply control in canal irrigation systems, in *Water Policy: Allocation and Management in Practice*, (eds. P. Howsan and R. Carter), E & FN Spon, London, pp. 186-193.
6.   Lankford, B. A. (1992) The Use of Measured Water Flows in Furrow Irrigation Management - a Case Study in Swaziland. *Irrigation and Drainage Systems*, Vol. 6: pp. 113-128.
7.   Korten, F. F. and Siy, Jr. R. Y. (eds.) (1988) *Transforming a bureaucracy. The experience of the Philippine National Irrigation Administration*, Kumarian Press.
8.   Diemer, G. and Huibers, F. P. (1996) *Crops, people and irrigation. Water allocation practices of farmer and engineers.* IT Publications, London.
9.   Meijers. T. (1990) The interaction between users and designer in the design process of village irrigation systems on the island of Ile a Morphil, in North Senegal, in *Design for Sustainable Farmer-Managed Irrigation Schemes in Sub-Saharan Africa,* Agricultural University, Wageningen.

CANAL INSPECTORS' KNOW-HOW IN WATER MANAGEMENT

R. BUSTOS and M. MARRE
School of Political and Social Sciences, National University of Cuyo, Argentina
J. CHAMBOULEYRON and S. SALATINO
National Institute for Water and the Environment, Andean Regional Center, Mendoza, Argentina

Abstract
In the oases of the province of Mendoza, Argentina, irrigation water management is decentralised and participatory. Users are consolidated in Canal Inspections, which are autonomous agencies with constitutional standing. Though the Inspector represents the users. Thus, the administration of the canals has become more complex on account of the struggle of interests and the development of new problems. Since the current administrative structure as well as canal inspectors' management capacity are being questioned, this project aims to: a) Learn how much know-how canal inspectors have acquired through their social practice, the incorporation of new uses, the autonomy by which this know-how is produced; b) Identify a set of conditions for improving and strengthening users' organisations management in order to ensure higher efficiency at a local level. The results show that canal inspectors' know-how is only applied to certain management areas where they are competent and effective, because the state has exercised the control over the social practice of the local management of canal authorities. There are also a set of institutional weaknesses which should be overcome, such as lack of training programs, low available resources and differences in water distribution and system's maintenance costs among users' organisations.
Keywords: Canal inspectors', decentralised and participatory administration, irrigation systems, operation, water management, water users' organizations

1 Introduction

Mendoza is located in the western arid region of Argentina. Its irrigated area is concentrated in four oases, out of which the Tunuyán River oasis is the largest in the province. In its middle and lower reaches, this water system covers an area of about 85,000 ha. Users' organizations are responsible for managing and operating secondary,

Water and the Environment: Innovative Issues in Irrigation and Drainage. Edited by Luis S. Pereira and John W. Gowing. Published in 1998 by E & FN Spon. ISBN 0 419 23710 0

tertiary and quaternary canals as well as ditches and drains.

After the enactment of the General Water Law in 1884, water management in the oases of Mendoza became decentralized and participatory. The water policy principles laid down by this law attained constitutional status in 1894. Since then, the provincial constitutions contain water management rules, especially as regards irrigation. The Constitution currently in force, which was enacted in 1916, in its Sixth Chapter on "Irrigation Department" sets forth the establishment of an autarchic management agency, fully independent from the provincial Executive, empowered to draw up its own budget, collect a charge for the use of public waters and manage the resource at river, dam and main canal levels.

Users' participation in water management is also laid down in the Constitution, which states: "The irrigation laws enacted by the Legislature of the Province shall not, under any circumstances, deprive interested parties in canals, ditches and drains of the power to elect their authorities and administer their respective revenues..". These users' organizations, which are called "Canal Inspections", are autonomous and autarchic organizations that elect their authorities every four years, draft and use their own budget, and collect the "canal pro rata" within their jurisdiction; it is with these resources that the Inspections' management activities are supported. They have an administrative structure made up of the Inspector (Water Manager), who is also the canal judge, three delegates who collaborate with him on water management, and the gate-keepers who are responsible for delivering water to different users.

Throughout this first century of operation, the decentralized and participatory administrative system has undergone substantial changes. Though the managers of the irrigation systems have not undergone great changes, the agricultural society has turned into an agricultural-urban-industrial society. This has given rise to problems not only related to the amount of water to be delivered but also to water quality and environmental equilibrium. Managers are not properly trained and are unable to grasp the changes that have taken place in connection with water allocation for industrial, recreational, public, urban and agricultural uses.

Some measures have been adopted to increase these organizations' management efficiency. For instance, small-sized organizations that reached low collection levels were consolidated in order to achieve economies of scale in the administration. At present, the number of Canal Inspections was reduced from 800 during the 80's to 157.

It is said that "Water Management" is the management of complexity. Are our managers capable of performing a task of this magnitude? Isn't society entrusting farmers with excessive responsibilities before providing them with adequate training? These are the questions that will be answered in this paper. To this end, the know-how developed by Canal Inspectors through their social practice, current uses, new characterization of uses and the structures that restrict their development shall be described. This paper also aims at identifying a set of conditions which are essential for improving water management in users' organizations in order to ensure greater efficiency and the achievement of goals at local level.

Sociological, legal and political principles endorse the existence of decentralized, self-managed users' organizations. Their institutionalization in modern times is the result of the development and progress attained by each community which, in turn, assigns them special powers according to the circumstantial reality and the prevailing uses in each of them [1]. In the water management field, these organizations elicit greater consideration for users' interests; reduce direct governmental management; make alternative use of financial, human and material resources; capitalize the community's efforts in water works; and generate economies of scale in the use of shared resources [2].

In order to deal with the real situation of decentralized organizations when they operate at al local level and are confronted with users' problems, it is necessary to know three fundamental aspects: the powers, the formally assigned resources, and the greater or smaller centralization that exist in the respective governmental organizations.

The requirements to exercise the powers formally conferred by legal rules are: financial resources and political representativeness of these organizations at the government's higher levels. The former will permit them to derive human, technical and material resources for their activities. The latter will enable them to manage the relations with other governmental organizational levels of the water sector and powerful groups of the local society. From this point of view, there are three fundamental issues which are necessary to understand the problems of local organizations: financial resources, degree of autonomy and users' participation through their natural representatives [3].

Water users' organizations are, from the management point of view, more flexible and versatile. The know-how, "knowledge", that water managers develop makes it possible to enforce regulations, to manage resources with more flexibility and to solve local conflicts.

Coward states that an irrigation system does not only consist of a physical infrastructure but also of a common ownership system of rights among irrigators that favors the establishment of a "sustainable" organization.

Brewer and Sakthivadivel consider that there is a relationship between water distribution rules and water resources performance. These rules define the users, the amount of water to be delivered, the period of time and the authorities. Some of these rules are formal and others are informal.

Perry sustains that in order to enhance performance, distribution rules must include multiple criteria [4]. In this sense, the know-how developed by Canal Inspectors to manage the irrigation system is the link between the rules that regulate under normal system operation conditions and the practice, which compels them to incorporate new "informal" rules that depend on their own technical skills. The formal or written rule does not stipulate how to solve water distribution problems when there is a great number of abandoned hectares with water rights, crisis due to droughts or floods, competition for new uses, or restrictions imposed by the existing irrigation infrastructure.

This know-how can be defined as the "own repertoire of knowledge socially attained, conveyed through work and transmitted from the interior and/or exterior of each generation, that endows individuals with the capacity to perform technical activities related to water distribution. This know-how is not static; it can incorporate new knowledge or techniques based on a new social practice and on the fulfillment of a new social need" [5].

One of the aspects that contributes to the operation of a hydraulic work is the development of this know-how since it reflects the variety of needs and strategies that are required to execute the tasks, the ability acquired to efficiently distribute water, the local priority objectives that users have established and which can be different at each system's level from the ones laid down by the General Irrigation Department [6].

2 Methodology

An intensive interview and a structured survey were carried out and secondary data was analyzed in order to collect information concerning the know-how and resources management of users' organizations of the middle and lower Tunuyán River Oasis (which covers about 85,000 irrigated hectares).   Out of the 21 Canal Inspectors representing all the organizations in the oasis that were surveyed, 62% manage areas larger than 5,000 ha while 38% manage areas of 5,000 ha or less.

By means of the documentary analysis technique, the Budget and Resource Calculations of each of them were analyzed in order to detect restrictions in the performance of their administrative-financial activities.

3 Results

The documentary analysis shows that there is a set of rules that stipulates the structure and operation of the irrigation system at users' organizations or Canal Inspections' levels.

First, there are all those water rights granted by the provincial legislation and registered by the General Irrigation Department. As these rights have originated from a constitutional rule, they are irremovable and pertain to the area to which they were assigned. Though these rights were allocated for irrigation purposes, in practice water rights have been incorporated for other uses.

Second, the Constitution stipulates "the autonomous administration of canals". This has given rise to a large number of laws concerning the powers and functional skills of Canal Inspectors:

- To equitably and efficiently administer and distribute water according to different uses and categories or registered rights.
- To prepare "rotation system schedules" for water delivery and to properly inform users of such schedules.
- To keep an updated register of water rights.
- To keep and update the official land register of the Inspection's command area.
- To execute the works and activities required for canal maintenance, conservation and improvement.
- To draw up the annual expense budget and calculation of the agencies' resources.
- To appoint or hire personnel.
- To control and prevent water and canal pollution (Constitution of the Province of Mendoza and Law N° 6405 of 1996).

The powers granted to Inspectors to operate and maintain the systems, together with those that allow them to exercise police power over pollution, have contributed to shape a real "environmental manager" of his command area.

In order to learn about the Inspectors' performance, their know-how ("knowledge") was qualitatively analyzed. The intensive interviews yielded four categories of know-how: 1) Know-how to identify the owner; 2) Know-how to meet users' needs; 3) Know-how to control water distribution; 4) Know-how to establish the irrigation charge.

From the analysis of the data collected in the survey, it was possible to identify the know-how that corresponds to each category and how Canal Inspectors use it.

Since Inspectors do not have access to updated information (which is monopolized by the General Irrigation Department) to be conveyed to plans and irrigators' registers, they resort to their empirical know-how and knowledge of the area to identify the owners by means of a sketch of the location of canal users'. 58% of the interviewed Inspectors use this sketch for water distribution and canal maintenance.

57,8% of the Inspectors use the irrigators' register to prepare the rotation system. This shows that they restrict their activities to water distribution using only the information contained in the register on the amount of hectares to be irrigated (Table 1).

In order to prepare the rotation system, which is essential for water distribution, the Inspector needs to know, among other things, the amount of water that shall be distributed that year. But in order to apply that know-how in the preparation of the rotation system schedule, the administrator must have profited from that know-how. There are at least two practical ways of determining the flow to be distributed and both have different precision levels. The first is qualitative and is expressed as a water height in the canal (in cm). The second is quantitative and is expressed in litters per second or cubic meters per second, which is an expression of flow. In the first case, the rotation system does not take into account the amount of water that is received at the Inspection's level and if that amount is correct. 42% of the Inspectors use the qualitative measurement, 32% of them do not know how much water the Inspection receives and 26% use the quantitative measurement. This clearly shows that water distribution is formal, not very accurate and follows uses and customs.

This type of management could be called "water distribution according to supply" because it does not acknowledge the user's demands and assumes that there is a solidarity system. The non-written rules set forth that due to water shortages, farmers distribute the available flow among themselves.

Table 1. Use of the register's information (in percentage and order of importance)

Uses*	1	2	3	4	5	6	NE, NS/C	Total Elections
A	57,80	10,50	15,50	-------	5,2	-------	10,50	100
B	21,05	36,80	-------	21,05	5,2	-------	15,70	100
C	5,20	21,05	15,78	10,52	5,2	5,2	36,80	100
D	-------	5,26	26,31	15,78	21,05	0,526	31,57	100
E	5,26	5,26	10,50	5,26	20,5	31,57	31,57	100

* uses: for preparing the rotation system schedules (A), for identifying the property's owner (B), for updating the register (C), for the distribution of the bill (D), for calculating default irrigators (E), not selected category (NE), does not answer (NS).

Water distribution is carried out according to a "rotation system schedule" which is prepared by the Inspector. Each farmer uses the water during the time (expressed in minutes) established in the schedule. Before preparing the rotation system schedule, the Inspector must know how much water is derived from the reservoir, the amount of irrigated areas and the way water is distributed ("bottom-up and top-down"). This rotation system is based on a solidarity system whose main assumption is that "irrigators can control water uses among themselves". The fulfillment of the rotation system is based on self-control.

Of all the resources that Inspectors manage, financial resources are the most critical ones since they are essential for efficiently performing the tasks assigned by the law and for meeting users' demands.

The collection of charges as compensation for water distribution services and system's maintenance is called "Canal's pro rata", which is in practice the only source of financial resources they have. Different factors, such as the economic crisis, the impact of agricultural activities on the oases and the existence of abandoned properties and/or default users, have gradually led to the financial deficit of these organizations.

The Inspections' collection levels are low and according to historical data they range between 50 to 60% of the amounts estimated in the Expense Budget and Resource Calculations they prepare every year. These values have slightly increased as from 1995 when the General Irrigation Department compelled Inspectors to cut off water supply to default users in compliance with the provisions of the Water Law [5].

At present, collection levels range between 60 to 70% of the estimated amounts (Table 2).

The source of the data in Table 2 is the Expense Budget and Resource Calculation of 21 Inspections of the Middle and Lower Tunuyán River.

- Operation and Maintenance: it includes the expenses of permanent and temporary personnel in charge of water distribution, control of the rotation system, canal cleaning and conservation, minor repairs in infrastructure, etc.
- Minor repairs: construction of intakes, flow dividers and bridges, repair of sluice gates and other tasks necessary for the operation of the infrastructure.
- Per diem expenses and transportation: since the post of Inspector is, in theory, honorary he is paid per diem and transportation expenses.
- Administrative expenses: administration expenses, accountant's fees, bank charges, office supplies, publicity, contributions to the Association, hardware, software and communications.
- Unforeseen expenses: purchase and maintenance of machinery, tools, well operation and maintenance expenses.

Table 2. Revenues and expenses of the inspections - fiscal year 1995

Inspection's area	Canal's pro rata (in $/ha)	Collection percentage	Budget Items (in %)					
			Operation and maintenance	Minor Works	Per diem expenses and transportation	Administrative expenses	Others	Total
Up to 5,000 ha	13	61	54	7	15	17	5	100
5,000 ha and more	12	65	61	6	13	10	7	100

As the investment percentage in works is rather small, little improvement is made in infrastructure. On the other hand, the so called "bureaucratic" expenses (per diem, administrative and minor expenses) represent more than 30% of the budget. The most important item in all Inspections is operation and management of the system which, in practice, is the most important task Inspectors perform (Table 2).

Management decentralization entails the possibility of constructing an autonomous decision-making space for Canal Inspections. It is based on users' participation and on the representation of interests. Users' participation has notably decreased due to the crisis that local, regional and national powers are facing [7]. The degree of autonomy by which they can take decisions is then reduced by their subordination to the General Irrigation Department. This is shown by the belief that "only the Subdelegation can update the irrigators' register". Though 47% of the Inspectors hold this position, 53% of them have updated the register. In spite of the fact that most of them are of the opinion that they have autonomy to take some decisions to overcome water shortages, they are aware of the fact that they cannot increase the number of rotation systems. This would call for a participatory form of management with the higher levels as regards the amount of water each should receive.

4 Conclusions

The causes for the system's weaknesses are: 1) the fact that Inspectors use their know-how only for water distribution and that it is poorly developed on account of the lack of adoption of technology and the limited access to new techniques; 2) the subordination to the General Irrigation Department; 3) the lack of financial resources and their allocation to expenses not related to the infrastructure's improvement; 4) the rigidity of the rules granting water rights. The main consequences of these weaknesses are that users' demands are not met and they, in turn, lack the incentives to pay the water charges. This gives rise to a financial deficit in the agencies and generates a vicious circle difficult to overcome. It is necessary to train water managers and to professionalize their activities. However, the incorporation of new technical-instrumental know-how shall depend on the transformation of the relations with users and of the subordinate relations with the General Irrigation Department.

It is also necessary to collect more financial resources and to allot them to expenses and investments defined by the users themselves. For instance, investment in infrastructure is a constant demand from the users and the main objective of the Inspections. If this demand is met, higher participation shall be achieved.

The budgets prepared by the Inspections are exactly the same for all of them and are suggested by the General Irrigation Department. This calls for the need to increase their autonomy. Training will enable Inspectors to know, interpret and enforce the rules in an effective way.

As regards rules, one of the rigidities of the water management system is the "water rights" that have been granted according to the area of each property but have not been divided according to uses. Socio-economic and environmental changes have rendered water allocation imprecise and inequitable. For instance, there are no differentiated water charges according to uses and it is the farmers themselves who have to bear the heaviest burden of the distribution system.

5 References

1. Matiello, H.(1992) *La administración del riego y el principio de participación de los usuarios.* CRA. INCYTH, Mendoza, Argentina.
2. FAO. (1982) *Organización, Operación y Mantenimiento de Sistemas de Riego.*Documento N° 40, Roma, Italia.
3. Clichevsky, N. et al.(1990*) Construcción y Administración de la Ciudad Latinoamericana* Instituto Internacional de Medio Ambiente y Desarrollo. Grupo Editor Latinoamericano. Buenos Aires, Argentina.
4. Brewer J.; R. Sakthivadivel, and K.V. Baju (1997) *Water Distribution Rules and Watwe Distribution Performance: a case study in the Tambraparani Irrigation System.* Research Report 12. IIMI,Colombo , Sri Lanka.
5. Chambouleyron, J et. al. (1996) *Evaluación del manejo y control de la calidad del agua de riego en Mendoza. Argentina.* Universidad Nacional de Cuyo, INCYTH, Mendoza.
6. Van deer Zaag, P. (1992) *Users, Operators and HydraulicStructures:A case of Irrigation Management in Western Mexico*, Network paper 19., Odi, London.
7. Bustos, R., (1997) *La Participación de los regantes en los recursos Hídricos: la constitución de espacios de lucha en los oasis de Mendoza*, Ponencia.46 Congreso Internacional de Americanistas, Quito, 7-11 de junio.

DEMAND MANAGEMENT BY IRRIGATION DELIVERY SCHEDULING

N. HATCHO
Department of International Resources Management, Kinki University, Nara, Japan

Abstract
Two major factors influencing the performance of in-field irrigation are volume and timing of water delivery. Both of these factors can be fixed or adjusted to meet crop water demands. Irrigation delivery scheduling with a supply-oriented approach often results in large losses of water because it does not reflect the actual crop water requirements. This paper discusses the application of the delivery scheduling module of SIMIS (for Scheme Irrigation Management Information System), developed by FAO, to irrigation systems in Argentina and Thailand, with their very different environments, in a check of its capabilities. Simulation of different delivery scheduling options was done for comparison of their results so that potentials and limitations of demand-oriented management can be clearly understood.
Keywords: Demand management, irrigation delivery scheduling, water saving

1 Introduction

From the point of water delivery to farmers' fields, it is very important to supply the appropriate amount of water at right timing so that optimum agricultural production can be obtained. The timing or interval and volume of water supply to the field can be defined by irrigation scheduling at farm level. Scheduling can be very flexible when both variables (timing and volume) are changeable, and can be very rigid when both are fixed. The complexity of operation and the burden of operator increase as the freedom of water use increases. Which scheduling methods to adopt in an irrigation system depends on the infrastructural setup and managerial capabilities as well as local conditions surrounding an irrigation system/irrigated agriculture.

Water and the Environment: Innovative Issues in Irrigation and Drainage. Edited by Luis S. Pereira and John W. Gowing. Published in 1998 by E & FN Spon. ISBN 0 419 23710 0

Four different types of delivery scheduling methods are included in SIMIS (Scheme Irrigation Management Information System); Supply Oriented, Fix, Rational, Optimal, and On-demand. With Fix method, both interval and volume of irrigation is fixed, while with Rational method, both interval and volume can be variable every month. Optimal method reflects changing soil moisture conditions and crop water requirements [1] and may be used for comparisons of what may be possible under most favorable conditions for water management. Under On-demand method, interval and application depth is requested by water users who judge their own irrigation needs.

A flow diagram of the Irrigation Scheduling Module of SIMIS [2] is shown in Figure 1. Weekly planning of irrigation scheduling is carried out based on the selected method and irrigation time ordering for each plot is converted to the flow requirements of each canal. The operation schedule of major intakes is calculated by adding up the flow requirement of different canals on an hourly basis. Possible crop damages due to water shortage and water losses by over-irrigation are also calculated.

Soil moisture balance (SMB) is used to monitor the effectiveness of water application and possible crop damage by water shortage. When water is applied more than the available volume of readily available moisture (RAM), the water is considered to be lost by over-irrigation. When all the RAM is depleted, possible crop damage is calculated by using yield response factor [3]. SMB at week t can be calculated as follows.

$$SMB_t = SMB_{t-1} + ERAIN_t + IRRIG_t - 7 \times CWR_t \tag{1}$$

where: SMB_t: soil moisture balance at the week t (mm)
SMB_{t-1}: soil moisture balance of the previous week t-1 (mm)
$ERAIN_t$: effective rainfall of the week t (mm)
$IRRIG_t$: net irrigation water application of the week t (mm)
CWR_t: crop water requirements of the week t (mm/day)

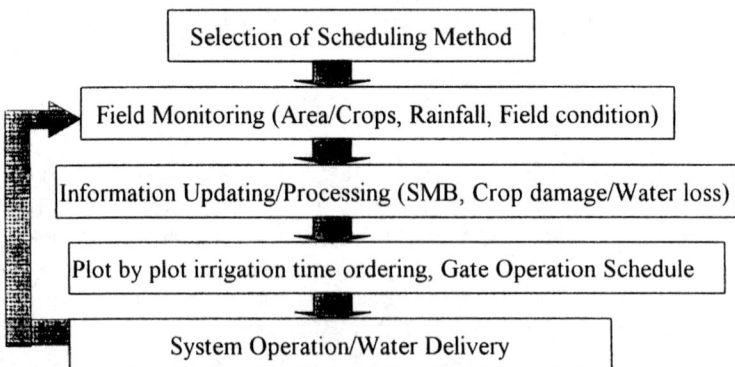

Fig. 1 Flow Diagram of Irrigation Scheduling Module of SIMIS

2 Application to the Mendoza project in Argentina

2.1 Project outline

Target irrigation system is in Mendoza Province, in the central west part of Argentina. The SIMIS irrigation scheduling module was applied to one tertiary unit, called the Hijuela (tertiary canal) Delgado, of the C. Cacique Guaymallen Irrigation System. The total irrigable area of the system is about 50,000 ha and the Hijuela Delgado has a total irrigable area of 386.5 ha, of which 332.6 ha is cultivated.

The climate of Mendoza is arid with a mean annual rainfall of 192 mm. There are two kinds of soil in this tertiary unit; one is sandy loam and the other is silt loam, with available soil moisture of 140 mm/m and 178 mm/m, respectively. Major crops grown in the area are vine, olive, fruits, alfalfa, tomato and garlic. In Mendoza, irrigation water is delivered by a supply-oriented approach, which distributes water by dividing the flow based on water rights or cultivated area.

2.2 Simulation of water management practices

Current water management practice was simulated for 16 weeks by use of the supply-oriented option (Supply Oriented method: SOM) of the irrigation scheduling module starting in August. 80 plots in the tertiary unit are grouped into 13 sections by the location of intakes. When the flow is too small, the tertiary unit is subdivided into two sub-units. The major difference between demand-oriented approach and supply-oriented one is whether there is a variation of water supply to each section or not. Under demand-oriented approach, it is assumed that each section has an access to the water supply whenever demand arises.

Comparisons in terms of total irrigated volume, water losses, and possible crop damage are carried out among the 4 methods of delivery scheduling. For the Fix method (Fix 1), fixed interval of 7 days and volume of 30 mm/application were used to every plot regardless of the type of crop or growth stage. The Rational method (Rat 1) used the same 7-day interval, but the application depth varied according to the crop water requirements (CWR). For the Optimal method (Opt 1), irrigation was planned when the soil moisture of a specific plot reached less than 20% of RAM and terminated at 80% of RAM.

2.3 Results of simulation in Mendoza

The results of simulation are shown in Table 1. Over Irrigation in the 4th column is the water lost due to over-irrigation beyond the limit of RAM, which is also a part of overall irrigation loss in the 3rd column. Among four scheduling methods, SOM and Fix 1 methods used a volume of about 2.4 million m^3 while Rat 1 and Opt 1 methods used only half as much. Delivery loss with SOM came from water allocated to plots which were not planted or cropped because water-right is the principle of water allocation. Nearly 15% of water withdrawn is lost by irrigating above RAM level and assigning water to non-irrigated plots. Overall irrigation efficiency, which was calculated by dividing the volume of net crop water use by total supply volume, was about 56% for Opt 1 and Rat 1 methods, while it was 35% for Fix 1 and 33% for SOM.

SMB calculated by Equation 1 was used to assess possible water losses and crop damage. SMB on a weekly basis is shown in Fig. 2 for a plot of garlic under different scheduling methods.

Table 1 Simulation resuluts different scheduling methods in Mendoza

Method	Total supply (m³)	Irrigation loss (m³)	Over- irrigation (m³)	Net crop use (m³⁾	Irrig Effic(%)	Crop damage(%)
SOM	2,375,655	1,582,666	(292,029)	782,990	33	0.4
Fix	12,357,500	1,540,931	(504,159)	815,570	35	0.04
Rat	11,160,659	511,092	(70)	649,603	56	0.02
Opt	11,232,147	541,836	(-)	689,310	56	0.03

For the SOM and Fix 1 methods, it is easy to see in Figure 2 how fields are over-irrigated beyond the level of RAM. When the crop had relatively small CWR at an early stage of growth and RAM was small with minimum root development, water could be easily wasted by the SOM or Fix 1 method because supply can not reflect CWR at a specific growth stage.

Crop damages are relatively small for all methods. This is because of the relatively short interval (7 days) adopted (except for Opt 1) method and the large application volume which resulted in water losses rather than crop damage. It is important to note that over-irrigation could result in a rise in the water table and waterlogging and salinization in arid and semi-arid areas

Based on the irrigation scheduling and irrigation time ordering of each plot, canal operation schedules were calculated. Operation requirements of quarternary canals were calculated by irrigation time ordering of each plot in the section. Quaternary canal operation was translated into operation requirements for a tertiary canal intake taking account of the time required for water to flow from the tertiary intake to the intake of each section.

2.4 Possible improvements of demand-oriented scheduling

Large water losses could be avoided by switching from a supply-oriented approach to a demand-oriented approach in this project. However, from the point of view of canal operation requirements, irrigation delivery scheduling based on a demand-oriented approach resulted in wide flow variations in the tertiary canals, necessitating frequent gate operations. In many developing countries, frequent and complex gate operations could pose heavy burdens on operation staff, and precise flow adjustment may not be

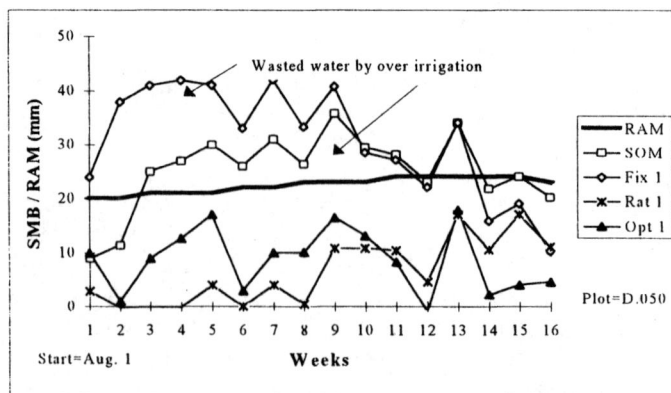

Fig. 2 Soil moisture balance under different delivery scheduling methods

possible to obtain.

To stabilize the canal flow on a daily basis, adjustments by applying the supply adjustment factor (SAF) which reduces the flow size by an assigned percentage (50-100%), and adopting rotational irrigation (14-day interval) were incorporated, and additional five simulations were carried out.

Simulation results of these adjusted scheduling methods for a 16-week period showed that total supply volume required to satisfy the crop water demands significantly declined as much as 25% in Fix 2 (7-day interval with 20-30mm application) and 38% in Fix 3 (14-day interval with 40-60mm application) methods compared with Fix 1, and came close to that of Rat or Opt method. Water losses for Fix 2 and Fix 3 methods due to over-irrigation declined as much as 20 to 30% over the original Fix 1method. The performances of Rat 2 and Opt 2 which used SAF of 50-80% were similar to those of original Rat 1 and Opt 1 methods.

The possibility of crop damage increased in the Fix 3 and the Rat 3 methods with 14-day intervals, because crops with small RAM could not store enough water and available water in the soil was depleted by the following irrigation. This also affected the application loss in Fix 3 and Rat 3 methods, because the application of larger volume simply exceeded the capacity of RAM and was lost. The use of appropriate interval and application depths which correspond to RAM and CWR is always important for better irrigation water management, as is demonstrated by the Optimal method.

The tertiary canal operation schedules under modified delivery scheduling methods showed that the fluctuation of flow became more stable compared with the original full flow case. As a result, gate adjustments and operation requirements will not be very high in these cases.

3 Application to Phetchabri project in Thailand

SIMIS was applied to the Phetchabri Project in Thailand to check the viability of the irrigation delivery scheduling module under paddy fields during land preparation and other growing periods. To allow SIMIS application under paddy cultivating conditions, calculation procedure for the specific water needs during land preparation and other periods were introduced.

3.1 Outline of Phetchabri irrigation project

Phetchabri Irrigation Project is located about 150 km southwest of Bangkok with a project size of 52,800 ha. The application of SIMIS was carried out in Zone 11 in the lowest reach of the Main Canal 3 with an irrigable area of 1,874 ha, of which 1,594 ha is cultivated under paddy rice. There are 1,161 plots in Zone 11 with a mean plot size of 1.35 ha. Soil types in Zone 11 are sand, sandy loam, loam, and heavy clay with irrigable areas of 127 ha, 604 ha, 31 ha, and 1125 ha, respectively. Since field conditions are homogeneous under paddy cultivation, and water is delivered by plot-to-plot flow after a field ditch, small plots were grouped into 49 large management units (MU).

There are three check-drop structures in Zone 11. Irrigation water is regulated by a check structure and diverted to secondary canals through an opening (orifice) of

D=300mm in a main canal which has the design capacity of 105 l/s under submerged condition. There are 34 secondary canals in the Zone. At present, the Zone is subdivided into three sub-zones. Each sub-zone uses the full main canal flow of 1.26 m^3, and rotated among 3 sub-zones on a weekly basis during land preparation period. Due to the lack of field ditches and different sizes of command area under each field ditch, irrigation water can not reach to the lower part under plot-to-plot flow.

Two major differences exit between paddy and upland crop irrigation water management. One is the additional water requirement by deep percolation and the other is the ponded water on a field surface. Deep percolation value, which varies with soil type, is subtracted from the weekly soil moisture balance. For upland crop water management, RAM is used as a criterion for judging shortage or excess of water, but in paddy cultivation, the variation of ponded water layer on the field surface is used as a criterion as if it were a RAM of upland crops. During SIMIS application, ponded water depth is monitored instead of RAM for possible shortage or excess of water on a weekly basis.

3.2 Scheduling methods

Five different delivery scheduling methods similar to the application in Mendoza were used to simulate irrigation water distribution for 16-week period from June to September. For Fix method, two different application volumes were used with 40mm and 60 mm. Zone 11 was grouped into 3 sectors and 12 sections. Criteria for grouping sections was the location of check structure, the plot location and the irrigation area of each section. Water intake from the main canal to the tertiary unit (section) was fixed with an opening of 105 l/s, and plots were grouped into equal section size as much as possible to allow better water allocation.

In case of the demand-oriented methods of delivery scheduling, it was assumed that intake gate of 105 l/s was opened and closed according to the calculated irrigation time ordering. Instead of adjusting irrigation depths by a flow size (gate operation), incoming volume of irrigation can be adjusted by controlling the duration of gate opening. All sections were assumed to receive the flow of 105 l/s which was used rotationally among each MU in the section. There are 42 intake gates to each MU to operate, and a zone man needs to know when to open and close these gates so that canal flow can be used most effectively.

3.3 Results and discussion of Phetchabri application

Results of simulating water distribution and allocation are shown in Table 2. In addition to demand-oriented approach, the performance of the Supply Oriented method (SOM) was tested, with an assumption that a proportional allocation of flow to each section was possible. Differences of total supply volume between SOM and the other four methods based on a demand-oriented approach was relatively small. However, the reduction of supply as calculated in Table 2 could only be possible when proper and precise operation is carried out, which requires skilled operation personnel and careful flow monitoring.

SOM showed the highest supply volume followed by Fix-60 and Opt methods. Efficiency values of all 5 methods showed relatively high values, suggesting that irrigation water loss by over irrigation was small in case of paddy fields because of large

Table 2 Simulation results of different scheduling methods in Phetchabri

Method	Total supply (m³)	Irrig loss (m³)	Over-irrigation (m³)	Net crop use (m³)	Irrig. Effic. (m³)	Crop damage(%)
SOM	15,482,880	6,068,649	(377,245)	9,414,231	61	5
Fix-40	9,632,716	3,496,012	(27,666)	6,136,704	64	24
Fix-60	13,131,431	4,854,364	(127,026)	8,277,067	63	9
Rat	12,149,754	4,378,993	(4,556)	7,770,761	64	7
Opt	13,090,954	4,721,695	-	8,378,250	64	0

storage capacity on the field surface.

It should be kept in mind that in some MUs the ponded water depth fell below 0 level. When the ponded water level on the field surface became less than 0, which was monitored by SMB (ponded water depth in case of paddy fields) on a weekly basis, a possible crop damage was calculated by applying a yield response factor. Possible damage was the highest for Fix-40 method, followed by Fix-60 method. It does not necessarily mean that 24% of paddy crop will be damaged for Fix-40 method, but it shows higher risk of crop damage due to water shortage. Thus very small application of water can increase irigation efficiency but could lead to crop damage.

MU with sandy soil could not maintain the ponded water layer, and the water depth dropped to 0 mm in the second week of July, which could lead to possible crop damage. SOM method could assure equal access of water with respect to the size of holding, but could not reflect variations in water demands by soil types, which lead to over-irrigation or possible crop damage.

In demand-oriented methods, frequent adjustment of check gate in the main canal would be necessary to achieve the performance shown in Table 2. Gate adjustment requirements of different scheduling methods under three sectors showed large fluctuations Opt and Rat methods. To minimize the operation requirements and achieve better performance, simulation results can be used to operate check gates and flow allocations. Check gates can be operated on a weekly basis and maximum or mean flow can be used as the actual operation flow for the entire week so that a daily gate adjustment would not be required as has been done in the case of Mendoza by SAF. This type of operational adjustment could be introduced easily by knowing how water should be controlled at each operating points.

It was found that under paddy cultivation with homogeneous field conditions the application of SOM or Fix method which does not necessarily reflect crop water demands can also be a viable delivery scheduling option, because the ponded water on the field surface can work as a buffer against the short-term shortage or excess of water.

4 Conclusion

Water management practices of two different irrigation systems have been simulated and analyzed by applying the irrigation delivery scheduling module of SIMIS. In the application in Argentina, a new option based on a supply-oriented approach has been added to simulate current management practice. It has been shown that this current practice based on a supply-oriented approach involves large water losses, because the water allocation is based on water rights and the supply does not reflect the actual crop water demands.

Scheduling methods based on CWR or SMB could save as much as 50% of water supply compared with a supply-oriented approach. However, the demand-oriented approach poses problems from the point of view of practical canal operation, because flow variations and hence the operation requirements in the tertiary canals will be high. Five additional simulations have been carried out to better adapt the method to practical management practice by introducing a supply adjustment factor and a weekly rotational irrigation. The results were very encouraging, further reducing the supply volume and improving the operational constraints caused by varying flow in tertiary canals.

In Thailand, the program was tested for its applicability under paddy cultivation condition, which has a distinct irrigation practice compared to upland crops, especially during land preparation period. Under upland crop cultivation, the shortage or the excess of water supply could directly lead to crop damage or water loss. However, in paddy fields the ponded water layer on the surface could work as a buffer against the limited supply, and minimize the impact of water shortage. Lowered water layer can be recovered when the supply becomes abundant or with big rainfall.

As has been discussed, irrigation performance can be different by adopting different scheduling approaches. It can not be improved only by operational measures, but associated infrastructural improvements are required. To determine which scheduling method to introduce to a specific irrigation system, local specific conditions such as the cost of labor and infrastructural setups, the capacity of staff to manage the system, and the value of water should be taken into account.

5 References

1.   Doorenbos, J. and Pruitt, W. O. (1979) *Crop Water Requirements* (revised). Irrigation and Drainage Paper 24. FAO, Rome.
2.   N. Hatcho(1995) Scheme Irrigation Management Information System (SIMIS) with Database Application. *Journal of JSIDRE*. Vol. 63. No. 4. pp.31-36.
3.   Doorenbos, J. and Kassam, A. (1992) *Yield Response to Water*. Irrigation and Drainage Paper 33. FAO, Rome.

PERFORMANCE ANALYSIS OF ON-DEMAND PRESSURIZED IRRIGATION SYSTEMS

N. LAMADDALENA
Centre International de Hautes Etudes Agronomiques Mediterranéenes, Bari Institute, Italy
L.S. PEREIRA
Department of Agricultural Engineering, Institute of Agronomy, Technical University of Lisbon, Portugal

Abstract
The spatial and temporal variability of flow regimes occurring in pressurized collective irrigation systems produces important variations on the pressure available at the hydrants. Both the on-farm system operation and the crop yield may be affected. The need for the analysis of collective irrigation systems, mainly where the spatial and temporal variability of flow regimes might be important, is a priority issue for the sustainable use of water. Under this perspective, a study on the analysis of a pressurized collective irrigation systems operating on-demand is presented. The maximum upstream discharge being known, a random model is used to generate scenarios of flow regimes. A model for analysis and reliability assessment is then utilized to identify where failures of the pressure at the hydrants would occur for those flow regimes. This approach leads to the identification of deficient hydrants and makes possible to identify the problem areas for which specific measures are required.
Keywords: Flow regimes, on-demand operation, performance analysis, pressurized irrigation systems

1 Introduction

Designers and managers of irrigation systems, to meet the needs of farmers, often prefer on-demand delivery schedules. These systems allow a wider freedom to farmers respect to the other types of delivery schedules.

In order to avoid oversizing the pipes of the network, the discharges are usually computed by using probabilistic approaches in which the Gaussian distribution of discharges is hypothesized [1]. In these methods a threshold of risk is accepted, i.e.,

Water and the Environment: Innovative Issues in Irrigation and Drainage. Edited by Luis S. Pereira and John W. Gowing. Published in 1998 by E & FN Spon. ISBN 0 419 23710 0

during the operation of the system, discharges higher than the designed ones may occur changes in cropping patterns make that operation conditions are often different from those expected at design. As observed in the "Sinistra Ofanto" irrigation scheme, managed by the Consorzio di Bonifica of Capitanata (Foggia-Italy), even when the designed discharges are not exceeded, very low hydraulic performances may occur due to important pressure deficits during the whole irrigation season, mainly during the peak periods.

These conditions pushed the authors to analyze the hydraulic behavior of one district of the "Sinistra Ofanto" system in order to identify the causes of the failures and measures which could be adopted for improving its hydraulic performance.

Several procedures for the hydraulic analysis of pressurized irrigation systems have been developed. Among these, the model ICARE [2] [3] is often utilized. However, the information provided by ICARE refers to the entire system and not to each hydrant, and a configuration is considered not satisfied when at least one hydrant has an hydraulic head smaller that the target one. To use the advantages of ICARE and enlarge the results of the analysis by evaluating the behaviour of each hydrant, a new model was developed. The application of this model and ICARE to the same system in Sinistra Ofanto is presented in [4]. This paper reports on the proposed approach.

2 Hydraulic and performance analisys

2.1 Model AKLA

The model AKLA [4] [5] is based on assumptions utilized in model ICARE, enlarging the analyzis to evaluate the hydraulic performance at the level of each hydrant of the network. This model, using a random number generator having a uniform distribution function, adopts a multiple generation of a pre-fixed number of hydrants simultaneously operating, which define a configuration.

Within each generated configuration r (r = 1, 2, ..., c), a hydrant j (j = 1, 2, ..., N) is considered satisfied when

$$H_{j,r} \geq H_{min} \tag{1}$$

where $H_{j,r}$ [m] is the hydraulic head at the hydrant j within the configuration r, and H_{min} [m] represents the minimum required head for the appropriate operation of the on-farm systems. The relative pressure deficit at each hydrant is computed as:

$$\Delta H_{j,r} = \frac{H_{j,r} - H_{min}}{H_{min}} \tag{2}$$

Assuming that each hydrant may withdraw the nominal discharge d [l s^{-1}] also when its head is lower than the minimum required (H_{min}), if the total discharge Q_r [l s^{-1}] is fixed at the upstream end of the network, the number R of hydrants simultaneously operating is:

$$K_r = Q_r / d \tag{3}$$

In the case of different hydrant discharges, random drawing will be performed to satisfy the relatioship $\mid Q_{tir} - Q_r \mid < \varepsilon$ where Q_{tir} is the discharge corresponding to K hydrants drawn at random ($l \ s^{-1}$) and ε is the accepted tolerance. In general, ε is assumed as equal to the value of the lowest hydrant discharge.

The following procedure is adopted for the computation. Once the available piezometric head at the upstream end of the network, Z_0 [m a.s.l.], is established, the set of discharges to be tested, Q_r, and the number C of configurations to be investigated for each discharge are selected. From the Eq. 3, the number of hydrants corresponding to each discharge Q_r is calculated. Later, by using a random generation model, the R hydrants simultaneously operating are randomly drawn. This procedure is repeated C times for each discharge Q_r. The number of configurations to be tested should be higher than the number of hydrants of the network (C > R)in order to obtain reliable results.

The discharges flowing in each section of the network for each discharge Q_r are obtained by aggregating, from downstream to upstream, the discharges delivered by the selected R hydrants. Then, starting from the upstream piezometric head, Z_0 [m], the head losses are calculated and, consequently, the head ($H_{j,r}$) available at each hydrant in each configuration is computed. Those hydrants having a head lower than the minimum pre-established (H_{min}) are identified. They will be defined as unsatisfied hydrants.

For each configuration, the range of variation of the head at each hydrant is identified. Indeed, the relative pressure deficit, $\Delta H_{j,r}$ (Eq. 2), may be represented in a plane where abcisses correspond to the hydrants numbers and the ordinates to $\Delta h_{j,r}$. In such a way, the hydrants which may be most subject to insufficient pressure head, and hence the critical zones of the network, may be clearly identified.

2.2 Reliability model

The system performance can be also described when analysing "how often the system fails". This is a measure of the system reliability [6].

The definition of this criterium is formulated assuming that the performance of an irrigation system can be described by a stationary stochastic process. It means that the probability distributions describing the time series (in this case the time series of pressure heads and discharges at the hydrant being considered) do not change with time. This hypothesis is only an approximation but, particularly during the peak periods, it is quite reasonable to be accepted.

Let X_t be the random variable denoting the state of the system at time t (where t may assume values 1, 2,......, n_t). In general, the possible values of X_t may be shared into two sets: S, the set of all satisfactory outputs, and F, the set of all unsatisfactory outputs (failure). At each instant t the system may fall in one of the above sets. The reliability of a system can be described by the probability α that the system is in a satisfactory state:

$$\alpha = \text{Prob } [X_t \in S] \qquad (4)$$

From the definition of reliability given above, the following relationship can be assumed:

$$\alpha_j = \frac{\sum\limits_{r=1}^{c} Ih_{j,r}\ Ip_{j,r}}{\sum\limits_{r=1}^{c} Ih_{j,r}}$$

(5)

where α_j is the reliability of the hydrant j and

$Ih_{j,r} = 1$	if the hydrant j is open in the configuration r
$Ih_{j,r} = 0$	if the hydrant j is closed in the configuration r
$Ip_{j,r} = 1$	if $H_{j,r} \geq H_{min}$
$Ip_{j,r} = 0$	if $H_{j,r} < H_{min}$

3 The case study

This study has been carried out in an irrigation district of the irrigation scheme Sinistra Ofanto, Consorzio di Bonifica of Capitanata, Foggia, Italy. The scheme (Fig. 1), covering an area of about 22500 ha, is approximately triangular-shaped, confined in the south by the Ofanto river and in the south-east by the town of Cerignola. The system is divided into seven irrigation districts (numbered 4 to 10) which are, in turn, subdivided into sectors with surface areas ranging from 20 ha to 300 ha. The irrigation districts are served by storage and daily compensation reservoirs supplied by a conveyance conduit which originates from the Capacciotti dam (Fig. 1). The pressurized irrigation network in each district is designed for on-demand delivery scheduling.

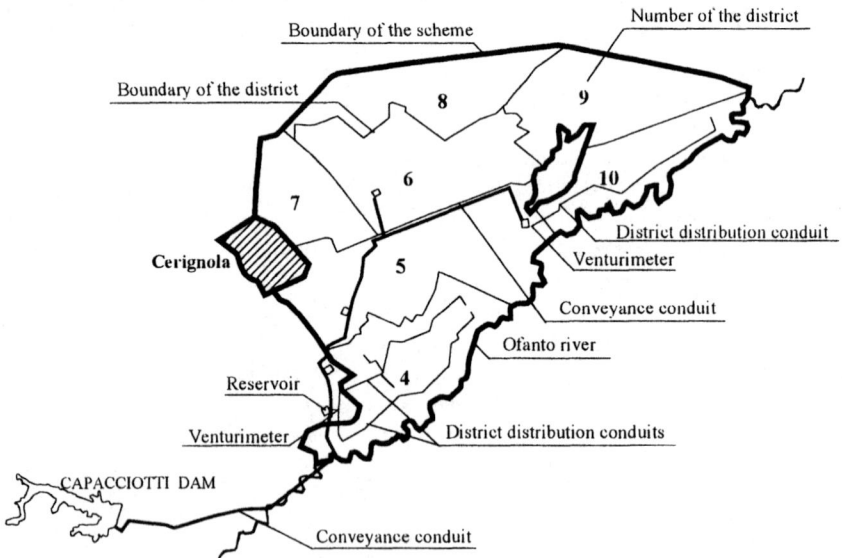

Fig. 1. The "Sinistra Ofanto" irrigation scheme, Foggia, Italy

The district 4 (Fig. 1) was chosen for the survey since a calibrated flow meter is available at the upstream end of the distribution network. District 4 has a topographic area of 3256 ha, and is supplied by a storage and daily compensation reservoir having a capacity of 28000 m^3, where the maximum water level is 143 m a.s.l. and the minimum water level is 139 m a.s.l.

The district distribution conduits consist of underground steel pipes starting with a diameter of 1200 mm. This conduit supply 32 sectors. A control unit is installed at the head of each sector and consists of a gate, a Venturi meter with recorder and a flow regulator. The sectors distribution networks serve the farms with hydrants having nominal discharge d = 10 l s^{-1}. The District 4 network is composed by 903 nodes of which 660 with hydrants. The discharges have been calculated with the Clément model and the diameters of the conduits have been calculated using linear programming techniques.

During the irrigation season 1991, the district 4 irrigation system operated continnously on-demand. It was therefore possible to perform a statistical analysis of the demand observed in the system using the records available from a Venturi-meter. During the peak period, it was obtained a good fitting between the theoretical Gaussian curve and the histogram of frequencies computed from observed field data during daytime water withdrawals (Figure 2). It means that the population of the discharges during this period is well represented by the Clément's formula, according to the design hypothesis [7].

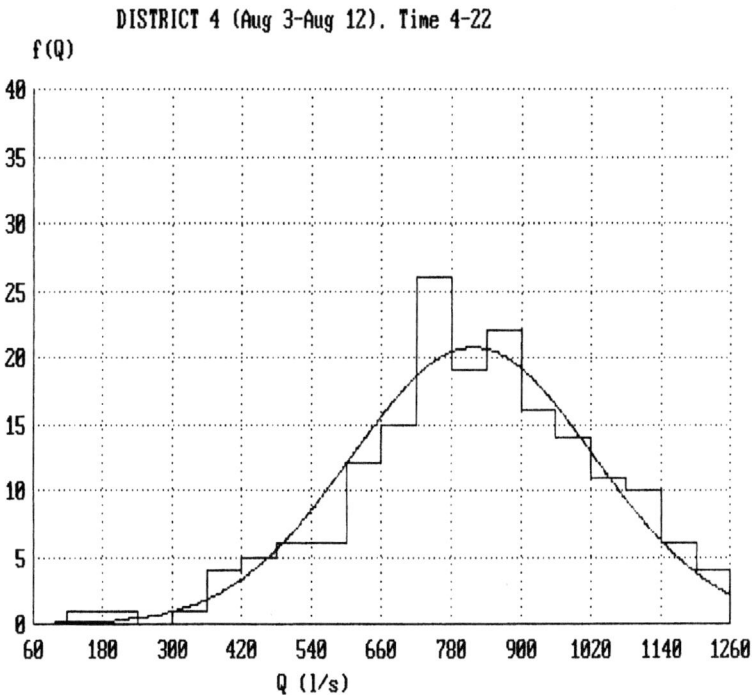

Fig. 2. Discharge frequency histogram. Peak period from August 3 to August 12.

Furthermore, the analyses relative to the peak period have shown that the maximum design discharge of 1200 l s^{-1} has been recorded with a frequency of 3%. Nevertheless, very important pressure head deficits occur for a large number of hydrants, as show below. Other information on this system, namely concerning cropping patterns and operation conditions during peak demand periods, as given in [8].

4 Application

The application of the model AKLA was performed taking into account an upstream piezometric elevation Z_0 = 140 m a.s.l. and the upstream discharge Q_r = 1200 l s^{-1}, corresponding to the design discharge. 1000 random discharge configurations have been generated in correspondence of that discharge. The model allows to calculate the relative pressure deficit for each hydrant in each configuration. They are represented in Figure 3 from which it is possible to see that the whole network falls in deficit conditions respect to the minimum required head (H_{min} = 20 m). Furthermore, from Figure 3 it is possible to observed that the relative pressure deficit, ΔH, is lower than -1 for several hydrants. This corresponds to a very dangerous condition because when negative pressure occur air may entry in the network and cause cavitation problems.

The performance improves when the discharge Q_r = 700 l s^{-1} is analyzed. The hydrants with negative pressure heads considerably decrease but, nevertheless, still cavitation problems persist in the network (Figure 4). This analysis shows that the system is undersized and requires that many sections of the network need to be reinforced. At collective level it is also possible to install in-line lifting units, or to

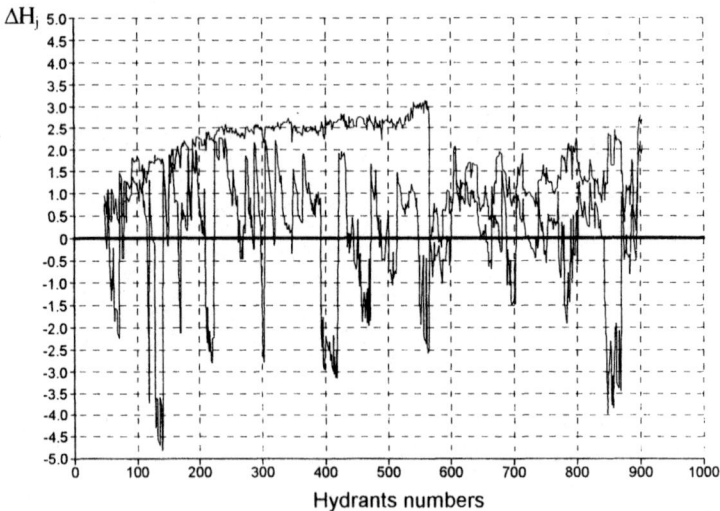

Fig. 3. Envelope curves for the relative pressure deficits at the hydrants of the district 4 when the upstream piezometric elevation is Z_0 = 140 m a.s.l. and the upstream discharge is Q_r = 1200 l s^{-1}. Simulations were performed for 1000 random discharge configurations.

impose limitations to the farmers' freedom to withdraw water during the peak demand hours.

This analysis is referred to on-demand operation. Previous analysis [9] showed that when arranged demand is adopted, the frequency of discharges higher than 700 l s^{-1} (causing very important failures in the system) increase. Therefore, hydraulic problems of the network increase when the on-demand operation is modified [8]. These results are confirmed from field surveys carried out through questionnaires and interviews to farmers. Farmers identified problems in functionning of the respective farm irrigation systems, sprinkler and microirrigation. Solutions at farm level include: to install booster pumps downstream of the critical hydrants, to optimize the design of the on-farm system in order to reduce head losses and improve operation, to suggest to the farmers in the critical areas to avoid irrigating during peak hours, or to suggest to the farmers to choose low pressure farm irrigation methods.

The reliability indicator (Eq. 5) has been applied for better quantify the performance of each hydrant of the existing system of district 4. It has been calculated from the results obtained by the model AKLA described above.

A set of 1000 random discharge configurations, corresponding to 120 hydrants simultaneously operating (Q_t = 1200 l s^{-1}), has been generated by using a random generation model. For each discharge configuration the analysis performed with the model AKLA gives the available head $H_{j,r}$ at each operating hydrant. Then, the indexes $Ih_{j,r}$ and $Ip_{j,r}$ (Eq. 5) may be easily calculated and Eq. 5 may be applied for calculating the reliability.

The reliability of the hydrants for the actual District 4 network is reported in Figure 5. It can be observed that there are areas where the hydrants reliability is very low (less than 0.7) and areas where the system is completely unreliable.

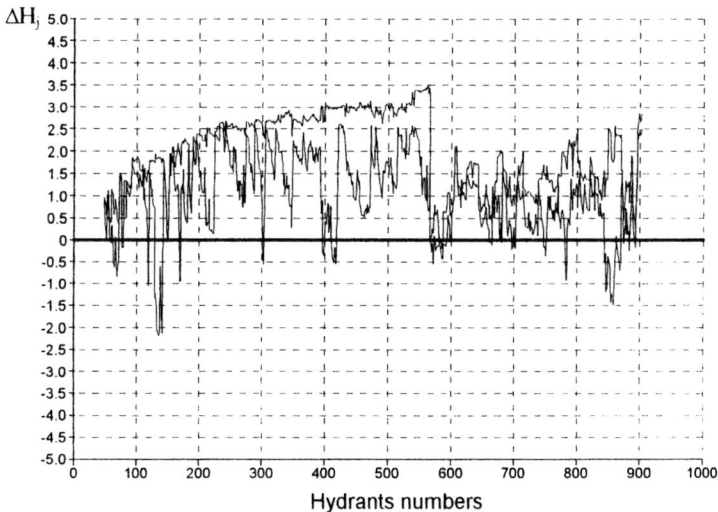

Fig. 4. Envelope curves for the relative pressure deficits at the hydrants of the district 4 when the upstream piezometric elevation is Z_0 = 140 m a.s.l. and the upstream discharge is Q_t = 700 l s^{-1}. Simulations were performed for 1000 random discharge configurations

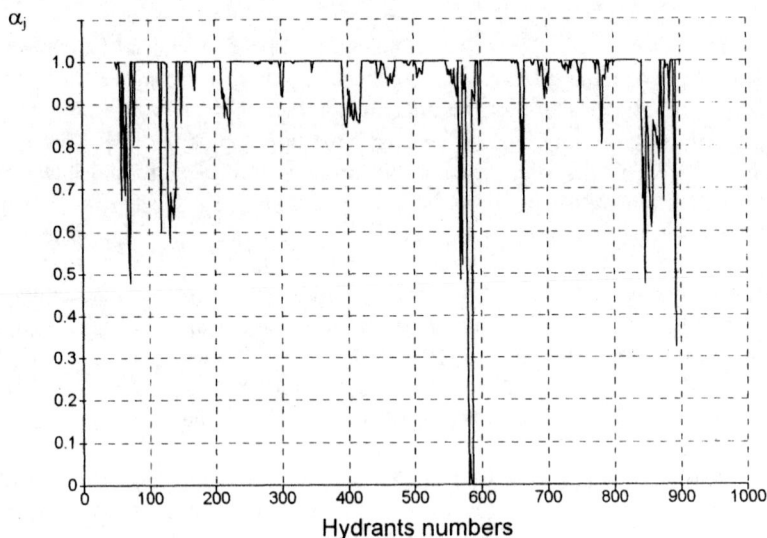

Fig. 5. Reliability of deliveries in Sinistra Ofanto (district 4), Foggia

5 Conclusions

This study shows that important failures may occur in a pressurized irrigation system network in terms of pressure deficits even when the design discharge is not exceeded. When these deficits are very large cavitation problemsmay occur. This condition should be always avoided. In all cases, very unfavorable conditions are created for on-farm water use, leading to low irrigation performances.

The application of the analysis models showed that the system is not able to supply the design discharge (1200 l s^{-1}). When lower discharges are withdrawn (i.e. 700 l s^{-1}) important pressure deficts also occur. Therefore, local rehabilitation is required by increasing the diameters of those sections upstream of the areas when failure occur. The analysis of the reliability of deliveries at hydrants level produce an appropriate system performance indicator. Because the analysis as proposed herein are performed for identification of problems at the hydrants scale, the information produced may support improvements both at system and farm scales.

6 References

1. Clément R., 1966. Calcul des débits dans les réseaux d'irrigation fonctionnant à la demande, *La Houille Blanche*, 5, 553-575.
2. Bethery J., Meunier M., Puech C., 1981. Analyse des defaillances et etude du renforcement des reseaux d'irrigation par aspersion, in *Onzieme Congres de la CIID*, question 36, 297-324.

3. Bethery J., 1990. *Reseaux collectifs d'irrigation ramifiés sous pression. Calcul et fonctionnement*, CEMAGREF, Etudes n. 6.

4. Lamaddalena N., 1997. Integrated simulation modeling for design and performance analysis of on-demand pressurized irrigation systems. Ph.D. Dissertation. Technical University of Lisbon. Lisbon (Portugal).

5. Lamaddalena N., 1995. Un modello di simulazione per l'analisi del funzionamento delle reti irrigue collettive, *Rivista di Ingegneria Agraria*, 4, 221-229.

6. Hashimoto T., Stedinger J.R., Loucks D.P., 1982. Reliability, Resiliency and Vulnerability Criteria for water resources system performance evaluation, *Water Resouces Research*, 18 (1), 14-20.

7. Lamaddalena N., Ciollaro G., 1993. Taratura della formula di Clément in un distretto irriguo dell'Italia meridionale, in *Atti del V Convegno Nazionale A.I.G.R. su "Il ruolo dell'ingegneria per l'agricoltura del 2000"*, AIGR, Ed. Europa (Potenza), 101-110.

8. Lamaddalena N., Ciolaro G., Pereira L. S. 1995. Effect of changing irrigation delivery schedules during periods of limited available water. *J. Agric. Engng. Res.* 61: 261-266.

9. Lamaddalena N., 1996. Problematiche connesse alla gestione ed all'esercizio di un sistema irriguo collettivo in periodi di disponibilità idrica limitata. Cinque anni di osservazioni, in *"Scritti dedicati a Giovanni Tournon"* Ass. Ital. Ing. Agr., Novara: 237-246.

NEW APPROACHES TO DESIGN AND PERFORMANCE ANALYSIS OF LOW PRESSURE DISTRIBUTION SYSTEMS

A. DOUIEB and R. BOUNOUA
Institut Agronomique et Vétérinaire Hassan II, Rabat, Morocco
L. S. PEREIRA and P. L. SOUSA
Department of Agricultural Engineering, Institute of Agronomy, Technical University of Lisbon, Portugal
N. LAMADDALENA
CIHEAM, Bari Institute, Bari, Italy

Abstract
Low pressure pipe distribution systems for surface irrigation provide for recognized environmental benefits at both off- and on-farm. However, expected benefits can only be attained when reliability, dependability and equity of systems are high enough to support appropriate conditions for water use at farm. An innovative methodology for design and analysis is proposed, which includes the generation of farm demand and flow regimes for the peak demand period, the optimization of pipe sizes using several flow regimes, and the performance analysis from simulations with several flow regimes. An application to one sector of the Sorraia irrigation system illustrates the method proposed.
Keywords: Dependability, design, equity, flow regimes, generation of farm demand, low pressure irrigation systems, performance analysis, reliability.

1 Introduction

Low pressure buried pipe distribution systems for surface irrigation constitute a valuable alternative to open channel distributors [1]. Under the environmental perspective, advantages relate to reduced water losses, more efficient use of agricultural land, reduced damage of land through waterlogging and salinity, reduced damage of water resources, greater transit efficiency, control of aquatic weeds and associated pests, control of disease vectors, and control of human water illness, namely schistosomiasis and malaria.

Benefits are particularly important at farm level because pipe systems enable greater flexibility and reliability of deliveries due to shorter transit times and smaller system

Water and the Environment: Innovative Issues in Irrigation and Drainage. Edited by Luis S. Pereira and John W. Gowing. Published in 1998 by E & FN Spon. ISBN 0 419 23710 0

losses than open surface systems. Pipe systems make easier to match water supplies to crop demand, providing conditions for more efficient water application at farm level, and contribute to eliminate tail-end equity problems. This results in improved conditions for controlling water wastes, for reducing the transport of solutes out of the root zone, and for increasing the efficiency of water in agricultural production.

Appropriate design is an essential condition for achieving the expected benefits. Associating the performance analysis with optimization of pipe sizes provides for the selection of the design alternatives that enable high performances associated with reduced costs. With these objectives, a new design methodology has been developed based on that used for collective pressurized systems. Pipe sizes are optimized using the iterative discontinuous method [2] applied to several flow regimes[3] [4], and the performance analysis is adapted from Bethery [5] and Lamaddalena [6] to derive appropriate estimators for reliability, dependability and equity.

2 Brief description of the model. Input data

The programme MSGOA has been developed to perform the generation of the flow regimes, the optimization of the pipe diameters, and the analysis of performance by simulating the network functionning at several flow regimes. MSGOA is written in Turbo Pascal 7.0 for MSDOS and WINDOWS. The model is limited to a system with a maximum of 150 outlets, 10 crops and 10 irrigation methods.

Input data are the following: (a) upstream discharge Q_0 [l s^{-1}], and hydraulic head H_0 [m]; (b) cropping pattern, with indication of the probability of occurrence of each crop, C_k [%]; (c) irrigation methods and the respective probability of occurrence, M_i [%], including the percentage of automation; (d) net irrigation requirements during the peak month, I_n [mm]; (e) average irrigation intervals m [days] and respective range a [days] according to the irrigation method and soil type; (f) application efficiencies relative to each couple irrigation method-soil type, e_a [%]; (g) day time hours for water supply as established by the management agency; (h) nominal discharge Q_n [l s^{-1}] and minimum head H_{min} [m] at the outlets; (i) soil type, land slope type and area served by each outlet, which provides the soil type distribution S_s [%]; (j) system layout with identification codes for each node and respective land elevation; (k) lengths of each pipe section between two successive nodes and, when analysing existing systems, the respective diameters; and (l) characteristics of the commercial pipes, mainly diameters and costs.

Data relative to items (a) through (h) are introduced with help of user friendly windows and can be modified from one session to another. System characteristics data are introduced through the sector files.

The operation of the model is commanded by a main menu through which the user can select: (1) the generation of flow regimes, (2) the optimization of pipe diameters, and (3) the analysis of performances. Generated flow regimes and pipe sizes (lengths and diameters) are stored in output files to be used in subsequent calculations. Model results concerning each one of the three main options are given as numerical or graphical outputs after selection by the user through an appropriate menu.

3 Generation of flow regimes

Each flow regime is defined as a combination of discharges flowing in the system in correspondance with each configuration of outlets simultaneously operating. In opposition to pressurized irrigation systems operating on demand, where each configuration can be randomly generated [4] [5], flow regimes have to be obtained from demand hydrographs which respect the arranged delivery schedules used in surface irrigation systems. Thus, the generation of flow regimes requires: first, the definition of the irrigation demand schedules for the areas served by each outlet; then, their aggregation at system level respecting the available upstream discharge; and, finally, the generation of the hourly demand hydrographs. Values for each hour, which correspond to the discharges at the outlets operating simultaneously at that hour, are then utilized to define the flow regimes.

Computations are made assuming that only one crop can be assigned to the area served by each outlet when the crop distribution over the total area is respected. Thus, knowing the percentage distribution of each crop in the project area C_k, it can be assumed that the probability for any crop k to occur in the area served by each outlet is equal to C_k. Therefore, adopting a random generation of numbers from 0 to 100 with uniform distribution, and assuming that each probability C_k [%] corresponds to a portion in the interval 0 to 100, it is possible to randomly select the crop k assigned to each outlet.

After this operation is concluded for all outlets, it is verified if the simulated crop pattern matches to the one proposed by summing to all area the surfaces assigned to each crop. When more than 10% differences are observed, the operation is repeated until satisfactory results are obatined.

The irrigation methods considered in this version are the traditional short blocked furrows, automated and non automated furrows and level basins, flooded rice basins, automated and non automated solid set sprinklers, drip irrigation and line source microirrigation. Low pressure pipes do not deliver water for pressurized systems but these can be supplied when appropriate pumps are available.

The probability for a given irrigation method i to be associated with a type of soil s is estimated by $M_i S_s$ [%], and the probability that it would be associated with a crop k corresponds to $M_i C_k$ [%]. Considering that the soil type s and the crop type k are known for the areas served by every outlet, it is possible to assign an irrigation method to each outlet area when the probability for an irrigation method i to be associated with the crop k and the soil s be known. This probability is estimated by

$$\left(S_s C_k\right)_i = \frac{\left(S_s M_i\right)\left(C_k M_i\right)}{\sum_{i=1}^{n}\left(S_s M_i\right)\left(C_k M_i\right)} \tag{1}$$

Using a procedure similar to that indicated for the random assignement of the crops to each outlet service area, the irrigation methods are also randomly defined for each outlet. It becomes then possible to associate a crop, a soil and an irrigation method to each outlet area.

For each couple irrigation method - soil type, the user selects the average interval between irrigations, m [days], and the range of variation, a [days], allowed for this interval. These data are then used by the model to randomly generate an irrigation interval F [days] for each outlet. The procedure consists in: (1) assigning to each value V_i [days], in the interval [(m - a), (m + a)] the lower and upper limits $R_i = V_i - 0.5$ and $R_{i+1} = V_i + 0.5$; (2) converting these real numbers R_i into the normal variables $X_i = (R_i - m) / \sigma$, where σ is the standard deviation of R_i (i = 1, 2,..., n'); (3) computing from the normal distribution the probabilities $P_i = P (X > X_i)$; (4) randomly generating a real number (0 to 100) which falls in one of the intervals $[P_i, P_{i+1}]$; (5) computing back, from these probabilities, the variables X_i, X_{i+1} and R_i, R_{i+1}; (6) determining the value V_i in the interval $[R_i, R_{i+1}]$, which is the estimator for F [days].

The average irrigation depths D_{av} [mm] during the peak period are computed from the input data relative to the net monthly irrigation requirements I_n [mm] and the application efficiencies e_a [%], and from the computed irrigation intervals F [days]:

$$D_{av} = 100 (I_n / e_a) (F / 30) \tag{2}$$

Actual irrigation depths D [mm] are computed from D_{av} considering that they can vary within the interval $[D_{av} (1 - d), D_{av} (1 + d)]$, where d is a fraction of D_{av}. Thus,

$$D = D_{av} (1 + d) - \alpha \left(D_{av} 2d\right) \tag{3}$$

where α is a random generated number [0,1]. The final result is rounded up.

For each outlet, the first day of irrigation during the peak month is randomly defined between 1 and the minimal value for the irrigation interval. Next irrigations are scheduled by adding the respective irrigation intervals F. The time duration of each irrigation is computed from the ratio irrigation volume/discharge available at the outlet. This allows to establish the daily schedule of the irrigations since the hours in the day when the irrigation management agency supplies water are known from input data. When automation is considered, irrigation is also considered during the night hours. Rice irrigation is performed using a constant discharge rate during the night hours or for the 24 hours.

After establish the daily irrigation schedules are established at each outlet, the discharges are summed up and a preliminary hourly hydrograph is obtained at the upstream end of the network. When the computed total discharge exceeds the upstream discharge Q_0, the model delays the operation of some outlets until this discharge Q_0 will not be exceeded. Using a simplified procedure based on the queeing theory, outlets are open only when the system do not become saturated. This procedure is applied to the full peak month, which allows to produce the hourly hydrographs for every day in the month, as illustrated in Fig. 1.

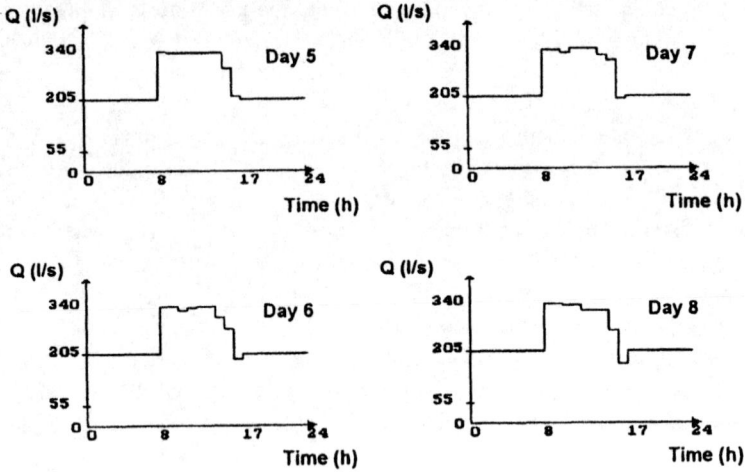

Fig. 1. Hourly discharge hydrographs for the days 5 through 8 of the peak month
(sector 11 of the Sorraia irrigation system)

4 Optimization of pipe diameters

The flow regimes obtained from the discharge configurations of outlets simultaneously
operating at each hour and defined from the hourly hydrographs are utilized to size the
network system.

The optimization is performed using the iterative discontinuous method of Labye [2],
as described by Pereira *et al.* [7], The computation uses the algorithm developed by
Lamaddalena [4].

The output files resulting from the optimization are utilized for the performance
analysis described below.

5 Performance analysis

Three performance indicators are utilized: reliability, dependability and equity. The
respective indicators are computed from comparing the nominal discharges Q_n [l s^{-1}]
and the minimum hydraulic head H_{min} [m] at outlets with the actual delivered
discharges Q_j [l s^{-1}] and the hydraulic head H_j [m] available at each outlet j. Q_n and
H_{min} are input design variables. H_j is computed from the hydraulic simulation of the
system for each configuration of outlets in operation, i.e. for each flow regime. Q_j are
calculated according to H_j:

$$Q_j = Q_n \qquad\qquad \text{if } H_j \geq H_{min}$$
$$Q_j = K_{j,o}\, H_i \qquad \text{if } 0 < H_j < H_{min} \qquad\qquad (4)$$
$$Q_j = 0 \qquad\qquad \text{if } H_j \leq 0$$

where $K_{j,o}$ is the discharge coefficient relative to the outlet j (j = 1, 2, ..., N).

The reliability represents the ability of the system to deliver the target design discharges at every operating outlet. It can be computed for each hour and each operating outlet by $p_a = Q_j / Q_n$. The system reliability P_A [0,1] is obtained by integrating over the time the average probability p_a relative to the R outlets simultaneously operating at each unit time, i.e., for each simulated flow regime. P_A is given by

$$P_A = \frac{1}{T} \sum_{t=1}^{T} \left(\frac{1}{R} \sum_{j=1}^{R} p_a \right)_t \qquad (5)$$

where T is the time [hours] corresponding to the number of flow regimes simulated.

The dependability illustrates the ability of the system to deliver the target discharge at each outlet along a given period of time, i.e., it mesures the temporal uniformity of deliveries at each outlet. If the performance analysis covers the time T [hours], the temporal uniformity at each outlet j will be $U_{t,j}$ [0,1] given by

$$U_{t,j} = \frac{1}{T} \sum_{t=1}^{T} \frac{Q_j}{Q_n} \qquad (6)$$

and the time variability of discharges at the outlet j is

$$V_{t,j} = \frac{1}{T} \sum_{t=1}^{T} \left| U_t - \frac{Q_j}{Q_n} \right| \qquad (7)$$

Extending to all the N outlets of the network, the dependability of the system is

$$P_D = 1 - \frac{1}{N} \sum_{j=1}^{N} V_{t,j} \qquad (8)$$

The equity measures the spatial uniformity of deliveries during the time T [hours]. The spatial uniformity of discharges delivered during each unit of time, corresponding to a configuration of R outlets simultaneously operating, can be defined by

$$U_{s,t} = \frac{1}{R} \sum_{j=1}^{R} \frac{Q_j}{Q_n} \qquad (9)$$

and the corresponding spatial variability of discharges is

$$V_{t,j} = \frac{1}{R} \sum_{j=1}^{R} \left| U_s - \frac{Q_j}{Q_n} \right| \qquad (10)$$

Considering the full time T under analysis, it results for the system equity P_E [0,1]

$$P_E = 1 - \frac{1}{T} \sum_{t=1}^{T} V_{s,t} \qquad (11)$$

Computations of these performance parameters are performed simulating any number of flow regimes during the peak demand period, in general between 240 (10 days peak period) and 744 (31 days). The analysis may be performed for an existing system or a system being designed. Flow regimes may differ from those used for the optimization when the demand hydrographs are generated using different scenarios relative to the cropping patterns and/or the irrigation management.

6 Application

The methodology described above has been applied to the sector 11 of the Sorraia irrigation system, Portugal. A buried pipeline system, designed and constructed in the years 1950, distributes water to 70 outlets, each one serving an area averaging 5 ha. Rice, representing 37% of the area, maize, tomato and sunflower are the main crops. Rice is irrigated by permanent flooding (paddy) and row crops are dominantly irrigated by short blocked furrows, with few areas adopting graded furrows and level basins. The maximal upstream discharge is $Q_0 = 340$ l s^{-1}. Nominal discharges at outlets are 10, 15 and 20 l s^{-1} according to the area served. The daily labour schedule, resulting from labour unions agreements, imposes that irrigation is practiced from 8.00 a.m. to 5.00 p.m. or, for rice, with a continuous flow for the 24 hours of the day. This is evidenced in the demand hydrographs in Fig. 1.

The study intended to assess the performance of the existing system, to develop design alternatives for the same system which could provide for the adoption of modern farm systems, including automation, and to evaluate these alternatives for different scenarios of crop patterns and on-farm irrigation management.

Under the present cropping and irrigation conditions, the reliability (Fig. 2a) averages 0.95 but shows poor values for the day time peak demand hours. Results for the equity (Fig. 2b) show that unequal service is provided among outlets during the same peak demand hours, with an average $P_E = 0.93$. Both indicators make evident that service is only excellent during the night hours, when rice is irrigated (cf. Fig. 1).

The dependability (Fig. 2c) also averages 0.93 but shows that service is not dependable for some outlets, which are mostly located in terminal branches of the network, having small diameters.

This analysis permits to identify some alternative solutions for improvement:(1) to increase the daily labour schedules, which could decrease the problems during the peak daily demand periods; (2) to enlarge the use of automated farm irrigation systems, which could allow for irrigation of row crops out of the normal labour hours; and (3) to reinforce the pipe network, including the respective layout and the outlet characteristics (for $Q_n = 30$ ls^{-1}).

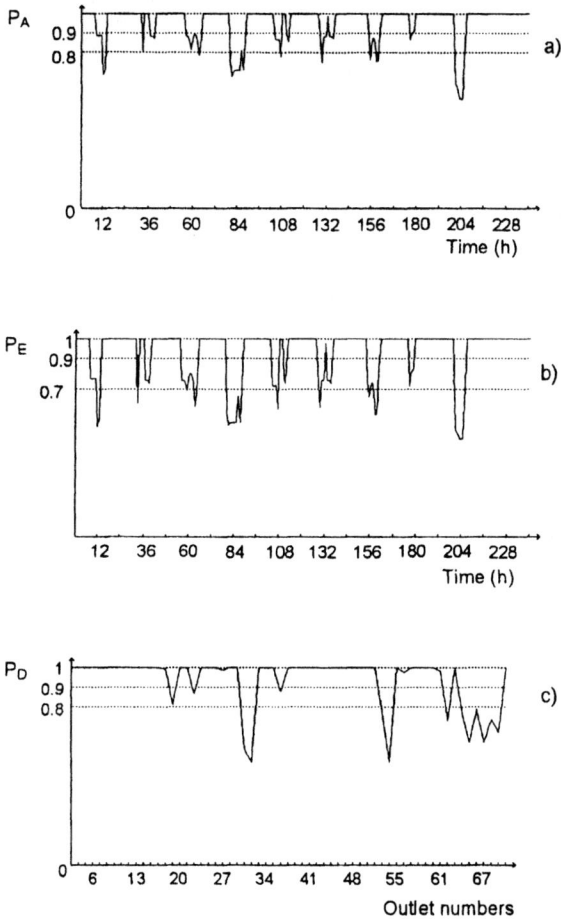

Fig. 2. Reliability, equity and dependability relative to the 10 days peak demand period for the existing network of sector 11, Sorraia system.

Alternative systems have been designed (using 240 flow regimes) to respond to issues indicated above but keeping the upstream discharge unchanged. These alternatives have been analysed using 744 flow regimes.

For scenarios where the cropping pattern remains the same as at present, the rice being the main crop, it is evidenced that average performances could be excellent if automation would be adopted by all farmers (Fig. 3a). However, not all problems could be solved because demand during daylight hours would remain the highest. Performances are lower when the percentage of automation would be smaller. Without automation, results are similar to those obtained at present.

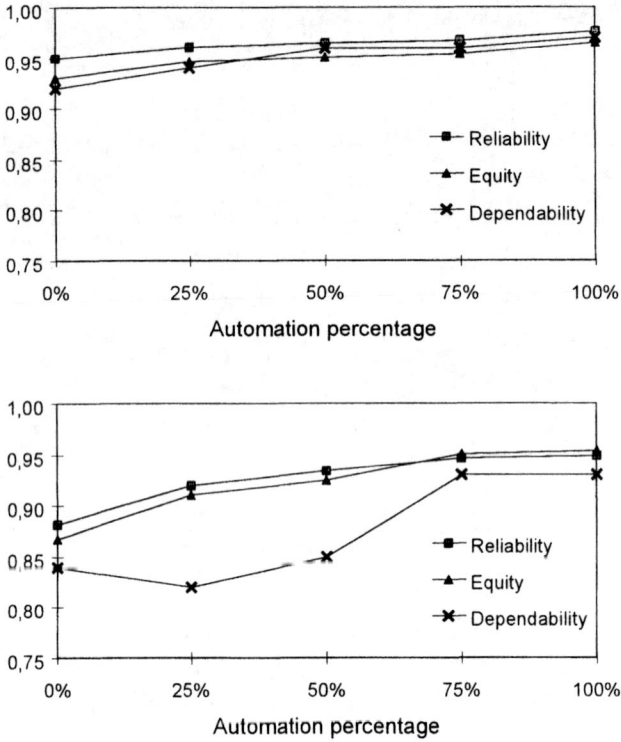

Fig. 3. Forecasted system performances in relation to on-farm automation: (a) without changing the cropping pattern, and (b) replacing rice by row crops.

For scenarios where the cropping pattern is modified (Fig. 3b), replacing rice by row crops, the attainable performances are smaller than those in Fig. 3a. This evidences the role of night time irrigation of the rice crop in decreasing the demand during the daylight time, which can not be compensated by automation only.

7 Conclusions

The methodology proposed for design and analysis of low pressure distribution systems for surface irrigation shows to be able to assess how systems would perform in practice. The procedure utilized to generate the demand hydrographs allows for simulating the demand under different scenarios for the cropping pattern and irrigation practices. This capability provides for assessing the performances of an existing or design system under different operational conditions.

The application of the methodology to an existing system and to alternative design networks shows that it is possible to antecipate the performance and benefits of the system. Further work is required to fully explore this methodology for decision making.

8 Acknowledgements

The support by the CIHEAM, IAM-Bari, and by the CEER (JNICT and ISA), Lisbon, is greatly acknowledged.

9 References

1. van Bentum, R., Smout, I. K., 1994. *Buried Pipelines for Surface Irrigation.* IT Publications, London, et WEDC, Loughborough.
2. Labye, Y. (1981) Iterative discontinuous method for networks with one or or more flow regimes, in *Proc. Int. Workshop on Systems Analysis of Problems in Irrigation Drainage and Flood Control.* ICID, New Delhi: 31-40.
3. Lamaddalena, N. (1996) Sulla optimizzazione dei diametri in una rete irrigua con esercizio alla domanda. *Riv. Ingegneria Agraria* 27 (1): 12-19.
4. Lamaddalena N. (1997); Integrated simulation modeling for design and performance analysis of on-demand pressurized irrigation systems. Ph. D. Dissertation, I.S.A., Univ. Técnica de Lisboa.
5. Bethery, J. (1990) *Réseaux collectifs d'irrigation ramifiés sous pression. Calcul et fonctionnement.* Études n°6, CEMAGREF, Antony.
6. Lamaddalena, N. (1995) Analisi del funzionamento dei sistemi irrigui collettivi. *Riv. Irrigazione e Drenaggio* 2: 18-26.
7. Pereira, L. S., Douieb, A., Bounoua, R., Lamaddalena, N., Sousa, P. L. (1998) A model for design of low pressure distribution in irrigation systems, in *Computers in Agriculture* (Int. Conf., Orlando, FL, Oct. 26-30, 1998). ASAE (in press).

EFFECTIVE MONITORING OF CANAL IRRIGATION WITH MINIMUM OR NO FLOW MEASUREMENT

B. A. LANKFORD
School of Development Studies, University of East Anglia, Norwich, UK

Abstract
The irrigation literature generally supports the idea that flow measurement is an important precursor to improved monitoring and management of surface irrigation schemes. However, the construction, maintenance and operation of flow measurement structures, coupled with the collection and analysis of data from them, create logistical problems which erode their effective use. In acknowledging this, but still aiming to raise the performance of irrigation, it becomes necessary to think of other ways that monitoring can be conducted while maintaining the value of the information collected. In this paper, three avenues are explored, based on; key-location flow measurement; passive flow measurement and no flow measurement. In the discussion under no flow measurement, applicable to rotational irrigation, the concept of irrigation progress per day (ha/day) is introduced. In addition, a brief discussion on the management of monitoring covers linkages with water management, use of computers, devolving responsibility and phased planning of interventions.
Keywords: Canal irrigation, flow measurement, irrigation management, monitoring, performance.

1 Introduction

Flow measurement on surface irrigation schemes is often stated to be an important precursor to monitoring and management, summarised in the saying "to measure is to manage." However, numerous problematic factors are involved, resulting in few successful monitoring programmes. Within this large subject-area, this paper discusses a few ways in which managers may introduce and sustain informative, cost-effective monitoring of irrigation management and performance.

The notion that flow measurement is an important part of monitoring of irrigation is

Water and the Environment: Innovative Issues in Irrigation and Drainage. Edited by Luis S. Pereira and John W. Gowing. Published in 1998 by E & FN Spon. ISBN 0 419 23710 0

widespread: "the existence and location of management tools such as flow measurement and water control structures can influence the managerial decision making process." [1]. "Usually water measurements should be planned at all points where it can be reasonably established that information on the flow rate will affect the management decisions." [2]. A survey among specialists on irrigation performance placed measuring devices within the top five along with performance data and performance criteria [3]. Measuring flows is also termed 'control for supervision' [4].

However, the commonplace situation on irrigation schemes is that flow measurement is not taking place, usually because of a combination of factors such as the difficulty of measuring varying flows; broken structures; difficult-to-operate structures; low staffing and training levels; weakly developed data collection procedures; a lack of motivation of staff and a lack of will on behalf of the managers. Nevertheless, irrigation schemes can and do manage without flow measurement.

2 An analysis of monitoring

Perhaps there is a need for effective monitoring rather than 'perfect' monitoring. The question is what is effective monitoring, what factors are involved, and how flexible, scheme-specific and appropriate are they for different levels of the scheme or organisation? Table 1 is an analysis of seven possible factors involved in irrigation monitoring. The left hand column gives the factor and sub-types, the middle column gives the definition and role of the factors, while the right column suggests the level of the irrigation scheme or organisation where the factor would be most appropriate.

Some important points arise from this analysis. Effective monitoring is about utilising each factor in the most cost-effective manner to arrive at available, up-to-date, relevant information. Effectiveness is also about building monitoring in stages; initially on simple questions (e.g. how many irrigations were achieved this season?) to create an interest in irrigation management. Then, monitoring of irrigation management (against management performance targets) can be introduced, and then monitoring of performance indicators (i.e. efficiency and equity) can be considered.

Secondly, flow measurement has a place in monitoring, but need not dominate the debate. The relative omission of flow measurement structures has been discussed by several authors, not least in respect of effective water pricing systems [5]. Further discussion on flow measurement is found in section 3 of this paper.

Thirdly, codification and up-to-date measurement of areas, which are often ignored, are precursors to well-planned (computerised), monitoring. Codification needs to recognise returns to water as well as other reasons for naming fields. Ideally, codes should be attached to the canals, fields and outlets with the lowest unit of flow. The author has visited a scheme where a single code in the record books was in fact four separate "fields" supplied by two separate tertiary canals and outlets. The managers said they found it difficult to determine how long each field took to irrigate. In addition, codes should reflect the connection of fields via the rotation of flows.

The fourth point is that the six factors are interrelated and without each addressed in some way, monitoring may be ineffective or collapse totally.

Fifthly, the success of monitoring is related to what the objective of monitoring is and how carefully monitoring procedures are thought out. Many factors are involved and each should be addressed with one question; how relevant is it? In other words,

Table 1. Analysis of factors involved in monitoring of surface irrigation schemes

Main factor and sub-types	Definition and role	Probable best location for cost-effectiveness
Codification	Naming and numbering of infrastructure and place. For assigning information to areas and locations.	Tertiaries, fields and farms.
Area measurement	Determining areas irrigated. For use in flow-area-time calculations	Fields and farms.
No. of irrigations measurement	Determining progress rates of irrigation. See section 3.4 of this paper	Fields and farms.
Time measurement	Measuring start and stop times of irrigation. For calculating durations and volumes applied	Fields and farms.

Flow and/or level measurement
    Automatic
    Passive manual
    Active manual
	Measuring flows and/or levels. For determining volumes applied and diagnosing problems.
	- No or few people involved (e.g. telemetry)
	- Flow often set by structure and is known
	- Flow is variable and has to be recorded
		(Passive and automatic can work at tertiary levels).
		Active and automatic measurement at main and secondary canals.

Organisational procedures
    Informal
    Formal
    Information
    Personnel
    Pathway
    Frequency
	Methods of monitoring. For allocating water and selecting method of monitoring.
	- Implicit monitoring against unwritten rules
	- Explicit procedures and rules for monitoring
	- What is collected (flow, time, place)
	- Who collects (farmers, gatekeepers, supervisors)
	- How collected (from what/whom to what/whom)
	- When collected (daily, weekly, missing data)
		Complex organisational levels involved; farmers, gatekeepers, junior and senior irrigation staff. 'Flow' of data is usually from many junior staff to fewer senior staff.

Evaluation and relevancy
    Method
    By/to whom
    To what stage and format
    Targets
    Irrigation
 	
    People
	Use of monitoring information. For demonstrating benefits and relevancy of monitoring in management.
	- Manual or computer
	- Who examines results, then to whom are they sent?
	- How analysed and what is final form (e.g. irrigation management measures or performance indicators)
	- What/how are targets set and who sets them
	- Can the results be used to identify zones or operational procedures resulting in poor irrigation?
	- Can the results be used to motivate and manage staff/farmers involved in data collection?
		All levels are involved.
		
		Also depends on stages of building of monitoring
		
		Traditionally, evaluated results moves from senior staff back to junior staff and farmers.

managers need to be wary of collecting information for the sake of collecting information, and ensure that steps along the way should quickly lead to other steps; it is just as important to add procedural stages as it is to delete or short-cut others.

Sixthly, apart from flow measurement which needs relatively high capital investment to construct and repair structures, other procedures involve mostly recurrent costs. Whereas this is advantageous, they involve person management which is problematic, and it is necessary to consider factors which affect the motivation of staff such as a sense of responsibility, of being listened and responded to, and salary levels.

Lastly, from the author's experience and from the literature regarding traditional irrigation, farmers manage irrigation using implicit forms of monitoring set against unwritten operational procedures. In an example in Sri Lanka, complex operational rules were adhered to by villagers [6]. On small-scale schemes, farmers readily know how long a field takes to irrigate thereby measuring performance against average performance. This works when fields are small, flows are circulated with few divisions and when farmers over time recognise spatial and temporal relationships between flow rate, time, area, crop, climate, losses, soil conditions and irrigation interval. However, such relationships are difficult to discern on larger schemes and it is on these that more formalised monitoring has the greatest potential impact.

3 Obtaining quantitative information for monitoring

3.1 Introduction
For irrigation managers, a high priority remains the need to obtain useful quantitative information in order to monitor irrigation management and performance. In this section three main ways are suggested. The first two address flow measurement and the third proposes a method of monitoring without flow measurement.

3.2 Monitoring using key-location flow measurement
One approach to flow measurement is to identify the key-location structures providing the most cost-effective information. The most elegant answer to this lies in the hierarchical nature of irrigation systems. Table 2 reveals for a 10,000 ha irrigation scheme, the number of turnouts required at different levels. The 23 structures down to the secondary level could be argued as being the key-location structures. However, it is not clear what can be done with the 8400 structures on the distribution system. Where rotational flows exist, one answer is to reduce the number of measuring turnouts by creating grouped canal networks, used for rotating the leadstream (main d'eau), centred around a single nodal turnout. The disadvantage is that canal networks of existing schemes would need to be altered. In the example in table 2, grouping four 25 ha tertiary networks results in a reduction of 400 to 100 turnouts requiring flow measurement. However, at the lowest level, grouping of farms still leaves high numbers of outlets requiring measurement. For example, grouping of 7 farms results in 1143 measuring outlets. At this level, the solution in section 3.3 is recommended.

Table 2. Numbers of command units, canals and turnouts on a large-scale irrigation scheme

Level to level interface	Usual name of structure	No. of divisions per canal	Command area per outlet	Total number required
River to main canal	Intake	1	10,000 ha	1
Main to main branch	Offtake	2	5000 ha	2
Main branch to secondary	Offtake	10	500 ha	20
Secondary to tertiary	Offtake/turnout	20	25 ha	400
Tertiary to farm outlet	Outlet/turnout	20	1.25 ha	8000

3.3 Monitoring using passive flow measurement
Another solution at the tertiary and farm turnout level is to use passive flow measurement (putting aside arguments for automatic means). Passive flow measurement can be designed into the turnout - a common design being the modular

gate. The flow is not necessarily observed or recorded but can be "measured" by simply opening shutters. The duration of flow is used to determine the volume discharged. Passive flow measurement is also possible where a flow previously measured upstream is split by a proportional divisor into two or more flows. In all such cases, tertiary head control is necessary to minimise fluctuations in water levels.

3.4 Monitoring using no flow measurement

A method is described, using easily recorded data, to monitor irrigation at the tertiary level without flow measurement. It allows irrigation managers to set targets and compare performance against them. The key concept is 'irrigation progress', in hectares/day, calculated for an individual field over time from the following equation:

$$\text{Irrigation progress (ha / day)} = \frac{\text{command area of field (ha)} \times \text{no. irrigation}}{\text{total duration of irrigation (days)}} \quad (1)$$

The number of irrigations is over the peak season and the total duration of irrigation is over the same period. The irrigation progress is used to calculate the potential interval period between irrigations which assumes no stops are made due to rain:

$$\text{Potential interval period (days)} = \frac{\text{leadstream duty area (ha)}}{\text{irrigation progress (ha / days)}} \quad (2)$$

The leadstream duty area is the nominal duty area served by a single rotating leadstream (see equation 4). The irrigation progress or potential interval of irrigation can be compared to calculated targets of the same, determined from the following:

$$\text{Gross daily crop water use (mm / days)} = \frac{\text{Reference crop ETo} \times \text{kc}}{\text{management allowed efficiency \%}} \quad (3)$$

$$\text{Nominal leadstream duty area (ha)} = \frac{\text{secondary unit command area (ha)}}{\text{no. of leadstreams}} \quad (4)$$

$$\text{Target irrig. prg. (ha / day)} = \text{duty area (ha)} \times \frac{\text{gross daily crop water use (mm / day)}}{\text{soil RAM (mm)}} \quad (5)$$

$$\text{Target (no rain) interval (days)} = \frac{\text{soil RAM (mm)}}{\text{gross daily crop water use (mm / days)}} \quad (6)$$

In equation 3, the gross daily crop water use includes the reference crop evaporation (ETo), the crop factor (Kc) for full canopy water use and the management allowed efficiency (MAE) for irrigation losses at the tertiary level (field and canal). Soil RAM (equation 5) is readily available moisture which includes the management allowed deficit. RAM is therefore the net target refill dose of irrigation in mm.

By comparing these results, indicators of performance may be determined, such as relative efficiency (set against management allowed efficiency) or a computed efficiency (recalculated against a management allowed efficiency of 100%):

$$\text{Computed irrigation efficiency \%} = \frac{\text{t arg et int erval (days)} \times \text{MAE \%}}{\text{potencial target (days)}} \quad (7)$$

An example clarifies the use of these equations. It is first necessary to calculate the target performance figures. The gross daily water use (equation 3) of 7.43 mm/day is from a ETo of 5.0 mm/day, a crop factor of 1.0 and an MAE of 70% (= 5.0*1.0/0.7). The nominal duty area (equation 4) for a secondary unit of 350 ha with 12 leadstreams is 29.2 ha per leadstream (= 350/12). The target progress rate (equation 5) is found to be 3.33 ha/day from the duty area, gross water use and a soil RAM of 65 mm (=29.2*7.43/65). The target interval (found from equation 6) is 8.75 days (= 65/7.43).

The next stage is to determine the actual irrigation progress rates which shows whether any particular field has too-quick or too-slow irrigation. As an example, one 7.8 ha field over a four month period (using sugarcane) receives 8 irrigations, and takes a total of 24 days. The irrigation progress (equation 1) is found to be 2.60 ha/day (= 7.8*8/24). The potential irrigation interval (equation 2) is found to be 11.2 days (= 29.2/2.60). By comparing the potential and target irrigation interval, and assuming both are correct, the computed irrigation efficiency is found to be 55% (= 8.75*0.7/11.2). In addition the following guidelines and assumptions apply:

1. The method is used during the peak period of water demand where and when managers schedule irrigation with a fixed dose per irrigation but alter the frequency.
2. It requires a constant secondary flow so that the number of leadstreams operating in equation 4 can be used to determine the nominal duty areas.
3. Actual progress is determined for individual fields as it may not possible to specify the precise duty areas that leadstreams repeatedly rotate around. If leadstreams do stick to set command areas then the method is applied to that group of fields.
4. It requires roughly even ratios between flow to area within the secondary area so that irrigation progress is accountable to losses rather than differing designs and l/sec/ha ratios; in such cases correction factors will be needed.
5. It is possible to determine weighted averages of performance and targets for the whole secondary unit using areas of the individual fields within the secondary unit.
6. Equity indicators (e.g. the coefficient of variation) may be calculated by examining variations in progress between fields, tertiary or secondary units.
7. Note that the irrigation progress is determined from averaging the duration of irrigation with the number of irrigations over the peak season. This smoothes out short term changes in irrigation progress due to early irrigation timing after rainfall.
8. The potential interval is compared to the target interval as given in equation 6, not to the actual interval which includes stoppages for rain.
9. The errors in this method arise when calculating meaningful targets from the values of ETo, RAM, efficiency and especially the number of leadstreams operating in the secondary canal which can change over time with differing volumes of water.

In summary, irrigation progress is a measure of the ability of the irrigation schemes to maintain the required frequency of irrigation, reflecting the way in which surface irrigation schemes can cope with periods without rainfall. Less efficient schemes will have a lower progress rate, and consequently a higher interval between irrigations.

4 The management of monitoring

This section discusses four issues regarding the management of monitoring:

1. *Irrigation management*: management of monitoring and irrigation management depend on and respond to each other in situations of both pre- and post introduction of monitoring. If monitoring is not present, informal water management rules are required against which operation may be informally gauged. In the second situation, if formalised monitoring is in place but irrigation water is not well managed (and is not improving) then the monitoring of water management (and performance) becomes a rather meaningless exercise.
2. *Use of computers*: in monitoring, computers are usually used for two main purposes; to enter, store and retrieve data for areas and to generate summary information for larger areas and for end-of-season reports. A third important use for them is rarely found; to add **relevancy** to the monitoring information, which is done in four ways. The first is to predict, from actual irrigation progress, the delay in scheduling and its effect on crop water stress and yield; the second is to locate problems and successes; the third is to identify the procedures and people involved in problems and successes, and the fourth is to do these three concurrently with the data collection stage rather than several months later. By setting up the relevant sub-routines within computer programmes and spreadsheets, managers can be directly involved in seeing the output of their irrigation management.
3. *Devolving responsibility*: in keeping with trends in irrigation management, ways should be found of allowing farmer associations and gatekeepers to monitor their own management while ensuring that it remains the irrigation managers' role to assist farmers in achieving their goals of higher productivity and saving water.
4. *Phased programmes*: interventions may be planned so that the success of each stage encourages the implementation of the next. Instead of beginning with a monitoring programme based on flow measurement, more modest aims might be appropriate. Although linear pathways of implementation are a contradiction in terms in irrigation management, goals can and should be prioritised, such as; identifying responsibilities, data collection, codification; time recording; computerisation; standardisation of record collecting; setting simple targets; and a phased introduction of key-location flow measurement and rehabilitation of turnouts.

5 Conclusions

This paper briefly presents some ideas on the monitoring of surface irrigation schemes. Some thoughts on the management of monitoring are discussed. The paper reminds irrigation managers of the following:

- It is important to distinguish between flow measurement and information on management and performance. Managers should seek ways of generating and encouraging the latter.
- Time and area measurements on irrigation schemes are of great value. They are, for example, the basis of monitoring of traditional warabundi systems.
- Climate, soil, crop and irrigation design data can be used to determine targets which in turn can be used to evaluate performance.
- Computers can generate relevancy between collected data and performance.
- Managers may accept an argument of no monitoring, but have to be much more proactive towards setting informal standards and operational rules.
- Flow measurement is an achievable 'high goal' and its introduction should be seen within the existing monitoring and irrigation contexts of each scheme. It will be most needed in areas where water is a scarce resource or where its apportionment to users is being accepted via pricing mechanisms or where diagnostic-orientated monitoring is required to raise productivity.
- It is concluded that flow measurement at key points, passive flow measurement where possible, and recording of irrigation progress, are cost-effective options for irrigation managers to consider. The management of monitoring should be clearly thought out from codification through to computerisation, and acknowledge the reinforcing linkages between monitoring and management of water.

6 References

1.   Walker, W.R. and Skogerboe, G.V. (1987) *Surface Irrigation. Theory and Practice*, Prentice Hall, New Jersey.
2.   Clemmens, A.J., Bos, M.G. and Replogle, J.A. (1993) *FLUME: Design and calibration of long-throated measuring flumes. ILRI Publication 54*, Wageningen.
3.   Abernethy, C.L. and Pearce, G.R. (1987) *Research needs in third world irrigation. Proceedings of a colloquium*, Hydraulics Research Wallingford.
4.   Wolter, H. (1994) Capacity building to implement supervisory control systems in China and India. Theme topic. *GRID Issue 4. March 1994. IPTRID Network Magazine*, Wallingford.
5.   Carruthers, I and Small, L. E. (1991) *Farmer-financed irrigation: the economics of reform*, Cambridge University Press, Cambridge.
6.   Leach. E. (1980) Village irrigation in the dry zone of Sri Lanka, in *Irrigation and Agricultural Development in Asia. Perspectives from the Social Sciences*, (ed. E.W. Coward, Jr.), Cornell University Press, Ithaca and London.

CANAL WATER DELIVERY SYSTEM AUTOMATION. A CASE STUDY

M. RIJO
Department of Agricultural Engineering, University of Évora, Évora, Portugal
V. PAULO
Hidroprojecto Engenharia & Gestão S.A., Lisbon, Portugal

Abstract

Upstream control in canals is efficient when associated with rigid water delivery methods. In practice, most of these systems work with flexible water delivery schedules and water operational losses become significant. Real-time technologies allow the water manager to continuously compare the actual management of the delivery system with its optimal management, and to take appropriate corrective steps as required. These innovations allow the manager to react rapidly and effectively to changing conditions. For these reasons, in Portugal, modernisation of upstream controlled canals uses remote monitoring and central control technologies. This paper presents the case study of the Mira Irrigation Project modernisation program.
Keywords: automation, irrigation canal modernisation, remote control, remote monitoring, upstream control, water saving.

1 Introduction

In many countries or regions, water is becoming the limiting factor for development. The marked competition between municipalities, industry and agriculture is pushing toward more efficient use of this limited natural resource.

Agriculture, the largest water consumer, has adopted modernisation programs based on new technologies to provide more reliable and flexible water delivery to farmers. An improved irrigation service will lead to high on-farm irrigation efficiency, less spillage and water losses within conveyance and distribution systems. These two factors will definitely decrease the water volumes required at the source.

The intervenients involved in water distribution must up-grade their service to

Water and the Environment: Innovative Issues in Irrigation and Drainage. Edited by Luis S. Pereira and John W. Gowing. Published in 1998 by E & FN Spon. ISBN 0 419 23710 0

improve water use efficiency. New service policies involve more complex delivery schedules and changes in flow and duration of the deliveries. Canal flow will never be in steady flow conditions. This new strategy may surpass staff capacity, therefore supervisory control and control theory can offer some alternatives to help management to accomplish the service goals.

Upstream control is the most widely used control methods for irrigation canal systems [1]. As there is no storage capacity within canals, as is the case of the downstream control with AVIS gates [1], water delivery must be predicted and programmed in advance. This control method is efficient when connected with rigid delivery methods, like rotation [2].

In Portugal, where all of the irrigation canals are upstream controlled with AMIL gates or duckbill weirs [1], social reasons, among others, led to the substitution of rotation by arranged delivery schedules [2]. Farmers got some freedom to use the turnouts, but distribution service became more complex, difficult and inefficient in the water use. Operation water losses within conveyance and distribution system became significant, sometimes up to 50% of the inflow volume [3].

Few years ago, Portugal began a policy of rehabilitation and modernisation of the main canal irrigation projects, most of them built in the 50's and 60'. The main purposes of this policy are saving water and installing more flexible water delivery rules, like on-demand schedule [2], in order to make possible the new irrigation methods and give some degrees of freedom for irrigation to farmers. For field implementation of this policy, usually the way is to maintain the system architecture, install supervisory control and data acquisition system and, when it is possible, install buffer and control reservoirs within hydraulic system.

Real-time monitoring and control can hold an important key to intensified management on these projects thereby improving economic output, implementing measures to ensure public safety, enhancing environmental conditions, conserving water, and reducing operational costs, all in a very cost-effective manner.

Remote monitoring and control systems are becoming cost-effective water management tools. This development results from major innovations and cost breakthroughs occurring in water management technologies, including computer hardware and software, remote terminal units, controllers, communications equipment and sensors.

Real-time technologies allow the water manager to continuously compare the actual management of the delivery system with its optimal management, and to take appropriate corrective steps as required. These innovations allow the manager to react rapidly and effectively to changing conditions, thereby accommodating both high and low flow conditions and reducing spillage and seepage. Usually, after one or a few irrigation seasons monitoring database, is possible to automate a few management procedures.

The paper presents the supervisory control and data acquisition system developed for the Mira Irrigation Project, in Portugal.

2 Main characteristics of the Mira Irrigation Project

Mira Irrigation Project is located near the Atlantic Ocean, in the Southwest of Portugal. It is an upstream controlled canal system, equipped with automatic AMIL radial gates, in order to provide potential good operation conditions for the Neyrpic orifice intakes to distributors and fields, and to control the hydraulic behaviour inside the canals. The dominated area is 12,000 *ha*.

The irrigation system is composed by (Fig. 1):

- One water source – Sta. Clara Reservoir, with a useful storage capacity of $240*10^6 m^3$.
- A main conveyor – Canal Condutor Geral, between Sta. Clara and Odeceixe reservoirs, designed for 11.2 m^3/s, with 38 *km* total length, 11 tunnels (7,375 *m* global length), 4 inverted siphons (2,491 *m* global length) and 13 canal-bridges (2,185 *m* global length). The main goal of this canal is to convey water to Odeceixe Reservoir, where the irrigated perimeter begins. Canal only delivery water to 1,000 *ha*.
- Two buffer and control reservoirs - Odeceixe and Milfontes, both connected, respectively with $230*10^3$ and $33*10^3 m^3$, located near the irrigation areas;
- The distribution system, with two main distributors, Milfontes and Odeceixe canals, and the corresponding minors, canals or low pressure buried pipes.

The system was originally designed for the rotation water delivery method. However, in practice, it was implemented the restricted arranged schedules, with fixed delivery rate and duration [2]. With these water delivery rules, most of the crops were irrigated during normal labour hours and labour days, according to established agreements between the manager and farmers. The operational water losses on the fields and in the conveyance and distribution system became very important.

The rehabilitation and modernisation program of the irrigation project defined:

- The substitution of furrow and other gravity irrigation methods for sprinkling irrigation for the entire perimeter and the substitution of the old low-pressure buried pipes for high-pressure buried pipes. The perimeter has only sandy soils.
- The substitution of the restricted arranged by demand, as water delivery method.
- The introduction of a buffer and control reservoir between the distributor and the pumping station of each new irrigation block defined, because the upstream control installed in the conveyance and distribution system doesn't allow a quick and efficient answer to the demand flow fluctuations in each block.
- 16 irrigation blocks for the entire perimeter, each of them with an approximate area of 840 ha.
- A supervisory control and data acquisition system and the possible evolution, after some field experience, to a central automatic control for the main network – conveyor, main distributors and all buffer reservoirs.

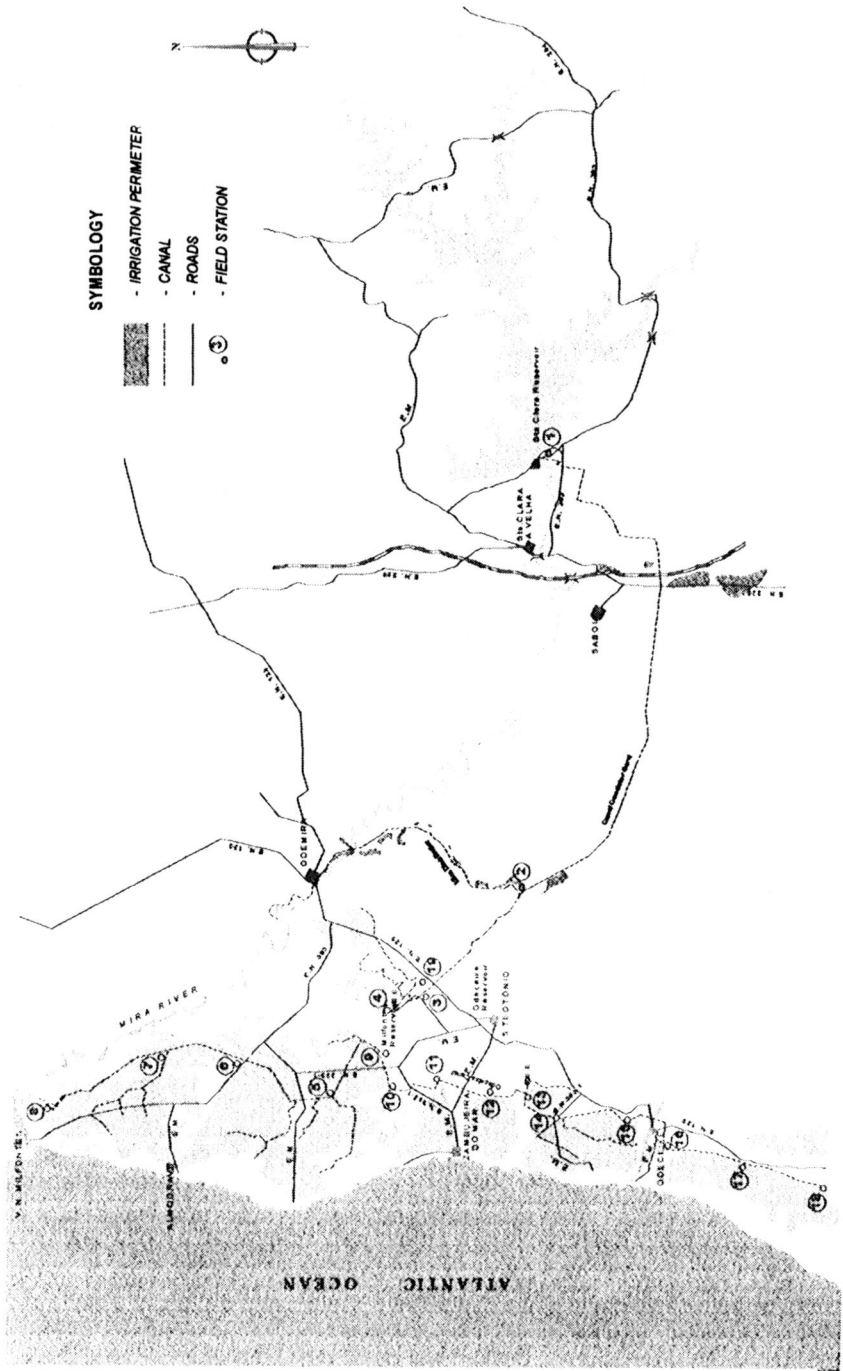

Fig. 1. Scheme of the Mira Irrigation Project.

3 Supervisory control and data acquisition system

3.1 General presentation

The central control of the conveyance and distribution network is only possible with reliable information about the hydraulic state of the system in real-time. Therefore, central automatic control of an open-channel system involves: i) a real-time remote monitoring action in order to keep abreast the canals conditions; ii) a remote control action in order to lead the system to desired state; iii) a management action to support operational decisions, ensuring the desired service performance and regarding the real and expected demands, the available storage volumes, and economic factors.

The central automatic control of an open-channel system is always supported by a remote monitoring and a remote control networks. Remote monitoring allows data acquisition, for example: water levels within canals and buffer reservoirs, gate positions, inflow and outflow determinations, making possible the diagnosis of the real hydraulic state. Remote control network allows the control orders sending, namely remote operation of gates, intakes and/or offtakes and valves, considering the hydraulic state diagnosis. These two networks, including their remote terminal units (RTU), and the communication and control software needed form the supervisory control and data acquisition (SCADA) system, are the first step of a central automatic control system.

RTU units are the interfaces between SCADA and the hydraulic systems. They are located inside field stations (for safety). RTU units main purposes are: control inputs and outputs of field devices (intakes, offtakes and gates); monitor conditions at field devices (water level and gate position sensors) and log alarms; stock data and report to the submaster or master station and carry out the commands they receive from these stations. The field stations can be of three types: control unit (RTUc), control and monitoring units (RTUcm) and monitoring unit (RTUm).

For the Mira Irrigation Project a tree structure for the SCADA system was defined (Fig. 2). With this format, the SCADA system can evolve easily, adding plus submaster stations (only one in this step), plus RTU units, or more sophistication at each RTU.

All the communications will be by radio (trucking system). The execution project of the SCADA system is already finished and it will be built soon.

The main goals of the presented SCADA system are:

- To control flows at main intakes - conveyor, two main distributors, distributors from the conveyor and buffer reservoirs at the head of the designed irrigation blocks (Fig. 1). In this first step, only one of the sixteen irrigation blocks will be built and equipped (Station 12, Figs 1 and 2).
- To monitor water storage volumes within all reservoirs, flows at each AMIL gate of the conveyor and main distributors, and flows at the outlets at the downstream end of two main distributors (Figs 1 and 2). Inflow and outflow of each pool within these three canals will always be known in real-time and it will be easier to the manager to define the best hydraulic management algorithms.

3.2 Field stations with control units (RTUc)

There will be two RTUc sites, stations 1 and 4 (Fig. 1 and 2), respectively, to control flow at the conveyor and Milfontes distributor Neyrpic orifice intake [1]. Both RTUc

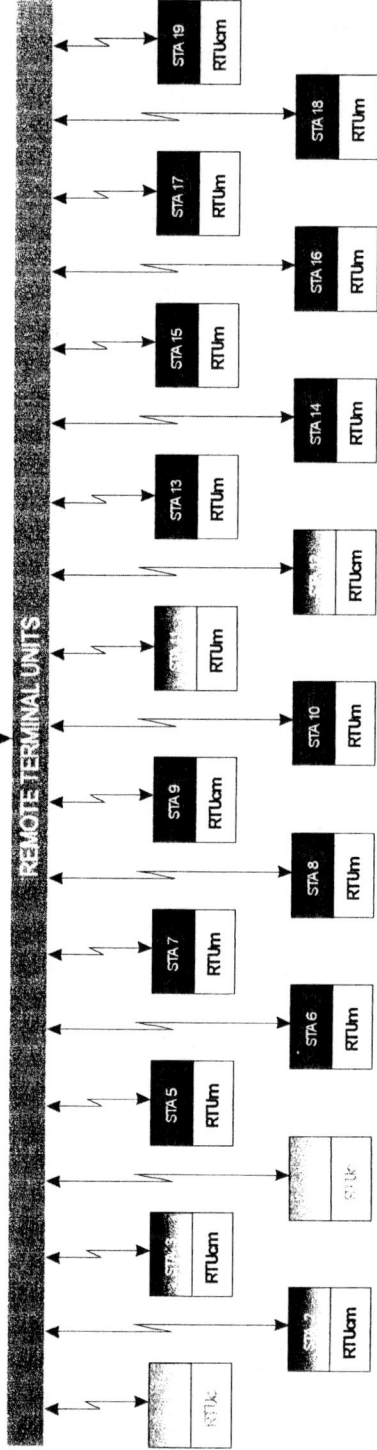

Fig. 2. Supervisory control and data acquisition system of the Mira Irrigation Project.

RTUc - Control Unit   RTUcm - Control and Monitoring Unit   RTUm - Monitoring Unit

will be equipped with a power supply, a radio transmitter/receiver and the corresponding antenna, a programmable computer controller, and module actuators.

3.3 Field stations with control and monitoring units (RTUcm)
There will be five RTUcm stations.

Stations 2 and 3 (Fig. 1 and 2) will control flows at a Neyrpic orifice intake to a distributor and will monitor the corresponding AMIL gate downstream (upstream and downstream water depths and gate position). At theses sites the delivered flows and the flow at the gates will be estimated.

Stations 9 and 19 will control flows at the distributors Odeceixe and Boavista Neyrpic orifice intakes and will monitor water depth within Odeceixe and Milfontes reservoirs (Fig 1 and 2), respectively.

The first sprinkling irrigation block designed will be controlled by station 12 (Fig 1 and 2). Field unit will control flows at the buffer reservoir inlet and at the pumping station and will monitor gate position and water depths immediately upstream and downstream (reservoir).

All these stations will have the same equipment as the RTUc stations plus the necessary water depth and gate position sensors.

3.4 Field stations with monitoring units (RTUm)
There will be thirteen RTUm stations.

Stations 5 to 7 and 10 to 17 will monitor AMIL gates (upstream and downstream water depths and gate position), the first three inside Milfontes Distributor and the others inside Odeceixe Distributor (Fig. 1 and 2).

Stations 8 and 18 will monitor flow at the downstream end of the distributors, respectively Odeceixe and Milfontes (Fig. 1 and 2).

All these stations will have the same equipment as the RTUc stations, without gate or module actuators, plus the necessary water depth and gate position sensors.

3.5 Master and submaster stations
At the present step of the SCADA system (Fig. 2), the constitution of the submaster and master stations are similar. However, the master station has the priority as system manager. The SCADA system can be monitored and controlled from these two stations, but master station orders are priority.

In the future, SCADA system will grow with new irrigation blocks built and new control functions. It will be necessary to have plus submaster stations, for example one submaster station integrating all pumping stations and another for all conveyance system. The SCADA system was already designed in order to support this evolution.

Presently, both stations will be equipped with a computer and the necessary software, printer, communication equipment, and power supply.

The master station will receive the relevant data from all RTU stations and the submaster station, will treat all the received data and will show it to the manager for control decision. In the future. control software will determine control actions.

3 Central automatic control system

It is already decided to obtain data during one or two irrigation seasons in order to automate the controlled flow devices and integrate a closed-loop control for the subsystem: main conveyor – Odeceixe and Milfontes reservoirs. Step by step, the SCADA system will evolve to a central automatic control system, with most of the control functions automated.

For the diversion flow control devices, automation control will be accomplished by installing enhanced software at the corresponding local programmable computer controllers. This way, instead of moving gates by remote control, the water manager sets flow targets and gates will automatically move to maintain the required flows. Target flows can be estimated considering the database obtained and other factors, as the day of the week, weather conditions and social reasons.

The closed-loop control for main conveyor will be obtained considering expected demands, storage volumes inside two downstream reservoirs and the dynamic response of the conveyor to the inflow variations. The database will be very useful for the closed-loop control algorithm design and calibration.

In order to analyse the dynamic behaviour of the main conveyor, the SIMCAR unsteady open-channel model [4] was calibrated. This model gives a particular attention to the simulation of the AMIL radial gates. Using this model, the response time to inflow variations for the conveyor was obtained [5]. Considering the several possible flow variations, the response times are:

- 16.1 ± 2.3 hours, for flow increments.
- 14.6 ± 1.8 hours, for flow decrements.

The exact values of the response time for all situations will be integrated in the closed-loop automatic control algorithm.

5 Conclusions

Low-cost, real-time (or near real-time) instrumentation promises a revolution in improved water management. Just as the rapid evolution in microcomputer hardware and software have expanded capabilities, so have the dynamic advances in sensors, telemetry equipment and programmable computer controllers. Today, real-time monitoring and control systems are within the cost range of almost all water user groups, including irrigators, canal companies and water districts.

Real-time technologies are, for these reasons, excellent tools for the upstream controlled canal modernisation.

4 References

1.   Kraatz, D.B. and Mahajan, I.K. (1975) *Small hydraulic structures.* Irrigation and Drainage Paper, n°26, FAO, Rome.

2.   Clemmens, A.J. (1987) Delivery system schedules and required capacities, in *Planning, operation, rehabilitation and automation of irrigation water delivery systems* (ed. Zimbelman, D.D.), ASCE, New-York, pp. 18-34.

3.   Rijo, M. and Almeida, A.B. (1993) Performance of an automatic controlled irrigation system: conveyance efficiencies. *Irrigation and Drainage Systems*, Vol.7, pp.161-172.

4.   Rijo, M., Almeida, A.B. and Pereira, L.S. (1991) Modelling automatic upstream control with SIMCAR, in *Irrigation and Drainage* (ed. Ritter, W.F.), ASCE, New-York, pp. 487-493.

5.   Rijo, M. and Almeida, A.B. (1993) Tempos de resposta às variações de caudal em canais de rega. *Recursos Hídricos*, Vol. 16, n°3, pp. 13-24.

WATERPROOFING MIRA, CAMPILHAS, SENHORA DO PORTO CANALS WITH DRAINED GEOMEMBRANES

J. L. MACHADO DO VALE
Technical Department, Tecnasol FGE, Lisboa, Portugal
A. M. SCUERO
Technical Department, CARPI Group, Arona, Italy
G. L. VASCHETTI
Technical Department, Geosynthetics Hydro, Torino, Italy

Abstract
To preserve and distribute a valuable, finite and vulnerable resource, canals must be maintained at the utmost efficiency. When the action of the environment deteriorates the structure, and decreases impermeability and water flow, the canal must be restored to its original capacity. Drained flexible PolyVinylChloride (PVC) liners provide long lasting watertightness, low hydraulic roughness to increase water flow, reduced installation times and costs, minimum maintenance. The paper describes application of a drained PVC system to Mira, Campilhas and Senhora do Porto canals, with increase in water supply from 50 to 90%. A system employing PVC geomembranes embedded in new concrete walls is also outlined.
Keywords: Canal lining, flexible liners, geocomposites, geomembranes, rehabilitation uplift.

1 Introduction

Protection of the quality and supply of freshwater resources is the issue of Chapter 18 of Agenda 21 for sustainable development in the next century, which defines freshwater as a finite and vulnerable resource, and recognises its economic value. Of all water consumed in the world, more than 70% is used for irrigation. As reported at the 15th Congress of the International Commission on Irrigation and Drainage, one of the main areas of recent design development of irrigation systems is the search for new or refined techniques for water control [1]. Canals can exert water control, and the maintenance of their efficiency must be a major concern in modern society.

Canals aim to modify the natural distribution of water in the environment, and to

Water and the Environment: Innovative Issues in Irrigation and Drainage. Edited by Luis S. Pereira and John W. Gowing. Published in 1998 by E & FN Spon. ISBN 0 419 23710 0

create a controlled water pattern transporting water from its source to a user which is generally far away. Canals must maintain over time the original controlled pattern, and avoid that by their deterioration a more random, diffuse and uncontrolled water distribution pattern is established, which can be dangerous to the environment.

To efficiently and safely preserve the controlled water transport pattern, canals must have and maintain very low permeability. Old unlined canals often experienced considerable water loss depending on the permeability of the native soil. In more recent times, canals have been lined with low permeability materials such as compacted clay, brick or rock masonry, concrete. All these materials deteriorate because of environmental aggression, mostly mechanical (pop off of sealing material of joints, undrained uplift pressure, temperature gradients, growth of vegetation, erosion, dynamic action of water, of transported sediments and materials, of animals' hoofs, action of ice and frost), or chemical (soft or sulphate water, acid rains). The results of deterioration are decreased impermeability and water flow: fissures form, the materials sealing the construction joints weather, overall porosity of the structure increases, hydraulic roughness increases, aquatic plants grow inside the canal. As the canal "ages", water is lost and the capacity of the canal decreases. When the entity of water and capacity loss is no longer acceptable, the canal must be taken out of service, vegetation removed, and watertightness must be restored.

2 Rehabilitation with rigid liners and flexible liners

Experience has shown that local repairs are a temporary measure. A complete rehabilitation installing a new liner must evaluate costs of out of service, costs of the decrease in water flow due to the reduction of the cross section of the canal if adding new material, and cost effectiveness of the new liner, which must consider the total cost of installation and the years of expected waterproofing life service.

The abovementioned traditional liners involve long construction times, significantly reduce the cross section, and are rigid and discontinuous, discontinuities being filled with impervious materials whose characteristics are different from those of the main liner. Discontinuity and rigidity increase susceptibility to deterioration so that after only several years the canal experiences the same deterioration problems which asked for its rehabilitation. The need for frequent chemical treatment to contrast the growth of aquatic plants is also prejudicial to the quality of water and can affect water table.

In this century flexible liners like bituminous geomembranes, in situ sprayed geomembranes, synthetic geomembranes, have been increasingly used in the place of rigid liners, especially in case of rehabilitation. Geomembranes are thin impermeable materials, which do not significantly alter the cross section of the canal, and have reduced installation times and costs.

The use of synthetic waterproofing geomembranes in hydraulic structures is now widely recognised as a dependable solution. In many countries national committees and agencies are involved in the study of these liners for application to canals. In the United States, who have been experimenting in the laboratory and in the field the application of geomembranes to canals lining since the 1940s, the Bureau of Reclamation, the Corps of Engineers and the American Society of Agricultural Engineers are very active in the field. In Europe, a pioneer country in the use of synthetic liners on dams since the late

1950s, a special committee (CEN Joint Working Group - Technical Committee 189/ Technical Committee 254) is preparing the future standards for use of synthetic membrane liners on canals.

The lining systems with flexible geomembranes can be divided into two main "families": the undrained and the drained systems. In undrained systems, there is complete adherence between the liner and the substrate on which it is applied. This is typical of in situ sprayed geomembranes and of glued bituminous geomembranes. Undrained systems, besides requiring perfect preparation of the subgrade, have a major disadvantage: the liner completely adheres to the substrate, there is no drainage, the liner is susceptible to detachment due to uplift by water impregnating the substrate.

In drained systems, the liner is mechanically fastened at points or lines. Two substantial benefits are provided: water behind the liner is constantly conveyed to bottom discharge and it cannot uplift the liner, seepage water impregnating the slopes is continuously intercepted and removed, with beneficial effect on the stability of the soil.

3 PVC drained geomembranes

Drained flexible PVC geomembranes, mechanically anchored, are the most widely used material for application on canals. Their main characteristics are:

- watertightness is continuous: a PVC liner, whose impermeability is several orders of magnitude higher than traditional rigid liners, has virtually no joints (welds tested 100% have the same watertight properties than the parent material). Due to elongation (>250% before rupture) and elasticity characteristics, it can bridge joints and large fissures in the deteriorated structure, and resist opening of new fissures in case of settlements in the natural slopes
- watertightness is durable: as the PVC geomembrane is engineered to resist mechanical and chemical aggression, and solar radiation, in particular UV, recorded experience on hydraulic structures has now surpassed 40 years
- PVC maintains a hydraulic roughness much lower than that of traditional materials: after years of exploitation, the global Manning-Strickler coefficient can remain in the order of 85 $m^{1/3}$/s, while experience shows that in concrete liners the coefficient will typically be 40% lower
- drainage decreases the water content in the natural slopes and in the structure
- growth of aquatic plants is hampered (Fig. 1 and 2)
- installation times and costs are lower than those of other rehabilitation systems, practically no maintenance is required.

Fig. 1. Mira distribution canal clogged by vegetation before rehabilitation.

Fig. 2. Mira canal after rehabilitation with a PVC drained geomembrane. Smoothness of the surface increases water flow and reduces growth of vegetation

4 Case histories

The following case histories illustrate how in Portugal the use of a PVC drained geomembrane system has allowed to improve the efficiency of water management: water is preserved, and the transport capability of the canals is increased. All installations have been performed by Tecnasol FGE and CARPI, a company specialising in the design, manufacturing and installation of flexible geomembrane liners.

4.1 Mira and Campilhas canals

The Obra de Rega do Mira irrigates an area of 12,000 ha and is operated by the Associação do Beneficiários do Mira. The distribution canal and the main canal, both lined with concrete and completed in 1970, were subject to deterioration resulting in increasing water losses. In 1991 the deterioration of most of the hydraulic structures, due either to their having reached the maximum service lifetime or to lack of maintenance, was such that it was deemed no longer acceptable from the point of view of watertightness and of stability. Among rehabilitation works, restoration of imperviousness was a major issue.

A flexible geomembrane system was deemed more adequate than a rigid new liner, as a flexible liner would assure longer term waterproofing of joints and cracks, would not reduce the cross section of the canal, and would imply lower costs. Among candidate geomembrane systems considered (PVC, PE, modified bitumen), a mechanically anchored PVC system was selected. Elements concurring to the choice were reliability, capability of drainage, durability of waterproofing life, and the additional advantage provided by the fact that design, manufacturing of the geomembrane and installation were covered by the same entity [2].

The Campilhas irrigation canal is owned by the Associação de Regantes e Beneficiários de Campilhas e Alto Sado. Characteristics and problems were similar to those described for the Mira canal.

4.1.1 Mira distribution canal and Campilhas canal

The Mira distribution canal and the Campilhas canals are both small canals with trapezoidal section and considerable length. Mira is 22,308 m long, wetted perimeter 1.56 m, maximum water velocity <1 m/s. Campilhas is 5,000 m long, wetted perimeter 3.2 m, maximum water velocity <1 m/s.

The capacity of the canals had decreased due to water loss at joints and cracks (seepage of 40 l/s was recorded at Mira), to the significant increase in roughness of the deteriorated concrete, to presence of vegetation and algae, to settlements which had further locally reduced the section. At Mira, of the available 210 l/s in its initial part only 170 l/s were delivered.

For both canals the main objective was to stop water leakage without further reducing a very small cross section, and at acceptable costs: cost of the installation, and costs of out of service, had to be kept to a minimum. The small cross section, low water head and low water velocity, allowed to choose systems simple in design (drainage, anchorage, drainage discharge) and in installation.

The Mira distribution canal was waterproofed by a flexible PVC geocomposite consisting of a 2.0 mm thick impervious PVC geomembrane heat-coupled during manufacturing to a 200 g/m^2 polyester geotextile. The geotextile provides drainage and antipuncture protection. The PVC geomembrane composition was engineered to make it capable of withstanding the estimated service conditions while at the same time being flexible enough to easily accommodate the small dimensions of the canal.

To allow an easy and quick installation minimising costs, the PVC geocomposite sheets were unrolled longitudinally to cover the entire cross section of the canal in stretches of 20 m. The low velocity of the transported water, and the substantial soundness of the concrete substrate, allowed mechanical anchorage to be made at crest by 30x2 mm flat stainless steel batten strips, and at the invert by impact anchors placed every 50 cm, covered by welded round patches of the same PVC parent material. Fig. 3 shows the installation.

In Campilhas canal, evaluation of the substrate, of service conditions and of budget requirements lead to the employment of a PVC geocomposite of the same type described above, but with a thinner (1.5 mm) PVC geomembrane. Layout of PVC sheets was made to accommodate at best the variations in the canal's cross section. A different anchorage system was adopted at crest: for the greater part of the canal, the PVC sheets overlapped the crest, were buried in a trench and then covered by natural soil, while at the flume anchorage at crest was made by rigid PVC shapes secured to the concrete elements. In both canals, junctions of all PVC sheets were made by manual heat-welding and tested 100%.

Drained water discharge was made by simple and effective one-way valves consisting of a square piece of the same PVC material, welded on three sides over the PVC geocomposite, and operated by the water pressure differentials (Fig. 4).

Watertight anchorage is always provided at the beginning and at the end of each lined area, to assure that the transported water does not infiltrate behind the liner.

The entire installation process at Mira, including preliminary cleaning operations, started and was completed in the Winter of 1994 (3 months, January through March, when the canals are not used for irrigation), for a total lined area of 25,300 square meters. At Campilhas, installation was performed in the Winter of 1997, and required 2 and a half months for 16,000 square meters. No outage was necessary for any of the

Fig. 3. Mira distribution canal: installation of the PVC geocomposite sheets to be left exposed to water flow.

Fig. 4. Campilhas canal: uplift relief by PVC one way valves.

two canals.

In the years from 1994 to 1996, the owner of Mira has been monitoring the behaviour of the waterproofing liner, with the following results:

- a capacity of 260 l/s is now possible, and the quantity of effectively transported water is now 255 l/s. This represents an increased water supply of 50%
- algae and silt have disappeared
- after three irrigation seasons, the geocomposite shows no alteration in physical or chemical characteristics
- even after some landslides occurred in the Winter of 1995, the geomembrane experienced no failure due to impact of rocks inside the canal.

4.1.2 Mira main canals

The main canals have a trapezoidal section, wetted perimeter varying from 7 to 9 m. Water velocity is <1 m/s.

In addition to decreased capacity due to water loss, high roughness of the concrete and formation of large algae, the water seeping from the canal had been impregnating soils with low coherence, with the result that sections with steep natural slopes were unstable. Capturing and discharging water already present in the soil, with the long term objective of improving its stability, was an important project issue. The use of a drained geomembrane was imperative.

The described PVC lining system was installed in the Mira canals in the Winter of 1997, after a 100 m long trial section had been waterproofed in 1994. The waterproofing geocomposite is 1.5 mm thick PVC + 200 g/m² geotextile on the slopes, 2.5 mm PVC + 500 g/m² on the invert. The anchorage and drainage systems are designed in function of the larger size of the canal, of the presence of high uplift from underground water, of the foreseen traffic of cleaning machines on the bottom.

Design had to assure that the liner and the anchorage system could resist drag forces acting on the liner. The high friction coefficient between the concrete substrate and the geotextile bonded to the PVC membrane allowed to achieve uniform distribution and transfer of forces to the substrate, so that they would not concentrate at anchorage lines. Anchorage at crest was made by stainless steel batten strips secured to the sound concrete of the crest. Anchorage at the invert had been made in the trial section (1994) by a ballasting concrete. After the canal was put back to service, algae formed again, so in 1997 the ballast layer was dismissed and substituted with anchorage by impact anchors, of the type which had performed satisfactorily in the distribution canal. The thicker geocomposite at the invert provides necessary resistance to mechanical damage, as proven by previous satisfactory results in similar service conditions.

Drainage capability was increased at the invert: a high transmissivity geonet was positioned under the PVC geocomposite. Discharge was made by one way valves. Total installed quantity at Mira main canals was 27,000 square meters. Results of monitoring show that:

- leaks have totally disappeared
- there is no formation of algae
- the geocomposite and all anchorage elements are totally unaltered, signalling no further increase in soil instability.

4.2 Senhora do Porto

Senhora do Porto canal is a 2,500 m long rectangular canal supplying water to the Senhora do Porto powerplant. The canal, with maximum bottom width of 5.1 m, height of 3 m, water velocity superior to 2 m/s, had lost capacity in respect to the need of the powerplant, due to large fissures located at the springing of the rock masonry walls, and to the roughness of the lining which had an estimated global Manning-Strickler coefficient of 55 $m^{1/3}$/s.

Rehabilitation aimed to watertight and upgrade the canal so that it could transport the 20 m^3/s flow necessary for maximum power output. This required installation of a new liner to restore impermeability, and heightening of the walls. The owner considered as reliable options a 15 cm reinforced concrete layer, and a PVC membrane. Hydraulic calculations comparing the two options evidentiated that the PVC membrane option allowed to attain, with the same heightening of the canal's walls, an average increase in capacity in the order of 50% higher than that allowed by a new concrete liner. The PVC option was selected. The objective of the intervention as specified in the contract was to attain a global Manning Strickler coefficient of 85 $m^{1/3}$/s, and to increase capacity from the 13 m^3/s of initial design, which had decreased over service, to the required 20 m^3/s [3].

To attain this objective, design and installation aimed to guarantee that the liner would have and maintain smoothness and planarity over the substrate. Anchorage and efficient drainage of uplifts (Fig. 5) were paramount.

The chosen waterproofing liner is a geocomposite consisting of a 2.5 mm thick PVC geomembrane heat-coupled during manufacturing to a 500 g/m^2 polyester geotextile (Fig. 6). The size and water speed in the canal required special mechanical anchorage to guarantee adherence of the liner to the substrate, so that stresses could be properly transferred and that no folds would form under the repeated action of flowing water. This was accomplished by increasing the anchorage lines, and by the use of special stainless steel coupled profiles which have the capability of tensioning the geocomposite. The tensioning profiles (125x1.5 mm) were installed at the base of the walls, stainless steel batten strips were installed at the invert, and 50x3 mm stainless steel batten strips provide anchorage at the middle of the walls and at crest.

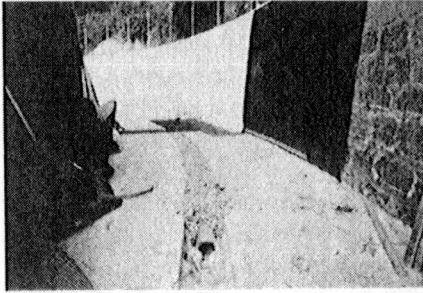

Fig. 5. Drainage discharge pipe, drainage layer, PVC geocomposite installed at Senhora do Porto.

Fig. 6. Waterproofing in canal is completed. The membrane can accommodate also passage of vehicles for cleaning, if required.

To take care of high uplifts in case of rapid drawdown of the canal, high transmissivity geonets were placed under the geocomposite on the walls and at the invert, and a collection system consisting of a fissured pipe was installed longitudinally in a trench excavated at the invert. Discharge of drained water is made by means of transverse outlets positioned at 100 m intervals and crossing the invert and the downstream wall.

In total, to perform civil works to heighten the walls, and to line 26,300 square meters, 6 months were necessary, in the two dry seasons of 1994 and 1995. Hydraulic measurements and calculations performed by the owner for acceptance confirmed that the global Manning-Strickler coefficient and the total capacity complied with specifications [3]. Comparing data for the same level in the forebay reservoir, and after 10% heightening of the canal's walls, increase in capacity amounts to 90%.

5 Heavy rehabilitation

The low hydraulic roughness of PVC can be an asset also when concern for the stability of the canal requires construction of new walls, reducing the cross section of the canal, and consequently its water flow. In application to hydropower supply canals in Italy, the reduction in cross section was counterbalanced by the gain in water flow obtained with the very low hydraulic roughness. Structural rehabilitation is made by new thin reinforced concrete walls embedding at casting special ribbed PVC geomembranes. PVC ribs on one side provide diffuse anchorage to the concrete, the side facing the flowing water is perfectly smooth. This system allowed an actual increase in water flow due to the favourable Manning-Strickler coefficient provided by the PVC membrane.

6 Conclusions

Conclusions of the owners [2, 3] show that the described drained PVC lining system is efficient from the point of view of waterproofing and of elimination of vegetation, that it constitutes a clear technical success (increase in maximum capacity of 50 to 90% were attained), that it is significantly less expensive than traditional systems, that

mechanical anchorage is adequate to resist water flow. In Senhora do Porto canal, power production increase 40%, corresponds to a gain of more than 200,000 US$/year.
The possibility of increasing water supply and minimising out of service time is an asset to consider when evaluating options for canals' rehabilitation. The described system is very promising to preserve and distribute water with the maximum efficiency.

7 References

1. *The International Journal on Hydropower & Dams*, Vol. 1, Issue 1, pp. 67-75.
2. Figueira, M. A. (1996) Utilização de geomembranas en obras hidráulicas e subterrâneas - Perímetro hidroagrícola do Mira. *Proceedings of the First Portuguese National Conference - Laboratorio Naciónal de Engenharia Civil*, APRH- ABRH, Lisbon.
3. Liberal, O. (1996) Remodelação do canal da Senhora do Porto - Estudos hidráulicos. *Proceedings of the First Portuguese National Conference - Laboratorio Naciónal de Engenharia Civil*, APRH- ABRH, Lisbon.

MINIMUM LEVEL OF WATER DEMAND FOR A PROFITABLE OPERATION OF AN IRRIGATION SCHEME

I. NICOLAESCU and E. MANOLE
Land Reclamation and Environmental Engineering Faculty, University of Agricultural Sciences, Bucharest, Romania

Abstract
Efficient operation of a large size irrigation scheme is strictly dependent on the actual water demand of the potential users within its command area. There is a certain minimum actual water demand level of an irrigation scheme under which, its operation will not be able to generate economic benefits. In order to evaluate this limit, it has been created a model on the base of benefit–cost rate analysis. This model was applied to a case study. The results indicate for this irrigation scheme that water demand minimum level varies in the range of 45 – 50%. There is not any economic reason to start in operation this scheme as long as the actual water demand is less than the minimum level evaluated by model running.
Keywords: Benefit–cost ratio, cropping pattern, economic input and output, irrigation water demand minimum level, water pumping efficiency, water use efficiency.

1 Introduction

In Romania, there are about one hundred irrigation schemes technically equipped for irrigation, covering almost three million ha; of this area, about 2.2 million ha are currently in operation and maintenance. The hydraulic schemes were designed to convey and distribute water among to irrigation plots that have a large range of individual size area (200 - 2000 ha/ plot). The irrigation plot consists of an underground pipe network where the water is pressurized from a pumping station. The majority of these schemes rely on hand-move lateral sprinkler system for irrigation of the main field crops: maize, wheat, sunflower, sugar beet, soybean a.s.o. More than 2.5 million ha has been developed to use medium to high pressure impact sprinklers. Each individual scheme has a set of specific operation and performance parameters, which determine water and

Water and the Environment: Innovative Issues in Irrigation and Drainage. Edited by Luis S. Pereira and John W. Gowing. Published in 1998 by E & FN Spon. ISBN 0 419 23710 0

pumping energy use efficiency and finally the economic viability. The values of these parameters are directly dependent on the size of the actual irrigated area and the net water volume applied to each crop within the irrigation scheme. In fact, these are related with the real demand of potential water users and climate evolution during the growing season. The irrigation water demand has dramatically decreased in Romania after 1991. This process has been generated by the land reform and dismantling of collective farms. The restitution of land to individual owners, to workers of the former collective farms and to other people qualified to clime the land, generated a great number of small land users who became new potential clients for the irrigation O&M units. The average size of the land ownership is around 2,5 ha, ranging from 0,5 to 10 ha. Land fragmentation increased tremendously the potential number of water users within the irrigation schemes, engendering difficulties in relations between the O&M units and irrigation water clients.

Another aspect is related to the fact that small landowners practice a subsistence farming which does not generate economic welfare in order to sustain irrigation scheme to run efficiently.

It is recognized that decreasing water demand within the irrigation scheme shall decrease its technical and economical performances.

Under such circumstances, this paper deals with a model in order to establish *the actual water demand minimum level (D_{min}) of an irrigation scheme* that is able to cover the total input expenses related to an increase of agricultural yield achievement.

2 The parameters involved in the model

2.1 Net irrigation requirement
Net irrigation requirement for each crop per month, corresponding to 50% probability, is noted with I_j [m^3/ha.month], where $j \in$ [April, September].

2.2 Irrigation water use level
Irrigation water use level within the scheme (D) is defined as:

$$\frac{V_{max}}{V_o} \geq D = \frac{V_r}{V_o} > 0 \tag{1}$$

In which:
V_r – net water volume applied to the crops within the scheme, during an irrigation season. This volume is considered to be consumed integrally by evapotranspiration;

V_o – actual net water volume required by the crops within the scheme area during the irrigation season, calculated for 50% probability;

V_{max} – V_o corresponding to 80% probability level, accepted on the scheme design phase.

2.3 Overall irrigation water use index

Overall irrigation water use index of the scheme (E_s) during a defined operation period can be evaluated by the general equation [1, 2]:

$$E_s = \left(\frac{1}{E_a \cdot E_t} + \frac{V_{kE}}{V_r} \right)^{-1}$$

(2)

E_a – field water application efficiency [%];
E_t – water distribution efficiency within the irrigation plot [%];
V_{kE} – volume of water that is lost along the conveyance network (seepage and operational mismanagement losses) during a defined period (month, season).
Involving the water use level (D) criteria, the equation (2) becomes:

$$E_s = \left(\frac{1}{E_a \cdot E_t} + \frac{v_{kE} \cdot \sum\limits_{m}^{n} T_j}{D \cdot S \cdot \sum\limits_{m}^{n} I_j} \right)^{-1}$$

(3)

Where:

v_{kE} – the monthly water volume that is lost along the conveyance network, considered to be constant for each month of the irrigation season (m^3/month).

$\sum\limits_{m}^{n} T_j$ – chronological months of irrigation schemes operation $\in$ [April, September];

m   – the start operation month;
n   – the end operation month;
j   – the month's index $\in$[m, n];
S   – total irrigation scheme area [ha].

2.4 Energy for pumping the water

The energy used for pumping the water in the irrigation scheme and plots, E [kWh/ha]:

$$E = e_o \cdot \frac{D \cdot \sum\limits_{m}^{n} I_j}{10^3 \cdot E_s}$$

(4)

Where:

e_o – unit pumping energy computed as the weighted average of all pumping stations within the irrigation scheme [kWh/1000 m^3 of pumped water], which is:

$$e_o = \frac{\sum\limits_{i=1}^{N} \left(s \cdot e_o \right)_i}{S}$$

(5)

N   – total number of pumping stations;

s_i – served area by „ i "pumping station [ha];

$(e_o)_i$ – unit energy consumed by „ i "pumping station [kWh/1000 m^3 of pumped water];

$$\left(e_o\right)_i = 2{,}725\frac{H_i}{E_p},\qquad (6)$$

H_i – total dynamic pumping head [m of water column];

E_p – pumping station operation efficiency [%].

2.5 Relation between the increment of crop yield and net irrigation application

Relation between incremental crop yield (y) and net applied irrigation water $\left(\sum_m^n I_j\right)$ using a power function type is:

$$y = \alpha \cdot \left(\sum_m^n I_j\right)^{\beta}\qquad (7)$$

Where:

y   – is expressed in kg/ha;

α, β – are individual crop statistical parameters which have been established on the basis of research data for at least ten years of investigation, where $\beta \leq 1$.

2.6 Economic input and output data

Economic input and output data involved in the achievement of increment of agricultural yield (y), for each crop, are:

- C_a – total supplementary agricultural inputs induced the crop irrigation by technology of irrigated crop, like: seeds, fertilizer, pesticides, extra yield harvesting and transportation a.s.o. [currency units/ha];
- C_i – total costs of irrigation O&M scheme, except the pumping energy cost [currency units/ha];
- C_E – total cost of energy for pumping [currency units/ha];
- p_e – electric energy unit price [currency units/kWh];
- p_c – unit price of the crop yield sold at farmgate [currency units/kg].

3 Model development

The model is based on the benefit – cost ratio concept analysis that could indicate the economic viability limits for a profitable operation strategy of a given irrigation scheme, which is:

$$\frac{B}{C} > 1 \tag{8}$$

B – incremental benefit obtained due to irrigation, by selling the incremental yield (y);
C – total cost of incremental inputs required for the incremental yield.
   In the model conception the following assumptions have been used:

- no government subsidies;
- total command area of the irrigation scheme is cultivated with a single crop.

   Taking into account the equations and economic data presented above, the combined relations were developed:

$$B = y \cdot p_c \cdot D = p_c \cdot \alpha \cdot D\left(\sum_m^n I_j\right)^{\beta} \tag{9}$$

$$C = C_a + C_i + C_E \tag{10}$$

$$C_E = e_o \cdot \frac{D \cdot \sum_m^n I_j}{10^3 \cdot E_s} \cdot p_e, \text{ or}$$

$$C_E = e_o \cdot p_e \cdot \frac{D\sum_m^n I_j}{10^3} \left(\frac{1}{E_a \cdot E_t} + \frac{v_{kE}}{D \cdot S} \cdot \frac{\sum_m^n T_j}{\sum_m^n I_j} \right) \tag{11}$$

$$p_c \cdot \alpha \cdot D\left(\sum_m^n I_j\right)^{\beta} > C_a + C_i + e_o \cdot p_e \cdot \frac{D\sum_m^n I_j}{10^3} \left(\frac{1}{E_a \cdot E_t} + \frac{v_{kE}}{D \cdot S} \cdot \frac{\sum_m^n T_j}{\sum_m^n I_j} \right) \tag{12}$$

   Rearranging in term of irrigation scheme water using degree (D), the last relation becomes:

$$D > \frac{\left(C_a + C_i\right) + \frac{e_o \cdot p_e}{10^3} \cdot \frac{v_{kE} \cdot \sum_m^n T_j}{S}}{p_c \cdot \alpha \cdot \left(\sum_m^n I_j\right)^{\beta} - \frac{e_o \cdot p_e}{10^3 \cdot E_a \cdot E_t} \cdot \sum_m^n I_j} > 0 \tag{13}$$

Passing over mathematical interpretations of relation (13) is more important to get the actual water demand minimum level (D_{min}) corresponding to the limit condition B equal C, such as:

$$D_{min} = \frac{\left(C_a + C_i\right) + \dfrac{e_o \cdot p_e}{10^3} \cdot \dfrac{v_{kE}}{S} \cdot \sum_m^n T_j}{\alpha \cdot p_c \cdot \left(\sum_m^n I_j\right)^{\beta} - \dfrac{e_o \cdot p_e}{10^3 \cdot E_a \cdot E_t} \cdot \sum_m^n I_j} > 0 \qquad (14)$$

In order to respect the basic condition ($D_{min} > 0$) it has to be fulfilled the following one:

$$p_c > \frac{e_o \cdot p_e}{10^3} \cdot \frac{1}{E_a \cdot E_t} \cdot \frac{1}{\alpha} \cdot \left(\sum_m^n I_j\right)^{1-\beta} \qquad (15)$$

This condition underlines the importance of crop yield selling price (p_c) concerning the establishment of cropping pattern in the proper way, according to actual operation parameters of a large irrigation scheme. It is also advisable to get the insurance of selling price before the cropping season starts.

An irrigation scheme is becoming more profitable as (D_{min}) value tends to decrease. Analyzing the model equation (14), this demand can be achieved by means of the proper agricultural practice's strategy, as well as by rehabilitation and modernization procedures. According to the model structure, the rehabilitation and modernization technologies should have essentially in view, to solve two main goals:

- to minimize the water losses along the hydraulic scheme network (v_{kc}) and specific energy required to pump water (e_o);
- to increase as much as possible water use efficiency inside the irrigation plots (E_a, E_t).

This model provides the user with the ability to evaluate the irrigation scheme management strategies, in correlation with agricultural needs.

In the situation of an actual $D < D_{min}$, the irrigation scheme should not be put in operation, as far as this one is not able to generate economic benefit.

4 Model application in a case study

The model has been applied to Mihail Kogalniceanu (MK) irrigation scheme that is stated on Central Dobrogea zone – one of the driest area in Romania.

This scheme is supplied with water from the Danube - Black Sea Canal by a main pumping station. Along the hydraulic open canals of the scheme there are six pumping stations for lifting water up to the land terraces. The scheme area is divided into 33 irrigation plots each one having its pressure pumping station. Since 1995, MK irrigation scheme is being under R & D project, so that the actual data for model application are as follows:

- $S = 23.141$ ha; $v_{kE} = 7,8 \times 10^6$ m^3/month;
- $E_a = 0,78$; $E_t = 0,92$; $e_o = 732$ kWh/1000 m^3; $p_e = 0,055$ $/kWh.

The rest of entry data used in model testing, according to the actual cropping pattern, are shown in Table 1.

The final results of model application are presented in Table 2.

Table 1. Main input data which were used in model testing

Crop	α	β	p_c ($/kg)	$C_a + C_i$ ($/ha)
1. Grain Corn	81	0.55	0.103	121.31
2. Corn Forage	234	0.65	0.020	122.10
3. Sugarbeet	90	0.72	0.042	104.50
4. Soybean	29	0.53	0.267	118.83
5. Sunflower	298	0.20	0.256	116.58
6. Wheat	47	0.54	0.140	115.12
7. Potatoes	55	0.74	0.156	106.05
8. Alfalfa	93	0.72	0.020	128.10

Table 2. Results of model application in M. Kogalniceanu irrigation scheme

Crop	Cropping Pattern (%)	ΣT_i	Continuous operation time (months)						
		2 (15 VI–15 VIII)		3 (1 VI–1 IX)		4 (1 V–1 IX)		5 (15IV–15 IX)	
		ΣI_j (m^3/ha)	D_{min} (%)	ΣI_j (m^3/ha)	D_{min} (%)	ΣI_j (m^3/ha)	D_{min} (%)	ΣI_j (m^3/ha)	D_{min} (%)
1.Grain Corn	40	1821	35,7	2463	34,2	2592	36,3	2728	37
2.Corn Forage	30	460	60,1	760	47,1	760	47,1	1261	36,6
3.Sugarbeet	6	1920	16,9	2752	14,6	2977	15,2	3245	15,5
4.Soybean	6	1355	55,9	1829	54,6	1953	57,8	2084	61
5.Sunflower	7	1585	59,9	2099	68,5	2262	75,7	2355	82,6
6.Wheat	30	500	75,3	700	68,2	1228	59,3	1366	59,6
7.Potatoes	1	1742	6,69	2394	5,86	2762	5,78	2903	6,04
8.Alfalfa	10	1441	60,3	2071	52,4	2478	50,7	2791	50,7
D_{min} (%) as weight average of cropping pattern			49,4		45,2		45,3		44,7

5 Conclusion

- The large size irrigation scheme should be defined and managed as an entire agriculture production system in order to generate economic benefits.
- The benefits can not be guaranteed if the water demand minimum level (D_{min}) is not fulfilled by the actual water requirement of the potential water users within the irrigation scheme. That is why this model, which has been developed on the benefit–cost rate analysis, provides a proper tool for decision markers.

- Applying the model to a case study, the following conclusions can be underlined:

  - this irrigation scheme should not be put in operation if the actual water demand level is not at least 45 – 50%, depending essentially on the cropping pattern and continuous operation time;
  - the crop profitability on this irrigation scheme area varies from the highest to lowest as: potatoes and sugarbeet; corn for grain and forage; alfalfa, soybean, wheat and sunflower.

- The optimization of the cropping pattern can be achieved using this conceptual model for any irrigation scheme.
- The model can also be successfully used in rehabilitation and modernization strategy of an individual large irrigation scheme.

6 References

1. Nicolaescu, I. (1992) Rehabilitation and modernization basis of large irrigation scheme in Romania. *Hydrotechnics Review*, No. 1 - 3.
2. Nicolaescu, I. (1994) Water efficiency – the base of irrigation system rehabilitation. *17th ICID European Regional Conference Proceedings*, Varna, Bulgaria, Vol. 2, pp. 173 – 180.
3. Moisa, M. (1995) Irrigation systems efficient utilization limits model. *ICID Special Technical Session Proceedings*, Vol.2, Rome, Italy.
4. Gilley, J. R. (1996) Sprinkler irrigation systems. *Sustainability of irrigated agriculture* (ed. L. S. Pereira *et al.*), NATO Series E, Vol.312, Kluwer Academic Publishers, pp. 193 – 209.

SPATIAL DISTRIBUTION OF IRRIGATION WATER REQUIREMENTS USING GEOSTATISTICAL ANALYSIS

V. SOUSA
University of Trás-os-Montes e Alto Douro, Vila Real, Portugal
L. S. PEREIRA
Department of Agricultural Engineering, Institute of Agronomy, Technical University of Lisbon, Portugal

Abstract
Contour lines of the net irrigation water requirements for the grass forage crop have been drawn using geostatistical analysis. The point values considered refer to 106 locations in the region of Trás-os-Montes, Northeast of Portugal, where rainfall data are available. Evapotranspiration is computed from data relative to 8 weather stations. At each one of the locations, the net crop irrigation water requirements are estimated using an irrigation scheduling simulation model, which was applied to the 19 years length climatic data series. Different water availability scenarios have been considered. For each one, the net irrigation requirements corresponding to the probability of 95%; 75% and 50% of not being exceeded were computed. It resulted in nine sets of 106 point values. The variogram analysis was applied to each of these sets providing the spatial estimation of net irrigation requirements. A cross validation prior to the application indicated that the average estimation errors range from 6.4% to 8.6% of the computed values. The contour lines have been drawn throughout kriging, which show appropriate correspondence to the variations induced by the relief on the rainfall and evapotranspiration patterns.
Keywords: Irrigation requirements, spatial variation, geostatistics, kriging.

1 Introduction

The present study deals with the spatial variation of the net irrigation water requirements of the grass forage crop in the Trás-os-Montes region. This is a mountainous region, located in the Northeast of Portugal, where irrigated agriculture is practiced after centuries, along the valleys of small rivers. Forage grass for pasture and hay is a main irrigated crop in the area.

Water and the Environment: Innovative Issues in Irrigation and Drainage. Edited by Luis S. Pereira and John W. Gowing. Published in 1998 by E & FN Spon. ISBN 0 419 23710 0

The mountain ridges are oriented from SW to NE and constitute successive barriers for the moist air coming from the seaside, predominantly from West. Rainfall increases with elevation and gradually decrease from West to East, being lower in the deeper valleys. Elevations of the irrigated fields vary along the region from 100m up to more than 1000m. These conditions lead to a large variation in the net irrigation water requirements along the region. The kriging estimation, widely used in earth and soil sciences [1] [2] [3], has been used for the spatial interpolation of that variable.

2 Material and methods

2.1 Climatic data and soil water balance
The climatic data available for the Trás-os-Montes region consisted of 106 rainfall data sets and 8 weather data sets with air temperature, relative humidity, sunshine duration and wind speed. The data sets refer to a ten-day time step and are 19 years long. All data sets have been tested for homogeneity and they were corrected when required. Missing values in the data sets were completed by a least square procedure using data from the nearest rainfall or weather station.

The model ISAREG [4] has been used for the simulation of the soil water balance in each one of the 106 locations where climatic data were available. Prior to the application, the model has been validated with field observations in the region. The reference evapotranspiration (ET_0), which is required for the model, is estimated from climatic data using the FAO Penman-Monteith method [5]. Parameters are computed as proposed by Allen *et al.* [6]. For the locations where only rainfall data is available, the ETo values of the nearest weather station are utilized, according to elevation and relief pattern influencing the local climate.

The model ISAREG has been selected because of its capacity to deal with several different irrigation constraints. In this work 3 different scenarios concerning the water availability are considered. These are defined from the ratios between actual and maximum crop evapotranspiration for the grass forage crop: $ET_a/ET_m=1$, $ET_a/ET_m=0.9$ and $ET_a/ET_m=0.75$. For each of these the net irrigation water requirements [mm] that are not exceeded for the probability of 50% ($I_{n,50}$), 75% ($I_{n,75}$) and 95% ($I_{n,95}$) are computed and constitute the regionalized variables under study. The soil water balance computations are performed for only one soil type but crop phenology is differentiated according the prevailing climatic conditions, mainly in relation to the elevation of the locations.

2.2 Spatial interpolation
The soil water balance simulation provides point values for the net irrigation water requirements of the pasture crop. Contour lines of this variable are drawn when one can estimate those values for points regularly spaced along the region. In the present work, kriging estimation was performed after variogram analysis.

Matheron [7] proposed the term regionalized variable to describe a spatially distributed phenomenon, i. e., a property witch magnitude depends on its location Examples of such variables are the rainfall depth, soil permeability or evapotranspiration [8]. In the present study, the regionalized variable is the net irrigation requirement of the grass forage crop.

Let us consider the regionalized variable z and $z(x_1)$, $z(x_2)$, ...$z(x_n)$ the realization of z at the points x_1, x_2, ..., x_n. Let us now consider every two points separated by the distance h. If we compute:

$$m(h) = \frac{1}{n} \sum_{i=1}^{n} \left[z(x_i) - z(x_i + h) \right] \qquad (1)$$

the expected difference in the magnitude of the variable z observed in two different locations separated by the distance h is obtained.

The semivariance of those values is computed from:

$$y(h) = \frac{1}{2n} \sum_{i=1}^{n} \left[z(x_i) - z(x_i + h) \right]^2 \qquad (2)$$

If the semivariance γ (h) is computed for different values of h and plotted against the distance h, an experimental semivariogram is obtained (see Fig. 1 and 2). The experimental semivariogram is not applicable in kriging estimation because it can not be represented by an equation. A semivariogram model must be fitted to the experimental one, as shown in Fig. 1, and 2, where the spherical model has been adjusted to the experimental points. The most common semivariogram models are described in Delhomme [2].

The semivariogram shows the spatial structure of the variable under study. Of particular importance in kriging estimation is the range of influence, i. e., the radius of the circle which contain the point values used for the estimation of the variable value relative to its center.

The UTM (Universal Transverse Mercator) coordinates were calculated for each location to compute the distance kilometer between locations to be used in the variogram analysis and in the kriging estimation. The results of kriging are presented as contour lines in a map format. A GIS software package was used for such mapping [9].

The estimator of the unknown value of the variable z at point x_p is:

$$z^*(x_p) = w_1 z(x_1) + w_2 z(x_2) + w_3 z(x_3) + \ldots w_n z(x_n) \qquad (3)$$

where w_1, w_2, w_3, ..., w_n are the weights assigned to the z (x_1), z (x_2), z (x_3), ..., z (x_n) point values within the range of influence. The weight factors w_i in equation (3) can be calculated from the following kriging system of equations [5].

$$\begin{cases} w_1 \gamma(h_{11}) + w_2 \gamma(h_{12}) + \ldots + w_n \gamma(h_{1n}) + \mu = 0 \\ w_1 \gamma(h_{21}) + w_2 \gamma(h_{22}) + \ldots + w_n \gamma(h_{2n}) + \mu = 0 \\ \cdot \\ w_1 \gamma(h_{n1}) + w_2 \gamma(h_{n2}) + \ldots + w_n \gamma(h_{nn}) + \mu = 0 \\ w_1 + w_2 + \ldots + w_n = 1 \end{cases} \qquad (4)$$

where the y (h) are determined from the semivariogram model adjusted to the experimental one, and μ is a lagrangian constant. A geostatistical software package was used to perform the variogram analysis and the kriging estimations [10].

3 Results

3.1 Irrigation requirements and variogram analysis

The descriptive statistics (mean, standard deviation, standard error and range) of the 9 sets of the 106 time series of net irrigation requirements for the grass forage crop are presented in Table 1.

Table 1. Descriptive statistics of the net irrigation requirements (mm) of the grass forage crop relative to 106 locations for the 50th, 75th and 95th percentiles and $ET_a/ET_m=1$; $ET_a/ET_m=0.9$ and $ET_a/ET_m=0.75$

ET_a/ET_m	Variable	Mean (mm)	Std. Dev. (mm)	Std. Error (mm)	Count	Minimum (mm)	Maximum (mm)
1	$I_{n,50}$	443	96	9	106	263	635
1	$I_{n,75}$	499	91	9	106	347	698
1	$I_{n,95}$	573	102	10	106	388	806
0.9	$I_{n,50}$	359	77	7	106	215	516
0.9	$I_{n,75}$	406	74	7	106	281	562
0.9	$I_{n,95}$	468	83	8	106	314	665
0.75	$I_{n,50}$	284	63	6	106	169	414
0.75	$I_{n,75}$	324	61	6	106	217	448
0.75	$I_{n,95}$	376	69	7	106	243	547

For all the 9 scenarios under study, one can observe that a large regional variation of the net irrigation requirements exists. For all cases, the range of variation is greater than the minimum observed depth. This variability follows the regional variation in rainfall and in the climatic variables determining ET_0.

For each set of the 106 point values a variogram analysis has been performed. Variogram models adjusted to each one of the scenarios studied are presented in Figures 1 and 2 for the conditions $ET_a = ET_m$ and $ET_a/ET_m = 0.75$, respectively. The variogram model that better fits the experimental variograms is of the Gauss type. When comparing the Figures 1 and 2 it can be observed that variograms are similar but the range of influence is smaller for $ET_a/ET_m = 0.75$ then for $ET_a = ET_m$.

A cross validation was performed to test the variogram models. It consists in estimating, one by one, all the observed values using the fitted semivariogram model. Estimated values are then compared with the observed ones. A summary of the results of the cross validation for all the scenarios is presented in Table 2. The estimation error observed in each one of the 106 locations is expressed as a percentage of the observed value. The estimation errors for main percentile distribution among the 106 locations are also given in Table 2.

The mean of estimates error ranges from 6.4 to 8.6% and the standard deviation ranges from 5.2 to 7.7%. Higher estimation errors correspond to $I_{n,50}$ and the smaller to $I_{n,75}$. The estimation errors for 50% of the locations (50th percentile) only exceed

a

b

c

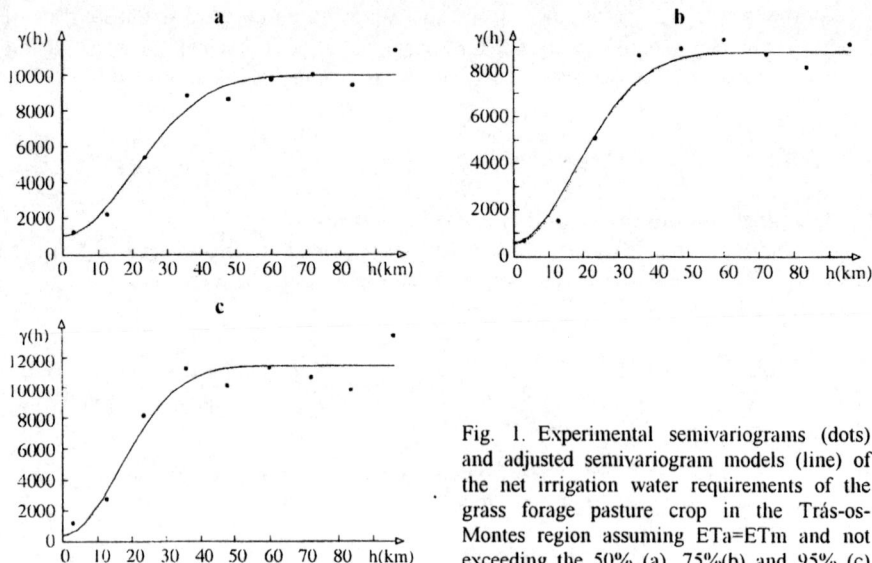

Fig. 1. Experimental semivariograms (dots) and adjusted semivariogram models (line) of the net irrigation water requirements of the grass forage pasture crop in the Trás-os-Montes region assuming ETa=ETm and not exceeding the 50% (a), 75%(b) and 95% (c) probability.

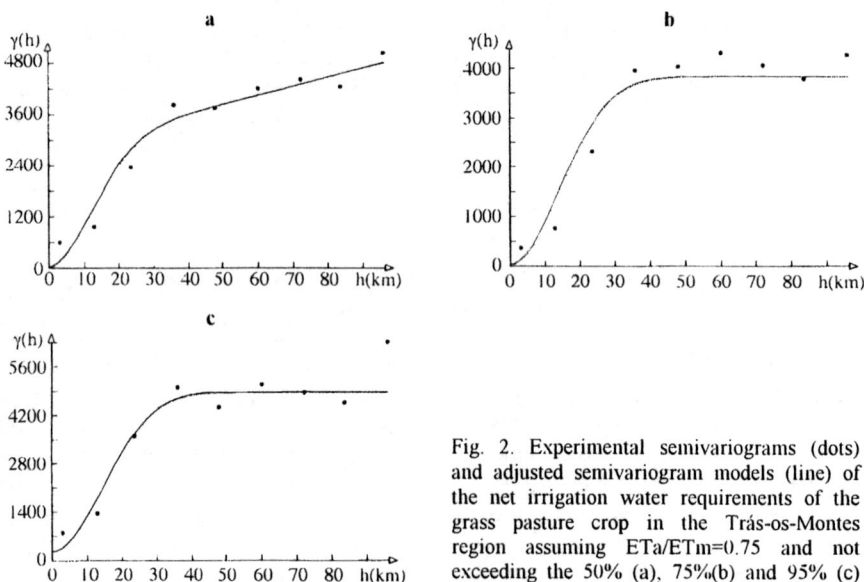

a

b

c

Fig. 2. Experimental semivariograms (dots) and adjusted semivariogram models (line) of the net irrigation water requirements of the grass pasture crop in the Trás-os-Montes region assuming ETa/ETm=0.75 and not exceeding the 50% (a), 75%(b) and 95% (c) probability.

6% for $I_{n,50}$ relative to the $ET_a/ET_m = 0.75$ scenario. For 75% of the locations, the estimation errors range from 9 to 12%. When 90% of the locations are considered, errors range from 13 to 18%. These results indicate that the procedure utilized is enough accurate to estimate the spatial distribution of net irrigation requirements.

3.2 Kriging estimation

The spatial interpolation of the net irrigation water requirements was performed, for each one of the scenarios considered using the kriging method and the semivariogram models presented above. Contour lines were drawn from the values estimated for the nodes of a grid 1 km wide. The results are presented in Figures 3, 4 and 5.

Figure 3 shows variation of the net irrigation water requirements of the grass forage crop in the Trás-os-Montes region, for the case of full irrigation, i.e., when $ET_a=ET_m$.

Table 2. Descriptive statistics of the cross validation error, expressed as percentages of the observed I_n values.

ET_a/ET_m	I_n	Mean	Std. Dev.	Std. Error	Percentile (%)				
		(%)	(%)	(%)	10th	25th	50th	75th	90th
1	$I_{n,50}$	8.1	7.2	0.7	1	3	6	12	17
1	$I_{n,75}$	6.4	5.2	0.5	1	2	5	9	13
1	$I_{n,95}$	7.1	6.7	0.6	1	2	5	10	15
0.9	$I_{n,50}$	8.0	7.2	0.7	1	3	6	11	18
0.9	$I_{n,75}$	6.4	5.3	0.5	1	2	5	9	12
0.9	$I_{n,95}$	7.1	6.7	0.6	1	3	5	10	15
0.75	$I_{n,50}$	8.6	7.7	0.8	1	3	7	12	18
0.75	$I_{n,75}$	6.6	5.6	0.5	1	2	6	10	13
0.75	$I_{n,95}$	7.4	7.0	0.7	1	2	6	11	17

For the average year (Fig. 3a) $I_{n,50}$ varies from less than 300 mm, in the West area, having the highest elevation, up to more than 600 mm, in the dryer zones and lower areas of the Douro valley. For the years with higher climatic demand (Fig. 3b and 3c)

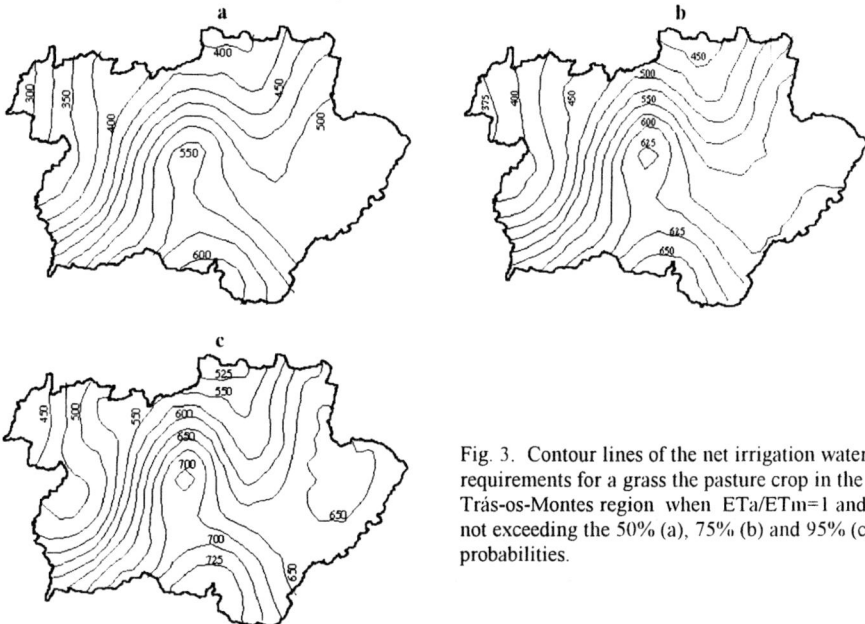

Fig. 3. Contour lines of the net irrigation water requirements for a grass the pasture crop in the Trás-os-Montes region when ETa/ETm=1 and not exceeding the 50% (a), 75% (b) and 95% (c) probabilities.

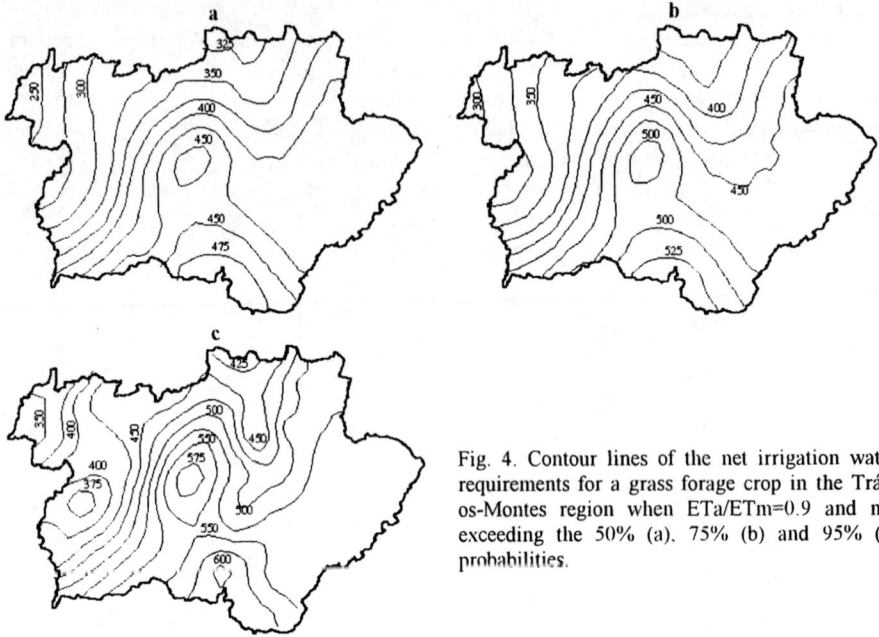

Fig. 4. Contour lines of the net irrigation water requirements for a grass forage crop in the Trás-os-Montes region when ETa/ETm=0.9 and not exceeding the 50% (a), 75% (b) and 95% (c) probabilities.

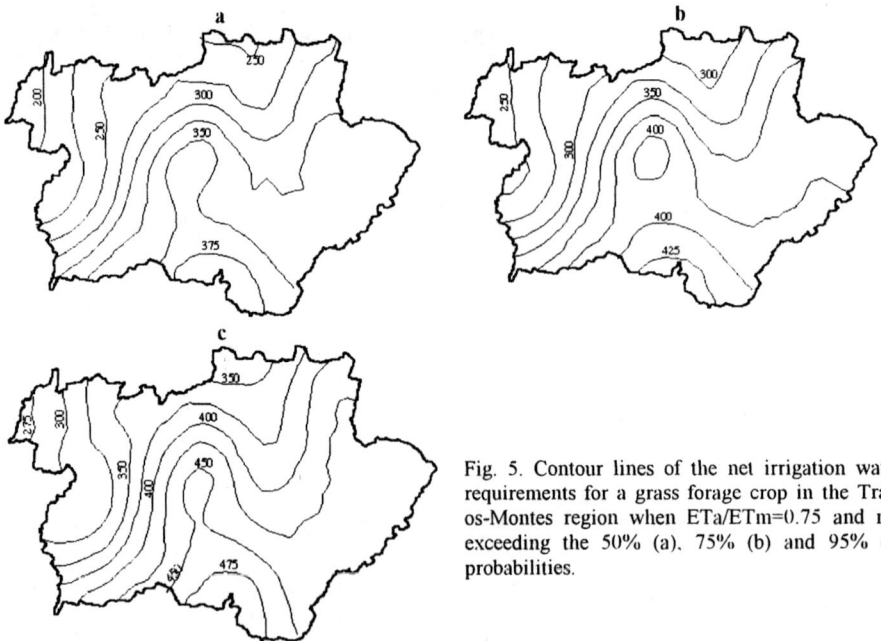

Fig. 5. Contour lines of the net irrigation water requirements for a grass forage crop in the Trás-os-Montes region when ETa/ETm=0.75 and not exceeding the 50% (a), 75% (b) and 95% (c) probabilities.

the spatial distribution of the net irrigation requirements of the crop has a similar pattern but yearly irrigation depths are higher, reaching more than 725 mm in some locations when the $I_{n.95}$ is considered (Fig 3c).

When crop irrigation requirements are not fully met (Fig. 4 and 5), the irrigation water requirements of the crop are naturally lower. As it could be expected, the spatial distribution exhibits a pattern similar to that Fig. 3, following the influences of the relief.

Lower I_n values correspond to the West and North areas of the region, where the elevation is higher and the rainfall is more abundant.

4 Conclusions

The spatial interpolation of the annual net irrigation water requirements of the grass forage crop through kriging shows to be appropriate to estimate their spatial distribution over a region. The contour lines drawn can be well explained by the influence of the relief and its orientation on rainfall and evapotranspiration.

Annual values of the net irrigation water requirements have been used in this study. Results encourage the use of a more detailed time scale like ten-day values in order to provide information for a better planning of water resources demand at the regional level. Using a short time scale, applications can also be done to provide information for farmers and to support regional irrigation scheduling programmes.

5 References

1. Bourgault, G.; Journel, A.G.; Rhoades, J.D.; Corwin, D.L.; Lesch, S.M., 1997. Geostatistical Analysis of Soil Salinity Data Set. *Advances in Agronomy*, **58**:241-292.
2. Delhomme, J.P., 1978. Kriging in Hydrosciences, *Advances in Water Resources*, **1**:251-266.
3. Goovaerts, P., 1997. *Geostatistics for Natural Resources Evaluation*. Oxford University Press Inc. 483 p.
4. Teixeira, J.L. , Pereira, L.S., 1992. ISAREG, an irrigation scheduling simulation model, *ICID Bulletin*, **41**: 29-48.
5. Allen, R.G.; Smith, H.; Perrier, A.; Pereira, L.S., 1994a. An update for the definition of the reference evapotranspiration. *ICID Bulletin*, **43**:1-34.
6. Allen, R.G., Smith, H.; Pereira, L.S.; Perrier, A., 1994b. An update for the calculation of the reference evapotranspiration. *ICID Bulletin* , **43**:35-92.
7. Matheron, G., 1963. Principles of geostatistics., *Economic Geology*, **58**:1246-1266.
8. Cuenca, R.H., Amegee, K.Y., 1984. Regionalized statistics applied to evapotranspiration, in: *Water Today and Tomorrow*, ASCE, N. York, pp. 330-337.
9. ComGrafix™ 1989. MapGrafix™. Geographic Information System. Users's Manual. ComGrafix, Inc., Clearwater, FL.
10. Englund, E., Sparks, A., 1991. *Geo-Eas. Geostatistical environmental assessment software. User,s guide*. U. S. Environemntal Protection Agency, Las Vegas, Nevada, 65 pp.

RICE PRODUCTION IN FLOODPLAINS: ISSUES FOR WATER MANAGEMENT IN BANGLADESH

J.J.F. BARR and J.W. GOWING
Centre for Land Use & Water Resources Research, University of Newcastle, Newcastle-upon-Tyne, U.K.

Abstract
Growth in food production in Bangladesh has kept pace with population growth due to i) a shift from a main crop of rain-fed rice to irrigated rice and ii) measures to mitigate the depth and duration of flooding. These changes have increased the competition for water by different users in the dry season and have limited the extent of open-water fisheries in the monsoon season. Studies from a floodplain depression waterbody show how these changes differentially affect particular social groups. The area is a closed basin system in the dry season, and an open system in the wet season. Analytical frameworks for better water management that can cope with this contrast are examined. A process that addresses the concerns of different stakeholders and leads to an integrated approach is proposed.
Keywords: Floodplains, open-water fisheries, rice irrigation, systems approaches, socio-economic environment, water resources management.

1 Water management and changing floodplain agriculture

Bangladesh has an high population, estimated to reach 140 million by 2000, living at a density of 850 people km^{-2} and growing at 2.17% per annum [1]. Nonetheless food production has managed to broadly keep pace with this growth, enabling Bangladesh to be close to meeting its objective of food grain self-sufficiency. The growth in agricultural production has been largely effected by dual-stranded agricultural development and water management policies. Firstly and of most importance, development of minor irrigation creating a shift towards intensive production of high yielding varieties (HYV) of irrigated *boro* (dry season rice) and wheat. This is at the

Water and the Environment: Innovative Issues in Irrigation and Drainage. Edited by Luis S. Pereira and John W. Gowing. Published in 1998 by E & FN Spon. ISBN 0 419 23710 0

expense of dry season oilseeds and pulses and to some extent monsoon season rice. Secondly through flood control and drainage works (FCD) designed to reduce the depth and duration of flooding, allowing a shift from traditional *aman* varieties (rice grown in the monsoon recession) to HYV *aman* [2]. There are however now indications of stagnant or declining productivity in both HYV *boro* and HYV *aman* crops, probably due to accumulating nutrient imbalances in this increasingly intensive agriculture [3].

The traditional pattern of agriculture on Bangladesh floodplains was based around broadcast-sown, local tall varieties of *aus* (monsoon rice) and *aman* grown in flooded paddies during the flood-vulnerable *kharif* (summer monsoon), and dryland crops such as pulses and oilseeds cultivated in the *rabi* (winter). The predominant pattern is now based on transplanted HYV *aman* and irrigated HYV *boro* [4].

The average *boro* area expanded from 1.12 to 2.56 million ha (mha) between 1979-82 and 1991-94, but the total rice area declined from 10.34 to 10.12 mha [3]. The increasing importance of *boro* rice means that there is an increasing dependence on irrigation for staple food supply. The area under irrigation has expanded rapidly from 1.04 mha in 1984 to 2.02 mha in 1987 and 2.48 mha in 1989, and this was predicted to reach 4.16 mha by the completion of the Fourth Five Year Plan in 1995 [5] [6] [7]. This represents 48.05% of the net cultivable area of Bangladesh.

In the *kharif*, the area of traditional (tall) *aman* has fell from 5.08 to 3.53 mha between 1979-82 and 1991-94, while the average area of HYV *aman* increased from 0.48 to 1.90 mha in the same period. This reflects the effect of flood control on the extent of deeply inundated floodplain. Nonetheless, Flood Action Plan (FAP) studies show that some farmers have preferred to allow water into FCD schemes and continue to grow less risky traditional broadcast *aman*, often followed by HYV *boro* [8].

2 Access to water

In addition to rivers, the floodplain in Bangladesh is characterised by an intricate pattern of waterbodies and land—water interfaces. These include *baors, haors, beels, khals*[1], ponds and ditches. These have a dry season extent of 5,480 ha, (*baors*) 2.8 mha (*haors*), and 114,161 ha (*beels*), but spread to cover up to 55% of the country in the monsoon. The larger bodies are CPRs; most are classed as *khas* (owned by the Government), and are leased for the purpose of fishing in them. However, *beels*, as with other floodplain waterbodies, have multiple uses and it is only the fishing right which falls to the lessee. Other uses, including rights to bathe, wash animals, navigate boats, and draw irrigation water are held communally [9]. The ability to exercise these rights is not straightforward. Access to the water resource may be limited by political, social, physical or economic factors, such as access to mechanical devices to extract irrigation water [5]. Where the resource is scarce, competing demands for water can result in conflict, denial of traditional access rights, diversion of water, and resource capture by powerful/influential sectors of the community. However there is also evidence of social organisation for co-operative water management in response to

---

[1] Ox-bow lakes; large floodplain depressions/small internal drainage basins; flooded depressions/back-swamps (perennial or seasonal); canals/channels

scarcity [10]. The *beel* may be considered a water resource endowment held by the very many floodplain dwellers around it, but the water resource entitlements (sets of utilities) that can actually be realised do not match the endowments for many disadvantaged social groups [11].

The burgeoning extent of irrigated dry season agriculture means that there is increasing competition for surface water from its multiple users. The impact of water extraction from *beels* by low-lift pumps in the *rabi* is currently uncertain [2]. However pumping reduces the size of dry season waterbodies, concentrating fish, making them easier to catch and stimulating increased involvement in fishing by farmers [12]. This has lead to further conflict between different user groups.

3 Water management and floodplain fisheries

In the monsoon season, extensive flood control measures, both large and small scale, together with structures such as all-weather roads, have modified the natural hydrological regime of large areas. The impacts of such modification have been studied under the Flood Action Plan. Despite some drainage congestion, impacts on agricultural production have been broadly positive due to the facilitation of the above changes in cropping pattern, whereas impacts on fisheries production have been generally negative. NEMAP estimates that FCDs have caused a 70% reduction in floodplain fisheries [13]. This is recognised by the Ministry of Water Resources who state that the main beneficiaries of water development projects are those that own assets (e.g. farmers), while there may be adverse impacts on specific occupational groups (e.g. fishers) and the poor generally, who are dependent on common property resources (CPRs) (e.g. floodplain fisheries) [14]. The identified impacts of FCDs on fisheries include loss of fish habitats due to reduction in extent of inundated floodplains and perennial *beels*, reduced fish abundance and biodiversity, blockage of fish migration routes, increased fishing effort and vulnerability to over-fishing, absence of integrated flood control and fisheries planning, and social conflicts between fishermen and farmers [2] [8]. In addition pollution and drainage of *beels* for agricultural use are further factors in the declining inland capture fishery [3]. There is an awareness that many *beels* are becoming increasingly ephemeral and such siltation has been recorded at the study site [15].

A common response to the decline in inland open water capture fisheries has been to develop Bangladesh's enormous potential for aquaculture. Aquaculture in homestead ponds and other small waterbodies, often with carp species, is popular and common. Technologies for integrated production of rice and fish (paddy aquaculture) are being widely promoted by agencies such as CARE [16] and enthusiastically taken up by farmers. Though these aquaculture systems are replacing some of the lost fisheries production, they are primarily constrained to privately owned ponds or paddies and are thus more suited to resource owners than the landless. This differs from open water floodplain fisheries in CPRs. The focus on carp species also contrasts with the diversity of small indigenous or non-economic species in *beels* which are significant in the diets of poor people, especially women and children [17].

4 The study area

4.1 Physical environment
The study centres on a medium—large *beel* and its expansion area; the perennial water in the *beel* covers 90 ha, its associated seasonal floodplain is a further 535 ha. A *beel* was chosen as it is a partly closed basin, and a common floodplain feature with key ecological, hydrological and socio-economic roles in floodplain production.

The area is located in Tangail District in north central Bangladesh. It is approximately 16 km west of the Jamuna river, and its agro-ecological classification is Young Brahmaputra and Jamuna Floodplains. It lies between two distinct river systems - to the west, the Dhaleswari, a major distributary of the Jamuna, with a peak flow of 3000 - 5400 m^3 sec^{-1} and carrying up to 13% of Jamuna flow as overbank spillage, and to the east, the Bangshi, which drains the slightly uplifted Madhupur Tract. The low-lying floodplain between these rivers is hydrologically complex, subject to seasonal cross-flows between the two systems, and becomes deeply inundated each year in the monsoon, acting as a natural flood water storage area. The *beel* is directly connected to a distributary of the Bangshi via a *khal*, on which there is a regulator gate set in a breached low embankment. The connection is thus open when flood levels permit, and the *beel* hydrology is closely related to peak river flows in the Bangshi distributary [15].

The soils are predominantly seasonally flooded, fine textured, non-calcareous soils developed in older Jamuna alluvium, although detailed soil survey within the study has revealed extensive areas of sandy soil around the *beel*, with important implications for irrigated dry-season cropping. In the adjacent *thana* (administrative sub-unit), high seepage and percolation losses from paddies of 27 - 29 mm day[-1] were measured [18].

The area falls between the 1750 and 2000 mm yr[-1] isohyets, with 85 - 90% of precipitation falling between mid-April to September - the *kharif* months. The *rabi* season experiences very litle rainfall and cropping is irrigation dependent. Although annually rainfall exceeds potential evapotranspiration (PET), PET exceeds rainfall by 250 - 400 mm in the *rabi* [19].

The predominant cropping pattern is fallow-mustard-*boro*, with a minority of *aman-boro* patterns. During *kharif*, 59.7% of land is fallow since the plots are deeply inundated (national flood classes F2 and F3 [0.9 - 1.8m and 1.8 - 3.0m]), mostly too deep for *aman*. Mustard is grown as a catch crop on 55.1% of plots once the monsoon recedes, for immediate sale to realise the cash needed to purchase inputs for the *boro* crop which follows. The cropping pattern is thus characterised by limited rice growing in the *kharif* due to excess water, and widespread rice production in the *rabi*, when water is at a premium.

Water resources audit found 91 functioning tubewell pumps used for irrigation, of which 15 are deep tubewells, the remainder shallow. Additionally there are 5 low-lift pumps, with capacities of ~50 1 sec[-1], drawing directly from the *beel*, and it is considered that these may have a marked effect on the dry season extent of the *beel*.

4.2 Socio-economic environment
An initial census of the 701 households (approximately 3900 people) in the five villages directly adjacent to the *beel* was used to group the households according to wealth (Table 1.). This stratification was based on land holding, dividing households between seven strata using Bangladesh Bureau of Statistics categories [20].

Table 1. Socio-economic stratification and land-holding

Group	Farm type (socio-economic category)	Amount of land owned (acres)	Proportion of households in group (%)	Proportion of land area owned by group (%)
1.	Landless - Categories I & II*	<0.049	17.83	0.07
2.	Landless - Category III	0.05 - 0.49	25.53	5.40
3.	Marginal	0.5 - 0.99	19.12	9.44
4.	Small	1.0 - 2.49	23.97	25.82
5.	Medium - I	2.5 - 4.99	8.56	22.10
6.	Medium - II	5.0 - 7.49	2.43	9.06
7.	Large	>7.5	2.57	28.10

*Landless categories: I - neither homestead or cultivable land; II - homestead but no cultivable land; III - homestead and cultivable land ≤ 0.49 acres.

Size of land holding is the factor normally used to denote socio-economic status in Bangladesh, and it has been shown to be a representative measure of wealth [21]. Nonetheless, a wealth ranking exercise [22] was carried out with a sample of households to validate stratification by land holding, since it is notoriously difficult to obtain accurate information about land ownership. Wealth ranking confirmed the importance of land as the main factor in wealth ('*shompod*'), with homestead size, family manpower, paid employment (remittances), ponds and trees, livestock, fishing equipment, cultivation equipment and irrigation equipment being other important factors. It also supported the ranking established by the stratification, except for Group 7 which ranked 4th. This is assumed to be due to more diversified wealth among the richer households, who, as proposed by Lewis *et al* now derive much income from their control of agricultural services such as irrigation and processing, rather than land *per se* [23].   Nonetheless, the stratification demonstrates that the poorest three groups represent 62.5% of households, but own only 14.9% of available land, whilst the wealthiest three groups make up only 13.6% of households, yet own 59.3% of the land area.

Floodplain agriculture does not consist solely of farmers cultivating their own land, there is an highly complex system of sharecropping, land mortgaging and labour purchase. It is estimated that 23% of the country's cultivable area is farmed by tenants or owner-cum-tenants, with a further 45% cultivated by wage labourers [24], and that 70% of sharecroppers farm the same plot for three years or less. However there is a broad relationship between socio-economic status and dependence on land and water resources. Drawing on households' self-assessed primary occupation, the census shows that overall 74.6% of households depend primarily on agriculture. Households with medium sized land holdings are the main farmers, whilst large land owners include the highest proportion of businessmen (17.7% of households). However 17.1% of Group 1 households are primarily dependent on fishing, compared to only 3.1% of fishing households overall. These households are 'professional' (full-time) fishermen. There are no Group 7 households dependent on fishing.

The low proportion of full-time fishermen belies the level of participation in subsistence fishing, which is practised by 73% of all households nationally [25]. The inexorable increasing pressure for land has resulted in increasing landlessness. One option for the landless is fishing, and this shift of poor households from agriculture to fishing is evident in the number of Muslim households now involved in what was previously an exclusively low-caste Hindu profession. The shallow and extensive nature of much of the floodplain means that entry into fishing does not necessarily depend on purchase of expensive capital equipment such as a boat [12].

5 Discussion

The study has shown two seasonally distinct sets of water management issues on Bangladesh floodplains. The dry season requires efficient use of surface irrigation water. Managing the use of this scarce resource by different farmers as well as fishermen requires an integrated approach in order to prevent a crisis of shortage. The additional layer of complexity in this case is the management of monsoon flooding, with the need to prevent crises of excess. The seasonally and temporally significant aquatic phase means there is a high dependency on fishing, particularly by the poor. Water management thus needs to address the competing and complementary demands between fishing and agriculture throughout the year. In the wet season agricultural interests aim to limit the extent of flooding and thus fish habitat, whilst in the dry season fish habitat is also threatened by extraction of surface water for irrigation.

As shown by the study site, the physical dynamic of the floodplain is overlain by the differential use of and access to land and water. Water management must therefore consider the normal physico-technical issues together with prevailing socio-economic patterns. Although these two dimensions are usually considered separately, combined approaches are being developed in land management [26]. The nature of the physical and social environment is characterised here by factors such as i) socially polarised land ownership patterns, ii) a majority of professional fishermen owning no land, iii) strong competition for dry season surface water, iv) high levels of involvement in subsistence fishing, and v) floodplain livelihood strategies which are co-dependent on a mosaic of aquatic and terrestrial natural resources. This interaction between the physical and social elements means a synthesis is necessary.

There is a growing consensus that an holistic or integrated approach is necessary in the management of floodplain natural resources. This comes not only as a consequence of the FAP [27] [10], but also as a result of concern about degrading wetland areas [28] and more efficient resource use and better government support to the natural resources sector [29] [24]. Integration is required between land and water resources - a systemic approach to floodplain management. The key question is whether the same analytical framework can be used throughout the year.

Analytical frameworks developed for irrigation management can contribute to addressing the wider pressures of integrated management of floodplain resources. The *beel* is effectively a small catchment and during the dry season it functions as a closed basin system. The pressures on surface water resources for multifarious uses are particularly keen in Bangladesh, with small waterbodies such as *beels* often being the focus of the pressure. The competing demands for water lend themselves well to Integrated Water Resource Systems (IWS) analysis [30]. Water management interventions in such closed systems, where water is a limited resource, tend towards zero-sum outcomes [31]. Implementation of irrigation practices with greater 'classical' efficiency in upper basin paddies is detrimental to water uses lower in the system such as paddy irrigation and activities in the *beel* sink. This is Seckler's "head-ender—tailender" problem [32], but in this context some of the tail-enders are fishermen who wish the tail-end sink to last as long as possible (even perennially to allow fish to breed), whilst some head-end farmers have access to the sink for irrigation and will abstract from it until it is exhausted. In addition to quantity implications, closed systems can lead to decreased quality for downstream users as pesticides and pollutants accumulate in *beels* in the dry season.

There are however possibilities for dry season water management in *beel* systems which could result in improvement in both agriculture and fisheries; positive-sum outcomes - Uphoff's "both-and" scenario [31]. There is anecdotal evidence from another study site of losses from ground-water irrigation *increasing* the dry season size of a small closed *beel*. Better 'effective' irrigation efficiency could be employed to systematically improve this. Identifying and working through the systems implications of a positive-sum change using stakeholder approaches and farmer participatory research methods will be the next phase of the research.

In contrast to being a closed basin in the dry season, the study *beel*, when flooded by river and rain water in the wet season, becomes an open system. Although definitions of 'open' and 'closed' are constructs based on decisions about the scale of analysis, open systems do present particular problems for analysis since they are always affected by influences beyond the scope of the analysis [31]. Most current water management measures are "either-or" options within the *beel*; either shallowly flooded agricultural land or deeper flooded open water fisheries. In the wet season, fishermen and farmers are essentially competing for the same space. At the study site, as on most of the floodplain, any point in space is likely to be in the aquatic phase for at least a third of the year, and in the terrestrial phase for the rest of the year. Interviews in the area indicate that farmers strive for a shorter flooding period (dry space) - land with a higher elevation being more highly valued due to its greater cropping potential, whilst fishers desire less restricted flood and more extensive fish habitats (wet space). The situation reflects competition between prevailing seasonal uses of land/water space, with potential for conflict over water management regimes which can affect whether an area is land or water at key times. Evidence for this comes from FCD areas, where conflict over operation of regulators is common [27].

The open nature of the system also leads to impacts outside the *beel*. FCD measures could be implemented to mitigate the depth and duration of the flood within the *beel* and enable the large areas of *kharif* fallow to be brought into HYV *aman* production, however FAP studies have shown that this would be exacerbate flooding outside the regulated *beel*. Agriculture inside the *beel* would be in conflict with agriculture outside the *beel*. The openness also permits reciprocal impacts; the open *beel* permits access to the *beel* fishery by those from outside the area, often facilitated by locally influential people (such as money-lenders) who can manipulate the environmental entitlements that different stakeholders can actually realise. Thus although the *beel*/basin conceptualisation and IWS analyses of scarce water use in closed basins is a useful construct for better water management in the dry season, it appears too restrictive for examining systemic water management in the monsoon season. Whole floodplain modelling and GIS have been used to examine these larger scale problems in open systems [33], but such approaches generally lack an integral socio-economic grounding. Only very recently has GIS been used as a facet of participatory approaches.

Participation in integrated water management requires a framework that elucidates the multiple perspectives that exist from different stakeholders [27]. Co-management approaches have been successfully used in CPRs in many places including Bangladesh, but need to be used with caution as they can be overly simple, especially if based on the premise of a now lost sense of community under which resources were sustainably managed [11]. Definitions of community and related issues of scale are germane since in the closed dry season *beel*, water management is essentially a local scale problem, whilst in the flooded open *beel*, the local scale must be taken with the broader floodplain management. Participatory approaches (albeit involving only farmers) based on water user groups have been trialed in Bangladesh, but the process has been found to be easily captured by locally influential people [10]. Though not simple or easy to achieve, it is clear however that better (i.e. more sustainable and more equitable) year-round management of floodplain water resources will only emerge from approaches based on strong principles of inclusivity. Approaches which have a grounding in mutuality, such as those used in the Gal Oya irrigation scheme [31], are the way forward. These approaches engender participation of all stakeholders and facilitate the expression of the multiple perspectives of different water users, leading to system-wide appreciation of other users and negotiated water management solutions.

6 Acknowledgements

This paper is based on work in Bangladesh funded by the Natural Resources Systems Programme of the UK Department for International Development, under contract number R6756. However the views expressed are not necessarily those of DFID.

7 References

1. BBS (1994). *Population Census, 1991.* Bangladesh Bureau of Statistics, Dhaka.
2. FAP 12 (1992). *FCD/I Agricultural Study. Volume 1. Main Report.* Bangladesh Flood Action Plan, Flood Plan Co-ordination Organisation, Ministry of Irrigation, Water Development and Flood Control, Dhaka.
3. Pagiola, S. (1995). *Environmental and Natural Resources Degradation in Intensive Agriculture in Bangladesh.* Land, Water and Natural Habitats Division, World Bank, Washington D.C.
4. Ali, A. M. S. (1995). Population Pressure, Environmental Constraints and Agricultural Change in Bangladesh: Examples from Three Agroecosystems. *Agriculture, Ecosystems & Management*, 55, 95-109.
5. Wood, G.D. and Palmer-Jones, R. (1991). *The Water Sellers. A Co-operative Venture by the Rural Poor.* Intermediate Technology Publications, London.
6. Rashid, H. R. (1991). *Geography of Bangladesh. 2nd Edition.* University Press Ltd., Dhaka.
7. BBS (1995). *1994 Statistical Yearbook of Bangladesh.* (15th edition). Bangladesh Bureau of Statistics, Dhaka.
8. FAP 17 (1995) *Main Volume.* FAP 17 Fisheries Studies and Pilot Project, Final Report. ODA, London.

9.   Toufique, K.A. (1997). Some Observations on Power and Property Rights in the Inland Fisheries of Bangladesh. *World Development*, 25 (3), 457-467.

10.   Soussan, J., Mallick, D, and Chadwick, M. (1998).   *Understanding Rural Change: Socio-Economic Trends and Peoples' Participation in Water Resources Management in Bangladesh.*   Environment Centre, University of Leeds & Bangladesh Centre for Advanced Studies, Dhaka.

11.   Leach, M, Mearns, R, and Scoones, I. (1997). *Environmental Entitlements: A Framework for Understanding the Institutional Dynamics of Environmental Change.* IDS Discussion Paper 359. University of Sussex, Brighton.

12.   FAP 17 (1994). *Thematic Socioeconomic Study.* Supporting Volume No. 19. FAP 17 Fisheries Studies and Pilot Project, Final Report (draft). ODA, London.

13.   NEMAP (1995). *National Environmental Management Action Plan.* Ministry of Environment and Forest, Dhaka.

14.   FPCO (1995). *Bangladesh Water and Flood Management Strategy (Draft).* Flood Plan Co-ordination Organisation, Ministry of Water Resources, Dhaka.

15.   EGIS & Delft Hydraulics (1997).   *Floodplain Fish Habitat Study.*   Water Resources Planning Organisation, Ministry of Water Resources, Dhaka.

16.   Kamp, K., Fahmida Begum and Setijoprodjo, A. (1996). Diversifying rice field systems in Bangladesh. *ILEIA Newsletter*, 12 (2), pp.22-23.

17.   Minkin, S.F., Rahman, M. M., and Halder, S. (1997). Fish Biodiversity, Human Nutrition and Environmental Restoration in Bangladesh.   In: Tsai, C. and Youssouf Ali, M. (eds.) p.75-88.   *Openwater Fisheries of Bangladesh.* University Press Ltd., Dhaka.

18.   Khan, L.R. (1990). Some Issues for Efficient Water Management in the Haor Area of Kishorganj and the Floodplain of Tangail. p. 19-39. In: Khan, M.A.A., Hossain, S.M.A, Nishimura, H., Kaida, Y. and Tanaka, K. (eds.) *Agricultural and Rural Development in Bangladesh.*   Proceedings of the Second JSARD Workshop. JSARD Publication No. 18. JICA, Dhaka.

19.   Brammer, H. (1996). *The Geography of the Soils of Bangladesh.* University Press Ltd., Dhaka.

20.   Hossain, T. (1995). *Land Rights in Bangladesh - problems of management.* University Press Ltd., Dhaka.

21.   Adams, A.M., Evans, T.G., Mohammed, R., and Farnsworth, J. (1997). Socioeconomic Stratification by Wealth Ranking: Is It Valid?   *World Development*, 25 (7), 1165-1172.

22.   Grandin B.E. (1988). *Wealth Ranking in Smallholder Communities : A Field Manual.* Intermediate Technology Publications Ltd, London.

23.   Lewis, D.J., Wood, G.D. and Gregory, R. (1996). *Trading the Silver Seed. Local knowledge and market moralities in aquacultural development.* Intermediate Technology Publications, London.

24.   Hasan, S. and Mulamoottil, G. (1994). Natural-resources Management in Bangladesh. *Ambio*, 23 (2), 141-145.

25.   DOF (1990). *Fish Catch Statistics of Bangladesh, 1987-1988.* Department of Fisheries, Dhaka.

26.   Collinson, M. (1998).   Social and Economic Considerations in Resource Management Domains.   p. 45-60. In: Syers, J.K. and Bouma, J. (eds.).

Proceeding of the Conference on Resource Management Domains. Kuala Lumpur, 26-29 August 1996. IBSRAM Proceedings no 16. IBSRAM, Bangkok.

27. Hughes, R., Adnan, S. & Dalal-Clayton, B. (1994). *Floodplains or Flood Plans? A Review of Approaches to Water Management in Bangladesh.* IIED, London and RAS, Dhaka.

28. Rahman, A.K.A. (1993). Wetlands and Fisheries. pp 147-161. In: Nishat, A., Hussain, Z., Roy, M.K. and Karim, A. (eds.). *Freshwater Wetland in Bangladesh: Issues and Approaches for Management.* IUCN, Gland.

29. Karim, Z. (1994). Needs and Priorities of NARS for the Management of Natural Resources: Bangladesh. p. 105-125. In: Goldsworthy, P. and Penning de Vries, F.W.T. (eds.), *Opportunities, Use, and Transfer of Systems Research Methods in Agriculture to Developing Countries.* Kluwer Academic Publishers, The Netherlands.

30. Keller, A,, Keller, J. and Seckler, D. (1996). *Integrated Water Resource Systems: Theory and Policy Implications.* Research Report 3, International Irrigation Management Institute, Colombo.

31. Uphoff, N. (1996). *Learning from Gal Oya. Possibilities for Participatory Development and Post-Newtonian Social Science.* Intermediate Technology Publications, London.

32. Seckler, D. (1996). *The New Era of Water Resources Management: From "Dry" to "Wet" Water Savings.* Research Report 1, International Irrigation Management Institute, Colombo.

33. Pandyal, G.N. and Syme, M.J. (1994). Profiling a flood management systems for Bangladesh: the strategy of the generic model—GIS connection. *Journal of Hydraulic Research,* 32 (extra issue), 21-34.

USERS' GOVERNANCE OF IRRIGATION WATER: ON-GOING REFORMS AND POTENTIALS

M. NAKASHIMA
Department of International Studies, Hiroshima City University, Hiroshima, Japan

Abstract
Increasing water demands urge both supplier's and users' institutions to evolve into a new stage, through a reform of irrigation management. This paper first presents a brief overview of two on-going institutional reforms. One is the case of Mexico's reform, started in 1989 and almost completed now. Another is the case of Pakistan's reform, started in 1997 on a trial basis. Considering the experiences in Mexico and arguments in Pakistan, conditions for a successful reform of irrigation management are discussed for designing a better frame of the institutional reform.
Keywords: autonomy, end-users' resource governance, institutional reform, irrigation charge, irrigation management transfer, participatory irrigation management, water market, water users' organization.

1 Introduction

End-users' governance of irrigation water or participatory irrigation management will contribute in solving the issues listed for this Conference. Water users can be most sensitive in combating water pollution and chronic water deficits as well as demand management. Although each water user may not be powerful, the users' collective voices and actions would help promote and achieve improvements in these issues.

This paper introduces and discusses the on-going institutional reforms toward physically and financially more sustainable irrigation management. The reforms have been initiated with a motivation to alleviate a financial burden of the government for irrigation facility operation and maintenance (O&M). Such a reform requires farmers' involvement in irrigation facilities and water management, necessitating formation and development of a water users' organization. At an administrative level, decentralization, public and private role sharing, and legalizing water right must

Water and the Environment: Innovative Issues in Irrigation and Drainage. Edited by Luis S. Pereira and John W. Gowing. Published in 1998 by E & FN Spon. ISBN 0 419 23710 0

accompany the reform. Many countries are in need of a reform of irrigation management, and this paper is intended to contribute to designing a frame of the institutional reform. First, I will introduce the Mexico's case of irrigation management reform, and next the case of Pakistan, which recently started the reform. Then, I will discuss potentials and conditions for a successful reform of irrigation management.

2 The Mexico's reform -- innovation for irrigation management

2.1 Deterioration of irrigation management
Irrigated agriculture in Mexico has a yearly production value of more than 50% of the total agricultural production. This is a significant contribution to the country's agriculture since the total farm land is 31.1 million hectors (ha) and the irrigated farm land is only 5.7 million ha out of the total area. The Irrigation Districts are large-scale irrigation created by the government investment. Currently, there are mutually independent 80 Irrigation Districts that irrigate 3.2 million ha. The government, before irrigation management reform, operated these Irrigation Districts.

The Mexico's economic crisis in 1980s affected the operation of Irrigation Districts. In 1988, irrigation users' payment covered only 15 % of maintenance cost, because water delivery was not made properly due to poor operation. The government could not continue spending considerable amounts of money that was necessary for proper O&M. Deterioration of the irrigation systems was accelerated. Eventually, it was considered that the only way, to recover proper operation of the Districts, was to make the irrigation users responsible for the management, as they are the ones who are benefited from irrigation works.

2.2 Irrigation management reform and water users' association
The government started from 1988 to transfer the irrigation management of the Irrigation Districts. The transfer program contained two stages. In the first stage, a water users' association (WUA) was organized by the farmers in order to operate and maintain a new irrigation unit. It was decided to artificially divide an Irrigation District into new units (irrigation subsystems), and they are called irrigation "modules." The WUA would be given a Concession Title for the use of water, secondary infrastructure (channels), and some of the machinery in the District. The essential condition for the transfer is that the WUA has to achieve a financial self-sufficiency to cover all the costs of O&M of the module. In the second stage, an organization -- Society of Limited Responsibility with Public Interest (SLR) -- was to be created by a group of WUAs. The roles of SLR are administration, O&M of the main channels, and coordination of the WUAs. The condition to establish a SLR is to show an institutional capacity of performing these roles to the government.

The transfer process started in 1988. However, an actual implementation was not until creation of the Federal authority, Comision Nacional del Agua (CNA), in 1989, and the national program for transferring the Districts to the WUAs was organized. The CNA was created by decentralizing a Ministry responsible for both irrigation and agricultural production. In order to repair the damaged irrigation and drainage infrastructure and to train and strengthen the WUAs, a credit with the World Bank and Inter American Development Bank was agreed.

2.3 Legal framework: the National Water Law

Legal and institutional arrangements were necessary to facilitate the irrigation management transfer. The National Water Law was promulgated in 1992 and made law in 1994 by abolishing the Federal Water Law of 1972. The Article 65 of Law states that the Irrigation Districts will be operated, maintained and conserved by the water users. The CNA provides the water and the necessary public infrastructure. It should be noted that although the users manage irrigation infrastructures, irrigation transfer is not exactly a privatization. Because, even after the transfer, infrastructures of irrigation, road, etc., are still the national properties. Also, irrigation water is not a private good, but a public and common good. The WUA operates secondary channels and the SLR manages main channels, however, a dam can not be transferred to the water users, since it has many functions such as urban water supply and flood protection.

When the transfer is completed, the Title of Concession is granted by the CNA to a WUA, which is the legal instrument that identifies and describes the duties and obligations of all parties involved in the transfer. The right of using infrastructures belongs to farmers for the period of at most 50 years. One of the most important features of the Law is that it allows trading of water rights. By the Law, water right is not attached to the land, and a water right can be conveyed to other users, regardless of the land being irrigated [1].

2.4 Formation of WUA and Irrigation management

In the beginning of the transfer process, the CNA officers of an Irrigation District convened meetings to explain the transfer program to farmer's leaders. The CNA recognized past mistakes such as bureaucracy and corruption and explained advantages of the transfer program. The irrigation transfer seemed a heavy burden for farmers, and some farmers were scared because they did not have any experience in irrigation management. They had many meetings (until 1992 there were more than 200 meetings in 2 years in a District) to discuss two major issues: defining a size of module and revising irrigation charge. The first issue was a size of module to create: for example, smaller sizes of 4,000 to 8,000 ha or larger sizes of 18,000 to 20,000 ha. A smaller size has an advantage of closer communications between a WUA and irrigation users and a disadvantage of higher production cost due to diseconomies of scale. To reduce production cost, the proposal of a larger module size won the majority in many Districts. Another issue was reaching an agreement with the irrigation charge; it was raised in most of the WUAs since financial self-sufficiency was a condition of irrigation transfer.

By the decentralization with irrigation management transfer, the government's role has been significantly reduced limiting to: water allocation for WUAs, administration of major facilities such as reservoirs, and supervising the WUAs and SLRs. The roles played by the farmers are to pay an irrigation charge and to attend General Assembly for electing the president of WUA. They also work with canal operators, WUA's employees, to irrigate their farms. WUA has a monthly meeting by its executives, and they hold an Assembly to decide important issues and to solve major disputes. The CNA and a WUA share the user's water charge, with the percentage of 25% and 75%, respectively. Financial self-sufficiency of Irrigation Districts is about 80%, as a whole in the country after great improvement [1].

There are cases of water transaction in an Irrigation District. According to a record of water delivery in the 1996-1997 crop year, water transactions were made among several WUAs. A record shows a water balance among WUAs in the District: for example, a module growing sugarcane used more water than allocated, and the delivery balance was negative [2]. If such a module with a negative balance can explain how (from which module) they obtained the extra water, then the overuse is acceptable for the CNA. However, if they can not explain it, they have to face a sanction. The modules of negative balances paid to the exporting modules with a rate of 25 pesos/1000 m^3. Those modules with surplus water saved water by a drip irrigation technology for vegetables.

2.5 Performance of the irrigation management transfer

After the irrigation transfer, water users (farmers) feel as if they have an ownership of irrigation facility, and it changed their attitudes of maintaining the facilities and saving water. Good results of the transfer include better water delivery to the farms and better maintenance of the channels. Farmers achieved such a technology advancement as drip irrigation, mainly in vegetable production areas, since farmers try to save water, as they must pay more for water. The major achievements are financial self-sufficiency, better finance with a gain from bank interest, new irrigation technique, and more second crop cultivation resulting from water saving.

There are varying cases, in the country, of success and no success yet in the process of irrigation transfer. Various factors have affected the performance, for example, economic, social, cultural, and political factors. It may be useful to examine the effects of such factors. Performance of irrigation transfer has been different among regions of north, central, and south of Mexico. Northern region completed the transfer sooner than the central and the southern region [1]. There are following reasons in such a varying performance.

The major reason seems to be the economic situations in the regions. Agriculture is more modernized and economy is better in the North. The farmers would be better prepared to create a business organization (WUA) and have a stronger capability to manage irrigation facilities of the module. The poor performance in the South may be due to its more traditional and small scale agriculture, where the farmers are not ready to manage a large Irrigation District. Another factor affecting the performance is cultural, for example a way of thinking. In the North they have more business-oriented mentality. The transfer was easier in the areas more developed with business orientation. By contrast, in a region where farmers have paternalistic thinking, that is, water should be provided by the government, the transfer process had to overcome such a mental obstacle.

Another factor is political situation of a region; e.g., opposing to a policy that comes from the central government. Such a political problem is often found in the South, which is caused mainly by a poor regional economy. Political problems make the government policy difficult to implement. Bad cases were those where political influences were made and sound management was hindered in important decisions in an Assembly. Land tenancy problems may complicate the political situations. A lesson was the importance of not to mix politics with administration of a WUA, since irrigation is for agricultural production and it can be purely a business matter. Politics tends to come in, when an increase of irrigation charge is proposed in Assembly.

Water scarcity should be another factor affecting the performance of irrigation transfer. In the South, they depend on rainwater for agricultural production; and irrigation is not always an essential agricultural input. The performance of irrigation transfer in the North and in the South may be related to the climate of less water resources in the North and more in the South. In a region with sufficient water resource, farmers may not have strong incentives to better manage their resource.

In 1997 the transfer was completed for 2.9 million ha, and only 0.4 million ha have remained not transferred yet, indicating the transfer is almost finished. However, such statistics do not reveal performance of each WUA and SLR. Although the transfer was almost finished at least in an official document, there are many problems that have to be resolved. The transfer is "a process," and the WUAs and the SLRs must continue to develop themselves in their capabilities. Some Districts may be struggling still at the first stage, and other Districts may be at the second stage. Each District and WUA should make progress in the transfer, according to its own pace most appropriate to them.

3 The Pakistan's reform -- trial for innovation

3.1 Deterioration of irrigation management and the proposed reform

The Pakistan's irrigation system was constructed by the British ruler in the Indus River basin more than 100 years ago. It is one of the largest irrigation system in the World, extending in the Indus Plain with irrigable area of 14 million ha. Pakistan's agriculture depends on the irrigation system, for more than 90% of total agricultural outputs.

The irrigation system was operated and maintained well for many years, and the irrigation charge collected from farmers was used to cover the necessary cost of O&M. Before deterioration in the system's O&M took place, there were increasing pressures of water demand to the irrigation system [3]. Irrigation water demands increased with needs of expanding agricultural production prompted by the economic development. Cash crop production and modern agricultural technology require more water than traditional agriculture for subsistence. Increasing population boosted the crop planting intensity, resulting another increase in water demand. Such increasing water demands were the results of modernization of the past decades observed in any country. In Pakistan, the irrigation system of Indus River was designed and built to meet only 60% crop intensity, more than a century ago; and therefore, water supply of the irrigation system could not meet the actual demands.

In this way, the hardware of the irrigation system could not meet the increased water demand, and software of the irrigation system also could not cope with it. The software may consist of institutions in both users' side and supplier's side. The water users' institution is called "Warabandi," that is a traditional water distribution system among farmers. The Warabandi had been working properly before the increases in water demand put pressures on the traditional institution. The supplier's side institutions are a century old rules and regulations of the government, enforced by Provincial Irrigation Departments. These institutions of both sides were not able to keep their own disciplines, when farmers faced with the needs of more production and government officers faced with the pressures from farmers demanding more water.

Since the early 1970s, the Irrigation Departments started to lose autonomy due to political pressures and influences demanding more water. Due to weakened disciplines and autonomy, the collection of irrigation charge from farmers declined, and the revenue fallen short of the government O&M expenditure. Deterioration of the physical system brought irregular water distribution to the canals, resulting unequal irrigation water available for farmers [4]. There are many cases reported that the tail end of distribution channels is dry and farmers do not receive water for many years. Such a physical deterioration has been acompanied by a social and administrative deterioration, manifesting itself as malpractice of both influential farmers and irrigation officers. Again, this augmented inequity in water distribution among farmers. The irrigation system is regarded now as facing a crisis, physically, socially, and administratively.

The World Bank has been financing in the Pakistan's irrigation and drainage infrastructure for the last three decades. The Bank made recommendations to remedy the irrigation system by the document [5]. The recommendations are basically oriented toward building a commercial relationship between water suppliers and users. The Bank recognizes the irrigation water in Pakistan as a private good but not a public good, as was treated by the Pakistan's government. This notion leads to the Bank's recommendation of establishing a public utility organization, which buys water from the government and sells it to the users' organizations. Together with this, the Bank recommended to set up a water market on the basis of water right as an individual property right. This is the only way, the Bank believes, to achieve equity and efficiency in water distribution.

3.2 Arguments about feasibility of the proposed reform
The Bank's recommendations brought about arguments within the country, with objections and partial agreements by various groups. The objections arouse from Pakistan's specific conditions or locality of social set-up. Probably most groups, if not all, were against the concept of water market, since this may be misused by influential farmers and landlords, creating a water monopoly. Also, water right as an individual property right, which is a basis of water market, was opposed with the reason that the water and land must be dealt together, in such a semi-arid land agriculture as in Pakistan.

There have been both pro and con arguments regarding decentralization of irrigation management. The resistance, if not a clear con, came from the Provincial Irrigation Departments and the large landholding farmers, and pros came from the Provincial Agriculture Department and the small landholding farmers. The resistance seemingly originates from distrust between "authority," Irrigation Department (and landlords), and "people," mainly those small landholders suffering from water shortage. The justification of the cons is lack of readiness in the rural society to take a responsibility of irrigation management. The cons claim the present rural people are uneducated and a rural social set-up would not allow democratic farmers' organizations, due to a kinship-based or a feudalistic society. The Agriculture Department has been responsible for organizing farmers to implement On Farm Water Management Project and promoting participatory irrigation management, which is essential for the decentralization.

3.3 The reform to be implemented on a pilot basis
The resistance to the World Bank's recommendations and prolonged discussions for more appropriate institutions in Pakistan deferred the initiation of reform for almost

two years. The Federal and Provincial governments and concerned groups have seemingly agreed on a Pakistan's model of institutional reform, and each Provincial Assembly has passed the bill in 1997 to establish the Provincial Irrigation and Drainage Authority (PIDA) in each Province.

The institutional frame consists of three entities: Provincial Irrigation and Drainage Authority (PIDA), Area Water Board (AWBs), and farmers' organizations (FOs). The PIDA, transformed from Provincial Irrigation Department (PID), is an autonomous body having independent revenue collection and spending authority with proper accountability. Below the PIDA in each Province, a financially self-accounting pilot Area Water Board (AWB) will be created on a trial basis around a canal command. And, below the AWB, FOs will be formed, again on a pilot basis, along a distribution channel (tertiary level). The AWB receives water from the PIDA and delivers it to FOs, and the FOs operate and maintain the distribution channel with the autonomy of financial self-sufficiency. The farmers' representatives will be on the Board of both the PIDA and AWB, to ensure participatory irrigation management. The pilot AWB will be evaluated. If it is successful, the reform will be extended to the nation wide scale.

The PIDA, at present, is legally the owner of irrigation channels and facilities. However, the PIDA exists only in a paper as law, since the transformation from the PID into the PIDA has not been materialized yet. The Provincial government is in the process of preparing rules and regulations to fill up the legal void, linking the PID with the PIDA. Also, the PID is to choose a pilot canal command to set up a pilot Area Water Board.

4 Potential and conditions of institutional reform and users' participation

4.1 Common problems after irrigation development

In the past decades many developing countries have invested in irrigation systems to achieve and sustain the food self-sufficiency for their populations. The irrigated agriculture also contributed to the country's economic development by earning the foreign currency. The governments, as investors, have operated and maintained the irrigation systems for many decades. Most of the cases, farmers as beneficiary managed only end-ditches running along their farms, and the governments had to take care of the primary and secondary canals. Such an official responsibility required a large number of government's employees, resulting in a country's largest agency in charge of irrigation. The financial burden to administer the irrigation system has become unbearable due to increasing needs of rehabilitation, O&M, and the government employees. This situation urges the central government seek a way to rectify the financial problem, as introduced in the case of Mexico and Pakistan.

Water demand always keeps increasing due to agricultural modernization and needs of production increase for economic development and population increase. High-valued cash crops often require more water, and high yield varieties usually require more inputs and water. Increased irrigation water supply, with a new reservoir, sometimes prompts more water demands, and total irrigation demand soon exceeds the system's supply capacity. It is difficult for the government to increase water supply, by new reservoirs, to satisfy increasing demand, since proper sites for a reservoir are limited and a construction cost and environmental problems become more sever constraints. Such a situation then requires a concept of demand management, in which both the government and end-users must contribute. It is not only the demand increase to make the situation difficult but also diversifying demand patterns of local needs and farmers. Particularly for a large irrigation system without a local farm pond, it is impossible for the water supplier to respond to diversified demand patterns. Rather, there has to be a local measure by irrigation users to respond to it and a local autonomy to facilitate such a measure.

4.2 Needs of evolving water users' institutions

When natural resources are abundant, no institution would be necessary for users to manage and sustain the resources. When the resource becomes scarce, then the resource users must establish rules (institutions) for them not to exhaust and destroy the resource. The "tragedy of commons" must be avoided. In order to create rules for the users and by the users, they must have autonomy in the users' community. With the absence of external interference, they can not rely on the policing of outside authority; instead, they have to create internal rules of the community to organize and sustain their own resource uses.

With the case of irrigation water uses, the governments have invested, operated and maintained the system. Such a presence of authority tends to weaken the internal rules of a water users' community. Government authority, in the case of Central Thailand, is not effective for the water users to observe the rules set by the government agency. In the case of Pakistan, the government authority is not effective for the water users to observe the traditional rules of water distribution. For various reasons, the government authority has been losing effectiveness for the equitable and sustainable use of irrigation water. In this situation, water users may be given an autonomy to develop their own internal rules, by limiting the scope of government authority. To cope with the new situations of increased water demand, irrigation water management institutions will have to evolve into those of a new stage.

4.3 End-users' participation as essential part of institutions

End-users' participation, in managing the irrigation facilities and water, is essential to remedy the present situations. The key to participatory irrigation management is the existence of farmers' organization. Utilization of water usually necessitates a users' organization for collective actions. Water users' organizations were formed in developing countries; however, there was a tendency for them to become inactive after a few years' operation of end-ditch maintenance. This is because most of the organizations were created to implement some projects, and they usually disappeared after the project was completed and all the incentives were eliminated. Therefore, sustainability is the most important issue with the organizations.

The primal importance for sustainability is to guarantee a power over irrigation water to its users. The power is the authority to control users' resources of water and physical facilities. Without having a power or an authority to govern the resources, there would be few benefits and incentives for organizing themselves. Such a power has to be institutionalized and protected by a law [6]. The benefit of users' autonomy is clearly observed by the case of Mexico's irrigation management transfer. The WUA most appreciated the autonomy with managing irrigation infrastructures.

There are some necessary conditions to make such a power more effective and to enhance sustainability of the users' organization: transparent information and skills training. Provision of information is necessary, since, for example, they first need to know how much water there is, how much water they are entitled to, and when they can receive it. With this information, they can plan for managing their scarce resource. Without knowing about their resource, they can not do much except just waiting for water. The key is transparency in information. It is usually a dictator's practice to make people unknowledgeable, with no transparency, and depend on him, which seems the principle for any ruler in the world. On the contrary, transparency in information is the basis for the democracy and the autonomy.

Users often do not have experiences in O&M of their resources. The skills in managing their resources and organization can be provided through governmental training facilities. A water users' organization often employs professionals to take care of technical matters such as O&M of the system. This will become necessary, when the organization takes over more responsibilities from the government, and a degree of technical complexity becomes higher.

There would be no universal model for the users' participation, applicable to all countries and all societies, since each society has a unique background of a tradition, a social set-up, an economic and political environment, and natural conditions. It is a tendency to try to replicate a successful model without considering differences in a social background. Instead, we must be flexible enough to adopt any local advantage.

As introduced in the previous chapter, the World Bank made a policy recommendation, based on the recognition that the irrigation water in Pakistan is not a public good but a private good, tradable among water users. Such a Bank's presumption is rather too simple minded, and it would be dangerous to apply the free market concept to a country only with an economic consideration. Water has the multiple aspects: not only a private good, but also a public good and a common good; and therefore, we can not confine water to just one function of a private good. Moreover, it is not only the human being that uses water, but also the living nature uses the water. This is why we save water also for the environment. It may be appropriate in a country to treat water as a tradable good but not in all countries and societies.

5 Conclusions

To cope with new situations of increased water demand, irrigation management institutions will have to evolve into those of a new stage. The Mexico's institutional reform has achieved financial self-sufficiency and more crop cultivation. Some Irrigation Districts, however, are still struggling for improvements in administration and O&M. Each District and WUA should make progress in the reform, according to its own pace and mode most appropriate to them. In Pakistan, it will take a few years to see how its reform will proceed, and even more years to see whether the pilot model will be extended to the nation wide scale. The key for its success would be a political will for the reform and a local creativity for the betterment. Water users' participation is essential for the reform. Participatory irrigation management would require a property right to guarantee a power to control users' resources of water and physical facilities. There are some necessary conditions to make such a power more effective: for example, transparent information and skills training. There are more experiences being accumulated in irrigation reforms; and it would be useful to exchange the

is essential for the reform. Participatory irrigation management would require a property right to guarantee a power to control users' resources of water and physical facilities. There are some necessary conditions to make such a power more effective: for example, transparent information and skills training. There are more experiences being accumulated in irrigation reforms; and it would be useful to exchange the experiences and learn the conditions for success. A unique background of a local tradition, a social set-up, an economic and political environment, and natural conditions is also important to understand together with the modes of institutional reform.

6 References

1. Gorriz, C.M., et. al. (1995) *Irrigation Management Transfer in Mexico*, The World Bank Technical Paper number 292.
2. CNA Gerencia Estatal en Sinaloa (1997) Reunion Ordinario del Comite Hidraulico.
3. Haq, A.U and Shahid, B.A. (1997) Public, Private or Participatory? Reforming Irrigation Management in Pakistan, *ICID Journal*, Vol.46, No. 1, pp.37-48.
4. Bandaragoda, D.J. and Rehman, S (1995) *Warabandi in Pakistan's Canal Irrigation Systems*, IIMI Country Paper, Pakistan No.7.
5. World Bank (1994) Pakistan Irrigation and Drainage: Issues and Options, Report No. 1184-PAK.
6. Ostrom, E. (1990) *Governing the Commons*, Cambridge University Press.

SECTION IV

WATER CONSERVATION AND HYDROLOGIC BEHAVIOUR

REUSE OF RETURN FLOWS AND LOCAL RUNOFF IN IRRIGATION SYSTEMS

J.W. GOWING and P.S. MAHEEPALA
Centre for Land Use & Water Resources Research, University of Newcastle upon Tyne, UK.

Abstract

In most conventional large-scale irrigation systems imperfect matching between water supply and demand is an inescapable fact of life that leads to operational spillages and contributes to low efficiency in use of water unless provision is made for reuse of return flows. Provision of auxiliary storage reservoirs at strategic points within the system can improve the recovery of return flows and absorb surplus supply that would otherwise be wasted. Other benefits also arise including more flexibility of supply within the distribution system and greater opportunity for multiple-use management, including fish production. A method of optimal operation of secondary storage reservoirs has been developed, which employs stochastic dynamic programming within a hierarchical multilevel framework. Application of the method was demonstrated using a case study in Sri Lanka. The proposed method provides higher utilisation of rainfall and lower operational losses than conventional operating methods, regardless of the size of the secondary reservoir. Further work on multiple-use management of systems incorporating such tanks is continuing.
Keywords: Irrigation systems, modelling, return flow, secondary storage, water saving.

1 Introduction

The evidence of poor performance of irrigation systems is overwhelming and the potential for improvement is undeniable. However, in seeking to bring about the desired improvement in performance it is essential to recognise the conflict that exists between two principal objectives:

Water and the Environment: Innovative Issues in Irrigation and Drainage. Edited by Luis S. Pereira and John W. Gowing. Published in 1998 by E & FN Spon. ISBN 0 419 23710 0

- to provide a reliable and flexible water delivery to farmers;
- to ensure highest possible efficiency and equity in water distribution.

It is possible to design a water delivery system that is capable of meeting both objectives operating in responsive mode under downstream control, but the vast majority of existing irrigation systems were designed to operate under upstream control. The case for modernising these systems is compelling and the literature on so-called 'modern' design is growing. Conversion to downstream control is expensive and may not be appropriate under conditions of water-shortage, therefore considerable attention has been directed towards improving the regulation of water delivery in systems operating under upstream control [1] [2] [3] [4].

Given that unsteady flow conditions generally prevail, it is difficult to design and to manage upstream control systems to provide a flexible, responsive water supply without excessive losses through operational spillage due to imperfect matching between supply and demand. Provision of secondary reservoirs at strategic points along the main canal system has been identified as one way of improving this situation. These secondary reservoirs provide on-line buffer storage which can absorb that part of the water supply which cannot be utilised immediately for irrigation and would otherwise be wasted. The advantages of incorporating such reservoirs have been advocated for some time [5] in the context of large-scale irrigated farming, but the contrasting case of irrigation systems serving many small-scale farmers has received little attention.

The purpose of this paper is to present a procedure for operating secondary reservoirs which have their own local catchments as well as main canal supply. The objective is to use irrigation water more efficiently in large irrigation systems by operating the secondary reservoirs in such a way that rainfall and local inflows are utilised as far as possible to meet crop water requirement. Secondary reservoirs which do not have a local catchment are studied as a special case of the general approach. The proposed operation method takes into account the stochastic behaviour of rainfall, crop evapotranspiration and local inflows. The applicability of the method is demonstrated using typical data from a rice irrigation system in Sri Lanka.

2 Model description

The operation of a secondary reservoir involves two management decisions:

- how best to allocate water from the reservoir to its command area and
- how best to request water from the main water source.

Both decisions are complex because they depend on uncertain hydrological variables (i.e. the rainfall and evapotranspiration over the command area and the inflow to the secondary reservoir) and on the difficult tradeoff between wasting water by over-irrigating and the loss of crop yield due to water shortages. Because of this complexity and because of the different timescales for the two decisions, the problem has been formulated as two distinct, but inter-related models.

- *an Allocation Model* to make short-term decisions on allocation of water from the secondary reservoir;
- *a Requisition Model* to make less frequent decisions on requisition of water from the main water source.

Both models contain variables which are unknown at the time the decisions have to be made, specifically the inflow to the reservoir and the effective rainfall during the period. Both models require decisions to be made regularly at the beginning of a fixed time period, and both models are seeking optimal solutions. For all these reasons stochastic dynamic programming was chosen as the most suitable solution technique.

For the Allocation Model the decision to be made at each time period is how much water to release for irrigation given the volume of water stored in the secondary reservoir (or known to be arriving from the main canal) and given the field water depth (or soil moisture content for crops other than rice). The objective is to irrigate in an economically optimal manner, that is to minimise the costs of irrigation, comprising the cost of the water and the cost of any loss of yield due to a shortage of water. Other costs of production such as the cost of labour, machinery and seeds, are assumed to be fixed and are not included in the objective function as they do not influence the decision. The objective function can therefore be written as:

$$F_m = \underset{IR_m}{\text{Min}} \sum_{m=1}^{M} \{(P_1)_m + (P_2)_m\} \tag{1}$$

where, $(P_1)_m$ is the cost of irrigation water applied during the decision stage m and $(P_2)_m$ is the cost of a reduction in crop yield due to any water deficit during the decision stage m.

The decisions are constrained by the volume of water stored in the secondary reservoir at the beginning of the decision stage, S_m, the depth of water in the field at the beginning of the decision stage, L_m, and the inflow from the main canal during this period, q_m, which are all known. They are also influenced by other variables which are unknown at the time the decision is made; the rainfall and evapo-transpiration during the decision stage (R_m and ET_m respectively), the net reservoir inflow (inflows minus losses such as evaporation, seepage and percolation), SP_m, the percolation losses from the fields, $PERC_m$, the overland drainage from the paddy, $DRAIN_m$, if the storage capacity of the paddy field is exceeded, L_{max}.

If the volume of water to be released for irrigation is called IR_m, then the constraints can be expressed, as two continuity equations (2) and (3) with volumes expressed as equivalent depths over the command area to be irrigated.

- For the reservoir:

$$S_{m+1} = S_m + q_m - IR_m + SP_m \tag{2}$$

- For the field:

$$L_{m+1} = L_m + IR_m + R_m - ET_m - PERC_m - DRAIN_m \tag{3}$$

This model is solved to determine the optimum release, IR_m, for each period from 1 to M, where Mt is the duration between requests for water from the main canal.

For the Requisition Model, the decision to be made at each time period $T (= Mt)$ is how much water to take from the main canal, Q_n, given the volume of water stored in the secondary reservoir, the field water depth, and the stage of development of the crop (rice in this case). It is assumed that there is no uncertainty in Q_n, (i.e. the volume requested is delivered).

The objective is again to irrigate optimally in an economic sense over the time horizon, but for this model the time horizon is to the end of the growing season. This season is divided into a number of decision stages of length T.

The objective function can be expressed as

$$G_n = \underset{Q_n}{Min} \sum_{n=1}^{N} \left\{ (Z_1)_n + (Z_2)_n \right\} \tag{4}$$

where, $(Z_1)_n$ is the cost of requested water during the decision stage n, and $(Z_2)_n$ is the expected cost of a reduction in crop yield during the decision stage n.

The constraints on this decision can again be expressed as continuity equations.

• For the secondary reservoir:

$$S_{n+1} = S_n + Q_n + IN_n - SIR_n - SPILL_n \tag{5}$$

where IN_n is the net inflow over the period n and SIR_n is total water released over period n. The reservoir storage is constrained by its capacity, S_{max} and $SPILL_n$ is the spillage if the reservoir overflows.

• For the field:

$$L_{n+1} = L_n + SIR_n + ER_n - PERC_n - DRAIN_n \tag{6}$$

where ER_n is the effective rainfall during this period.

This model is solved to determine the optimum volume of water to request, Q_n, for each period from 1 to N, which covers the whole growing season from land preparation to harvest.

3 Model application

A rice irrigation system located in the Dry Zone of Sri Lanka was selected to demonstrate the applicability of the methodology. The year is conventionally divided into two seasons: wet (from October-March) and dry (from April-September). Generally water is conserved in the wet season for use in the dry season. In principle, the models are applicable to both dry and wet seasons. However, in most rice irrigation systems in the Dry Zone the water use efficiency in the wet season is very low compared to that of the dry season [6], and therefore, these models will be much more useful in the wet season. It is this season which has been chosen to demonstrate the effectiveness of the approach.

The rice variety grown in the Dry Zone has 135 days growth period. Land preparation was assumed to start in the first week of October, and to last two weeks. During this period irrigation is applied at a constant rate of 12.5 mm/day, net of losses, evaporation and effective rainfall. The scheduling process was started from the third week of October with transplanting. The last 16 days were allowed for drying-out the fields for harvesting. Therefore, the time horizon of the Requisition Model was 15 weeks, starting from the third week of October (i.e. N = 15 and T = 1 week). Allocations from the reservoir were made on a daily basis with a time horizon of 1 week (i.e. M = 7 and t = 1 day).

Both objective functions (equations 1 and 4) are made up of two terms - the loss of yield due to water deficits and the unit cost of the water. A model [7] was used to assess the yield loss (ΔY_m) due to a water deficit, such that:

$$\Delta Y_m = K_y.SD/100 \tag{7}$$

where, K_y is the percentage reduction in yield per day stressed, SD is the number of 'stress days' during the total time period under consideration. A stress day is defined as any day on which the water level in the field falls below a 'critical level' at which the crops are likely to experience water stress. In the study the critical level was taken as the saturated soil moisture condition and K_y was assigned the value 2.0.

In the study area the dry season crop is entirely dependent on the water conserved from the wet season; therefore, an opportunity cost can be assigned to the water used in the wet season in terms of the cost of reduction in crop yield during the dry season. The main crop grown in the dry season is rice and the price of the dry season rice crop was taken as the same as that of the wet season. If w units of water are required to obtain the potential yield from a unit area of irrigated land in the dry season, and if there is a deficit of one unit of water at the beginning of the dry season, the area should be reduced by (1/w) in order to obtain the potential yield from the rest of the area. The cost of reduction in yield due to this reduction in area is $(1/w)Y_{max}PC$. This loss has been caused as a result of using an extra unit of water in the wet season. Therefore, the opportunity cost of the irrigation water in the wet season can be expressed as

$$c = (1/w)Y_{max}PC \tag{8}$$

In order to get an idea about w, a simulation model of crop growth of the dry season rice crop was run for 7 years (1978-1984).

Daily data of pan evaporation and rainfall for 32 years for this climatic zone were obtained from [8]. Measured daily inflows to the secondary reservoir were not available but were generated to be consistent with the rainfall and evaporation data [9]. All the variables were expressed as a depth of water per unit area of irrigated land. The reservoir storage and field water depth were discretised in 5 mm steps in the Allocation Model. In the Requisition Model, the reservoir storage was discretised in 10 mm steps and the field water depth in 25 mm steps.

By considering the design capacities of the canals of the selected scheme, the daily irrigation decisions in the Allocation Model could range from 0 to 25 mm/day in 5 mm/day steps, and the water requisition in the Requisition Model could range from 0 to 70 mm/week in 10 mm steps.

To determine the transition probability matrix in the Allocation Model, an extreme value type 1 distribution was fitted to the natural logarithm of net effective rainfall (i.e. rainfall-evapotranspiration-percolation) offset by a constant amount to avoid negative values of the random variable. The distribution changes slowly through the growing season: this was accommodated by combining the 15 weeks of the growing season into 5 groups with a slightly different distribution for each group. Using these distributions and equation (3), the transition matrix for field water depths was derived.

The transition matrix for the requisition model was more difficult as it was a function of two random variables. The procedure chosen for determining the joint probability density function pdf used conditional and marginal distributions of the weekly inflows, IN, and the weekly values of net effective rainfall f(RF-ET). For the marginal distribution, the two parameter Weibull distribution fitted to (RF-ET+γ), where γ was a type of displacement parameter, was found to be the most suitable distribution. For the conditional distribution a two parameter Weibull distribution was found to fit the weekly inflows with the parameters of the distribution α and β conditional on the net effective rainfall. This conditional relationship was found by regression to be

$$\alpha = A(RF\text{-}ET)^B \text{ and } \beta = C(RF\text{-}ET) \tag{9}$$

where A, B, and C are the regression parameters. If the distribution adequately describes the data, the residuals defined by $(RF\text{-}ET)\beta^\alpha$, should be exponentially distributed by a mean of 1.0; this was tested using a Chi squared test [12]. Again, the distribution changes through the growing season, which was accommodated as before but with six groups in this case.

4 Results of case study

Six different reservoir capacities were studied: 50, 80, 100, 150, 200 and 250 (all in mm per hectare of land irrigated). The allocation schedule of each of these capacities was derived and these schedules were then used to derive the corresponding requisition schedules. These schedules define an optimum operating policy for the secondary reservoirs, to be referred to as the 'proposed method' of operation.

In order to check the performance of the proposed method, a simulation model of water management was developed for the wet season rice crop. In this model, the rice field and secondary reservoir water balance was carried out on a daily basis throughout the growing season. The procedure was repeated for 10 years, using a period which was not used to define the policies, and the performance was studied with respect to the 10 year average.

The performance of the proposed method was compared with three other operating policies (A, B, C) which are often used in practice (see Table 1). Operating mode C can be thought of as a good pragmatic policy: try to maintain a field water depth of 50 mm and fill the reservoir at the beginning of each week. This, of course, takes no account of likely rainfall or natural inflows. It should be noted that the requisition rule for mode A was the same as the proposed method. Table 2 shows the comparison of

Table 1. Requisition and allocation rules of different operation methods

Operation method	Requisition Rule (Q_t in mm)	Allocation Rule IR_t in mm)
Proposed method	Requisition schedule derived using the SDP	Allocation schedule derived using the SDP
Operating Mode A	Request mean weekly crop ET plus losses	Irrigate 10 mm per day whenever possible
Operating Mode B	Maintain reservoir full	Irrigate 10 mm per day whenever possible
Operating Mode C	Maintain reservoir full	Maintain field water depth at 50 mm

costs obtained from each operating method for the different reservoir capacities. The performance of the four modes of operation were also compared according to how well they utilised the natural rainfall and the natural inflows to the secondary reservoir.

The proposed method gives the lowest average annual cost irrespective of the capacity of the secondary reservoir. For this method, the total cost decreases with an increase in the reservoir capacity and becomes almost zero at a reservoir capacity of 200 mm. The main reason for this behaviour is the increased use of local inflows for the irrigation water which becomes 100% at a reservoir capacity of 200 mm. The other modes are relatively insensitive to the reservoir capacity.

Under the proposed method, crops are allowed to undergo a degree of water deficit (i.e. yield reduction) during the wet season in order to save irrigation water for the dry season (see Table 2). The other operating modes do not consider the long term benefits of saving irrigation water, and they also lead to a rather inefficient use of rainfall and local inflows. The proposed method utilises the rainfall and local inflows far more effectively and as a result purchases much less water from the main irrigation system (P_1 in Table 2). However, it incurs some loss of yield (P_2 in Table 1) which the other modes do not.

In terms of costs, mode C is the best of the traditional methods tested. This is the same as B in terms of purchases from the main system, (i.e. it buys each week enough water to fill the reservoir regardless of any anticipated inflows), but it allocates the water each day to bring the field water depth up to a chosen operating depth - in this

Table 2. Cost elements P_1 (water purchase) and P_2 (loss of yield) under the different operation methods.

R	Cost elements P_1 and P_2 in pounds per hectare							
	Proposed method		Operating Mode A		Operating Mode B		Operating Mode C	
	P_1	P_2	P_1	P_2	P_1	P_2	P_1	P_2
50	12.4	13.8	100	0.0	66.8	0.0	39.0	0.0
80	7.7	8.0	100	0.0	87.1	0.0	43.8	0.0
100	5.9	5.5	100	0.0	87.1	0.0	44.6	0.0
150	1.6	2.6	100	0.0	87.1	0.0	44.6	0.0
200	0.0	0.6	100	0.0	87.1	0.0	44.6	0.0
250	0.0	0.0	100	0.0	87.1	0.0	44.6	0.0

Note: R = reservoir capacity in mm.

case 50 mm. Again, no allowance is made for anticipated rainfall. Alternative operating depths were tested and mode C was even better with a depth of 25 mm.

All the results presented so far are for the case of equal size catchment and command area. To study the effect of catchment size, seven different values of catchment command area ratio (C_a) were tested. The values chosen were 0, 1, 3, 4, 6, 10 and 15. When the secondary reservoir does not possess a local catchment, (i.e. when $C_a = 0$), the Requisition Model had to be modified. For all the other values of C_a, the corresponding parameters of the conditional pdf of inflows were re-estimated. The least total costs were obtained with the proposed method irrespective of the catchment size. However, as the catchment size increases, the total costs obtained with both operating modes B and C become closer to those obtained with the proposed method [9].

All the results presented above were obtained assuming the water was supplied at the beginning of the week, but this type of delivery method is not always practicable. An alternative procedure provides the water continuously at a uniform rate over the week. With this method the term q_m (equation 2) is taken as the constant daily supply from the main canal, (i.e. the decision variable in the Requisition Model). For each decision q_m, there is a separate allocation schedule. The results show that there is no significant difference in the average operating cost obtained with both supply methods for all operation options [9].

5 Conclusion

Provision of auxiliary storage reservoirs at strategic points within an irrigation system can alleviate problems that arise from attempting to meet the conflicting aims of improving flexibility of supply to users, whilst also reducing water losses through operational spillages. If such reservoirs also recover return flows from local runoff and drainage, then they can provide a means of increasing the effective efficiency by adopting a more integrated approach to managing water distribution. They may also provide additional opportunities for multiple-use management of the water resource.

A case study in Sri Lanka has demonstrated a method of managing the operation of such a reservoir. It is shown that significant water savings can be made during the wet season thus allowing increased irrigation during the dry season. Further work is needed to test the feasibility of this approach in a practical situation, but it is anticipated that such innovation will bring about the desired operational objectives without need for complex water control technology.

6 References

1.   Zimbelman, D.D. (Ed.) (1987) Planning, operation, rehabilitation and automation of irrigation water delivery systems. *Amer. Soc. of Civil Eng.*, New York. 380pp.
2.   Plusquellec, H., Burt, C. and Wolter, H.W. (1994) Modern Water Control in Irrigation. *Technical Paper No. 246*, World Bank, Washington DC. 98pp.

3. Goussard, J. (1993) *Automation of Canal Irrigation Systems*. ICID, New Delhi, India. 116pp.
4. Malatere, P. (1995) Regulation of irrigation canals. *Irrigation & Drainage Systems*, Vol. 9, pp. 297-327.
5. Merriam, J.L. (1987) *Introduction to the need for flexibility and automation*. Proceedings ASCE Symposium on Planning, Operation, Rehabilitation and Automation of Irrigation Water Delivery Systems, pp.1-17.
6. Weller, J.A. (1986) *Water losses during irrigation operation: Irrigation Water Management*. Overseas Development Unit Bulletin (October), Hydraulics Research Ltd., U.K.
7. Green, A.P.E. (1988) *A productivity indicator for paddy rice*. Technical Note, OD/TN 45, Hydraulics Research Ltd., Wallingford.
8. Siddeek, F.Z. (1987) *Water Management for Lowland Rice*. Ph.D. Thesis, Virginia State University, U.S.A.
9. Maheepala, P.S. (1990) *Operation of Secondary Reservoirs in Irrigation Systems*. Ph.D. Thesis, University of Newcastle upon Tyne, U.K.

REMOTE SENSING FOR RAINWATER HARVESTING IN MEDITERRANEAN AGRICULTURE

D. PRINZ
Dept. of Rural Engineering, University of Karlsruhe, Karlsruhe, Germany
T. OWEIS and A. OBERLE
International Center for Agricultural Research in the Dry Areas (ICARDA), Aleppo, Syria

Abstract
Since time immemorable, farmers in the Mediterranean have collected surface runoff, using various types of 'water harvesting', which is the collection of surface runoff mainly for agricultural and domestic purposes. Water harvesting can be a valuable technique to supplement the other sources of water for irrigation. However, the selection of appropriate sites and the determination of suitable methods of water harvesting on a large scale present great challenges, since the necessary data on hydrology, soils, etc. are often lacking. A research work at ICARDA, Aleppo, Syria was aimed at developing a methodology for the application of remotely sensed data and GIS for identifying appropriate sites and methods of water harvesting in dry areas in West Asia and North Africa. A pilot project site was selected in central Syria, covered by a whole LANDSAT TM scene. The image processing software ERDAS IMAGINE and GIS software ARC/INFO were used to process the images and to establish a geo-information system. Using these tools it was possible to identify areas generally suitable for water harvesting and to determine water harvesting techniques for those sites. The developed methodology is also applicable in other regions with similar conditions. When heading for the application of water harvesting techniques, sustainability concerns have to be paid due attention to.
Keywords: Geographical information system, mediterranean agriculture, remote sensing, water harvesting

Water and the Environment: Innovative Issues in Irrigation and Drainage. Edited by Luis S. Pereira and John W. Gowing. Published in 1998 by E & FN Spon. ISBN 0 419 23710 0

1 Introduction

Only a small fraction of the rainfall falling in dry (non-urban) areas percolates into deeper soil or rock layers to recharge an aquifer. Another small fraction is used for the transpiration of vegetation or of agricultural crops. The majority of the precipitation evaporates from the often bare soil or from surface depressions. To feed the growing population in the Mediterranean basin (with special reference to North Africa and the Middle East), more irrigation is needed, but the quantity of irrigation water is extremely limited. The "classical" sources of irrigation water are often at the break of overuse and therefore untapped sources of irrigation water have to be sought for. To increase agricultural production in the Mediterranean, the necessity exist to think on alternatives, e.g. the utilisation of the evaporative portion of precipitation to be used for agricultural purposes before it is released to the atmosphere [1].

Since time immemorable, farmers in the Mediterranean have collected surface runoff, using various types of '(rain) water harvesting' [2], [3]. Rainwater harvesting' or 'Water Harvesting' is here defined as the collection of surface runoff mainly for agricultural and domestic purposes. In spite of several thousand years of experience in rainwater harvesting in the Mediterranean, a number of open questions remain, e.g. how to increase the water yield of a given catchment, what impact on ecology may occur, or how to identify areas suitable for certain techniques of water harvesting. The latter question was in the focus of a research project carried out by ICARDA (International Center for Agricultural Research in the Dry Areas) and the University of Karlsruhe, Germany.

Some years ago, the University of Karlsruhe was already engaged in the development of a methodology to identify areas suitable for water harvesting in West Africa [4], [5]. This methodology was developed under tropical conditions in a zone with precipitation amounts of about 500 mm/annum, in a summer rainfall area and a region with a sedentary farming population. The situation in the Middle East, were the new methodology was developed, is rather different: the region in question belongs to the winter rainfall area, experiences only 100-300 mm of rainfall/annum and the rural population is nomadic or semi-nomadic. Therefore, a totally new approach has to be used and a new methodology to be developed.

In rain water harvesting three different groups of techniques are distinguished:

- Flood water harvesting from far away, large catchments,
- Macro-catchment systems utilising the runoff from a nearby slope for agricultural purposes (with or without interim storage) and
- Micro-catchment water harvesting, where the water from an adjacent, small catchment is used for cropping [6], [7].

It is evident, that all three groups of water harvesting techniques need different environmental settings to be implemented. Besides the topography, the runoff conditions of the surface, the infiltration rates, the soil types of the run-on areas and the depth of the soil layer in the cropping areas are among the most important natural parameters for the implementation of any water harvesting system. Additionally, socio-economic factors have to be taken into due consideration.

For relatively small areas (in the range of several hundred hectares) a ground truth carried out by a number of experienced people will be the best technique to identify suitable areas for water harvesting. For medium range sizes of areas, the use of aeroplanes equipped with photographic equipment could be suitable and for even larger areas the application of remote sensing by using satellite images could be the most relevant means of identification of areas suitable for certain techniques of water harvesting. For any of the above mentioned techniques, the application of a suitable GIS (Geographic Information System) is indispensable. It has to be mentioned however, that the application even of the best GIS will not guarantee the success of any water harvesting scheme, as a number of external factors such as water and land rights, macro-economic conditions, traditional rules and believes can hardly be incorporated into such a GIS, but might influence the development of the water harvesting scheme.

2 Objective

The general objective of the study was an improved agricultural production in the steppe zone of the Middle East which helps combating desertification by more efficient utilisation of rainfall through proper water harvesting planning. Specific objectives include:

- Identification of potential areas which are suitable for given methods of water harvesting in a pilot area in Central Syria,
- Development of a methodology through which remotely sensed data together with a minimum of field investigations and the application of a GIS can be used to assess the potential for water harvesting and advise on the appropriate water harvesting methods in those areas.

3 Project pilot area

For this study, an area in Central Syria covering about 33,000 km^2 between latitudes 33,8° N and 35,5 °N and between longitude 36,8° E and 38,8° E was chosen. This area is covered by a scene of path 179, row 36 of the LANDSAT Thematic Mapper. The area is sparcely populated with major settlements being Salamiyeh, Palmyra and Quaryatain. Small villages and minor settlements are scattered mostly over the southern part of the project area. The infrastructure is rather poor and no detailed information on topography, soils, climatic conditions and socio-economic data is available for the area.

The climate in the Syrian steppe is characterised by cold, rainy winters and dry and hot summers. Mean monthly temperatures vary from 6°C in January to about 32°C in August. Precipitation occurs during the months November to April, this means annual ranges between 250 mm in the north-west to less than 100 mm in south-east of the project area. Rainfall is highly variable in space and time and falls as storms with often high intensity causing substantial runoff. The low precipitation amount contrasts sharply with the evaporation demand of the atmosphere which is calculated to be in the range of 2200-2300 mm /annum [8].

The soils of the area are classified as aridisoils and entisols according to the FAO classification system. The soils are weakly developed and low in organic matter (1-

2%). One common feature is the presence of a surface crust, consisting of pavement of embedded pebbles the so called 'desertic pavement' which impedes infiltration. Some wide plains are often inundated after rainstorms in springtime and the lower slope of the mountain valleys are often covered with a thick horizon of alluvial sediments of loamy find sand. These soils are richer in organic carbon and often used for the cultivation of barley.

Shrubs and grasses dominate the vegetation of the area, which is associated to the Irano-Turanian zone. In earlier days *Pistacia atlantica* was widespred but can rarely be found nowadays. Bushes like *Artemisia herba-alba* and *Haloxyletum articulatum* and annual herbs such as *Astralagus sp.* can be found along wadis and in local depressions. The natural vegetation is heavily degraded due to overgrazing; especially the disappearance of legumes affects the productivity of the pastures and illustrates the degradation of the land. As a consequence of the degradation of the vegetative cover, a severe soil degradation is visible in many locations. The land is mainly used for grazing; about 90% of the project area is rangeland. In low laying, fertile areas some barley and forages are cultivated by Beduins; in some isolated areas orchards of pomegranates, apricots and olives can be found, often irrigated with groundwater. In areas favourable in terms of infrastructure, soils and groundwater resources, vegetable cropping was started, leading to a significant lowering of the ground water table due to overpumping.

4 Methodology

The core data for the determination of suitable water harvesting areas were taken from satellite images which were provided by the LANDSAT Thematic Mapper. For the evaluation of land vegetative cover and geological /geomorphological features the near and mid-infrared channels were used. As the spatial variability of the features in question is very low in this sparsely populated arid region, the resolution of 30 m x 30 m on ground is apparently quite sufficient. Satellite scenes, shortly taken after the rainy season (April 1994 and 1995) were selected to assess the water availability and to identify regions with sufficient runoff.

For reasons of comparison with the dry summer period, satellite scenes of July 1994 and October 1993 were also evaluated. To insure the exact fitting of the pixel grid of the image to a map projection system, a geo-information system data base was established. The pixel grid of the satellite data is usually distorted due to instrument errors or earth rotation effects. Rectification was conducted by using roads, landmarks and the Global Positioning System (GPS). The complete satellite scene was divided into four sub-sets for easier handling of the data. The image was processed on a DEC workstation, using the software ERDAS IMAGINE. Thirteen land cover units were distinguished (from 'Scree slopes' to 'Orchards of palm trees, olive trees and grenadiers') and classified (Code 10: suitable for catchment, Code 20: suitable for cultivated area, Code 30: not suitable). The classification of the images was carried out by using the maximum likelihood procedure. The classification was checked by field investigations were the typical characteristics of each class were investigated. The separability of the various classes was checked by different statistical approaches. The classified image was incorporated into the Geo-Information System ARC-INFO.

Topographic maps at a scale of 1:100.000 were digitised and a Digital Terrain Model was developed. The most important parameters were determined as follows:

Vegetative cover: The vegetative cover is determined by classifying satellite scenes taken after the end of the rainy season. 13 different landcover units were determined and grouped into 3 classes, according to their suitability for (1) catchment, (2) cultivated area or (3) not suitable. The high degree of degradation of the vegetation made it difficult to conclude from the density of the vegetation to soil features. In the study carried out in West Africa [4] there a high degree of congruence between density of vegetation and suitability of the soil to be used for cropping areas was detected.

Topography: The land cover map was indexed by the digital elevation model, calculated by applying the triangulation method to digital data of the topographic maps (scale 1:100,000) by use of the TIM module of ARC INFO package (Table 1).

Table 1. Terrain classification

Land form	Slope Gradient (%)	Relief intensity	Different units of land form
Level land	0 to 8	50 m/2 km	Intermontane plains
			Flat saline plains
			Slightly undulating plains with dry wadis and desert type outliers
			Plateaux
			Depressions
			Low-gradient footslopes
			Wadi beds
Sloping land	8 to 30	> 50 m/2 km	Medium gradient mountain with undulating hilly relief
			Ridges
			Plains dissected by wadis
Steep land	> 30	>600 m/2 km	High gradient mountains
			High gradient hills (badlands)

The geomorphological classification involved the general classification into (1) level land (0-8% inclination), (2) sloping land (8-30% inclination) and (3) steep land (above 30% inclination). Further on 15 different units of land form were distinguished. The terrain analysis was done by visual analysis of the LANDSAT TM scene in combination with interpretation of the digital elevation model.

Length of slope: The terrain analysis was also used for the determination of the length of slope, a parameter regarded of very high importance for the suitability of an area for water harvesting. With a given inclination, the runoff volume increases with the length of slope, but conveyance losses have to be taken into account. The slope length was also one parameter used for differentiation between the suitability for macro-, micro- or mixed water harvesting systems.

Soil depth: Wet surfaces after several rainless days, shown by difference in colour, indicate larger soil depths. Further on, certain assumptions could be made on soil depth by classifying the topographic situation (foot of large slopes, depressions). The minimum requirement for soil depth was considered to be 1 m; soil depth of less than 1 m was regarded as suitable for catchments only.

Drainage system: The drainage system was classified according to the density (average channel length divided per unit area of land). According to the SOTER data base, three classes were defined: (1) slightly dissected (< 10km/km²), (2) dissected (10-25 km/km²) and (3) highly dissected (more than 25 km/km²). The drainage system was digitised on the LANDSAT TM satellite image and the areas with high drainage density were ranked higher in suitability for cropping area of a water harvesting system. By determination of the location of the farest point contributing to runoff, it was tried to define the runoff contributing areas of the various catchments.

Precipitation: Climate data were incorporated in terms of isohyetes, showing the average number of rainfall events exceeding 5mm /day. This quantity of rainfall has proved in several locations (Jordan, SW U.S.A.) to be the minimum amount of rainfall causing runoff.

Table 2 shows the coverage and raster units used for the setup of the GIS.

Most of the data layers were created in vector format; to analyze data within the GIS module of ERDAS IMAGE, the vector layers were converted to raster format.

Table 2. Coverage and raster units used for the setup of the GIS

Raster units for the setup of the GIS
LANDSAT TM. False colour image (453)
Land cover map (classified satellite image)
Inclination of slope (DEM)
Length of slope (DEM)
Land unit map
Drainage system
Water catchment area
Isohyetes of number of rainfall events exceeding 5 mm/d
Map of soil depth
Soil map according to SMSS classification
Geological map

5 Results and discussion

In a subset of about 50 to 60 km of the study area in a zone with about 100 mm precipitation/annum about 11% of the area were considered to be unsuitable for water harvesting systems. About more than 70 % of the area are suitable for the application of microcatchment systems e.g. like contour strips. Sites which may be utilised for macrocatchment systems are found in 6% of the area.

Flood water spreading systems can be applied in other parts of the study area but they need normally rather large constructions to guarantee the safety in case of flash floods. In a nearby project of UNDP in the Mihassa valley, (below 100 mm rain per year) dams and spillways were washed away during one big flood recently.

In general the area can be used for tree and shrub cultivation or range land improvement. The decision making process concerning the best method applicable also depends on the kind of crop and on economic and social factors. Microbasins like semicircular hoops are more convenient for tree cultivation than e.g. contour strips.

Labour is often the most important economic factors if local material only is used. The construction of contour ridges of 0.2 m height with an horizontal interval of 1.5 m needs 90 man days (MD) per ha in the first year and 50 MD in the second year (Experience from Kenya; [7]).

There are a number of reports that water harvesting can be economically very profitable; Rodriguez [9] e.g. showed that wheat grown under microcatchment water harvesting in highland Balochistan is more viable and profitable than any of the traditional methods.

One of the crucial social aspects for the success is the participation of the beneficiaries. The Syrian steppe is mainly used for livestock production: groups of Bedouin families, (semi-)sedentary or migratory, with herds of small ruminants are users of the range land, which is regarded as an open access system. Data layers showing accessibility to the land and migratory pathways in the area can be included in future in the geographical information system and they will alleviate the decision making process.

The study in question has proved that satellite images can be a valuable tool in assessing the water harvesting potential of a region with low infrastructure.

As a first step a qualitative approach was practised, which excludes the quantification of runoff volumes.

To quantify runoff volumes, more field data like measurements of the permeability of the soil and determination of runoff coefficients are needed.

Additionally, a rainfall-runoff model can be introduced, simulating the different runoff volumes according to the precipitation amount.

The methodology developed gives the opportunity to assess most of the parameters important for water harvesting systems, but socio-economic parameters as well as ecological features are not yet included. Flora and fauna can suffer, if e.g. a slope is cleared for higher runoff coefficient or if water is diverted from wetlands for agricultural use [10].

The presented methodology can be easily applied and adapted to other regions with similar climatological (and socio-economic) conditions. Although the current study was carried out with rather expensive equipment (workstation under UNIX), the same procedure can be done with a PC workstation under Windows NT, too.

6 Acknowledgement

The project was funded by BMZ (Federal Ministry for Economic Cooperation and Development), Federal Republic of Germany.

7 References

1.   Boers, Th.M. (1994) Rainwater Harvesting in Arid and Semi-arid Zones. Doctoral Thesis, Wageningen Agricultural University, Wageningen, The Netherlands.
2.   Critchley, W. and C. Siegert, (1991) *Water Harvesting Manual.* FAO Paper AGL/MISC/17/91, FAO, Rome
3.   Oweis, T. and A. Taimeh (1996) Evaluation of a Small Basin Water Harvesting System in the Arid Region of Jordan. *Water Resources Management*, 10, 21-34
4.   Prinz, D. (1995) Water Harvesting in the Mediterranean Environment - Its Past

Role and Future Prospects, in *Water Resources Management under Drought or Water Shortage Conditions.* (Tsiourtis, N.X., ed.) Proceedings EWRA 1995 Symposium, Nicosia, Cyprus, pp. 135-144. Balkema, Rotterdam

5. Prinz, D., W. Tauer, and Th. Vögtle (1994) Application of remote sensing and geographic information systems for determining potential sites for Water Harvesting. in *Proceeding, FAO Expert Consultation on Water Harvesting for Improved Agricultural Production.* Cairo, 21 - 25 Nov. 1993. FAO, Rome, 135-144

6. Reij C., P. Mulder and L. Begemann (1988) *Water Harvesting for Plant ProductionTechn.* Paper 91, World Bank, Washington D.C.

7. Rodrigues, A. (1996) Sustainability and Economic Viability of Cereals Grown Under Alternative Treatments of Water Harvesting in Highland Balochistan, *Pakistan. J. of Sustainable Agriculture* 3: 305-315

8. Soumi, G. (1991) Supplemental Irrigation Systems of the Syrian Arabic Republic (SAR), in: *Supplemental Irrigation in the Near East and North Africa.* (Perrier, E. R. and A.B. Salkini, ed.).Proceedings, Workshop on Regional Consultation on Supplemental Irrigation, ICARDA and FAO, 7-9 December 1987, Rabat, Morocco, pp. 497-511. Kluwer Academic Publ., Dordrecht, NL

9. Tauer, W. and G. Humborg (1992) *Runoff Irrigation in the Sahel Zone.* CTA (Technical Centre for Agricultural and Rural Cooperation), Ede/Wageningen, NL

0. Prinz, D. (1994) Water Harvesting and Sustainable Agriculture in Arid and Semi-arid Regions. In: Hamdy, A. (ed.), *Land and Water Resources Management in the Mediterranean Region.* Proceedings, Intern. Conference Valenzano/Bari, Italy, 4.-8. Sept. 1994 Instituto Agronomico Mediterraneo, Valenzano/Bari

A MODEL TO OBTAIN THE HYDROGRAPH OF SURFACE RUNOFF IN TERRACED AREAS

F.F. PRUSKI, J.M.A. SILVA, L.N. RODRIGUES and D.D. SILVA
Dept. of Agricultural Engineering., Fed. Univ. of Viçosa – MG - Brazil.

Abstract
Estimates of surface runoff are often needed for projects in urban and rural watersheds. The amount of sediment transported by surface runoff depends on the maximum value as well as the surface runoff variation with time. The difficulty in applying the available procedures to estimate the temporary variation of the flow rate and its maximum value is due to the inaccuracy of some methods that are usually used for this purpose and because of the big variability in the results that can be obtained by different professionals that use the same procedure. Knowledge of the hydrograph of surface runoff and of the corresponding maximum peak flow is necessary for the design of drains and in projects against flood and water erosion. For this reason the investigation of a method that produces reliable estimates of these parameters is of great interest. To reach this objective the area between two serial terraces was subdivided in a grid. Using the intensity-duration-frequency equation of the rainfall and the characteristics of infiltration of water in the soil, the necessary time for the beginning of surface runoff was quantified. Applying the Manning's equation to the flow of each one of the cells of the grid, the hydrograph of the surface runoff could be obtained, for some point located on the surface of the land as well as for some section along the channel of the terrace.
Keywords: Hydrograph, overland flow, peak-flow rate, soil and water conservation, surface drainage, surface runoff.

1 Introduction

After precipitation begins, the vegetal cover intercepts water until the potential interception storage is met. When the interception capacity is exceeded, infiltration into the soil begins. Since infiltration rate decreases exponentially as the soil water storage increases, a point is often reached when the rainfall rate exceeds the infiltration capacity.

Water and the Environment: Innovative Issues in Irrigation and Drainage. Edited by Luis S. Pereira and John W. Gowing. Published in 1998 by E & FN Spon. ISBN 0 419 23710 0

When this occurs, water begins to stand on the surface in micro-depressions. Once the micro-depression storage (surface retention volume) is filled, runoff begins [1]. The overland flow is usually a very complex hydraulic and geometrical phenomenon [2], and it occurs as a mixture of broad sheet flow which is found on interill areas and concentrated flow which occurs within rills. The quantity of flow within a particular rill is influenced by rill drainage area. Due in part to differences in microtopography across a slope, variations in flow rate frequently occur between rills [3].

Most flow equations are used for estimating discharges in pipes and open channels. Suitable equations for overland flow with its shallow depth, as in border irrigation, are not available, although such equations are important in models used in designing and managing irrigation systems [4].

Water erosion is the removal of soil from land's surface by running water. For this reason, the risk of erosion is significantly decreased by minimizing surface runoff [5]. In the models to describe the erosive process it is essential the knowledge of the overland flow inside the considered area. The amount of removed and transported particles by the surface runoff depends on that parameter.

Land terracing is one of the best known and most widely used method to control water erosion. The design of a terrace system involves proper spacing and location of terraces, the design of a channel with adequate capacity, and development of a farmable cross section. For the graded terrace, runoff must be removed at nonerosive velocities in both the channel and the outlet [6].

With graded terraces the rate of runoff is more important than the total runoff, whereas both rate and total runoff influence the design of level and conservation bench terraces. Graded terraces are designed as drainage channels or waterways, and level terraces function as storage reservoirs. The terrace channel acts as a temporary storage reservoir subjected to unequal rates of inflow and outflow. Inflow is affected by variables given in the runoff equation; outflow is influenced by the grade in the channel as well as by the inflow rate [6].

The first step to determine the project discharge of a surface drainage or graded terrace system is to calculate the rainfall fraction that is transformed in runoff. The application of empiric methods in the prediction of the runoff produced by a precipitation can be considered as a first approach, that should be corrected later with base in the evaluation of the system in operation [7].

In watersheds without instrumentation, the determination of the surface runoff is more difficult and less accurate than in watersheds with instrumentation. The rational method is the simplest method available, being recommended only for small basins and with the following restrictions: a) every area of the basin should contribute with runoff in the outlet cross-section in small intervals of time; b) the rainfall should be of high intensity and short duration; and c) in small intervals of time, the variation of the infiltration velocity should be small.

A preliminary study by the "Water Resources Council", mentioned by [8], documented the difficulties in applying peak-flow procedures, including the wide variability in estimates that can be obtained by different professionals using identical procedures, and the gross inaccuracy of some of the methods that are commonly used. Consequently, further investigation into a procedure for estimating peak runoff rates, which utilizes present knowledge of flood-generating mechanisms and analyses of data, is needed.

[9] developed a procedure to determine the maximum surface runoff volume

employing the constant rate of infiltration after prolonged wetting of the soil which is applicable to places where the rainfall intensity-duration-frequency equation is known.

The objective of this paper was to develop a model to estimate the overland flow inside terraced areas, the flow rate in channels of graded terraces, and the corresponding hydrographs of surface runoff in these situations.

2 Methodology

The following assumptions were made to develop the model:

- precipitation is uniform over the whole area analyzed and the intensity-duration-frequency equation is known.
- when the design rainfall occurs, the soil moisture is at field capacity and the infiltration rate close to the constant rate of infiltration after prolonged wetting of the soil.
- evaporation is zero during the design rainfall.
- infiltration rate through the channel surface remains constant during the accumulation of water into the channel terrace.

The initial abstraction (rainfall before surface runoff starts) depends upon the interception, depression accumulation, and infiltration prior to surface runoff. The value of initial abstraction was calculated by the curve number method, using the equation recommended by [10]:

$$Ia = 50,8(\frac{100}{CN} - 1) \tag{1}$$

where Ia = initial abstraction (mm); and CN = curve number, which describes the potential watershed storage (dimensionless).

To determine CN, the criteria of the Soil Conservation Service (SCS) were used. It was considered that when the rainfall occurs, the soil moisture, as defined by the curve number, was the maximum antecedent moisture condition (AMC III), i.e., the rainfall accumulated during five days prior to the design rainfall was equal to or higher than 52.5 mm.

The time corresponding to the initial abstraction was obtained using

$$t_{Ia} = \frac{Ia}{i_m} \tag{2}$$

where t_{Ia} = the time interval from the beginning of the rain to the beginning of runoff (min).

The average maximum rainfall precipitation (i_m), in mm h^{-1}, was obtained from the rainfall intensity-duration-frequency relationship by the equation

$$i_m = \frac{K\,T^a}{(t+b)^c} \tag{3}$$

where T = return period (years); t = rainfall duration (min); and K, a, b, c = parameters for a given geographic location.

The average maximum rainfall precipitation (i_m) was substituted in equation 2 with t = t_{la} and to solve the resulting equation the Newton-Raphson method was used.

To obtain the hydrograph of surface runoff the area between two successive terraces was subdivided in a grid composed by i lines and j columns and the analysis was made for two different conditions: a) overland flow: flow that occurs following the land slope direction; and b) flow rate: flow that occurs concentrated into channel cross section.

2.1 Overland flow conditions

For this condition it was considered that the overland flow occurs exclusively in the land slope direction and increases up to the when the contribution that came from line 1 reaches the considered line. After this time the overland flow decreases with the time.

Under overland flow conditions the overland flow value was equal to zero for all cells of the grid until the time t_{la}. After that time the overland flow value for each cell [i,j] was obtained by summing the overland flow produced in the specific cell and the overland flow produced by cells that contribute with runoff to the considered cell using the equation

$$q_t[i,j,t+t_d] = q_t[i-1,j,t] + \frac{(i_i[i,j,t+t_d]-f_c)W_s\,L}{3.6 \times 10^6\,r\,c} \tag{4}$$

where $q_t[i,j,t+t_d]$ = overland flow presented by cell [i,j] for a time equal to $t+t_d$ (m^3 s^{-1}); $t_d[i,j,t]$ = time that the overland flow that occurs in line i-1 in the time t takes to reach the line i for a given column j (min); $q_t[i-1,j,t]$ = overland flow presented by the cell [i-1,j] for a time equals to t (m^3 s^{-1}); $i_i[i,j,t+t_d]$ = instantaneous rainfall intensity into cell [i,j] for a time $t+t_d$ (mm h^{-1}); f_c = constant rate of infiltration after prolonged wetting of the soil, mm h^{-1}; W_s= horizontal spacing between channels (m); L = channel length (m); r = number of lines; and c = number of columns.

In the equation 4 it was considered that $i_i[i,j,t+t_d]$ is larger or equal to f_c.

The instantaneous rainfall intensity was obtained using the equation 5 proposed by [9]

$$i_i = i_m(1-\frac{c\,t}{t+b}) \tag{5}$$

Among the empirical equations developed for open channel flows, the Manning's equation is often used to describe surface irrigation flows [4,11], condition in which the depth of water is small. For this reason the calculation of the time of the overland flow was made using the Manning's equation, being the $t_d[i,j,t]$ obtained by the equation:

$$t_d[i, j, t] = \frac{W_s\, n_s}{S_s^{1/2}\, y_t[i-1, j, t]^{2/3}\, 60\, r}$$ (6)

where n_s = Manning hydraulic roughness coefficient of the soil surface (s m$^{-1/3}$); S_s = land slope (m m^{-1}); and $y_t[i-1,j]$ = flow depth under overland flow conditions, for the line i-1 of a given column j in the time t (m).

The $y_t[i-1,j]$ value was obtained by the equation

$$y_t[i-1, j, t] = (\frac{q_t[i-1, j, t]\, n_s}{S_s^{1/2}})^{3/5}$$ (7)

2.2 Flow rate into channel

For this condition the calculation of the flow rate in the cross section of the channel was made by the equation

$$q_c[j, t + t_{can}] = q_t[r, j, t + t_{can}] + q_t[r, j-1, t]$$ (8)

where $q_c[j,t+t_{can}]$ = flow rate in the channel for the column j and the time $t+t_{can}$ (m^3 s^{-1}); $t_{can}[i,j,t]$ = time that the flow rate that occurs in the column j-1 in the time t takes to reach the column j (min); $q_t[l,j,t+t_{can}]$ = overland flow in the line r for a column j in the time $t+t_{can}$ (m^3 s^{-1}); and $q_t[r,j-1,t]$ = overland flow in the line r for a column j-1 in the time t (m^3 s^{-1});

The time corresponding to the movement of water and the water depth in the channel were calculated using Manning's equation. The time that the flow rate that occurs in the column j-1 takes to reach the column j for a channel with triangular section was determined with the equation

$$t_{can}[j, t] = \frac{L\, n_c\, (2\, S_s\, S_f\, (\sin(aa) + \sin(bb)))^{2/3}}{S_c^{1/2}\, (y_{can}[j-1, t](S_s + S_f)\, \sin(aa)\, \sin(bb))^{2/3}\, 60\, c}$$ (9)

where n_c = Manning hydraulic roughness coefficient of the channel (s m$^{-1/3}$); S_f = ridge front slope = tg (aa), (m m^{-1}); S_s = land slope = tg (bb), (m m^{-1}); and S_c = channel slope (m m^{-1}); and y_{can} = water depth in the cross section of the terrace channel (m).

The schematic cross-section for a terrace with triangular section shape is presented in the Fig. 1.

The value of the water depth in the cross section of the terrace channel for channels . with triangular cross section was calculated by the equation

$$y_{can}[j-1, t] = [\frac{q_c[j-1, t]\, n_c\, (2\, S_f\, S_s)^{5/3}\, (sen(aa) + sen(bb))^{2/3}}{S_c^{1/2}\, (S_s + S_f)^{5/3}\, (sen(aa)\, sen(bb))^{2/3}}]^{3/8}$$ (10)

where $q_c[j-1,t]$ = flow rate for column j-1 in the time t (m^3 s^{-1}).

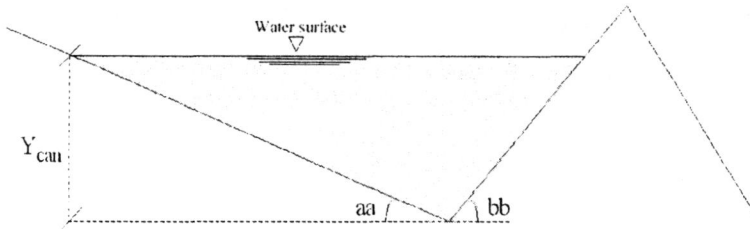

Fig. 1. Schematic terraces cross-section with triangular shape

3 Practical application and comments

An application of this methodology is presented for the conditions of rainfall typical of the Cascavel county, Paraná, Brazil, where the rainfall intensity-duration-frequency equation is expressed by the equation

$$i_m = \frac{1062.92\, T^{0.141}}{(t+5)^{0.776}} \tag{11}$$

The other data used for this practical application were:
CN (bare soil of group B and AMC III) = 94; T = 10 years; f_c = 10 mm h^{-1}; L = 400 m; W_s = 40 m; c = 100; r = 100; S_c = 0.001 m m^{-1}; S_f = 0.16 m m^{-1}; S_s = 0.10 m m^{-1}; n_s = 0.025; n_t = 0.10.

Figure 2 shows the hydrograph of the overland flow for line 75, column 50. As it can be observed in this hydrograph, after the beginning of the precipitation there is a time (corresponding to t_{1a}) along which the overland flow is zero. This time corresponds to duration of the initial abstraction, that is the time before surface runoff starts. After this time the overland flow increases gradually with the time until time t = 3.02 min, corresponding to the time of concentration of the overland flow arising from the cells situated above the considered cell. The value of the overland flow in this cell for this time is 0.00828 m^3 s^{-1}. After this time the overland flow begins to decrease because the water arising from the most remote cell already reached the considered cell and the instantaneous rainfall intensity decreases with time.

Figure 3 is shows the hydrograph of the flow rate for column 50, where it can be observed that the runoff flow for the cross-section considered increases until the moment when the rainfall which occurred in the most remote of cells reaches the considered cell. After this time the runoff rate diminishes with the time because the instantaneous rainfall intensity decreases with the time as occurs for overland flow conditions too.

4 Conclusions

The methodology developed allows the determination (with base in physical principles) of hydrographs of surface runoff for overland flow conditions as well as for

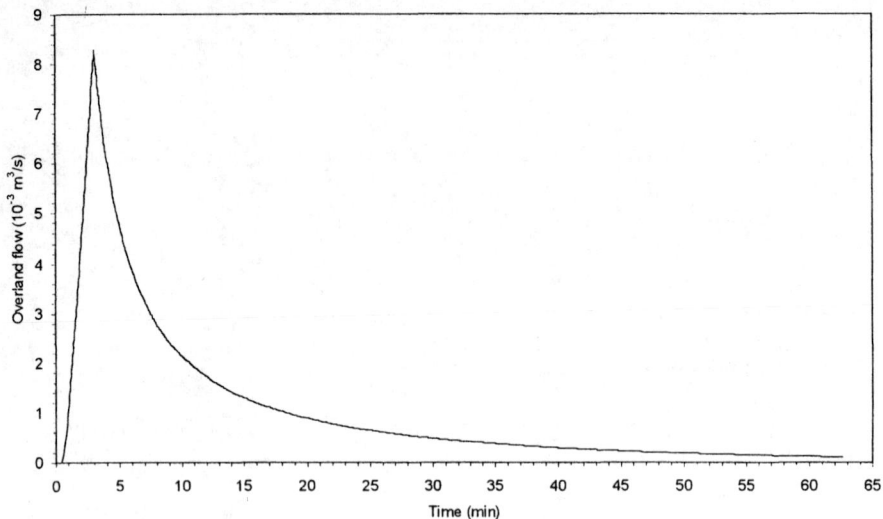

Fig. 2. Hydrograph of the overland flow for line 75, column 50.

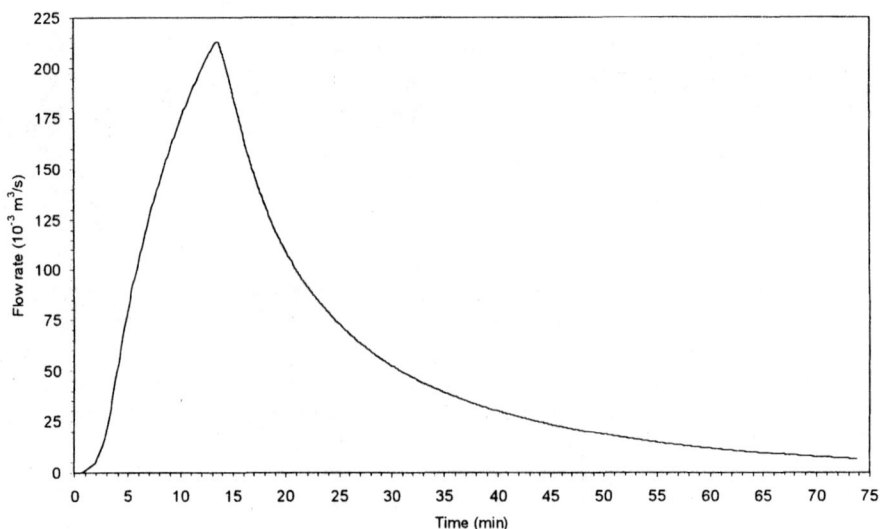

Fig. 3. Hydrograph of the flow rate for the column 50.

cross-sections located along the terrace channel. These hydrographs allow, when associated with the critical shear stress of the soil, the estimate of soil losses that occurs as a consequence of the erosive process. The methodology developed also allows quantifying the peak-rate flow that is essential to design the cross-section of graded channels. Although the methodology developed was not tested under real conditions the results obtained approach the physical behavior observed in the field. To use this methodology it is necessary to know the rainfall intensity-duration-frequency equation for the place considered; the topography of the land (land slope); the conditions of the land surface (roughness of the land and of the channel and the constant rate of infiltration after prolonged wetting of the soil); and the following characteristics of the terracing system: spacing, length, shape and slope of the channel.

5 References

1. Beasley, D.B.; Huggins, L.F. and Monke, E.J. (1980) ANSWERS: the model goes watershed planning. *Trans. ASAE*, Vol. 23, No. 4. pp. 938-944.
2. Engman, E.T. (1986) Roughness coefficients goes routing surface runoff. *J. Irrig. and Drain. Engrg*, Vol. 112, No. 1. pp. 39-53.
3. Gilley, J.E., Kottwitz, E.R. and Simanton, J.R. (1990) Hydraulic characteristics of rills. *Trans. ASAE*, Vol. 33, No. 6. pp. 1900-1906.
4. Maheshwari, B.L., and McMahon, T.A. (1992) Modeling shallow overland flow in surface irrigation. *J.Irrig. and Drain. Engrg*, Vol. 118, No. 2. pp. 201-217.
5. Derpsch, C.H., Roth, C.H., Sidiras, N., and Köpke, U. (1991) *Control of the erosion in Paraná, Brazil: systems of covering of the soil, direct plantation and I prepare conservacionista of the soil.* Eschborn, Deutsche Gesellschaft für Technische Zusammenarbeit (GTZ). (in Portuguese).
6. Schwab, G.O., Frevert, R.K., Edminster, T.W., and Barnes, K.K. (1981) *Soil and water conservation engineering*,3rd. Ed. John Wiley & Sons Inc., New-York, N.Y.
7. Beltran, J.M., Sanchez, I.G. and Fruk, M.P. (1988) Agricultural Drenaje. In: *I study international of specialization in irrigation engineering.* Brasilia, s.ed. (in Spanish).
8. Bonta, J.V., and Rao, A.R. (1992) Estimating peak flows from small agricultural watersheds. *J. Irrig. and Drain. Engrg*, Vol. 118, No. 1. pp. 122-137.
9. Pruski, F.F., Ferreira, P.A., Ramos, M.M., Cecon, P.R. (1997) Model to design level terraces. *J.Irrig. and Drain. Engrg*, Vol. 123, No.1. pp. 8-12.
10. Soil Conservation Service. (1972) *National engineering handbook, section 4, hydrology*, USDA.
11. Skogerboe, G.V. and Walker, W.R. (1987) *The theory and practice of surface irrigation*, Logan, Utah State University (in Spanish).

DYNAMIC FLOOD CONTROL ALONG ARTERIAL DRAINAGE NETWORKS, A CASE STUDY

Y. NEDELEC, D. ZIMMER, C. CHAUMONT and M.D. PORCHERON
Cemagref, division ouvrages pour le drainage et l'étanchéité, Antony, France

Abstract

Sub-surface drainage peak flow rates are in general smoothed at the plot scale because of an increase of local infiltration capacity and a reduction of surface runoff, whereas transfers are accelerated by the arterial drainage system, which is often over-designed due to the depth of the collector drains. This flow acceleration can be reduced by letting the network overflow in its less critical zones. This solution, which can result from adequate reduced cross sections such as crossovers for roads, is referred to as dynamic flood control. But very few methods are available to correctly define how to obtain these overflows and to afford maintenance problems. In order to describe the peakflow transfer through a small agricultural catchment, an hydraulic model based on St-Venant's equations was used, and applied to an experimental catchment near Paris (Orgeval, Seine-et-Marne), in which runoff comes essentially from subsurface drainage. The experiment and model results are used to build recommendations on dynamic flood control at a small catchment scale, by means of a correctly designed surface network, especially its reduced cross sections. These recommendations are now to be validated by multi-scale experiments, and design of effective equipment.

Keywords: Arterial drainage, dynamic flood control, flood routing, modelling, peak flow, subsurface drainage

1 Introduction

Many floods occurred in France during the past five years, and agriculture intensification was accused by media and public opinion, of being largely at their origin. The development of agriculture may lead to a greater influence of human activities on

Water and the Environment: Innovative Issues in Irrigation and Drainage. Edited by Luis S. Pereira and John W. Gowing. Published in 1998 by E & FN Spon. ISBN 0 419 23710 0

water regime, because of larger plots or specific hydraulic equipment, but responsibilities are not so easy to evaluate.

Subsurface drainage is one of the structures implied in the discussions about aggravating factors of floods. But if subsurface drainage may locally facilitate the soil infiltration capacity and reduce surface runoff, its combination with arterial drainage leads to easier transfers from plots to catchment outlets. Nevertheless, arterial drainage networks interact with many other structures, such as cross-overs for roads or pathways. These interactions, or any ways to dispatch water retention on multiple parts of the catchment, and at simultaneous times, are referred to as dynamic flood control. They can play a major role in the mitigation of flow regime during average floods.

The present work studies such interactions between drainage discharge, channel transfer, and reduced cross section influence. In order to collect and use information from an experimental site mostly subsurface drained, a simplified model was built to simulate the behavior of such catchments, and to test management scenarios to optimize the influence of cross sections on floods.

2 Description of the experimental catchment

The « Orgeval » experimental catchment has been monitored since 1962 by the *Cemagref* (Groupement d'ANTONY, division qualité des eaux, hydrologie) in order to study hydrological processes in a context of intensified agriculture. It is located about 70 km east of Paris, in the Brie region. The total watershed area is 104 km² and the average altitude 150 m, ranging from 75 m to 180 m with gentle slopes. The main part of the catchment is covered with a thick table-land loess (up to 10 m thick), characterized by low permeability. Soils are leached brown soils (alfisols), with a silty loam texture in its top layers. The Orgeval catchment is situated in a rural area : 80 % of the total area is covered by crops, principally wheat, corn, colza, peas. A large part (more than 50 % of its agricultural area) of the catchment is subsurface drained to enhance crop production.

A particular sub-catchment (drained by the « Ru du Rognon ») ending at Melarchez (Fig. 1, this catchment will further be referred to as « Melarchez catchment ») was selected to study peakflow transfers in an arterial drainage channel. The network was chosen quite simple (first order stream), and crossed by 3 crossovers. This sub-catchment has an area of 7,2 km². The channel length is 3.900 m (1.600 subterranean, 3.200 open channel). Its width varies from 4 m upstream (open air part), to 7 m downstream. Water inflow in the channel comes from groundwater table (the vast « Brie » groundwater system), and from subsurface drainage networks, which cover almost the whole catchment area (75 %). Therefore, subsurface drainage appear in this case to play a major role in flood generation, and precise geometrical characteristics of subsurface drainage networks (Area, positions of main collectors) were collected on the experimental site.

3 Simulation model and approach

3.1 Global approach
Because isotope studies showed that groundwater discharge could become predominant in the recessing part of the hydrographs [1], the model focused on the channel bed behavior during intense drainage inflow and flood generation. Thus model validation criteria were restricted to the peak characteristics.

Figure 2 shows the global representation of the experimental catchment, combining plots subsurface drainage, main drainage collectors, arterial drainage network with its major bed and cross-overs.

In this approach, the following information are collected, and associated in a global hydraulic scheme:

- subsurface drainage hydrographs, measured out from an experimental plot, and used as production function,
- collector drains geometry,
- arterial drainage network geometry: bed slope, cross-sections, major-bed perpendicular slope, hydraulic characteristics of reduced cross-sections, locations of collectors outlets where the results of the transfer function above are applied,
- downstream limit condition : the actual relation between flow rates and water level is taken into account.

The simulations performed according to this scheme give predictions of catchment outlet flow rates, during flood events, characterized by drainage outflow.

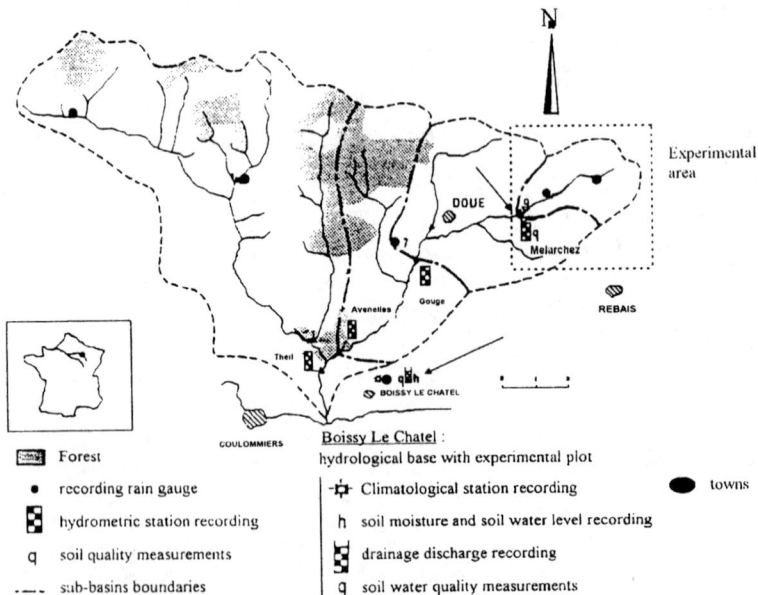

Forest	Boissy Le Chatel : hydrological base with experimental plot
recording rain gauge	Climatological station recording towns
hydrometric station recording	h soil moisture and soil water level recording
q soil quality measurements	drainage discharge recording
sub-basins boundaries	q soil water quality measurements

Fig. 1 The Orgeval catchment location and experimental equipment. Arrows show the location of flow rates measuring equipment: Boissy-le-Châtel (Subsurface drainage) and Melarchez (Catchment outflow).

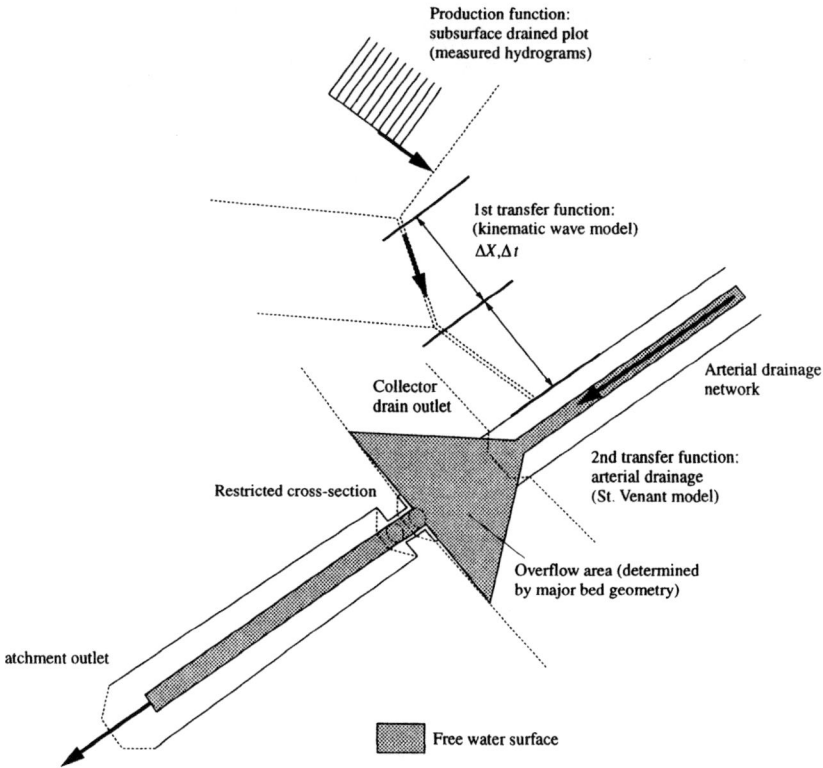

Fig. 2 Global approach of the complete drainage system

3.2 Subsurface drainage
Though the subsurface drainage hydrographs could not be measured at the drainage network scale on the Melarchez site, it was observed that peak flow prediction at the catchment outlet was impossible without applying a delay to the contributions of distant parts of networks. This delay, due to long subsurface drainage systems, was obtained by discretizing networks into successive parts of the same length, and multiplying the corresponding distance to channel with a same mean celerity. Such a discretization is equivalent to a kinematic wave approximation [2] of the subsurface drainage network.

3.3 Free surface network
An hydraulic model based on St. Venant's equations [3] was used (MAGE [4]) to simulate water transfers in the main channel. MAGE, developed at the *Cemagref*, solves this system of equations by the finite difference method between interpolated cross sections. The geometry of channel and major bed was adapted from topographical data and calibration of major bed perpendicular slopes. Drain discharges were introduced as point hydrographs, calculated by multiplying the drained plot observed flow rates by the subsurface drained area. These hydrographs were introduced at the actual locations of drain outlets.

3.4 Reduced cross-sections and overflow areas

Reduced cross-sections were modeled by classical hydraulic relations, according to different flow regimes. These relations are listed in table 1.

Table 1: Reduced cross-sections equations

Flow regime	Conditions	Equations	
Free discharge	$h_1 < w$ and $h_2 < \frac{2}{3} h_1$	$Q = \mu \sqrt{2g} \, Lh_1 h_1^{1/2}$	(4
Free discharge, control at outlet	$h_1 < w$ and $h_2 \geq \frac{2}{3} h_1$	$Q = \mu \frac{3\sqrt{3}}{2} \sqrt{2g} \, Lh_2 \sqrt{h_1 - h_2}$	(5)
Submerged discharge	$h_1 \geq w$ and $h_2 \leq \frac{2}{3} h_1$	$Q = \mu \sqrt{2g} \, L \left[h_1^{3/2} - (h_1 - w)^{3/2} \right]$	(6)
Submerged discharge, partial control at outlet	$h_1 \geq w$ and $\frac{2}{3} h_1 \leq h_2 < \frac{2}{3} h_1 + \frac{1}{3} w$	$Q = \mu \sqrt{2g} \, L \left[\frac{3\sqrt{3}}{2} h_2 \sqrt{h_1 - h_2} - (h_1 - w)^{3/2} \right]$	(7)
Submerged discharge, control at outlet	$h_1 \geq w$ and $h_2 \geq \frac{2}{3} h_1 + \frac{1}{3} w$	$Q = \mu \frac{3\sqrt{3}}{2} \sqrt{2g} \, Lw \sqrt{h_1 - h_2}$	(8)

Notations:
h1 : upstream water height above hole bottom
h2 : downstream water height above hole bottom
L : orifice width (supposed rectangular)
w : orifice height
μ : discharge coefficient

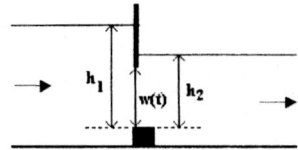

4. Data collection

Three types of data were necessary to study the influence of subsurface drainage on flood generation and transfer: input data, model parameters, and measured catchment outflow for model calibration and validation.

Measured drainage flow rates were used as input data for simulations. In this study concerned by subsurface drainage influence, sets of data were limited to floods occurring between approximately 10/01 and 04/30, at times when subsurface drainage discharges occur immediately in response to rainfall events. This particular period is referred to as the « intense drainage season ».

Measured drainage flow rates are available at the experimental plot of Boissy-le-Chatel (Fig. 1). These flow rates are calculated from variable timestep data (timesteps are as short as flow rates are high) collected by a water level recorder. No drained plot was equipped to measure drainflow rates on the Melarchez catchment, and therefore drainage data is influenced by a short distance (6 km) from the catchment.

The model was validated by comparisons to measured flow rates at the catchment outlet. These flow rates are calculated from variable timestep data (timesteps are as short as flow rates are high) collected by a water level recorder. Eight flood events which occurred during the intense drainage season, were selected between 1981 and

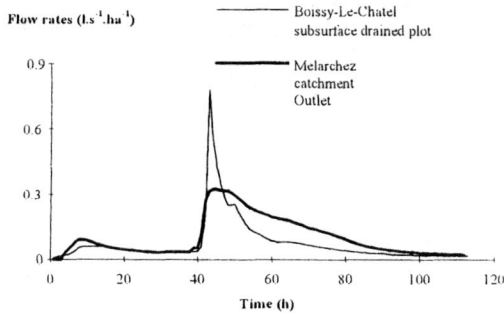

Fig. 3 Specific flow rates in l.s^{-1}.há$^{-1}$ from subsurface drained plot and Melarchez catchment outlet, between 1982/12/15 and 1982/12/20.

1989. Peakflow rates range from 1,2 m^3.s^{-1} (0,17 l.s^{-1}.há$^{-1}$) to 2,3 m^3.s^{-1} (0,32 l.s^{-1}.há$^{-1}$). Fig. 3 shows the transformations of specific flow rate between drained plot scale and catchment scale.

5 Results

The model was first calibrated, in two steps: transfers in the subsurface drainage system, and transfers in the arterial drainage network and cross-overs hydraulic characteristics.   Because of our interest in peakflow mitigation during subsurface drainage discharge, calibration focused on peak simulation accuracy.   Na application was then performed to study flood transfers, and how peakflows can be dynamically mitigated by this system.

5.1 Model calibration : transfers in the subsurface drainage system
The eight hydrographs measured at the subsurface drained plot outlet during the selected events were introduced as production functions.

Introduction of the subsurface drained plot hydrographs directly as lateral inflow in the arterial drainage channel can not predict proper catchment hydrographs, because of flood transfer velocities far too high.   A peakflow mitigation as can be seen on figure 3 can not be explained by the arterial drainage alone, but is better predicted by the application of slower transfers between subsurface drained plots and drainage channel. This has been done by the kinematic wave simulation [2].

It can be noticed that delays involved in these two steps in flood transfer appear to be quite different : whereas peakflows in channel can be transferred in less than 1 hour (na order of usual value for mean velocity in channels is 0.91 m.s^{-1} [2]), the longest subsurface drainage and collector network in the Melarchez catchment appear to induce delays of 8 hours.

Predicted peakflow rate (m³.s⁻¹)

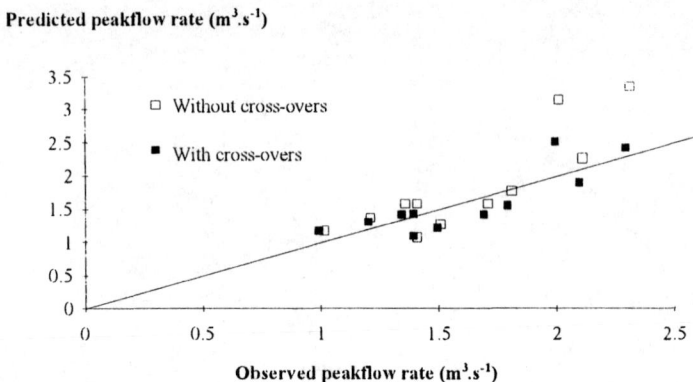

Observed peakflow rate (m³.s⁻¹)

Fig. 3 : Comparison between predicted and observed peakflow rates in m³.s⁻¹ at catchment outlet, with and without modelized cross-overs.

5.2 Model calibration : transfers in the arterial drainage network and cross-overs hydraulic characteristics

The arterial drainage network model, with existing reduced cross sections was calibrated after having set the subsurface drainage parameters. Though a topographic survey gave indications on the actual major bed geometry, it was simplified to make computation easier. Thus, a single major bed perpendicular slope was calibrated.

A good correspondence could be obtained between observed and predicted flow rates, as can be seen on fig. 3. In general, introducing culverts into the system reduces simulated peakflow rates. The graph shows also that during 2 of the most intense events, cross-overs played a significant role. Consequently, these particular events allowed to calibrate parameters related to cross-overs hydraulics (μ) and overflow dynamics (K and s). Criteria for this calibration were peakflow rates and the very first part of recession (Fig. 4). The second part of recession, not properly explained by our model, could be related to groundwater discharge from the Brie groundwater system, but will not be taken into account in model application.

Table 2 shows the values of the main parameters after calibration

Table 2 : Main hydraulic parameters for arterial drainage model after calibration

Parameter	Value
Manning-Strickler coefficient (Equation 3)	15 I.S.U.
Major bed perpendicular slope	2 %
Discharge coefficient (Equations 4 to 8)	0.65 I.S.U.

5.3 Model application: influence of cross-overs and reduced cross sections on flood control

After calibration, considered satisfactory as far as it is concerned by peak flow proper prediction, the model was used to study the effect on flood control of cross section local management. The study analyzed the effect of cross-overs sizes, number and distribution on the global peakflow mitigation ratio and local storage conditions.

Considering agricultural land tolerance to overflow, actual culverts dimensions were

Flow rate (m^3.s^{-1})

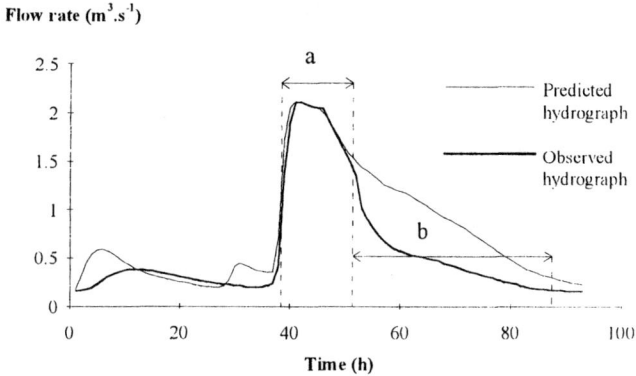

Fig. 4 : Observed and predicted flow rates in m^3.s^{-1} at the catchment outlet, between 1982/12/15 and 1982/12/20 (a : subsurface drainage discharge period, b: groundwater discharge period).

first modified in the model, in order to create a simultaneous overflow as soon as flow rate exceeds the value corresponding to the catchment daily flood at 2-year return period, further referred to as the reference event. The peak flow numerical value for this reference event was calculated from observed data at the catchment outlet, and extended to its other parts where new cross-overs were to be tested, by mean of the QdF synthetic model [5].

Up to 9 cross-overs were tested along the arterial drainage network, almost regularly located. It was a theoretical configuration as far as no topographical nor geographical considerations were taken into account. The results showed that more than 7 cross-overs on this catchment could not bring significant amelioration in terms of storage height and area.

On figure 5 five successive compared configurations (3 actual cross-overs, the corresponding theoretical flow rates during the reference event, 3 cross-overs with new dimensions, 7 cross-overs, and one single downstream storage) are schematized. The corresponding predicted values of three indicators of the mitigation effect for a particular event (1982/12/15 to 1982/12/20) are reported in table 3 : mitigation coefficients, storage areas and storage volumes.

For the flood observed between 1982/12/15 and 1982/12/22, the model shows that the mitigation ratio is about 28 % with the three initial structures, and reaches 34 % with 7 structures which distribute equally overflow volumes. The gain in mitigating coefficient does not result from higher spilled volumes, but from longer retention times. Mitigating is in fact the result of dynamic storage distribution along the channel, more than its cumulative value. Water stored at a first culvert located up stream can be stored again downstream. This can be confirmed by the comparison between distributed structures and a single reservoir, needing a greater available volume for the same mitigation ratio: 13.990 m^3 are needed by the major bed equipped with 7 reduced cross-sections, whereas 23.500 m^3 may be required if a single reservoir is constructed.

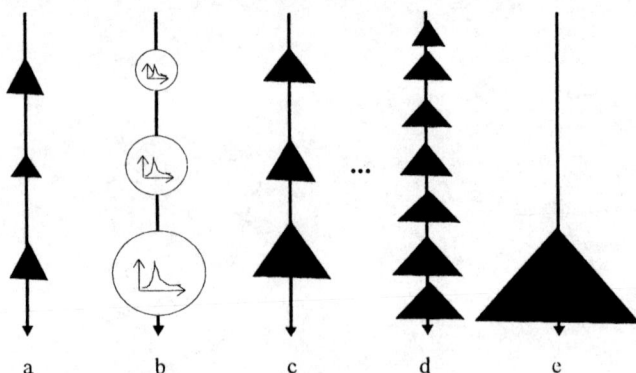

Fig. 5 : Successively tested storage configurations : (a) 3 actual cross-overs, (b) the corresponding theoretical flow rates during the reference event, (c) 3 cross-overs with new dimensions, (d) 7 cross-overs, (e) one single downstream storage

Whereas transfer times in open channel network appeared to be far shorter than in subsurface drainage systems, dynamic storage in the major bed can contribute to flood mitigation with significant efficiency.

Table 3 : Comparison of predicted values of mitigation coefficients, storage areas and storage volumes for a particular event (1982/12/15 to 1982/12/22), between 3, 7, and 1 storage structure.

ross-over n°	3 cross-overs			7 cross-overs			Single water storage		
	Mitigation coefficient (%)	Storage area (ha)	Storage volume (m³)	Mitigation coefficient (%)	Storage area (ha)	Storage volume (m³)	Mitigation coefficient (%)	Storage area (ha)	Storage volume (m³)
	17	0.92	2970	17	0.92	2970	34	1-2	23500
	26	0.88	4930	25	0.83	2770			
	28	0.97	5630	26	0.38	1140			
				31	0.83	3310			
				34	0.47	1760			
				37	0.13	270			
				34	0.45	1770			
otal		2.77	13530		4.01	13990		1-2	23500
utlet	28			34			34		

6 Conclusions

In order to promote dynamic flood control along arterial drainage networks, tools should be developed to guarantee enough protection for downstream exposed areas, together with limited restrictions to the use of major beds for agriculture. The model developed here in the case of intensively subsurface drained catchments, shows the effective dynamic storage in the major bed of an arterial drainage network. Reduced cross sections such as cross-overs for roads or pathways may play an active role in flood mitigation, as far as they introduce delays in flood routing, and may store water during longer and more efficient periods.

The effect of dynamic flood control, allowing mitigating ratios up to 34 % during events at about 2-year return period, adds to the already strong mitigation effect of the subsurface drainage system. Comparison with a single reservoir constructed downstream, which is a current practice, show that volumes to be stored may be nearly 2 times lower in case of dynamic flood control.

Numerical results could be exploited to introduce a more general method available for engineers responsible for rural land management. It replaces measured flow rates available in this experimental cases, by reference flow rates calculated by QdF synthetic hydrological model [5] for a reference event. Hydraulic properties of culverts are also simplified for easy computation. This method needs now to be completed to cases of partially drained catchments, and needs also further validation by convenient experiments and measurements.

After having studied the hydraulic possibilities of dynamic flood control, a more global approach should try to analyze their economical acceptability, and make sure that their general principles will be followed. Potential obstruction of minor bed at reduced cross sections because of reduced water velocities and sedimentation, must also be studied in terms of risks and maintenance costs.

7 Acknowledgments

Experimental data and information about the Orgeval catchment could not have been collected and exploited without the financial contribution of Region Ile-de-France.

8 References

1. Blavoux B. (1978) Etude du cycle de l'eau au moyen de l'oxygène 18 et du Tritium. Possibilités et limites de la méthode des isotopes du milieu en hydrologie de la zone tempérée. , Thèse de Doctorat d'Etat, Université P. et M. Curie, Paris, 333 p.
2. Ponce, V. M. (1986) Diffusion wave modeling of catchment dynamics., *Journal of Hydraulic Engineering*, 112 - 8, 716-726
3. Cunge, J. A. , Holly, F. M. Jr and Verney, A. (1980) *Practical aspects of computational river hydraulics.* , Pitman Advanced Publishing Program, 420 p.
4. Giraud, F., Faure, J.B., Zimmer, D., Lefeuvre, J.C. and Skaggs, R.W. (1997) Hydrologic modelling of a complex wetland. , *Journal of Irrigation and Drainage Engineering*, Sept-Oct 1997, 344-353
5. Galea, G. and Ramez, P.(1995) *Maîtrise du ruissellement et de l'érosion en vignoble de coteau-Guide à l'usage des aménageurs.*,Cemagref Editions, 112 p.

RIVER MEADOW PROGRAMS - PLANNING IN PARTNERSHIP

H. PATT
University of Essen, Essen, Germany

Abstract
Streams and rivers in Europe have been altered by past human activities. Sections of rivers that were once meandering and lined with vegetation are today very often straight. The connection to their natural floodplains is either reduced or no longer exists. Changes in agricultural and urban land use have resulted in erosion and sediment problems. The loss of retention volume has generally increased the flood risks. The ecological consciousness of the population and the changing of environment policy over the last 20 years has led to a complete rethinking of water resource engineering. River meadow concepts were developed to facilitate the realization of measurements for the ecological improvement of rivers.
Keywords: Public participation, rehabilitation of rivers, river meadow concepts, river meadow programs.

1 Introduction

The protection of waters and ecologically orientated rehabilitation measurements are a precaution for future generations. Within river meadow programs some concepts for selected rivers were installed to facilitate the coordination and realization of all measures through integrating different experts (e.g. hydraulic engineers, water resource management engineers, landscape engineers, biologists) and social groups (e.g. land owners, nature protection associations). Such concepts can help to avoid some of the serious conditions in rivers and finally save a lot of public money (e.g. drinking-water purification, flood protection).

Water and the Environment: Innovative Issues in Irrigation and Drainage. Edited by Luis S. Pereira and John W. Gowing. Published in 1998 by E & FN Spon. ISBN 0 419 23710 0

2 Aims and concepts

River meadow programs are usually sub-divisions of nature protection programs from the federal states. These nature protection programs include such different sub-programs as river meadow programs, programs for the implementation of bank strips, rehabilitation of urban areas, biotope integration and protection of endangered fauna and flora or rare habitats.

Within the river meadow programs of the states themselves there is a further sub-division into the rivers concerned (e.g. the River meadow concept of the river Sieg [1]). River meadow concepts have the intention of simplifying the formation and re-alization of all measures within the planning close to the particular river. This is achieved by an early incorporation of all participants into formation of assigned scopes, which includes politicians, public authorities from different specializations, land-owners and concerned citizens as well as representatives of agriculture, nature protection and leisure associations.

3 Legal aspects

In Germany the Wasserhaushaltsgesetz (Federal Water Law of Germany), the Bundes-naturschutzgesetz (Federal Nature Protection Law) and the Raumordnungsgesetz (Fed-eral Regional Planning Law) are the main legal basis for all measures concerning river meadow programs or river meadow concepts. Within the framework of the federal laws the 15 states of Germany have the right to install further regulations due to their given regional circumstances and political intentions. The development of river meadow programs, or general water and nature policy in Germany, is therefore very much influenced by state policy.

In 1996 the Wasserhaushaltsgesetz (Federal Water Law) was modified. The law now includes a much stronger order to improve the ecological conditions for rivers and to protect natural floodplains against further urban development.

Important changes of the law for the rehabilitation of rivers concern the question of whether the river development has to be part of a extensive approval procedure (the so-called "Planfeststellungsverfahren") or can be realized after simple permission from the appropriate authority (the so-called "Plangenehmigung").

The large-scale approval procedure requires the participation of different public authorities (e.g. water authorities, nature protection authorities) as well as the inclusion of certain representatives of interest (e.g. nature protection, agricultural, fishing, bird protection). The result is very often a compromise between all existing interests in-cluding indemnifications for restrictions. It is important to known that the final result of the approval procedure (the so-called "Planfeststellung") has a legal binding for each following decision.

Following the regulations of the new water law the approval procedure is especially necessary when water discharge is negatively influenced. This must usually be taken into consideration when a change influences flood security or the imperative minimum flow for ecological reasons.

The commitments to a river meadow concept (e.g. in the state of Nordrhein-Westfalen of the Federal Republic of Germany) are legally only advice from the state government to the subordinate public authorities (province and municipal administrations). This implies that within the German law system river meadow programs are only obligatory for public authorities but not for private planning.

All measures within a river meadow concept must be realised voluntarily, accommodate, and accord with different types of agreements (purchase of the land, agreements over specific modes of land use (e.g. extensive greenland use) or forms of maintenance (e.g. spatially and temporally) or indemnifications in the case of restricted utilization). It is a political decision to avoid antagonism of public authorities.

4 Measurements affecting agricultural areas

A river meadow concept involves the river bed and the river meadow laterally as well as longitudinally since all influences on the river stretch from the source to the mouth. They consist of different types of measures that generally improve the ecological situation of a river [2]).

There are measurements like the construction of fishways [3] that are important and relatively easy to realize because it mainly influences areas that are often in public property. But in most cases the areas that are affected are in agricultural utilization and therefore the central point of many conflicts.

4.1 Re-opening of natural flood plains
The natural development of rivers and flood plains that is suggested in the river meadow concepts require much more area that is often not available because of the human activities which have expanded into the flood plains. These regions are protected by flood protection installations (e.g. dykes, flood walls) and have no direct contact to the water level in the river. In the case of agricultural use of these areas there is at least a possibility to partially reobtain the land and to improve the discharge characteristics of the rivers especially during flood.

Special investigations must include the affects of re-opening the former flood plains on the travel time of a flood wave, water depth etc. as well as the groundwater situation and the sedimentation in the flooded area.

4.2 Bank strips - Riparian areas
Bank strips could be a first step for conservation and rehabilitation designs. Obviously, as more area is at a rivers disposal, it can develop its natural profile so much the better. The main parameters are slope, sinuosity, stream entrenchment ratio, width-to-depth ratio and land form feature.

In small to mid-size streams forested riparian zones can moderate water temperature and provide important sources of organic matter. The stream bank cover increases bank stability, introduces shade and buffers the soil from the eroding forces of flowing water (Fig. 1). In the case of new stream bank vegetation, the bank stability will develop only after several years. Growing bank vegetation can be protected through biological engineering scopes [4].

There are direct entries by the land-use, transport of soil by wind or floods or existing drainage systems. One important reason for bank strips is simply the distance factor, that protects a river against direct intakes and its functions as buffer and filter for injurious substances. New investigations show that the filter function leads to decreasing nitrate contents, while the phosphate standard does not change.   The

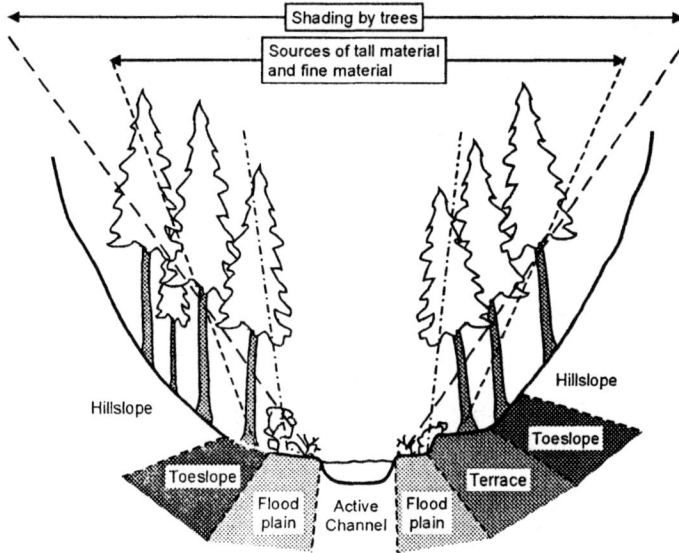

Fig. 1. Significance of bank strips for rivers (after [5])

reduction of nutrients depends on the kind of transition into the bank strips (e.g. continuous, over an even surface or local through furrows). Bank strips can not prevent entries through the ground water body.

Based on existing legal regulations in the states of the Federal Republic of Germany the minimal width of bank strips has to be 5 m. This is in many cases not enough for an ecological improvement [4]. In spite of the outlined distances a complete release of the natural flood plains from each kind of load that negatively influences a river is desirable so that a river can exhibit its natural structure (Fig. 2).

If there is a lot of handwork, the maintenance costs can be much higher than those of lined channels. With increasing width of the bank strips the maintenance of rivers may be reduced in many cases.

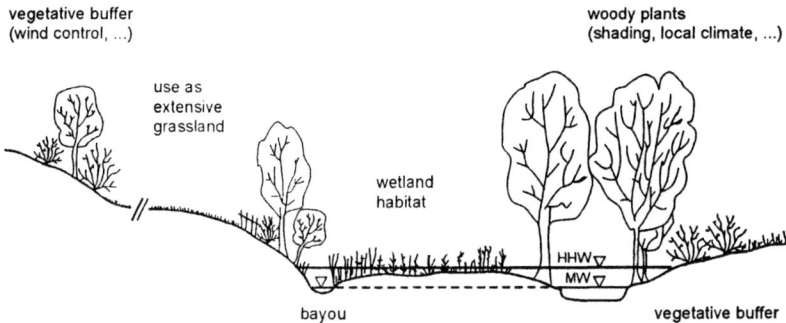

Fig. 2. Structure of bank strips

4.3 Ecologically orientated maintenance

The development of natural structures is also very much influenced by the maintenance work. It is necessary to distinguish between regular and irregular maintenance measures. Typical regular maintenance includes mowing and weeding (especially in the case of low-land rivers) and the maintenance of trees and woods. It must be mentioned that for the development of rivers the presence of dead wood is very important

Maintenance work must be done in accordance with biological time scales. This includes the brooding times of birds and the flourishing time of the vegetation. A comprehensive instruction book with different types of maintenance devices and their applications already exists [6]. However there are still intensive scientific investigations on test stretches to find out the influence of different types of maintenance on the development of rivers. The investigations not only take into account temporal variations of e.g. mowing and weeding, but also different spatial dependences.

4.4 Leisure and recreation on rivers

An rising problem for the protection of nature concerns the increasing number of leisure activities close to a river. A balance between areas with preference for nature conservation and those for recreation in free nature has to be found (Fig. 3).

Catalogues of assessment factors for different kinds of biotopes on one side and leisure activities on the other side allow a first description of occuring interactions and influences. Areas with absolute priority for nature conservation (e.g. through a legal appointment) may not be inserted into the weighing process.

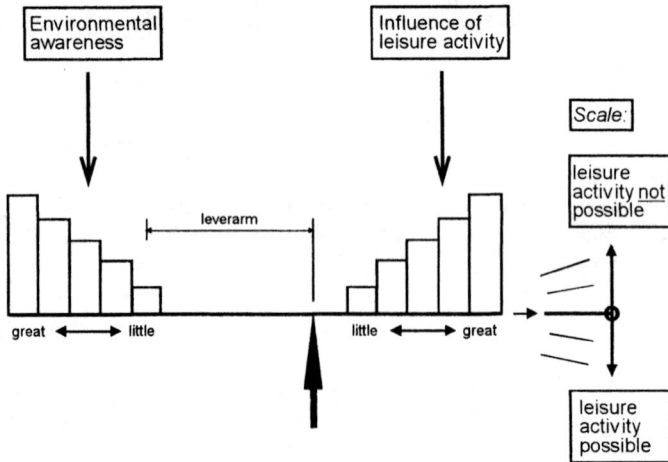

Fig. 3. Balance between leisure and recreation activities and environmental awareness

5 Steps to Realization

The implementation of the river meadow concepts requires a lot of different planning steps and patience during realization. Therefore the planning consists of short term, middle term and long term measures. The legal side (e.g. land ownership, land use, road allowances or existing water rights), social-economic aspects (mainly economic problems of agriculture), ecological risks and last but not least the constitutional right to recreation in open country in Germany requires a constant weighing of each parameter against the others.

An important first step is a continuous conservation of nature to avoid further irreversible changes. Such changes in flood plains are generally forbidden by the new federal water law, although there are still some exceptions that can influence the concept of a river rehabilitation. If there is a higher weighted public interest and the intended plan is therefore unavoidable it has to be compensated.

Beside the technical work the planning team has to think about the non-technical questions concerning the implementation of a concept. Investigations show that there is very often an general misunderstanding between the experts of the public authorities and the farmers concerned [7]. It is provable that there is more often a general consent of the land owners for ecological improvements especially in the river meadows but steps and goals are quite often non-identical. This is mainly due to the fact that for an biologist e.g. biodiversity or range of species is different from the number of spezies (e.g. hares) that can be observed by a layman. This is only one example but it is clear that this source of misunderstanding can easily abolished by a comprehensive information policy.

Due to the good soils in the river meadows many farms are not interested in giving up those areas for ecological reasons because this would imply heavy losses of income and endanger the existence of many farms. In spite of all political promises of the voluntary realization of the concepts many land owners and farmers do not trust the authorities. They are uncertain, because sometimes it has happened that after ecological improvement had reached a certain degree, the authorities saved the areas for nature protection reasons.

There is a wide range of possible results of the negotiations between the persons concerned but it is obvious that monetary compensation for disadvantages especially in agriculture plays an important part. It is worth knowing that the modification of the Bundesnaturschutzgesetz (Federal Nature Protection Law) failed a few times because of the unsolved problem of monetary compensation for disadvantages due to nature protection measurements.

6 Conclusions

Realisation of river meadow concepts seems to be daily work for all persons concerned. But the intentions and goals of the very often ecologically orientated planning are in many cases far away from practice because of many constraints that appear during the whole realization process.

The implementation of river meadow concepts is influenced by a lot of emotions that depend on changing public interests, political structures, giving up of farming or financing problems. A good information policy that incorporates all persons concerned improves the acceptance of a proposal and is sometimes decisive for the whole project.

7 References

1. Städtler, E. (1997) Das Gewässerauenkonzept Sieg. *Wasser und Boden*, 49. Jahrg., Nr. 10/1997.
2. Patt, H., Jürging, P., Kraus, W. (1997) *Der naturnahe Wasserbau - Entwicklung und Gestaltung von Fließgewässern*, Springer Verlag, Berlin, Heidelberg, New York.
3. DVWK (1996) *Fischaufstiegsanlagen*. DVWK-Merkblätter zur Wasserwirtschaft 232/ 1996. Wirtschafts- und Verlagsgesellschaft Gas und Wasser mbH, Bonn.
4. DVWK (1997) *Uferstreifen an Fließgewässern - Funktion, Gestaltung und Pflege*. DVWK-Merkblätter zur Wasserwirtschaft 244/1997. Wirtschafts- und Verlagsgesell schaft Gas und Wasser mbH, Bonn.
5. Maser, C. and Sedell, J.R. (1994) *From the Forest to the Sea*. The Ecology of Wood in Streams, Rivers, Estuaries and Oceans, St. Lucie Press, Delray Beach.
6. DVWK (1992) *Methoden und ökologische Auswirkungen der maschinellen Gewässerunterhaltung*. DVWK Merkblätter zur Wasserwirtschaft 224/1992. Verlag Paul Parey, Hamburg und Berlin.
7. Luz, F. (1996) *Zur Akzeptanz und Umsetzbarkeit von landschaftsplanerischen Projekten zur Extensivierung, Biotopvernetzung und anderen Maßnahmen des Natur- und Umweltschutzes, Europäische Hochschulschriften, Reihe 42, Ökologie, Umwelt und Landespflege*, Bd. 11, Peter Lang GmbH, Europäischer Verlag der Wissenschaften. Frankfurt.

EFFECTS OF RIVER DIKING ON THE WATER AND SALT REGIMES OF FLOOD PLAIN SOILS

F. STATESCU, A. NICOLAU and V. TOBOLCEA
Hydrotechnical Faculty, Technical University "Gh. Asachi", Iasi, Romania

Abstract
The paper presents the results of research on modifying several characteristics of alluvial soils, due to hydromeliorative works in the investigated area. The evolution of these soils is different from that of the soils found in the typical alluvial soils areas. The development of a superphreatic layer over some impervious horizons, with a high content of soluble salts has been found here, which makes it an essential element in the specific evolution of these soils.
Keywords: flood plains, river diking, soil salinity, water quality, water regime.

1 Introduction

It is well known that technical procedures applied to regulate the hydric regimen of soils determine important modifications of their physical, chemical and biological characteristics. The direction and amplitude of these modifications are, however, very different from one pedoclimatic area to another, depending on numerous factors. For this reason, the research in Romania becomes more complex and more substantiated as soil pollution, under various forms, becomes more and more aggressive.

The results of some experimental research, on the field and in laboratories, on the modifying of very sensitive pedologic indices, generated by human intervention in the hydric regimen of some alluvial soils are being presented in what fallows.

Water and the Environment: Innovative Issues in Irrigation and Drainage. Edited by Luis S. Pereira and John W. Gowing. Published in 1998 by E & FN Spon. ISBN 0 419 23710 0

2 Investigated area

The investigated area (18.386 ha) is situated in the East-North-East of Romania in the Prut river flood plain, with a semidry climate. The land is generally flat, with a low longitudinal slope, ranging between 0,015% and 2.3% and small transversal subsident areas.

The flood plain of Prut river is made up of sediments that can reach 14 m in thickness, laid over sarmatian clays, covered by coarse gravel and sand deposits as well as thinner sand layers towards terraces over which fine surface clays, silty clays and clayey silt, thicker in the central area, have been added. The make up of the fine deposits is not homogeneous, displaying variations in the their granulometry and thickness, both on the horizontal and the vertical.

The average phreatic level varies between 0.8 and 3.3 m being lower in the bank ridge area and ever lower in the terrace area.

The climate is characterized by mean annual rainfalls of 373 mm and an average annual temperature of 10.8 °C. Annually a moisture deficit ranging between 40 and 280 mm is recorded. Even in the rainy years a moisture deficit is recorded during the months of July, August and September.

The alluvial soils in the Prut river meadow display a great diversity of pedologic units on quite narrow areas, with an alternance of different textures and chemical compositions and of buried soils (Fig. 1) situated at different levels up to a 3 m depth. Their physical and chemical properties are determined by the high content in swelling clay (55-73%) and higher amount of soluble salts, as compared to the rest of the profile. The presence of buried soils and transitory alternances, to and from them, which have almost identical physical properties, determines an independent dynamics of moisture and soluble salts from the phreatic acquifer.

3 Hydromeliorative works

The alluvial soils from the investigated area have evolved naturally until 1966 when the works for their banking and drainage started. These hydromeliorative works stopped the salts influx from surface waterflow, with a high mineralization degree (higher than 1 g/l, exceeding 5 g/l in some places) and the influx sediments enriched in salts, removing the effect of leaching due to periodical floodings of the Prut river. Gradually, the percolation hydric regimen has ben replaced by a periodical exudative hydric regimen and it has triggered the secondary soil salination.

Since 1971, these soils have been irrigated both to ensure the soil moisture required by the crop plants growth and to establish a hydric regimen that would lead to soil desalination. However, due to the formation of a superphreatic layer, the secondary salination processes have evolved which justified buried drainage works to eliminate this superphreatic layer.

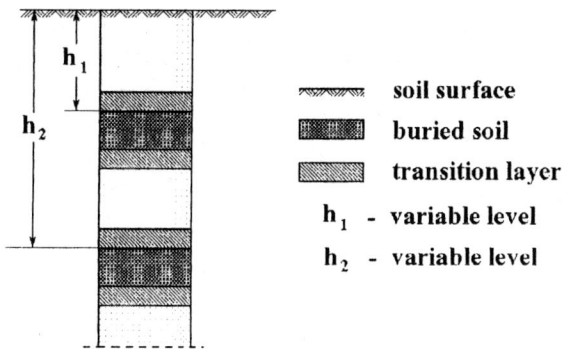

Fig.1 Characteristic profile of aluvial soil of river meadow Prut

4 Analysis and results of regulating hydric regimen of soils

The investigations have been carried out in 184 pedologic sites. Specimens of soil and phreatic water were taken from soil profiles and drillings and were analysed in laboratory by means of current methods [1].

4.1 Salinity modification

The secondary salination mechanism is due to the formation of a superphreatic, highly salinized layer and not to the rising of the phreatic level. This can be seen in Figs. 2 and 3, where the variations of two indices, depth of phreatic level (H) and total soluble salt content (TSSC) are presented for two situation: unirrigated (S1) and irrigated regimen (S2).

The average mineralization of the phreatic layer is higher by 2.636 g/l than that of the superphreatic one, as it result from the data presented in Table 1.

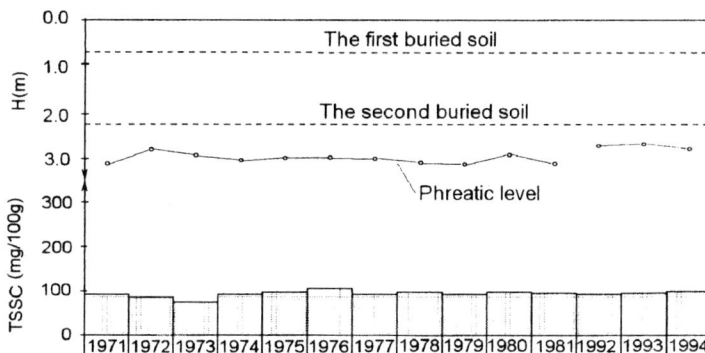

Fig.2 Dynamics of phreatic level (H) and of Total Soluble Salt Content (TSSC) in permanent site S1

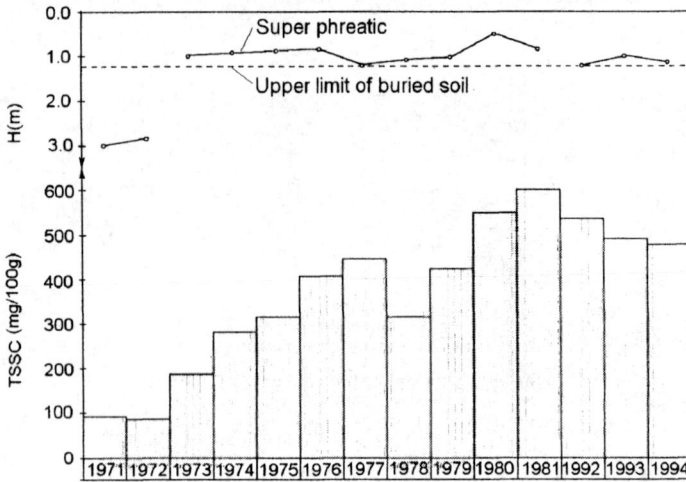

Fig.3   Dynamics phreatic level (H) and of Total Soluble Salt Content (TSSC) in permanent site S2

Table 1. Average composition of water in the superphreatic and phreatic layers, pedologic site S2

Nature of phreatic layer	TSSC (g/l)	Cl⁻ (g/l)	SO_4^{-2} (g/l)	HCO_3^- (g/l)	Ca^{+2} (g/l)	Mg^{+2} (g/l)	Na^+ (g/l)	K^+ (g/l)
Super-phreatic layer	14.681	2.799	4.416	0.967	0.409	0.712	3.010	0.004
Phreatic layer	17.317	4.001	3.414	0.901	0.406	0.798	3.985	0.003

The secondary salination process had set up before applying irrigation, immediately after the banking and drainage works (Table 2).

The mechanism of this process consists in the appearance of a superphreatic water layer above buried soils, during heavy rainfall. This dissolves the soluble salts,which, by capilarity, migrate and settle in the surface layers during drier periods.

The drainage applied on these soils in order to drain the superphreatic layer has evolved in a varied soil salinity, function of the drain location as against the impervious layer on which the superphreatic water stagnates.   Three distinct situations have been found, as one can see in Table 3.

To calculate the evolution rate the following relation was used:

$$Re = \frac{Im - Ipr}{|Ipr - Ii|}$$ (1)

where: Im - momentary value (of soil salinity),
       Ipr - prognosis value,
       Ii - initial value.

Table 2. Modification of TSSC over radicular layer depth

Pedologic	TSSC (mg/100 g sol)		
Site	1968	1969	1970
S1	63	115	276
S2	108	139	239

Table 3. Rate of soil salinity evolution as influenced by drainage

Case	Drainage conditions	Evolution rete (Re)
I	Drain located under a layer of very low permeability with a depth in excess of 45 cm	2.24
II	Drain located under a layer of very low permeability with a depth smaller than 20 cm	0.27
III	Drain located above the layer with a very low permeability	0.15

The evolution rate values, closer to or equal to the unit indicate relatively stable behaviour of the developed soils. The sub-unitary values show a positive process of soil evolution, with the present day indices tending to reach the value of the prognosis index. The super-unitary values show a negative process of soil evolution. We have taken 100 mg/100 g soil as a prognosis value which is the higher limit for the soil unaffected by salination.

4.2 Modification of chemical composition of soil solution

After applying irrigation, the dynamics of some soil solution compounds such as: Cl^-, SO_4^{-2}, HCO_3^-, Ca^{+2}, Mg^{+2}, Na^+ and K^+ has been investigated. The research done during 1973-1981 showed that the type of salination did not change after applying irrigation, being a sulfate salination. The ratio Cl^-/SO_4^{-2} had an average value of 0.42, with small deviations from this value for the whole period of research (Fig. 4).

Much more significant was the evolution of the Ca^{+2}/Na^+ ratio, which witnessed a continuous drop from 5.48 to 0.97. This means that in the soil solution, after irrigation, the quantity of Ca^{+2} ions dropped by aproximately 5.5 times as compared to that of Na^+ ions.

Knowing the Ca^{+2} and Mg^{+2} and the higer toxicity of Mg^{+2} compared to Na^+ and Ca^{+2}, it was interesting to study the evolution of Ca^{+2}/Mg^{+2} ratio, as well. Fig. 5 shows that, under irrigations, the content in Ca^{+2} ions of the soil solution descreased as compared to that in Mg^{+2} ions, the ratio of these two cations descreasing continually from 14,23 to 2,72.

The research showed that the crops in this area absorb higher amounts of Na^+ from the $NaHCO_3$ solutions as compared to the $NaCl$ and Na_2SO_4, the HCO_3^- anion being found in higher amounts that the Cl^- and SO_4^{-2} anions. It was concluded that irrigation did not modify significantly the soil contant in HCO_3^- ions.
Fig.5 Dynamics of Ca^{+2}/Mg^{+2} ratio

4.3 Modifications of hydrophysical characteristics in soil

The measurements for the investigation of the evolution of hydrophysical properties of the soil under the influence of irrigation consisted in determining the soil-water characteristc curves of the unirrigated (witness) soil as well as of the same one after

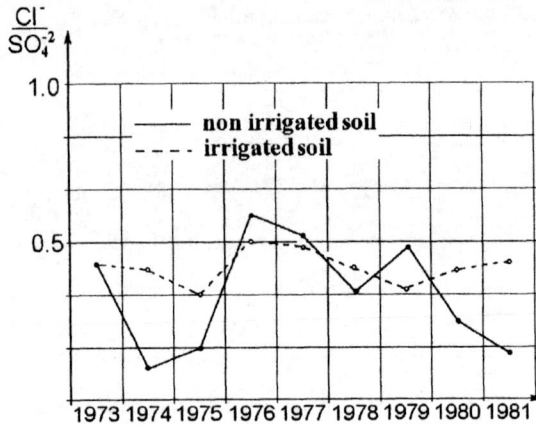

Fig.4 Dynamics of Cl^-/SO_4^{-2} ratio

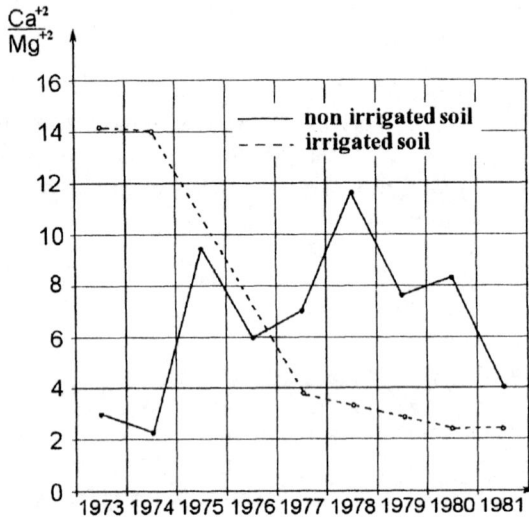

Fig.5 Dynamics of Ca^+/Mg^{+2} ratio

five years of irrigation if the drainage was adequate (Fig.6).

The processing of data grounded in the Brooks and Corey model resulted in a decrease of air-entry (bubbling) pressure in soil by aproximately 2.5 times and an increase of the pore size distribution index ($\lambda=0.176$ for the unirrigated soil and $\lambda=0.230$ for the irrigated soil). Thus the soil aeration improves increasing, at the same time. the range of available water.

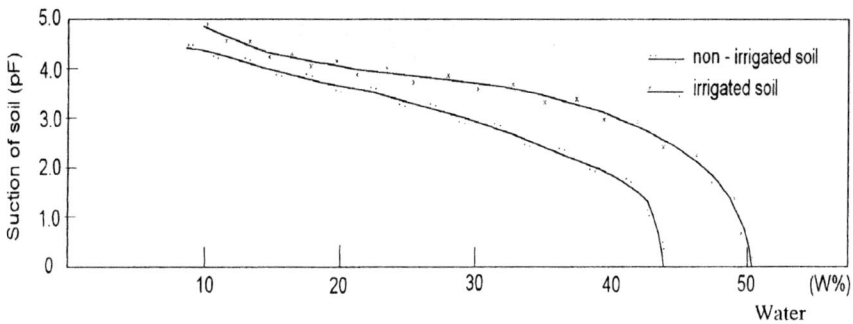

content

Soil type	Granulometric composition (%)			Bulk density (g/cm³)	Total porosity (%)
	Sand	Silt	Clay		
Aluvial soil	34.42	32.03	33.06	1.27	52.07

Fig.6 Soil-water characteristic curve

4.4 Modifications of soil fertility

The experimental technology used comprises the following stage: a soil specimen, positioned under optimum conditions for microorganisms development enables the measurement of the extent to which they can turn to acount the organic and mineral matter, without plants intervening in any way.

The working procedure consists in: weighting of 10 g soil which is introduced into an Erlenmayer vessel; 50 ml NaCl solution is added in physiological concentration and is well stirred; then is kept 48 hours in incubator at 27 °C after which is filtered; the live cells count is done afterwards (on a Thoma plate three average squares are counted). Table 4 shows the result obtained on samples from an unirrigated soil as well as from an irrigated one for 7 years.

Table 4 shows that by irrigation the soil fertility has decreased. This is a consequence of soil salination.

Table 4. Evolution of live cells counts in an experiment with unirrigated and irrigated soil samples, (0-10 cm depth)

Type of Sample	No of live cells/average square			
	Experiment initiation	After 4 weeks	After 3 months	After 1 year
Unirrigated soil	18	22	11	2
Irrigated soil	8	10	4	-

5 Conclusions

The research have shown that the control of water regime in the soil, in the investigated area, has had different efects, as follows:

1. Where drainage was adequate and superphreatic water layer was not formed as a consequence of irrigation, the hydrophysical and chemical properties have evolved positively, leading to an increase in fertility.
2. There, where the soil was not drained or the drainage was not adequate, due to the formation of a superphreatic water layer subsequent to irrigation, the physical and chemical properties have had a negative evolution, leading to a decrease in soil fertility.
3. The evolution of these soils is determined, especially, by the specific profile and the mineralogic composition of clay.
4. During the research period, some results, different from those found in the literature, have been obtained, and they will have to be checked by more thorough experiments.

5. References

1. Intitutul de Cercetari pentru Pedologie si Agrochimie (1986) *Metode de analiza chimica a solului*,Editura Academiei de Stiinte Agricole si Silvice.Bucuresti.

INFLUENCE OF TILLAGE SYSTEM ON WATER REGIME IN IRRIGATED AND RAINFED SUNFLOWER PRODUCTION

G. BASCH, J.P. MENDES, M.J.G.P.R. CARVALHO, F. MARQUES and M.J. SANTOS
Department of Crop Science, University of Évora, Évora, Portugal

Abstract
Recently, sunflower production increased substantially and the crop can be found on many soils and crop rotations normally associated to cereal production. The costs for ploughing, usually performed before the sunflower, are attributed partly to the wheat crop which, however, provides similar yields using no-tillage. Reduced and no-tillage (NT) systems have proved to be feasible alternatives to the traditional tillage methods (TT) based on the mouldboard plough, both regarding economical and environmental aspects. Nevertheless, the results obtained with direct drilled sunflower in the past were less favourable as those observed with other crops, although the water saving through no-tillage is supposed to favour crop growth under a strongly limited water regime. The improvement of the performance of sunflower in relatively wet springs or when early sowing was applied indicates that there exists an interaction between the tillage system and the existing water regime. In order to study this interaction a trial with two tillage systems and 3 levels of water regime was carried out. The results are discussed on the basis of the crop parameters, soil moisture content, root length and distribution, bulk density and penetration resistance, showing that sunflower performance was affected by no-tillage both rainfed and irrigated conditions.
Keywords: Irrigation, leaf area, no-tillage, penetration resistance, root growth, soil moisture, sunflower, traditional tillage.

1 Introduction

Sunflower is one of the few annual spring crops capable to be grown under rainfed conditions in typical Mediterranean environments with very pronounced summer dryness. Remarkable breeding progresses, which increased the productivity of the crop and its adaptation to various ecological conditions guaranteed sunflower its place

Water and the Environment: Innovative Issues in Irrigation and Drainage. Edited by Luis S. Pereira and John W. Gowing. Published in 1998 by E & FN Spon. ISBN 0 419 23710 0

in many crop rotations in Central and South Europe during the last two decades.In the South of Portugal it replaced safflower (*Carthamus tinctorius*, L.) and chickpea (*Cicer arietinum*, L.) as traditionally cultivated spring crop under rainfed conditions and on soils with a comparatively good water storage capacity. During the last few years, sunflower spread to even marginal soil conditions, a fact, which cannot be explained by further breeding advances or low cost production systems, but purely by the attribution of considerable subsidies.

Regardless of the reasons, which make farmers cultivate sunflower, this crop is certainly a reasonable alternative on many soils in Southern Portugal in order to break up the cereal dominated crop rotation, depending, however, on the production system used. Due to the great variability of the precipitation pattern in spring and the deficient water storage capacity of most soils the productivity of the crop is very uncertain and can range from complete failure to over one ton per hectare. This means that production costs have to be very low to make the crop economically feasible and less risky and that water saving production techniques should be used, once water is the main limiting factor.

In this context reduced and especially no-tillage techniques can contribute to both a considerable decrease of production costs, once the costs for soil preparation make up a high percentage of the total production costs of this crop, and the reduction of water loss through evaporation [1] and [2].

However, results obtained in previous experiments indicate that sunflower, especially under dry conditions, performs worse under no-tillage as compared to the traditional tillage system based on the mouldboard plough [3]. In order to understand the reasons for this apparent interaction between the tillage system and the water regime a trial with these two extremes regarding tillage intensity and 3 different water regimes was installed.

2 Material and Methods

The trial was carried out on an experimental farm of the Direção Regional de Agricultura do Alentejo (Herdade da Revilheira) near Reguengos de Monsaraz. The soil used can be classified as a Luvisol with the characteristics given in table 1 and 2.

2.1 Treatments and experimental design
Although the complete trial consisted of 5 tillage treatments only the two extremes – traditional tillage (**TT**) based on the mouldboard plough and disking and no-tillage (**NT**) – were used to study the interaction with the water regime. This treatment was installed as a split-plot on the main plots for the tillage treatment, which was laid out in a complete randomised block design. The three levels for the water regime were the following:

Level 1 – rainfed (without any irrigation)
Level 2 – irrigation until beginning of flowering (based on the potential
            evapotranspiration (pan evaporimeter))
Level 3 – irrigation until the beginning of maturity (maintaining at least two
            hirds of the field capacity)

The size of the main plots for the tillage treatment was 21 x 6 m and 6 x 6 m for the sub-plots.

Table 1. Soil characteristics

Horizon	Depth (cm)	Sand (%) coarse	fine	Silt (%)	Clay (%)
A	0-30	27.2	50.8	9.1	12.9
B	30-70	13.9	37.7	11.6	36.7

Table 2: Pore space characterisation (%)

Tillage system	Total porosity	pF = 1	pF = 1.8	pF = 2.54	pF = 4.2
NT	35.8	34.1	29.3	23.0	14.4
TT	37.4	35.2	28.0	21.4	14.3

2.2 Trial installation and irrigation

Due to the lack of precipitation during spring the seeding of the crop was realised only on April 23rd. The seeding density of the variety "Florasol" was around 5 seeds per m2 with a line spacing of 75 cm.

Irrigation was performed by drip irrigation having the tubes placed close to the crop row and in the middle of the rows in order to obtain a well distributed watering of the crop. The distance between nozzles was 32 cm. The schedule and amount of irrigation depended on the precipitation, which was exceptionally frequent and high in spring of 1997, and on the evapotranspiration determined by the class A pan evaporimeter in the field. Figure 1 shows the irrigation scheme and precipitation before and during the trial period. The evapotranspiration curve, as calculated according Penman-Monteith, shows however, that none of the water regimes satisfied the real evapotranspiration of the crop.

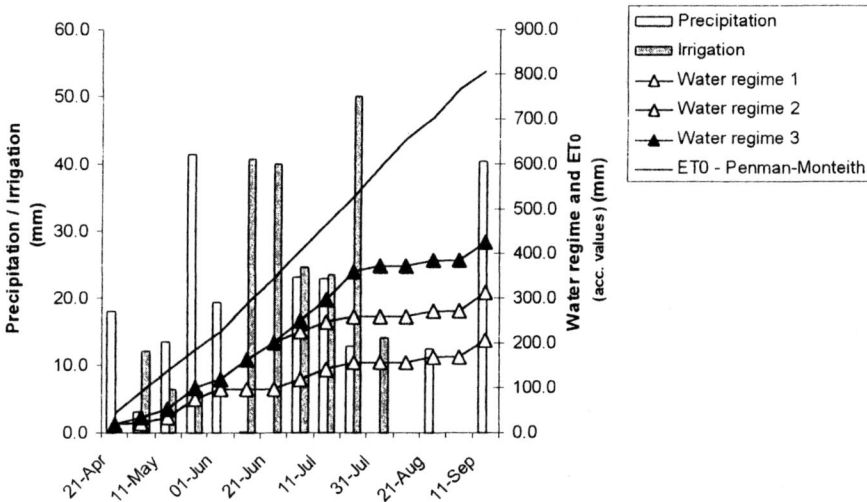

Fig. 1. Irrigation and precipitation before and during the trial period

2.3 Field and laboratory measurements

Soil parameters: During the trial period the soil moisture was measured with a neutron probe having one access tube installed per sub-plot close to a crop row and between two adjacent plants. The readings were done before each irrigation event and transformed to %Vol of H_2O-values per soil layer through a calibration curve established for this field site.

This information was used to calculate directly the water consumption by the crop and indirectly the water stress of the crop through a method proposed by [4] based on the potential evapotranspiration calculated according Penman-Monteith.

As soil tillage interferes strongly with soil physical characteristics and as sunflower is known as a crop with a reduced growth pressure of its roots [5], bulk density and penetration resistance of the soil were determined before and after crop emergence, respectively. Penetration resistance was measured by the penetrograph (Eijkelkamp) using a 1 cm^2 cone with an angle of 60°.

Crop parameters: In the beginning of the vegetation cycle the evolution of dry matter production was determined using the leaf area of the plants, which was found to be closely correlated with the above ground biomass of the crop during this early period ($r > 0.9$).

The leaf area, measured during the whole vegetation cycle, was based on leaf width measurement once a high correlation ($r = 0.977$) was found by [6] for this variety. In addition to the leaf area its duration between two measuring points was calculated according [7].

The production parameter recorded were the following:

- total dry matter yield at harvest $(g \cdot m^{-2})$
- number of seeds per plants
- number of seeds per m^2
- thousand kernel weight (g)
- seed yield $(g \cdot m^{-2})$

Another important crop parameter, which reflects the performance of a crop, especially when tillage systems are involved, is the root system. Although two methods in different periods were used for the monitoring of the root growth, only the results of one period will be presented as the second one is still subject to methodological processing. As the irrigation scheme of level 2 and was still the same at the time of the first root sampling only the results for the root distribution under the water regime 1 and 2 will be presented and discussed. The first monitoring scheme, carried out 2 weeks before flowering, consisted of the sampling of 8 soil cores per 10 cm layer, four of which were taken directly at four plants and four in the middle of the rows next to the plants selected for this purpose. The core diameter and height were 9.8 and 10 cm, respectively. Sampling depth was until 50 cm. After sampling the cores were deep frozen and processed according the capacity of the root washing machine. The washed and cleaned roots were stained with methylene-blue and then submitted to a scanner based detection of root length and diameter. This method based on an improved intersection model allows a fast processing of the samples and gives additional information in relation to the traditional intersection method. A full description will be provided in a separate publication.

3 Results and Discussion

As root elongation of sunflower is very sensitive to soil compaction [5] bulk density and penetration resistance were measured right before and after the emergence of the crop. Table 3 confirms that the lack of tillage leads to a higher bulk density. Astonishing seems the fact that even after soil preparation by ploughing, the bulk density shortly after seeding already reaches values of over 1.5. The differences between tillage treatments are decreasing with depth.

This observation is reflected by the results of the penetrometer resistance measurements (figure 2). Whereas under traditional tillage the penetration resistance increases steadily and considerably with depth no-tillage offers almost the same resistance from the soil surface down to 40 cm, increasing at lower depths at a similar rate as under traditional tillage.

These results may help to explain differences in the performance of the crop and especially in what root growth and distribution is concerned.

Based on the moisture profile, which was calculated for periods of ten days, the water use and the water stress of the crop for each treatment were established (figure 3). Both parameters show a strong effect of the water regime whereas the tillage treatment had a smaller and timely limited impact. Due to the wet spring the first readings during end of March reveal small differences even between water regimes, which increase considerably with the development of the crop and the higher temperatures in early summer. After the end of irrigation in water regime 2, water use and water stress reach the levels of water regime 1 within 20 days. Only with irrigation (water regime 3) high water use rates of above 4 mm·d^{-1} could be maintained and water stress kept at a small level.

Analysing the different periods it seems that no-tillage without irrigation had a reduced water uptake before the flowering stage, which may indicate that root development was somehow affected by this treatment as compared to traditional tillage. On the contrary, in water regimes 2 and 3, there are periods with a higher water use and reduced water stress under no-tillage. This could be the result of either a

Table 3. Bulk density of the soil layer 0-30 cm under traditional and no-tillage

Tillage system	No-tillage			Traditional tillage		
Depth (cm)	0-10	10-20	20-30	0-10	10-20	20-30
Bulk density	1.73	1.72	1.75	1.54	1.63	1.67

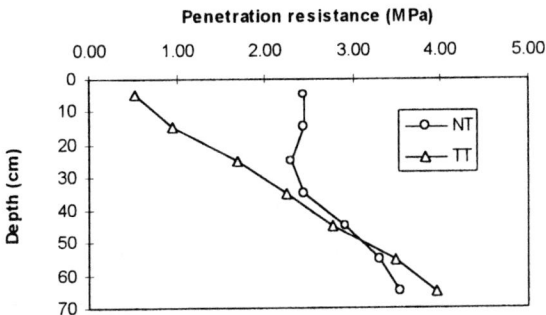

Fig. 2. Penetration resistance under traditional (TT) and no-tillage (NT) after crop emergence

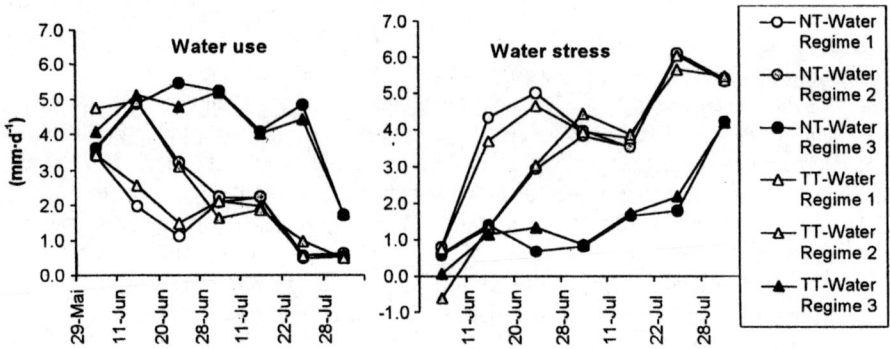

Fig. 3. Daily water use and stress of sunflower under different tillage systems and water regimes

better-developed root system and/or more water available in the soil.

Regarding crop performance figure 4 reveals that already 4 weeks after seeding marked differences in dry matter production existed between tillage treatments. Those persisted almost until head appearance of the crop, being the differences between water regimes until then very small, which is certainly due to the wet spring. Only at flowering the first differences in DM production could be noticed between water regimes. The considerable differences between tillage systems during the early crop stages disappear almost and especially for the two lower water regimes.

This tendency for a recovery of the sunflower under no-tillage in comparison to traditional tillage can be explained and is confirmed by the evolution of the leaf area of the crop (table 4). The big differences between tillage treatments in the beginning of the season reduce steadily until flowering and the leaf area becomes even greater under

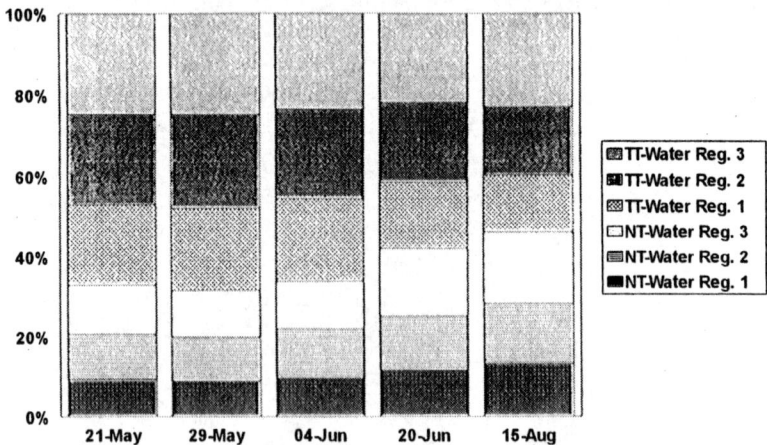

Fig. 4. Relative total dry matter evolution of sunflower under different tillage systems and water regimes

no-tillage at the end of the cropping season. Therefore the leaf area duration calculated for the critical period around flowering and over a considerable period of the crop cycle reveals almost no differences between tillage treatments.

With regard to leaf area, however, no-tillage responds more to irrigation than traditional tillage. A possible explanation for this obvious discrepancy may be the fact that the late recovery in terms of leaf area came too late to be adequately transformed into biomass.

Table 4. Leaf Area Index and Leaf Area Duration of sunflower under different tillage treatments and water regimes

Treatment	Leaf Area Index							Leaf Area Duration	
	21-May	29-May	04-Jun	12-Jun	20-Jun	17-Jul	04-Aug	20/06-17/07	04/06-04/08
No-tillage	0.03	0.15	0.35	0.85	1.35	1.51	0.74	38.7	72.6
Traditional tillage	0.06	0.27	0.59	1.22	1.57	1.37	0.61	39.6	75.8
Water regime 1	0.04	0.19	0.44	0.99	1.26	1.14	0.39	32.4	60.9
Water regime 2	0.05	0.22	0.49	1.07	1.54	1.26	0.47	37.7	69.9
Water regime 3	0.05	0.21	0.48	1.05	1.59	1.92	1.17	47.4	91.8
NT – WR 1	0.03	0.12	0.31	0.76	1.09	1.17	0.41	30.4	56.4
NT – WR 2	0.04	0.17	0.39	0.94	1.48	1.33	0.51	37.9	69.4
NT – WR 3	0.04	0.15	0.36	0.86	1.50	2.03	1.31	47.6	92.0
TT – WR 1	0.05	0.25	0.58	1.22	1.42	1.12	0.36	34.3	65.4
TT – WR 2	0.06	0.27	0.59	1.20	1.60	1.18	0.43	37.5	70.3
TT – WR 3	0.06	0.27	0.59	1.23	1.68	1.81	1.03	47.1	91.6
Mean	0.05	0.21	0.47	1.03	1.46	1.44	0.68	39.2	74.2

Looking below the soil surface the root length and its distribution along the soil profile at the beginning of flowering show a considerable effect both of soil tillage and water regime. As water regime 2 and 3 started to differ from that time on, only regime 1 and 2 were compared for its root development at this crop stage. In both tillage treatments the highest root density could be found in the upper soil layer, being reduced by approximately 5 and only 2 times in the 10-20 cm layer under no-tillage and traditional tillage, respectively. Traditional tillage had significantly higher root densities until the depth of 30 cm, whereas the water regime showed significant differences only for the top layer.

This root density and distribution certainly reflects the differences in crop performance between the tillage treatments during the crop stages until head appearance and helps to explain the final results at harvest (table 5) in a sense, that only small differences between tillage treatments were obtained for water regimes 1 and 2. The lower root density under no-tillage and rainfed conditions might also have recovered in later crop stages.

Although significant, the differences in grain yield between the two tillage systems were rather small in absolute terms. Unexpectedly this is due to the differences in the highest water regime and not to the results obtained under rainfed conditions. Only water regime 3 was able to increase grain yield not only significantly but also in a pronounced total grain yield. This was obtained both through a considerably higher number of seeds per m^2 but also through a high kernel weight. In all water regime treatments the kernel weight was higher under no-tillage, which compensated the significantly lower seed number especially under rainfed conditions. This also

Root density (cm· cm⁻³)

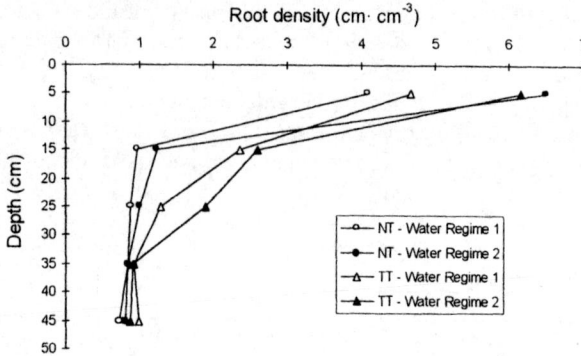

Fig. 5. Root density and distribution along the soil profile

indicates that under no-tillage the crop suffered less stress during the final stages, which was not detectable through the soil moisture measurements but through the leaf area index.

Table 5 Grain yield and production parameters of sunflower under different tillage treatments and water regimes

Treatment	Seeds·m⁻²	1000 kernel weight (g)	Total dry matter (g·m⁻²)	Grain yield (g·m⁻²)
No-tillage	5257 b	41.6	569 b	221 b
Traditional tillage	6417 a	38.5	664 a	252 a
Water regime 1	5002 c	36.5 b	501 c	182 c
Water regime 2	5588 b	37.5 b	589 b	210 b
Water regime 3	6922 a	46.1 a	760 a	317 a
NT – WR 1	4626 c	38.4	479 d	178 d
NT – WR 2	5296 b	38.2	558 cd	204 cd
NT – WR 3	5850 b	48.0	671 b	281 b
TT – WR 1	5379 b	34.6	523 d	187 cd
TT – WR 2	5879 b	36.7	620 bc	216 c
TT – WR 3	7994 a	44.2	850 a	353 a
Mean	5837	40.0	617	236
CV (%)	6.1	4.6	7.1	6.4

4 Conclusions

Due to the relatively wet spring only the highest water regime led to a considerable increase in the grain yield of sunflower. The differences observed between traditional and no-tillage were significant but not very pronounced as it was expected especially under the rainfed conditions. This different behaviour of the sunflower crop when compared to previous results could be the consequence of the wet spring, which allowed the recovery of the crop under NT at the end of the season.

On the other hand, the crop was much more able to take advantage of the highest water regime under traditional tillage. This indicates that there were probably other limiting factors under no-tillage besides the water availability. The more than 2 times higher root density under TT and irrigation in the soil layer between 10 and 30 cm could be an explanation for a better nutrient availability.

The crop parameters studied during the trial indicate that sunflower under no-tillage suffers a considerable delay in its development right after emergence, but is able to recover at the end of the season if the hydrological conditions are favourable as it was the case in this year even without irrigation. However, this delay, which undoubtedly is the result of a deficient initial root development, explains the reduced performance of NT in unfavourable years and stresses the impact of the soil structure for the success of NT in sensitive crops in terms of root pressure and elongation. Thus the economic and agronomic advantages of NT for these crops seem somehow restricted to well structured soils.

5 References

1. Basch, G., Carvalho, M. and Marques, F. (1997) Economical considerations on no-tillage crop production in Portugal, in *Proceedings of EC-Workshop - IV - on experience with the applicability of no-tillage crop production in the West-European countries*, Boingneville, May 12-14, 1997, (ed. F. Tebrügge and A. Böhrnsen), pp. 17-7.
2. Giráldez, J.V., González, P. and Fereres, E. (1986) Aprovechamiento del água del suelo en distintos sistemas de laboreo, cinco anos de experiencia en el Guadalquivir, in *I simpósium sobre mínimo laboreo en cultivos herbáceos*, Madrid 1-2 October, pp. 11-6.
3. Carvalho, M.J.G.P.R., (1997) Unpublished data.
4. Allen, R.G., Smith, M., Pereira, L.S. and Perrier, A. (1994) An update for the calculation of reference evapotranspiration. *ICID Bulletin*, Vol. 43, No. 2, pp. 35-59.
5. Misra, R.K., Dexter, A.R. and Alston, A.M. (1986) Maximum axial and radial growth pressures of plant roots. *Plant and Soil*, Vol. 95, pp. 315-11.
6. Carvalho, M.J.G.P.R., Basch, G., Azevedo, A.L. and Machado, L.C.R.T. (1991-95) Efeitos de datas e densidades de sementeira na cultura do girassol, em solo de barro preto (Bp). *Agronomia Lusitana*, Vol. 45, No. 1-3, pp. 137-21.
7. Roderick, H. (1978) *Plant growth analysis*. Studies in biology 96, The Camelot Press Ldt., Southhampton.

MEASUREMENTS OF SOIL HYDRAULIC PROPERTIES IN AN OLIVE ORCHARD UNDER DRIP IRRIGATION

F. MORENO, J.E. FERNANDEZ, M.J. PALOMO, I.F. GIRON, J.M. VILLAU and A. DIAZ-ESPEJO
Instituto de Recursos Naturales y Agrobiología de Sevilla, Seville, Spain.

Abstract
The increase of process modelling of water balance in irrigation and tillage experiments has imposed a demand of accurate measurements of the hydraulic properties of soils. The objective of this work was to determine in situ the hydraulic conductivity and sorptivity, in the range near saturation, of the soil of an olive orchard with drip irrigation. These provide data to predict the extension of the wetted zone into the soil, and thus to manage drip irrigation in a better way. Experiments were carried out on an olive orchard with drip irrigation. A tension-disc infiltrometer was used to determine in situ the hydraulic conductivity and the sorptivity in the range near saturation. The hydraulic conductivity of the soil surface layer showed relatively high values (0.030 mm s^{-1}) for $\psi > -10$ mm, and decreased with the decrease of ψ. In contrast, the hydraulic conductivity values at the depth of the plough pan layer was lower than those of the surface (0.014 mm s^{-1} for $\psi > -10$ mm). The variability of the hydraulic conductivity and sorptivity of the soil surface layer was relatively low, the coefficients of variation being always lower than 30%. For the plough pan layer the variability was much higher (coefficients of variation were between 40% and 60%).
Keywords: Drip irrigation, hydraulic conductivity, olive trees, soil hydraulic properties, sorptivity

1 Introduction

The increase of process modelling of water and solute transport in the soil has imposed a demand of accurate measurements of hydraulic properties of soils. For given climatic conditions and soil type, tillage methods and irrigation practices are the main factors

Water and the Environment: Innovative Issues in Irrigation and Drainage. Edited by Luis S. Pereira and John W. Gowing. Published in 1998 by E & FN Spon. ISBN 0 419 23710 0

that can alter the soil structure of top layers and consequently the hydraulic properties, [1], [2], [3]. For cultivated soils, the transport properties of the soil surface can change during the growing season, [4]. However, in modelling solute transport through unsaturated soils, it is usually assumed that the characteristics of the soil remain temporally without changes.

An important field characteristic of drip irrigation is the size of free-water ponds under drippers, [5]. Hydraulic properties of the soil and the drip discharge can be used to predict the size of free-water pond under drippers. This can help to define appropriate irrigation strategies when drip irrigation is used.

The objective of this work was to determine in situ the hydraulic conductivity and sorptivity, in the range near saturation, of the soil of an olive orchard with drip irrigation.

2 Materials and methods

Experiments were conducted in an olive orchard at the experimental farm of the Instituto de Recursos Naturales y Agrobiología (IRNAS-CSIC), which is located at Coria del Río close to the city of Seville in SW Spain (37° 17' N, 6° 3' W, elevation 30 m). A 0.5 ha plot containing 28-year-old olive trees (*Olea europaea* L., var. *Manzanillo*) that are planted at a spacing of 7 x 5 m was selected for the experiments. The soil of the plot is a sandy loam (Xerochrept) over 1.5 m deep, with clay, silt, fine sand and coarse sand percentages of 14.8, 7.0, 4.7 and 73.5 respectively. Olive trees are drip irrigated using a single pipe placed on the soil surface in each tree row, with five 3 L h-1 emitters 1 m apart per tree. Irrigation was applied during the dry season of 1997, from middle of March to the beginning of October.

The tension disc infiltrometer [6] was used to determine in situ the hydraulic conductivity and sorptivity in the range near saturation. Experiments were carried out with a disc infiltrometer of 125 mm radius. Several sites were chosen within the plot to measure the infiltration at several pressure potentials. The pressure potentials (ψ_0) chosen varied from -100 to 0 mm. Infiltration tests were carried out on the undisturbed soil surface of the selected sites and the depth of the plough pan. The hydraulic conductivity, $K_0 = K (\psi_0)$, and the sorptivity, $S_0 = S(\psi_0)$, were obtained using the mono-disc multiple-head method described by Ankeny et al., [1]. This method is based on Wooding's [7] equation for the steady-state asymptotic flux.

From the measurements with the tension disc infiltrometer, a soil structure index can be described by the frame-weighted mean pore size, [8]:

$$\lambda_m = \frac{\sigma}{\rho g} \frac{(\theta_0 - \theta_n)K_0}{bS_0^2} \tag{1}$$

where σ is the surface tension of water, ρ is the density of water, g the acceleration due to gravity, θ_n is the initial volumetric water content, and θ_0 the volumetric water content at the imposed water pressure potential ψ_0. Usually the parameter b for a field soil can be taken to be 0.55, [8], [9]. This length scale defines a mean characteristic width of the pores that are hydraulically functioning at the imposed water pressure potential ψ_0.

3 Results and Discussion

The hydraulic conductivity for the soil surface and for the soil at the depth of the plough pan changes very little across the range $-120 < \psi_0 < 0$ mm (Fig. 1). The difference in K_0 values between the soil surface and the depth of the plough pan is due to the consolidation of the soil at that depth. This in agreement with higher bulk density (1.49 Mg m^{-3}) in the plough pan than in the soil surface (1.31 Mg m^{-3}). The hydraulic conductivity was significantly higher in the soil surface than in the plough pan at the pressure potentials > -50 mm. At $\psi_0 = -120$ mm, the hydraulic conductivity was also higher in the soil surface than in the plough pan but not significantly different.

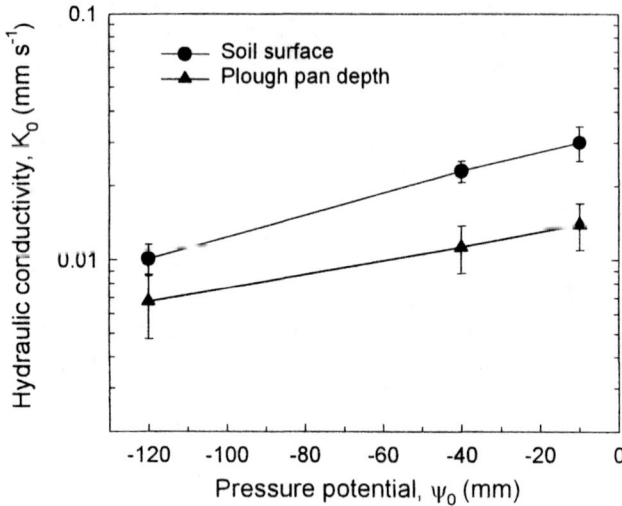

Fig. 1. Variation with the imposed water pressure potential of the hydraulic conductivity (vertical bars are standard errors).

Sorptivity values are shown in Fig. 2. As in the case of hydraulic conductivity, sorptivity decreases with the decrease of the pressure potential, and shows similar pattern in both soil layers. The sorptivity values in the range $-120 < \psi_0 < 0$ mm were significantly lower in the plough pan than in the soil surface. This was due to a higher initial water content in the plough pan than in the soil surface.

The variability of hydraulic conductivity and sorptivity of the soil surface was relatively low (coefficients of variation were lower than 30%). In contrast, the variability of these parameters was higher in the plough pan layer (coefficients of variation ranged between 40% and 60%).

Fig. 3 shows the change of the characteristic mean pore radius, λ_m, with the pressure potential imposed in the range near saturation. On the basis of the Student's *t*-test to both the mean and standard deviations at the 95% confidence interval, λ_m is practically the same in the range $-120 < \psi_0 < 0$ mm for the soil surface. In contrast, at the depth of the plough pan an slight increase of λ_m was observed between $\psi_0 = -10$ mm and $\psi_0 = -120$ mm. The characteristic mean pore radius was higher at the soil surface than at the

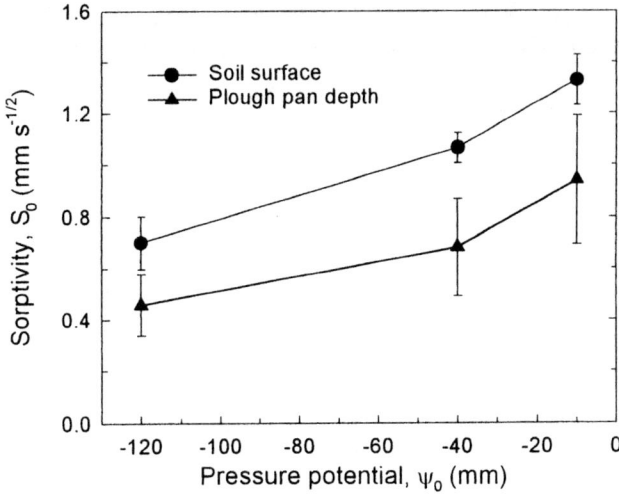

Fig. 2. Variation with the imposed water pressure potential of the sorptivity (vertical bars are standard errors)

plough pan, but only significantly different for ψ_0 = -40 mm. The smaller hydraulic conductivity, in both soil surface and plough pan, for ψ_0 = -40 and -120 mm is in apparent contradiction to the increase of λ_m, and suggests a lack of interconnected pores as reported by Angulo-Jaramillo et al., [4]. For ψ_0 > -40 mm, the reduction of λ_m does not follow a decrease of sorptivity. This apparent discrepancy could possibly be explained both by some fraction of the smaller pores (Beven and Germann, 1982) becoming hydraulically isolated and by the deposition of eroded small easily transportable particles.

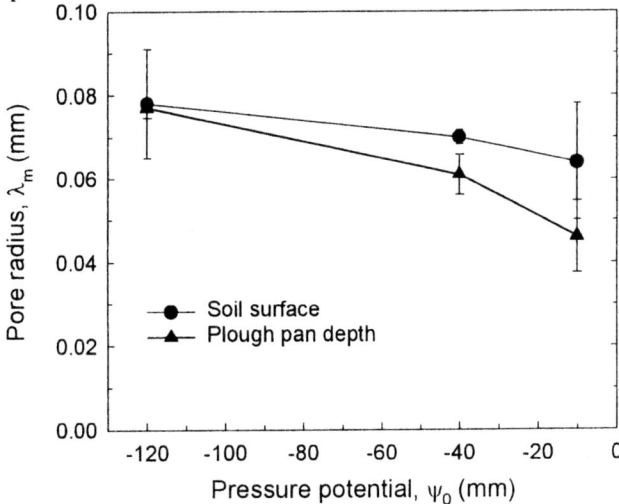

Fig. 3. Variation with the imposed water pressure potential of the characteristic mean pore radius (vertical bars are standard errors)

According with Philip, [10], during one-dimensional infiltration the time (t_{grav}) from which the gravity is the dominant factor that controls the infiltration is given by:

$$t_{grav} = \left(\frac{S_0}{K_0}\right)^2 \qquad (2)$$

where S_0 and K_0 are the sorptivity and the hydraulic conductivity, respectively. From the measurements carried out in this work we have calculated the gravity times that are shown in Table 1.

Table 1 Gravity time (t_{grav}) at different pressure potentials (ψ_0)

ψ_0, (mm)	Soil surface t_{grav} (min)	Plough pan t_{grav} (min)
-10	40.4a	69.1a
-40	36.8a	55.7a
-120	83.9a	77.0a

Values in each line followed by the same letter are not significantly different (P<0.05)

These results clearly show that the infiltration in this soil is controlled mainly by the gravity.

These basic soil hydraulic properties obtained in this work and the drip discharge rate can be used to predict the size of free-water pond under drippers as reported by other authors, [11]. This can help in improving the management of drip irrigation.

4 Conclusions

The tension disc infiltrometer is an easy and practical tool for hydraulic characterisation of soils in the range near saturation under field conditions. Both the tension disc infiltrometer and the approach of Ankeny et al., [1], used in this work allow a high number of replications over the field with low time and labour requirements. Results that are obtained with this device provide with a high information content on soil hydraulic properties.

The basic soil parameters obtained in this work are very useful to predict the size of free-water pond under drippers in our olive orchard under drip irrigation. This can help to define the appropriate irrigation strategies.

5 Acknowledgements

The authors wish to thank Mr. J.P. Calero for help with measurements in the field. This study was supported with funds of the Spanish CICYT, project HID96-1342-CO4-01, and the Junta de Andalucía (Research Group AGR-151).

6 References

1. Ankeny, M.D., Ahmed, M., Kaspar, T.C. and Horton, R. (1991) Simple field method determining unsaturated hydraulic conductivity. *Soil Science Society of America Journal*, Vol. 55. Pp. 467-470.
2. Messing, I. and Jarvis, N. (1993) Temporal variation in the hydraulic conductivity of a tilled clay soil as measured by tension infiltrometers. *Journal of the Soil Science*, Vol. 44. pp. 11-24.
3. Somaratne, N.M. and Smettem, K.R.J. (1993) Effect of cultivation and raindrop impact on the surface hydraulic properties of an Alfisol under wheat. *Soil and Tillage Research*, Vol. 26. pp.115-125.
4. Moreno, F., Pelegrín, F., Fernández, J.E. and Murillo, J.M. (1997) Soil physical properties, water depletion and crop development under traditional and conservation tillage in southern Spain. *Soil and Tillage Research*, Vol. 41. pp. 25-42.
5. Angulo-Jaramillo, R., Moreno, F., Clothier, B.E., Thony, J.L., Vachaud, G., Fernández-Boy, E. and Cayuela, J.A. (1997) Seasonal variation of hydraulic properties of soils measured using a tension disk infiltrometer. *Soil Science Society of America Journal*, Vol. 61. pp. 27-32.
6. Revol, P., Clothier, B.E., Vachaud, G. and Thony, J.L. (1991) Predicting the field characteristics of drip irrigation. *Soil Technology*, Vol. 4. pp. 125-134.
7. Perroux, K.M. and White, I. (1988) Designs for disk permeameter. *Soil Science Society of America Journal*, Vol. 52. pp. 1205-1215.
8. Wooding, R.A. (1968) Steady infiltration from a shallow circular pond. *Water Resource research*, Vol. 4. pp. 1259-1273.
9. White, I and Sully, M.J. (1987) Macroscopic and microscopic capillary length and times scales from field infiltration. *Water Resource Research*, Vol. 23. pp. 1514-1522.
10. Warrick, A.W. and Broadbridge, P. (1992) Sorptivity and macroscopic capillary length relationships. Water Resource Research, Vol. 28. pp. 427-431.
11. Philip, J.R. (1969) Theory of infiltration. *Advances in Hydroscience*, Vol. 5. pp. 215-296.
12. Revol, P., Clothier, B.E., Kosuth, P. and Vachaud, G. (1996) The free-water pond under a trickle source: a field test of existing theories. *Irrigation Sciences*, Vol. 16. pp. 169-173.

EFFECT OF MULCHING WITH BLACK POLYETHYLENE SHEETS ON SWEET PEPPER EVAPOTRANSPIRATION LOSSES

G. GHINASSI and L. NERI
Department of Agricultural and Forestry Engineering, University of Florence, Italy

Abstract
The influence of black polyethylene mulching on the evapotranspiration losses of irrigated sweet pepper (*Capsicum annuum* L.) was studied in an inland area of central Italy during the growing seasons 1996 and 1997. The influence on the agronomic length of the cycle, the crop yield and quality was also considered, comparing the results of plots with and without mulching. Reductions in water losses of 92 and 84 mm (28.3 and 24.3%) were measured for the mulched treatment in the first and the second year respectively. The total amount of such reductions, as well as the distribution during the growing cycle, has been observed to be connected with the seasonal evolution of the climate and the related crop response. The influence of mulching on the evapotranspiration rate is greater in the early vegetative period, when the crop cover is negligible. The 68% (63 mm) and the 49% (42 mm) of the total reduction of the irrigation water has been measured from transplanting to 10% of soil cover in 1996 and 1997 respectively. Such distributions are due to the different climatic conditions under which the crop developed. Fully satisfactory productions have been obtained in both years, without significant differences between treatments. A 10 days advance in harvest time was allowed by mulching in 1997.
Keywords: Crop response, evapotranspiration, irrigation agronomy, microlysimeters, mulching, black polyethylene, rainfall, water saving.

1 Introduction

Research related to water saving is probably the most important item for agriculture in many Countries. The general aim of any irrigation management is the efficient use of

Water and the Environment: Innovative Issues in Irrigation and Drainage. Edited by Luis S. Pereira and John W. Gowing. Published in 1998 by E & FN Spon. ISBN 0 419 23710 0

the water, both natural and applied, taking into consideration the increasing scarcity of the water resource associated with the intersectorial competition for its use.

In the mediterranean environment evaporation from the soil can be the major sourceof water losses. Several authors estimated such losses ranging from 25% to 66% of the total evapotranspiration [1]. When soil surface is wet, evaporation occurrs whether a crop is present or not. For a bare soil in optimal water conditions, losses due to direct evaporation can roughly be quantified nearly 40% of potential evapotranspiration [2].

Some agronomic practices can allow substantial reductions in soil evaporation losses; among them, mulching can be an effective tool in water saving strategies [3, 4, 5], permitting also the efficient use of the natural water supplies [6] as well as the control of weeds in the competition for water and nutrients [7]. In spite of this, few quantitative results of field experiences are available about this peculiar characteristic of mulching.

2 Materials and methods

A research financed by the Regional Agency for Development and Innovation in Agriculture (A.R.S.I.A.) has been carried out in the growing seasons 1996 and 1997. The aim was to evaluate quantitatively the influence of black polyethylene mulching on the evapotranspiration losses of irrigated sweet pepper, taking also into account the related influence on the qualitative aspects of the marketable yield [8] and the lenght of the growing cycle [9]. The trials were carried out at the Regional farm of Cesa in Valdichiana (Arezzo), placed in a fertile plain of the inland Tuscany. The physical and chemical properties of the soil at the experimental site are reported in table 1.

Table 1. Soil physical and chemical properties at the experimental site (10-30 cm depth)

Soil property	Value	Method
Clay (%)	35.0	Hydrometer
Sand (%)	45.0	Hydrometer
Silt (%)	20.0	Hydrometer
pH	8.3	Water 1:2.5
Organic carbon (%)	1.04	Walkey-Black
Total N (%)	0.08	Kjeldahl
C/N ratio	13.05	
Available P (ppm)	51.0	Olsen
Available K (ppm)	196.93	CH_3COONH_4
Available Na (meq $100g^{-1}$)	0.299	CH_3COONH_4
Active limestone (%)	0.5	Droineau-Gehu
C.E.C. (meq $100g^{-1}$)	24	$BaCl_2$+TEA

Capsicum annuum L., cv. *Heldor*, was transplanted in coupled rows 150 cm spaced, on the 10th of june in 1996, and the 29th of May in 1997. The transplanting layout was 50 cm between the rows, and 35 cm along the row, with a resulting density of 38,090 plants ha^{-1}.

The sheets of black polyethylene, 120 cm large and 0.05 mm thick, covered the 80% of the soil surface.

The water consumptions of the crop, corresponding to the maximum evapotranspiration rates, were measured by means of microlysimeters.

2.1 Microlysimeters

A microlysimeter is a small and regular (parallelepiped shape) hole, dug in the field, lined with an impermeable sheet of synthetic rubber, filled up with the same soil, and regularly cultivated among the open field crop (figure 1). A floating sensor connected with a valve allows to maintain the water table inside the microlysimeter at a constant level (figure 2), replacing the evapotraspiration losses of the crop [10].

Fig. 1. Microlysimeters in 1996

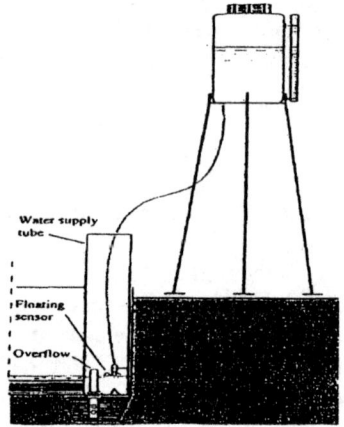

Fig. 2. Water replacement system

Microlysimeter size depends on the distinctive features of the crop. For sweet pepper, two microlysimeters 1.5 m width, 1.2 m length, and 0.8 m depth were prepared for each treatment (mulched and not mulched). Plants cultivated in the microlysimeters can satisfy the maximum evapotranspirative demand (ET_M). Therefore, without significant* natural rainfalls, the total depth of water supplied in a certain period is equal to the maximum evapotranspiration losses of the crop during the same period, that is the application rate (I) for the plots where $I=ET_M$. For each treatment, the microlysimeters were cultivated according to the agronomical practices (weeds, pests and diseases control) currently adopted in the zone. The average total amount of applied fertilizers is reported in table 2.

2.2 Data collection

Water consumptions of the microlysimeters have been measured in different ways during the two years of research. In 1996, the replacement of the crop water

---

* The water balance of a microlysimeter can roughly be written as $ET_M=I+R$, were I=irrigation, and R=rainfall, both expressed in mm or mm/d. The equation can be satisfactory for a range of R -mainly depending on the management of the microlysimeter (depth of the water table)- whose lower value is enough great with respect to the fraction of the total rainfall depth not affecting ET_M, and the upper limit is quite little to be completely stored inside the microlysimeter without overflowing.

Table 2. Average total amount of applied fertilizers

Element	Amount (kg ha^{-1})
N	200
P_2O_5	110
K_2O	200
MgO	20

consumptions was provided with the mechanical system illustrated in figure 2. Measurements were done every 2÷5 days, according to the crop evapotranspiration rate. In 1997, water supply was automatized using a PC-programme for a daily replacement of the evapotranspiration losses. However, in order to comply with the data measured during the previous growing season, the average values related to periods ranging from 2 to 5 days have been considered.

Sweet pepper harvesting normally requires a number of interventions depending on the characteristics of the cultivar, the prevailing use of the marketable product ("green" or "physiological" maturation), and the lenght of the growing season, which in turn is mainly affected by the evolution of the late-season temperatures. In the mediterranean environment the number of harvest interventions can vary from 5 to 15 [11].

In 1996, three interventions only were effected for the crop harvesting (1-23/10 and 26/11), because of the rainy weather from the end of august to the end of november. In 1997, the harvest time developed under favourable climatic conditions, allowing five interventions from the end of august to the end of october.

Whereas the measurements of the ET losses could only be carried out on the microlysimeters, the evaluation of the crop yield could significantly be observed on the plots where I=ET$_M$.

3 Results

3.1 Water consumptions

A significant reduction of the total evapotranspiration losses has been allowed by the black polyethylene mulching during the two years of research activity, as reported in table 3. The distribution of such reductions during the growing seasons has been greatly affected by the different evolution of the climatic conditions, illustrated in table 4. Since the irrigation management was based on the measured water consumptions, the irrigation season could begin only when the optimal hydrological conditions inside the microlysimeters took place.

In 1996 the measurements regularly started toward the middle of june, and extended until the beginning of september, when a prolonged rainy period obliged to end the irrigation water supply.

Table 3. Measured total water consumptions during the irrigation seasons 1996-1997

Year	Total consumption (mm)		Water saving				
	Mulched	Not mulched	Total		Until 10% soil cover	
			mm	%	mm	%
1996	232	324	92	28.4	63	68
1997	268	354	86	24.3	42	49

Table 4. Meteorological data at the experimental site during the growing seasons 1996-1997

Month	Decade	Temperature (°C)				Rainfall (mm)	
		Max		Min			
		96	97	96	97	96	97
May	1st	20.7	20.8	7.6	6.2	37.0	31.0
	2nd	22.3	27.6	8.6	9.6	27.0	0.4
	3rd	24.8	25.3	7.7	8.5	0.6	16.4
June	1st	29.4	24.6	12.9	12.2	4.2	157.0
	2nd	29.6	29.2	12.0	13.8	13.6	19.4
	3rd	25.7	27.3	10.3	12.6	9.4	3.6
July	1st	28.4	28.1	11.7	11.2	12.6	6.0
	2nd	29.9	30.1	12.8	13.0	4.2	15.8
	3rd	30.8	31.5	13.4	14.4	1.0	3.8
August	1st	32.3	31.0	13.6	15.3	0.8	1.4
	2nd	29.1	30.1	13.3	14.6	60.4	10.0
	3rd	27.5	28.4	13.2	13.7	52.6	29.0
Sept.	1st	23.1	31.5	9.6	14.0	27.6	0.0
	2nd	22.0	25.8	9.1	11.8	34.8	27.6
	3rd	21.0	24.9	10.0	9.8	44.4	0.0
Oct.	1st	19.3	25.7	10.4	11.4	22.8	2.4
	2nd	18.3	20.7	8.5	5.7	28.4	8.0
	3rd	17.7	13.8	3.8	6.2	1.6	32.4
Nov.	1st	17.2	16.6	4.5	6.9	1.4	70.0
	2nd	16.9	12.0	9.7	2.5	84.8	22.8
	3rd	9.0	11.2	0.9	6.1	46.0	136.4
Average/total		23.6	24.6	9.7	10.5	515.2	593.4

From transplanting to ~10% ground cover (~40 days) evaporation from the soil has been the major source of water loss for the not mulched treatment. The evolution of the evapotranspiration rates during the first irrigation season is illustrated in figure 3.

As reported in table 3, the greatest influence of mulching on water savings occurred in this period, decreasing as the crop soil cover went on and the relative weight of

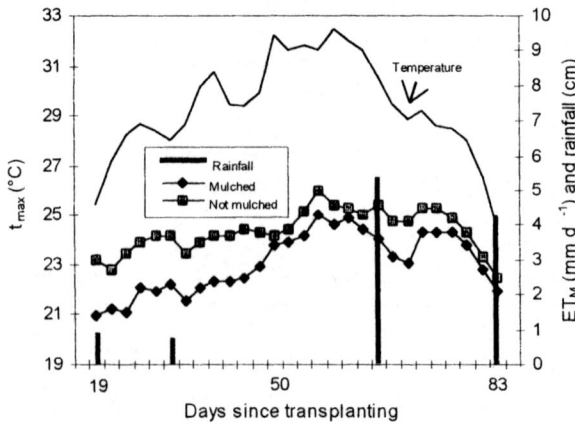

Fig. 3. Sweet pepper evapotranspiration losses during the irrigation season 1996.

transpiration on the total water losses increased.

With regard to the physiological development of the plants, no differences have been observed between mulched and not mulched treatment. In the early vegetative growing period, under the climatic conditions reported in table 4, average evapotranspiration rates of 1.7 vs. 3.7 mm d^{-1} have been measured for mulched and not mulched microlysymeters respectively. During the following period of higher evapotranspirative demand (40÷70 days after transplanting), the same average daily use rates increased up to 2.7 and 3.9 mm d^{-1}, with maximum values of 5 and 6 mm d^{-1}.

High precipitation depths occurred immediately after the planting in 1997 (table 4), allowing field capacity conditions on the microlysimeter topsoils. Such situation delayed the beginning of the irrigation season for the mulched treatment, on account of the favourable influence of mulching on the conservation of stored rainwater. A better vegetative response of mulched crop has been also observed until the middle-late growing season, probably due to the different evolution of the initial hydrological conditions. The greater early vigor can explain the behaviour of the seasonal evapotranspiration rates illustrated in figure 4, and the differences in comparison with figure 3. A swift decrease in crop water supplies took place from the last decade of august to the first days of september (80÷95 days), on account of the combined effect of temperature drop and 28 mm rainfall, as showed in figure 4.

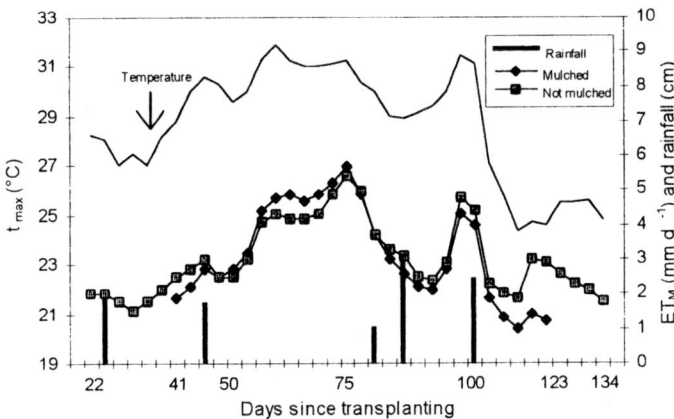

Fig. 4. Sweet pepper evapotranspiration losses during the irrigation season 1997.

The favourable climate in summer and mid-autumn allowed the not mulched crop to recover the vegetative vigor and extend the growing season until the last decade of october. Figures 3 and 4 are both made by using weighted 3-terms movable averages.

3.2 Yield response

In 1996 crop yield was heavily affected by frost, which twice injured the fruits just few days before the last harvest intervention. The effect of frost was lighter on the mulched crop, on account of the effect of the black polyethylene on the radiation flux and the thermic exchange between soil and atmosphere, as well as the better heat conduction in the wetted soil [7]. No differences among treatments have been observed in the cycle lenght, as well as in the quality of the marketable yield. The evolution of the unusual management of crop harvesting, due to the forementioned adverse climatic conditions, is illustrated in table 5.

Table 5. Yield response in 1996. Few days before the last harvest intervention, fruits have been damaged by frost, especially in the plot without mulching

Date	Mulched		Not mulched	
	q/ha	%	q/ha	%
01/10	59	24.6	76	32.6
23/10	100	41.7	104	44.6
26/11	81	33.7	53	22.8
Total	240	100.0	233	100.0

In 1997 crop harvesting was carried out according to the usual schedule of the area. An advance in the yield formation was observed for the mulched treatment. Such advance, which can be quantified in 8÷10 days, allowed the yield pattern reported in table 6.

Table 6. Yield response in 1997

Date	Mulched		Not mulched	
	q/ha	%	q/ha	%
28/08	87	23.0	39	11.8
09/09	165	43.5	96	29.1
22/09	69	18.2	68	20.6
09/10	34	9.0	84	25.5
28/10	24	6.3	43	13.0
Total	379	100.0	330	100.0

Not significant differences among treatments have been observed with regard to the yield and the characteristics of the marketable product. The crop performances can be considered in full accordance with the usual response of the crop in the zone.

4 Discussion and conclusions

The quantitative analysis carried out during the two years of research clearly shows that mulching can allow water savings in different and/or combined ways: directly reducing the evapotranspiration losses of the crop, and indirectly affecting the length of the crop cycle and improving the conservation of the stored soil water. In both 1996 and 1997, the effect of mulching has been greater in the first vegetative period. The total amount of saved water, 92 and 86 mm, has been quite similar in absolute values but very different with regard to the seasonal evolution, depending on the climatic response during the growing cycle of the crop. In 1996 the total water saving developed mainly as direct reduction of the water losses from the soil-plant system, decreasing from transplanting to the maximum soil cover. The influence on the rainfall storage in the soil is not evident, and could be limited to the single event occurred 63 days after transplanting. In 1997 the reduction of the total irrigation water requirement was achieved according to a combined way, as illustrated in figure 4. A better conservation of the rainwater stored in the soil was due to the effect of mulching on the reduction of the evaporation from the soil. This situation allowed a delay in the beginning of the irrigation season with respect to the not mulched treatment. According to the better hydrological conditions of the soil in the early vegetative period, the mulched treatment gained an advance in the crop development which brought the length of the irrigation season into a reduction of about 10 days.

On the ground of the positive hydrological results, and the fully satisfactory yield response of the crop, it can be stated that black polyethylene mulching can be an effective tool in water saving strategies under different climatic conditions.

5 Acknowledgements

The Authors wish to thank prof. Mario Falciai and prof. Antonio Giacomin for the helpful comments in reviewing the manuscript.

6 References

1. Turner N.C. (1997) *Further progress in crop water relations*, Advances in Agronomy, vol. 58, pp. 293-338, Academic Press.
2. Caliandro A. (1979) *Consumi idrici delle colture e fattori climatici*, Agricoltura Ricerca, pp. 3-26, anno II-n. 6.
3. Jones T.L., Jones U.S., Ezell D.O. (1977) *Effect of nitrogen and plastic mulch on properties of troup loamy sand soil and on yield of 'Walter' tomatoes*, J. Amer. Soc. Hort. Sci. 102(3), pp. 273-275.
4. Doorenbos J., Pruitt W.O. (1984) *Guidelines for predicting crop water requirements"*, Irrigation and Drainage Paper, n. 24, FAO-Rome, Italy.
5. Bhella H.S. (1988) *Tomato response to trickle irrigation and black polyethylene mulch"*, J. Amer. Soc. Hort. Sci. 113(4), pp. 543-546.
6. Caliandro A., Catalano M. (1991) *Principi di aridocoltura*, Rivista di Agronomia, anno XXV, n. 3, pp. 372-386.
7. Landi R. (1997) *L'agronomia e l'ambiente*, pp. 315-316, Firenze.
8. De Pascale S., Barbieri G. (1992) *Effetti della distanza tra linee erogatrici e della pacciamatura sulla produzione di pomodoro da industria irrigato con manichette forate*, Irrigazione e drenaggio, n. 2, pp. 23-30.
9. Silvestri G.P., Siviero P., Marasi V. (1990) *La pacciamatura per ampliare il periodo delle consegne di pomodoro alla rasformazione*, L'informatore agrario, n. 1, pp. 35-39.
10. Pardossi A., Bertolacci M., Gemignani S. (1993) *L'uso di un microlisimetro per l'automazione dell'irrigazione del melone in coltura protetta"*, Irrigazione e drenaggio, n. 1, pp. 14-19.
11. Pimpini F., Chillemi G. (1994) *Il peperone: aspetti biologici e tecnico-colturali*, Il Ponte del C.I.A.G., n. 5, pp.60-70.

SECTION V

COPING WITH WATER SCARCITY AND DROUGHT

REDUCED DEMAND IRRIGATION SCHEDULING UNDER CONSTRAINT OF THE IRRIGATION METHOD

R.M. FERNANDO and L.S. PEREIRA
Department of Agricultural Engineering, Institute of Agronomy, Technical University of Lisbon, Portugal
Y. LIU, Y.N. LI and L.G. CAI
China Institute of Water Resources and Hydro-electric Power Research, Beijing, P.R. China

Abstract
Approaches currently adopted to develop a water saving irrigation management program in the North China Plain are presented. They include the installation of an irrigation scheduling model, the improvement of the surface irrigation practices, and the establishment of an irrigation calendar which responds to both irrigation scheduling and surface irrigation constraints. The application to the Xiongxian area, Hebei Province, P. R. China, shows that when the basin irrigation practices could be improved, the seasonal irrigation depth could be greatly decreased.
Keywords: Irrigation scheduling, surface irrigation, water saving.

1 Introduction

In the North China Plain, winter wheat and summer maize are planted successively in one year for more efficiently using the limited land. Improved irrigation management and water savings are required to face water shortage due to uneven distribution of the annual rainfall and the competition for water by non-agricultural users, and to control the groundwater depletion. This is particularly important for low elevation lands and coastal areas due to sea water intrusion. Improvements include rational irrigation scheduling and betterments in field irrigation practices which could lead to more efficient water use, and provide for decreasing field water losses.

Computer models have been widely utilized for irrigation scheduling because they easy allow to develop and evaluate alternative strategies [1]. The model ISAREG [2] was selected to evaluate and support improved irrigation scheduling programmes because it is simple and accurate enough. The model performs the soil water balance with variable time scales and simulates alternative irrigation scheduling practices for a

Water and the Environment: Innovative Issues in Irrigation and Drainage. Edited by Luis S. Pereira and John W. Gowing. Published in 1998 by E & FN Spon. ISBN 0 419 23710 0

given soil-crop-climate system. The model has shown capabilities to select the most appropriate irrigation schedules under limited water supply and drought [3]. The validation of the ISAREG model for the North China plain has been performed using data from irrigation experiments in Wangdu and in the Xiongxian pilot area , Hebei Province. The exploration of the model to compare several irrigation scheduling strategies for the wheat-maize cropping sequence is presented. Irrigation depths utilised in the model result from field evaluations of surface irrigation and further simulations with the surface irrigation simulation model SRFR.

2 Models calibration and validation

The model ISAREG was first calibrated with data from experimental plots of winter wheat and maize at Wangdu, Hebei Province. Data refers to several irrigation treatments, ranging from severely to non stressed, and were collected independently of the model validation purposes. The calibration consisted of selecting the more appropriate crop coefficients (K_c) and soil water depletion fractions for non-water-stress (F_{ns}). The procedure is described in detail in [4] [5]. The water balance was performed with the reference evapotranspiration (ET_0) computed from the FAO Penman-Monteith equation [6].

The version of ISAREG utilized in China is modified from the original one to solve two peculiar situations: the long soil freezing period, from December to March, and the sudden rise of the watertable to near the soil surface when heavy summer monsoon rains occur.

In order to validate the irrigation scheduling model, several irrigation experiments of winter wheat and summer maize were performed in the Xiongxian Experimental Station Hebei Province, from 1994 to 1996. Results of this validation trials [7] confirmed the K_c and F_{ns} values adopted, and created the information required for exploring the model for different scenarios, including deficit irrigation. These validation trials were also utilized for the calibration of the capillary rise function. For completing the installation of the model to be used in the North China Plain, main results are being published in Chinese. This is the case for the ET_0 calculation procedure [8] and for the water balance [9].

Surface irrigation studies comprise: (a) field evaluation of basin irrigation in farmers fields, in the Xiongxian project area; (b) obtention of the infiltration parameters relative to the Kostiakov equation, and the roughness parameter of the Manning equation by the inverse solution of the simulation model SRFR [10]; and (c) exploring the model SRFR to determine the basin irrigation design parameters.

Field evaluations [11] provide information for initializing the iterative search of the infiltration and roughness parameters, and for characterizing the advance and recession along the basins. Observations also include the geometry and microtopography of the basins. This model is then run and parameters are modified progressively until advance and recession can be well described by the model. In order to further explore the model for design in the region, the parameters are not selected by event but for all events in a similar soil and for the same seasonal period (first irrigation, after planting, early October; second irrigation, before soil freezing, early December; third irrigation, early spring, by March-April; fourth irrigation at flowering, early April; and fifth irrigation, at milking, in May). Since infiltration and roughness parameters are available, the model SRFR is utilized to determine the design parameters relative to inflow rate and cuttoff time which allow appropriate irrigation performances for different basin sizes (length and width) and land levelling conditions.

3 Irrigation scheduling strategies

The simulation of alternative irrigation scheduling strategies was performed for the years representing 50%, 75% and 90% of probabilities for the irrigation demand not being exceeded, thus corresponding to the average, dry and very dry years. These years were selected from performing the frequential analysis of the net irrigation requirements of the wheat-maize cropping sequence, computed with the model ISAREG using daily meteorological data of 22 years (1975-1996) observed at the Xiongxian Meteorological Station.

Local climate is dry during the winter wheat crop season, making irrigation necessary for the wheat crop from planting to the end of the crop season. Near 80% of the precipitation occurs during the maize crop season, which is in general not irrigated.

At present local farmers adopt an irrigation schedule with well defined irrigation timings but variable and excessive application depths. Farmers cut the application when the advance is completed. Advance varies with roughness, soil infiltrability, inflow rate per unit width, basin length and field levelling conditions. The following irrigations are currently practiced:

- Planting irrigation (25 Sept.-5 Oct.), which is practiced when the available soil water (ASW) is depleted by 60 to 70% by the precedent maize crop.
- Winter irrigation (25 Nov.-5 Dec.), applied before freezing. The best application time is when the water freezes by night and melts during the day. The winter irrigation is practiced to store water in the soil for a long period since there is almost no soil water flux when the upper layer of the soil is frozen, and to improve soil porosity by the freezing and melting effects of the soil water stored in the upper layers.
- Spring irrigation (30 Mar.-10 Apr.). This is practiced when the wheat crop develops into the jointing stage. In order to get higher yield and earlier maturing, top fertilizer should be applied before rapid growth starts. Thus the spring irrigation not only refills the soil reserve but is required to ensure appropriate effect of the fertilizer.
- Irrigation at the heading stage (28 Apr.-8 May). The leaf area of the wheat crop attains the maximum development at heading and flowering, thus crop evapotranspiration becomes maximal for the next period. This reproductive period is also critical in relation to water stress, which also justifies this irrigation.
- Irrigation at the filling stage (25 May-5 Jun.). This irrigation is required for the wheat crop to provide for full grain filling, and for the summer maize, ensuring enough soil water for seeding timely and achieving an high emergence rate.

Traditionally, no irrigation is applied for summer maize because it grows during the rainy season. However, the frequency analysis for the 22 years shows that, although the total amount of precipitation during the maize growing season is larger than the evapotranspiration of maize, since most of rainfall occur in July and August, periods of water deficit occur in June and Sptember. If no irrigation is applied for summer maize, among the 22 years there are 10 years where the maize crop is slightly or severely stressed during September. Therefore, one irrigation for maize at the filling stage in such dry years could be required, also providing for enough soil water for the seeding and emergence of the wheat crop.

The following irrigation strategies have been simulated for the wheat-maize cropping sequence:

- IS1: Refilling the soil reservoir to the field capacity: Irrigations are applied when the average soil water content falls to the non-water-stress threshold (defined from F_{ns}). Irrigation depths are those required to refill soil moisture to the field capacity. Two restrictions were adopted: no irrigation is applied during the freezing period (5,Dec.-15,Mar) and during the last 10 days before harvesting.
- IS2: Refilling soil moisture to 85% of total available soil water (ASW) also for maximizing yields: The irrigation threshold is the same as for alternative (a) but application depths are smaller. Restrictions are the same as for (a). Keeping the soil moisture contents below 85% of ASW accommodates better for rainfall occurring after irrigation and decreases the probability for percolation losses.
- IS3. Irrigation at fixed dates: the irrigation depths are calculated by the model to refill the soil moisture reserve. The irrigation dates are fixed according to the current irrigation schedules as indicated above.
- IS4: Irrigation with fixed application depths: The selected depth is 80mm and the optimal irrigation dates are calculated by the model to maximize the relative yield.
- IS5: Irrigation with fixed dates and application depths: the irrigation dates are selected as for (c). The application depths for each irrigation are fixed considering both the requirement to refill the soil and the potential capabilities of the irrigation method. Based on results from surface irrigation studies, and assuming improvements in the basin irrigation practices, the application depths for the five irrigations are, respectively 70mm, 80mm, 70mm, 90mm, and 90mm.

Alternatives for deficit irrigation [12] are not considered because farm sizes are extremely small and yield deficits could produce an important reduction of the farmers income.

Results for these five alternative strategies relative to the probabilities 50, 75 and 95% for net irrigation demand not being exceeded are presented in Table 1. The analysis of these results provides some relevant information:

- The strategy aiming at highest yields by refilling the soil reservoir at field capacity, IS1, leads to the highest total irrigation depth and the highest water (irrigation and precipitation) losses;
- The alternative IS2, reducing the refill volumes to 85% of ASW, produces similar yields, lower irrigation water demand and less water losses. However, the number of irrigation events increase, becoming higher than the common practice. Computed irrigation depths, mostly below 70 mm, are often much smaller than those required by the irrigation application method. In basin irrigation, when land is uneven the

Table 1. Main results from simulations of alternative irrigation schedules

Irrigation Strategies	Year Prob. (%)	Irrigations			Irrigat. percolat. losses (mm)	Rainfall percolat. losses (mm)	ET_c (mm)	Relative Yield (%)	ASW end season (%)
		Crop	Num.	Depth (mm)					
IS1:	50	Wheat	3	279	0	0	425	99.6	73
Refill to		Maize	2	156	0	151	267	100	72
field			5	435	0	151	692	---	---
capacity	75	Wheat	4	341	0	0	453	98.5	37
		Maize	1	94	0	161	275	99.6	40
			5	435	0	161	728	---	---
	90	Wheat	5	452	0	1	502	98.9	60
		Maize	---	---	---	307	256	100	58
			5	452	0	308	758	---	---
IS2:	50	Wheat	4	278	0	10	425	99.6	66
Refill to		Maize	2	111	0	110	267	100	62
85% of			6	389	0	120	692	---	---
ASW	75	Wheat	6	400	0	0	458	99.7	74
		Maize	---	---	---	122	275	99.6	40
			6	400	0	122	733	---	---
	90	Wheat	7	451	0	0	508	100	56
		Maize	---	---	---	301	256	100	58
			7	451	0	301	764	---	---
IS3:	50	Wheat	4	288	0	1	427	100	76
Fixed		Maize	---	---	---	68	245	91.4	42
irrigation			4	288	0	69	672	---	---
dates	75	Wheat	5	391	0	7	457	99.4	64
		Maize	---	---	---	107	275	99.6	40
			5	391	0	114	732	---	---
	90	Wheat	5	434	0	0	489	96.1	57
		Maize	---	---	---	303	256	100	58
			5	434	0	303	745	---	---
IS4:	50	Wheat	3	240	0	0	426	99.8	46
Fixed		Maize	1	80	0	102	245	91.4	42
irrigation			4	320	0	102	671	---	---
depths	75	Wheat	4	320	0	0	441	95.8	32
		Maize	1	80	0	138	275	99.6	40
			5	400	0	138	716	---	---
	90	Wheat	4	320	0	0	438	85.6	15
		Maize	1	80	0	320	256	100	58
			5	400	0	320	694	---	---
IS5:	50	Wheat	4	330	44	0	427	100	76
Fixed		Maize	---	---	---	68	254	91.4	41
dates and			4	330	44	---	681	---	---
depths	75	Wheat	5	400	23	7	457	99.4	54
		Maize	---	---	---	93	275	99.6	40
			5	400	23	100	732	---	---
	90	Wheat	5	400	1	0	488	95.9	35
		Maize	---	---	---	271	254	98.9	58
			5	400	1	271	742	---	---

advance can only be completed with large application depths or relatively large discharges per unit width of the basin. These constraints oppose to consider irrigation depths below 70mm.

• The alternative IS3, for irrigations at fixed dates, leads to the lowest irrigation requirements and minimized water losses, but computed irrigation depths are often not appropriate considering the constraints relative to the irrigation method. Results confirm that current irrigation timings are appropriate.

• The alternative IS4, for irrigations with constant depths of 80 mm, lead to the less good yield results and, generally, to the lowest ASW percentages at the end of crops seasons, not favoring the establishment of the next crop. Seasonal irrigation depths are relatively high as well as rainfall losses comparatively to other alternatives.

• The alternative IS5, when both dates and depths are fixed, produce good results, similar to alternative IS3 in which concern yields. Water losses are the lowest and the irrigation demand is relatively low.

Under a practical perspective, the alternative IS5 is the most useful because it adopts irrigation timings close to those being practiced and irrigation depths respect the constraints of the irrigation method when application practices would be improved. This alternative was selected to build new irrigation scheduling calendars which could be adopted to advise farmers. Their implementation would require information on the soil moisture at planting the wheat crop to decide when the irrigation at planting is necessary, and a follow-up to adjust the depths of the last irrigation when the climatic demand would be higher, as well as to advise farmers when irrigations are required to the maize crop.

4 Establishment of an improved irrigation calendar

Table 2 gives the current applied water depths, obtained from field evaluations, and the optimal ones, produced by the simulations analysed above. Observations correspond to five field evaluations per each irrigation period.

Data in Table 2 shows that currently applied depths exceed those required by about 200 to 250 mm. The excess water application is mostly due to uneven basin surface and relatively small inflow rates per unit width of the basins. The unit inflow rate can be improved when the width of basins would be reduced or when the outlets would be

Table 2. Comparing observed and optimal irrigation depths

Irrigation timings	Observed depths (mm)		Optimal depths (mm)		
	Range	Average	Average year (P=50%)	Dry year (P=75%)	Very dry year (P=90%)
At planting	90 - 229	156	-	78	80
Winter	161 - 142	129	61	54	74
Spring	116 - 140	124	80	71	65
At heading	119 - 143	133	58	84	91
At filling	84 - 117	97	88	104	123
Total		639	287	391	433

modified. Land leveling can only be applied to a limited extend because the soil is without crops for very short periods. However, it could be necessary to correct inverted slopes near the downstream end. These improvements are required to reduce the time of advance and therefore allow for smaller application depths.

Table 3. Irrigation calendar when current application constraints could be solved

Irrigation timings	Minimum irrigation depths (mm) when inflow rates increase		Targets irrigation depths (mm)		
	Favorable field slope	Unfavorable field slope	Average year (P=50%)	Dry year (P=75%)	Very dry year (P=90%)
At planting	70	75	-	70	70
Winter	105	115	90	90	90
Spring	100	110	80	80	80
At heading	80	85	80	80	90
At filling	70	75	80	90	100
Total	(425)	(460)	330	410	430

Results from the analysis of field evaluation data using the model SRFR show that it is possible to reduce the application depths under constraints of the irrigation method when the unit inflow rates could be increased to 3.5 l s^{-1}m^{-1} without land levelling. Assuming that inverted slopes could be modified, unit discharges increased, and basin lengths adjusted, then improved target irrigation depths could be applied, as shown in Table 3. Improving the basin irrigation method could lead to a water saving irrigation scheduling calendar which could provide for near 200mm reduction in the demand for irrigation water.

5 Conclusions

The case study referred above shows that the sole improvement of the irrigation calendar would not produce the desirable results in terms of decreasing the demand for irrigation water. Results indicate that irrigation timings currently applied are appropriate. Implementing an irrigation scheduling programme under the perspective of water saving definitely requires that the improvement of the irrigation method be dealt together in order to allow reducing application depths to requirements.

Computed irrigation depths are much smaller than current ones. In practice no differences occur in seasonal irrigation depths for average and very dry years. This is due to constraints in water application; mainly insufficient inflow rates and uneven land surfaces. When these constraints could be overcome by adopting appropriate measures an improved irrigation calendar could be implemented leading to water savings without affecting yields.

6 Acknowledgments

This research is developed in the framework of the cooperative research contract TS3*CT93-0250, funded by the STD programme of the European Union.

7 References

1. Pereira, L.S., B.J. van den Broek, P. Kabat and R.G. Allen (eds). 1995. *Crop-Water-Simulation Models in Practice*, Wageningen Pers, Wageningen, 339p.
2. Teixeira, J.L. and Pereira L.S. (1992) ISAREG, an irrigation scheduling simulation model. *ICID Bulletin*, 41(2): 29-48.
3. Teixeira, J.L., Fernando R.M., and Pereira L.S. (1995) Irrigation scheduling alternatives for limited water supply and drought. *ICID Bulletin* 44(2): 73-88.
4. Teixeira, J.L., Liu Y., Zhang H.J., Pereira L.S. (1996) Evaluation of the ISAREG irrigation scheduling model in the North China Plain, in: *Evapotranspiration and Irrigation Scheduling* (ed. Camp, C.R., E.J.Sadler, and R.E.Yoder), ASAE, St.Joseph: 632-638.
5. Liu, Y., Teixeira, J.L., Zhang H.J. and Pereira L.S. (1998) Model validation and crop coefficients for irrigation scheduling in the North China Plain. *Agri. Water Manag.* (in press).
6. Allen, R.G., Smith M., Pereira L.S. and Perrier A. (1994) An update for the calculation of the reference evapotranspiration. *ICID Bulletin* 43(2):35-92.
7. Liu, Y., Fernando, R.M., Li, Y. and Pereira, L.S. (1997) Irrigation scheduling strategies for wheat-maize cropping sequence in North China Plain, in *Sustainable Irrigation in Areas of Water Scarcity and Drought* (ed. J.M. de Jager, L.P. Vermes and R. Ragab). British Nat. Com. ICID, Oxford, pp. 97-107.
8. Liu, Y., Pereira L.S., Teixeira J.L.and Cai L. (1997) Update definition and computation of reference evapotranspiration. Comparison with former method. *Journal of Hidraulics Engineering*, 6: 27-33 (in Chinese).
9. Liu, Y., Teixeira, J.L., Pereira, L.S. and Zhang H.J. (1997) Simulation of crop water requirements and irrigation scheduling. *J. Water Resources and Hydropower Engineering* 28 (4): 38-43 (in Chinese).
10. Strelkoff T SRFR 20.5: a Computer Program for Simulating Flow in Surface Irrigation. USDA-ARS,U.S. Water Conservation Laboratory, Phoenix. AZ.
11. Walker, W.R. and Skogerboe, G. (1987) *Surface Irrigation: Theory and Practice*. Prentice-Hall, englewood cliffs, NJ.
12. English, M.J., Musick J.T. and Murty V.V.N. (1990). Deficit irrigation, in *Management of Farm Irrigation Systems* (ed. G.J. Hoffman et al.), ASAE, St. Joseph, pp. 631-655.

WATER AND SALT MANAGEMENT STRATEGIES IN THE ARAL SEA BASIN

V.A. DUKHOVNY and V.I. SOKOLOV
Scientific-Information Center of the Interstate Commission for Water Coordination in the Aral Sea Basin, Tashkent, Uzbekistan

Abstract

At the time the five countries of the Central Asia became independent from the Soviet Union, they found themselves in very complicated conditions in the field of water and environment management. The primary focus of past interventions has been on the development and utilization of the water resources of the Aral Sea Basin. While major drainage and land reclamation programs were also conceived, progress was limited and the environmental consequences of development were not fully anticipated. Since salinity represents the critical threat to the long-term sustainability of the systems created, cost-effective programs to address salinity will be an essential element of the strategy, the basic premise of this presentation being that water and salinity management are inter-dependent and that one cannot be conceived without the other.
Keywords: Aral Sea Basin, salt accumulation, salt mobilization, transboundary water resources, water conservation, water management.

1 Introduction

The Aral Sea Basin is shared by five countries of the Former Soviet Union (southern Kazakhstan, southern Kyrgyz Republic, most of Turkmenistan, and all of Tajikistan and Uzbekistan), which together account for about 1.55 million km^2.

The rivers originate in snow melt and rainfall in high mountains that reach up to 7,500 m to the west and south of the region. Two main rivers cross the Aral Sea Basin: the Amu Darya river and the Syr Darya. The Amu Darya originates in Afghanistan and Tajikistan, and the Syr Darya in the Kyrgyz Republic. They fall rapidly to the desert plains of downstream riparian: Uzbekistan and Turkmenistan on the Amu

Water and the Environment: Innovative Issues in Irrigation and Drainage. Edited by Luis S. Pereira and John W. Gowing. Published in 1998 by E & FN Spon. ISBN 0 419 23710 0

Darya; Uzbekistan, Kazakhstan and - to a small extent - Tajikistan on the Syr Darya.

Central Asia has a continental climate, typified by low and irregular precipitation, and by cold rainy winters and hot dry summers.

The oases spread out along the old caravan routes exploited the rivers from ancient times, and irrigation has a history of more than two thousand years. However, it was only during the Soviet period that water was diverted on a large scale from the river valleys to the intervening steppes and deserts, primarily for cotton cultivation. Huge dams were constructed (Toktogul on the Syr Darya and Nurek in the Amu Darya basin are among the largest in the world); and massive diversion structures, pumping stations and canals took the water to large-scale settlement schemes. The total irrigated area rose from an estimated 2.5 million ha at the turn of the century, to 4.5 million ha in 1960 and 7.95 million ha in 1995.

The impacts of these diversions were striking and are well known. Large areas of the sea bed were exposed, becoming open desert, and the local climate - which had been moderated by the effects of the Sea - may have been affected, with hotter summers and colder winters. Elsewhere in the basin, irrigation under arid conditions was associated with rising water-tables, salt accumulation and secondary salinisation of the land, and large desert lakes were formed by drainage water that had previously returned to the rivers and Sea.

2 The key principles of the water and salt management strategy

Among the major objectives within the Aral Sea Basin Program the one is to prepare a joint Water and Salinity Management Strategy for the water and associated land resources that provides for the balanced economic and social development of the countries of the Aral Sea Basin over the short, medium and long-terms while protecting and enhancing environmental conditions and improving the quality of life of the region's inhabitants.

A number of important considerations should be borne in mind in carrying out this complex and multi-faceted assignment:

1. The primary focus of past interventions has been on the development and utilization of the water resources of the Aral Sea Basin. While major drainage and land reclamation programs were also conceived, progress was limited and the environmental consequences of development were not fully anticipated. Since salinity represents the critical threat to the long-term sustainability of the systems created, cost-effective programs to address salinity will be an essential element of the strategy, the basic premise of which being that water and salinity management are inter-dependent and that one cannot be conceived without the other.
2. The scale of potential interventions will be directly dependent on macro-economic developments in the region, as well as on developments in each of the major sectors concerned. The programs to be supported under the strategy will need to reflect such considerations, with priorities spelt out in ways that facilitate a flexible response to developments in the wider macro and sectoral context in the short, medium and long-terms respectively.

3. Successful implementation of the strategy will be a formidable undertaking, and can only be achieved based on: (i) a clear picture not only of the actions proposed but also of the probable consequences of these actions; (ii) an understanding of these actions and their consequences on the part of the many participants in the management system, extending to the general public and the private and voluntary sectors; and (iii) effective and close collaboration between each level of the management hierarchy ranging from the basin agencies, through the district and sectoral agencies, down to the end users.

4. Besides detailing specific proposals, the critical additional element under the future investigations will be the *analytical* justification for the water and salinity management interventions under the strategy. The aim is to develop a coherent and flexible family of inter-related tools and analytical approaches, adapted to the needs of the basin agencies and of national agencies responsible for complementary action. These tools will cover simulation of the natural system as well as the evaluation and selection of development interventions. They should be adopted to conditions in Central Asia and should be installed within local agencies for subsequent use and development.

5. Preparation of a water and salinity management strategy is an iterative process in a long-term perspective. Such a strategy will necessarily evolve: as physical and socio-economic conditions change; as additional data are collected and analyzed; as the understanding of the consequences of different lines of action increases; as the capacities of management at each level in the hierarchy are strengthened; and in the light of numerous other factors.

3 Criteria of the environmental safety

The ecological sustainability of water bodies and connected with them landscape zones should be considered from the position that natural processes are constantly changing depending on common laws of natural biogeochemical territorial cycles which are typical for all types of landscapes, and there are specific peculiarities for each taxon. Common laws of hydrological, hydrogeological, hydrochemical and other natural processes can be systematized through application of balance methods and methods of finite differences; and differentiation of conditions on the basis of a landscape method will give classification of both irrigated and non-irrigated lands, and water bodies; at the same time it will determine independent ecosystems within the landscapes which are the least liable to artificial influence; then transit and partially accumulative systems and, at last, accumulative and less often transit ecosystems with the most complicated conditions of their stabilization (for example, deltas, desert lowlands etc.).

That is why examination of ecological sustainability should follow determination of ecological zones and consideration of those data of changing natural-artificial processes which are changed and transformed more than others.

Doing this, on the basis of the correlation «inflow-outflow of waters» (surface and underground waters), «inflow-outflow-transportation of substances (ingredients, pollutants, suspensions etc.) changing of the main factors of instability under the influence of water regimes, and possibility of their stabilization by engineering and other nature-protection measures can be considered. The main condition for obtaining sustainability of natural and natural-artificial cycles is minimization of interaction between a river and irrigated (non-irrigated) lands, minimum interaction between surface and underground waters.

Naturally, there are different criteria of sustainability for different systems. For instance, for mountains, foothill area, tops of alluvial cones, slopes of wavy plateau the most important factor of sustainability is minimum discharge of both surface and underground waters and transportation of waters and masses from those zones, which cause either erosion or loss of particular biogeochemical elements.

The basis is consideration of two principal nature-protection aspects connected with each other: water quality in a river and salt accumulation in irrigation areas. From the position of ecological sustainability criteria of safety for these parameters are conceived as follows:

1. Salinity in irrigation areas and adjacent territories should not exceed permissible limits, and intensity of salt accumulation should be negative i.e. there should be gradual reduction of salt content in irrigated areas (and adjacent territories).
2. Salt content in river water in all reaches (from upper reaches to delta) should not exceed permissible content for all water users using water from that river (or a stream).

An assessment of salinity and other water quality issues should be undertaken on the basis of these criteria, and as they may need to be reflected in international agreements, recognizing that this is to provide a framework for evaluating alternative strategic mechanisms and not to reflect detailed technical investigations. This should include but not necessarily be limited to:

1. A broad evaluation of the sources of salinity in the rivers and irrigated lands of the Aral Sea Basin, the identification and description of trends in salinity in rivers and irrigated lands, and--in the light of international experience--an assessment of alternative approaches to considering the issue of salinity in the context of an international agreement in the light of the proposed criteria.
2. A broad evaluation of the other water quality concerns in the rivers and irrigated lands of the Aral Sea Basin, the identification and description of trends in water quality parameters, and--in the light of international experience--an assessment of alternative approaches to considering the issue of water quality in the context of an international agreement in the light of the proposed criteria.

4 The water conservation is the vital law in the region

Acceptance of a water saving approach as a basis for the regional water strategy and all the activity on future development and water resources management and its elaboration in the Aral Sea Basin Program requires purposefulness of work, which concerns most of all developers of the program. Developers of both national and regional strategies should make detailed analysis for each planning zone within the country and after that for the whole country within the basin:

- Potential land and water efficiency on the basis of available practical information (pilot projects, tests, WUFMAS, advanced experience, especially in dry years).
- Specific consumption of minimum water discharges for production of biological goods, using common methodical approaches determined according to CROPWAT-FAO, which will be adopted.
- Analysis of production shortages caused by reclamation and water factors and possibility of elimination of these causes with estimation of priority of undertaken measures.
- Assessment of salt and water balance for planning zones using recent data, possibility of its bringing to conformity with ecologically sustainable parameters (minimum salt exchange between a river and an irrigated zone, and between an aeration zone and underground waters with gradual reduction of salt content in an aeration zone and in the whole planning zone); possibility of maximum involvement of its own return waters and their use directly near local formations.
- Assessment of an opportunity to use waste and underground waters at each level of a supply system which are being lost now.
- Assessment of inefficient water expenditures at each level of irrigation water use, in the first place, in the field, which allows to determine low capital-intensive elements of water saving.
- Determination of zones with high infiltration on slope lands, steppes and high meadows, which cause not only water losses, but increased expenditures for pumping water supply and negative meliorative influence on lands located lower; their technical and economic estimation.
- Assessment of water saving influence on reduction of return waters discharge into rivers and water reservoirs and improvement of water quality.

Informing of the population, public awareness and direct involvement of society and water users into water saving should become a very important component of the water saving program. Development of the special social company in this direction is very important and requires detailed elaboration, as it should involve various strata of society and population, both decision-makers forming public opinion, mediators and professionals, direct water users - in the rural area (in irrigated agriculture), in municipal economy, industry and other branches of water use.

Various forms should be used here, but, with due regard for possible opposition of decision-makers and politicians not considering such problems as important under the conditions of economic depression of local leaders taking into account interests of 1-2 years; water users with their living and economic difficulties; water officials, which can hardly find funds for support of their structures with low payment of their staff's work.

Nevertheless the society should receive the following:

- We can leave for our children, grandchildren and next generations a desert like once fertile Mesopotamia, or open for winds and jackals ancient Gavkhare, if we continue unwise water use like now.
- To organize permanent agitation press on society through all the media: «Do you like your country and your nature? Or you are their enemy?»
- It should be shown, that only through common efforts we can save a small place in

the world, for which the five countries are responsible - through the efforts of all the states, all the strata of society and all the water users.

This company should be elaborated, planned and organized thoroughly in all the countries of the region. And prestige of separate politicians should reflect degree of their involvement and effectiveness in this process. Social questioning on water saving and management issues is a very good tool for the development of the company. Effectiveness of the company should be directed on the interests of each stratum of society which does not seem to have any necessity for water saving; moral pressing should be permanent and purposeful, which can and must break inertia of even deep-rooted indifference.

Naturally, the role of upbringing of society is high, beginning from a kindergarten and school, vocational and high schools. From childhood the idea about high value of water under the conditions of arid climate of our countries should be cultivated in everyone. Biology and Geography should show exceptional importance and irreplaceable role of water for forming and support of life, History and Literature should emphasize the attitude of our ancestors towards water and tell how water united and fed them, how thousands of people worked together on water, how they saved water for us, which we have spent unwisely during the life of one generation. It is clear, that at first this idea should be suggested to teachers, and then - to children.

5 Regional water and salinity management strategy preparation

Presentation of a coherent conceptual and analytical framework for the technical and economic evaluation of alternative strategies for water and salinity management in the short, medium and long-terms for each of the regional rivers of the Aral Sea Basin should be done, so as to avoid the danger that partial solutions will be adopted and to ensure balanced approaches to the management of water quantity and water quality (especially of salinity), taking into account the following considerations:

- With respect to water management, the conceptual framework should accommodate the need to satisfy present and possible alternative future water demands - in both consumptive and instream uses - taking into account: the potential contribution of water savings programs: future environmental and other needs, seasonal, annual and inter-annual operating issues; and in the light of agreed and possible future allocations and agreements between the riparian states.
- With respect to salt management, the conceptual framework should accommodate the need to manage the interactions between the rivers and irrigated lands, between surface water and groundwater, and between surface water, groundwater and irrigated lands, with a view to managing the storage and disposal of the salt in the system, and to meeting alternative salinity objectives.

In the light of the screening exercises described above, coherent alternative combined water and salinity management options should be specified for the *short-term, medium-term* and *long-term*, reflecting alternative macro-economic and sectoral development perspectives. These alternatives should include but not necessarily be limited to:

- For the short-term, revised operating rules and water management practices (supported by complementary incentive and related mechanisms) broadly appropriate to existing reservoir, conveyance, utilization and disposal systems, so as to satisfy realistic short-term water demands and salinity objectives in the light of investment programs that are at a level consistent with continued economic and financial stringency.
- For the medium-term, revised operating rules and water management practices (supported by complementary incentive and related mechanisms) for existing and feasible new combinations of reservoirs and water conveyance, utilization and disposal systems, so as to satisfy realistic medium-term water demands and salinity objectives in the light of investment programs that are at a level consistent with stabilized and steadily expanding macro-economic conditions.
- For the long-term, perspective operating rules and water management practices for alternative combinations of reservoirs and water conveyance, utilization and disposal systems, so as to satisfy realistic long-term water demands and salinity objectives in the light of investment programs that are at a level consistent with tackling the major problems associated with the sustainable development of the Aral Sea Basin.

The last step should be the evaluation of the alternative options specified for the *short-term*, *medium-term* and *long-term*, to demonstrate the technical and economic advantages and disadvantages of alternative lines of action, and facilitate coherent choices between the specified options.

6 Concluding remarks

The strategies proposed should reflect regional and national goals and objectives, and be realistic in terms of macro-economic and sectoral prospects. The proposed strategies should also provide for regional balance, subject to existing and future regional allocations, and the need to establish sustainable management practices and to protect the natural environment. They should be specified in relative detail over the short and medium-terms, and in outline terms for the long-term. The strategies should be presented in a format suited to further review and formal approval by the Governments of the riparian countries concerned.

It is necessary to analyze the impacts of the proposed strategies in the short, medium and long-terms with a view to checking the consistency of the overall physical impacts and presenting an systematic justification for the strategies proposed.

DROUGHT FORECASTING AS AN AID TO IRRIGATION WATER MANAGEMENT

I. PÁLFAI
Lower-Tisza Region District Water Authority, Szeged, Hungary
GY. SZILÁRD and L.M. TÓTH
National Water Authority, Budapest, Hungary

Abstract
The climate of Hungary is very changeable. Sometimes there is more water than necessary, but very often it's not enough which causes many problems, especially in the agriculture. So it's very important in the practice for people concerned to know the actual hydrometeorological situation and the probability of drought before the beginning of the irrigation period. This paper contents a drought-forecasting method made for this purpose. The Pálfai drought index is a relative number which is based on the sum of precipitation have been fallen down from last year's October and the depth of the groundwater (these are known at the time of making forecast) and takes into account the sum of precipitation falling down till the end of actual summer and the mean of temperature with their values belonging to different probabilities. These indexes can help the comparison of strength of drought one can take into account. The method worked out at the beginning of 1990s and practically have been applied in Hungary since 1995 using data of 68 PAI-stations. The forecast received by the different regional water authorities, Ministry of Agriculture and also the press (National Press Agency) are published by the National Water Authority.
Keywords: Droughts, drought indices, drought forecasting, irrigation planning, water management

1 Introduction

The climate of Hungary has a continental character, however in the country - and even in the whole Carpathian Basin - oceanic and Mediterranean effects prevail too

Water and the Environment: Innovative Issues in Irrigation and Drainage. Edited by Luis S. Pereira and John W. Gowing. Published in 1998 by E & FN Spon. ISBN 0 419 23710 0

sometimes. The amount of precipitation is very variable. In certain years there is too much of water appearing, another time the trouble is caused by the insufficient water quantity mainly in agriculture and especially on the flatlands of the country. Therefore it has a great practical significance that the concerned persons (mainly the farmers and the regional water agencies) should get appropriate information about the current hydrometeorological conditions and the expectable turn of water conditions. All these have a special importance in those zones which have been established for irrigated farming.

The National Water Authority initiated the elaboration of a drought forecasting method at the beginning of the eighties, then financed it later. The initial results of this research were presented at the XIV. Congress of the ICID [1]. Now, we are going to expound its form which has been developed further respectively which has been transformed, including the obtained experiences with it.

The forecast is focused on the rough estimated determination of the values of a complex index, the Pálfai drought index (PAI). As a preliminary we are going to communicate the calculation method of this index.

2 The calculation method of the drought index

There are indexes for the characterisation of the severity of a situation (drought) by single digit derived from only few meteorological and/or hydrological parameters. The great advantage of such indexes is that long-term data series could be produced by them.

The formula to calculate the base-values of the aridity index is:

$$PAI_0 = \frac{t_{IV-VIII}}{P_{X-VIII}} \cdot 100 \qquad (1)$$

where PAI_0 = base-value of the aridity index (°C/100 mm),

$t_{IV-VIII}$ = mean value of the temperature of the air in the period April-August (°C),

P_{X-VIII} = precipitation depth summed up by the monthly values of precipitation of the period of October-August (mm).

Monthly weights for the precipitation values were based on the conditions of moisture-storage and on the changing water demand of the crops. Estimates of the weighing factors are the following (with due regard on the overall natural conditions of the Carpathian Basin):

0.1 in October
0.4 in November
0.5 from December to April
0.8 in May
1.2 in June
1.6 in July
0.9 in August

It is evident that month July is the most critical period from the water supply point of view.

When the values PAI_0 were compared to the well-known Palmer's "drought-index" strict correlation was found between them [1][2].

For the more accurate expression of aridity the base value of PAI_0 should be corrected by the following factors:

- temperature (hot days) correction factor:

$$k_t = \sqrt[6]{\frac{n+1}{\overline{n}+1}} \qquad (2)$$

where k_t = temperature correction factor

n = number of the hot days ($t_{max} = 30$ °C) in the period of June-August (d)

$\overline{n}$ = multi-annual national average of the "n" value (d); in Hungary this value is 16 days.

- precipitation correction factor:

$$k_p = \sqrt[4]{\frac{\tau_{max}}{\overline{\tau}max}} \qquad (3)$$

where k_p = precipitation correction factor

τ_{max} = the longest precipitation poor period (if the sum of precipitation in the successive days do not exceeds max. 5-6 mm) between middle of June and the middle of August (d),

$\overline{\tau}max$ = multi-annual national average of τ_{max} (d); in Hungary this value is 20 days.

- groundwater correction factor:

$$k_{gw} = \sqrt{\frac{H}{\overline{H}}} \qquad (4)$$

where k_{gw} = groundwater correction factor

H = mean depth of groundwater table below ground level in March (m),

$\overline{H}$ = multi-annual value of "H" on the given area (m).

The use of correction factor is important on plain area. Practically it is best to use the data of the nearest 2 or 3 groundwater wells in the surrounding of the meteorological station or observation point.

The final value of the aridity index - defined as PAI - is obtained from the base-value (PAI_0) by corrections

$$PAI = k_t \cdot k_p \cdot k_{gw} \cdot PAI_0 \qquad (5)$$

where the correction factors are those described above.

According to the Hungarian experiences the threshold value of the Pálfai Aridity Index shall be at PAI = 6.0. Smaller values at a particular site - are for wet years, larger values would indicate the different severities of drought. These may be categorised as follows:

PAI = 6 - 8   moderate drought
       8 - 10 medium drought
      10 - 12 heavy drought
       > 12   extremely drought.

For the territorial distribution of the drought index within Hungary and for its annual changing we have presented an example, among others, during the ICID meeting held in Oxford [3]. On one of the measuring stations of the Great Hungarian Plain (at Szarvas municipality) the smallest value of PAI was 1.5 °C/100 mm between 1931 and 1996, its major value was 14.0 °C/100 mm. These values illustrate well the extremities of a great extent which characterises the Hungarian climatic conditions.

3 The presentation of the forecasting method

The essence of the method is, that those factors of the drought index which we know already in the time of the forecast (fact data) should be simply substituted to the index formula which we do not know yet, we take them into consideration according to the three supposed variants and thus we might calculate the numerical value of the index. However it is not about a traditionally interpreted forecast but in the case of certain conditions. It is about the numerical characterisation of the drought situation, which is called, for the sake of simplicity the forecast itself.

When preparing the forecast, from among the unknown factors, in the case of precipitation we reckon with the value of a 10, 50 and 90 % of appearing probability (statistical data) of the weighted precipitation sum of the period we are facing, that is to say from the time of preparing the forecast till the end of August, while the missing parameters of temperature and the length of the period having lack of precipitation - depending on the supposed precipitation - are determined from regressive equations (derivated data). We have seen in the previous item that the many years national average of heat days is constant ($n = 16$) in the case of all measuring stations, such as the national average of the period having lack of precipitation ($\tau_{max} = 20$).

It is worthy preparing the first forecast by the end of March, at the beginning of April and publishing it. This is the time when we know already the quantity of precipitation fallen in the winter semester (from the 1st of October - till the 31st of March), from which the weighted sum of precipitation of this period could be calculated (P_{X-III}). We also know the medium subsoil water level of March (H) from which - and from the many years average of the subsoil water level of March (h) - the correction factor k_{gw} could be determined.

The values of a 10, 50 and 90 % of appearing probability of the weighted precipitation sum for the period between the 1st April till the 31st August ($P_{IV-VIII}$)

were determined by the examination of distribution of the 62 years series of data (1931-1992) - starting from the monthly values.

The average temperature of the period between April-August ($t_{IV-VIII}$) e.g. in the case of the measurement station in Túrkeve municipality is calculated - in three variants - from the below regressive equation:

$$t_{IV-VIII} = 19,44 - 0,005.P_{IV-VIII} \tag{6}$$

The number of heat day (n) is calculated from the weighted sum of precipitation ($P_{VI-VIII}$), the duration of the period having a precipitation lack (τ_{max}) is calculated from the July precipitation (P_{VII}) also according to the 10, 50 and 90 % probability variant:

$$n = 31,02 - 0,047.P_{VI-VIII} \tag{7}$$

$$\tau_{max} = 29,39 - 0,134.P_{VII} \tag{8}$$

The forecast can be repeated monthly up to the beginning of August. Thus we shall have known data at disposal from longer periods (beside precipitation we shall have data of temperature too) and thus we have to produce statistical respectively derivated data for always shorter periods. Parallel with this the time advantage of the forecast would be gradually reducing moreover we could forecast only the final development of the meanwhile eventually evolving drought.

4 The application of the forecasting method

We are pursuing operative drought forecasting with the presented method since 1994. The calculations were carried out upon 68 meteorological stations. The first forecast is usually issued in early April. The results are published partly in numerical form (in tabular form) partly as a map and we even prepare textual evaluation too, underlining those regions which are supposed to be the droughtiest ones and we also communicate the national average of the forecast PAI values. Although from the three forecasting variants the occurring of that one has the major prospect to come about, which calculates with additional precipitation having 50 % of probability, however we recommend to take into consideration that variant calculating with 10 % of precipitation which could be considered as a pessimistic one, for the preparation for irrigation.

In Figure 1. we demonstrate the cartographical result of that forecast which reckons with the 10 % of the occurrence probability of the weighted precipitation sum for the period of April-August, issued on the 6th of April, 1994; that is to say the territorial distribution of the PAI values. The Figure 2. was set up on the basis of the factual PAI values of the year 1994. When comparing the figures it becomes visible, that the issued forecast is quite well informed about the presumable extent of the drought, respectively of its territorial distribution.

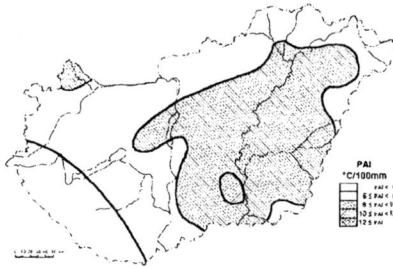

Fig. 1. The territorial distribution of the values concerning the index of droughtiness (PAI) in Hungary forecast on the 6th April,1994 (Supposing a weighted precipitation sum of April-August with a 10 % of occurring probability)

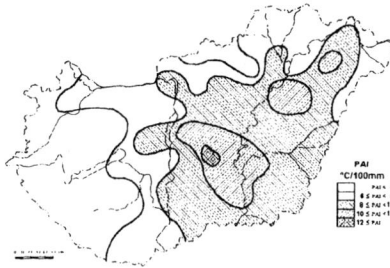

Fig. 2. The territorial distribution of the effective values of the droughtiness index (PAI) in Hungary, in 1994. (1: moderate drought, 2:medium drought, 3:heavy drought, 4:extremely drought)

The national averages of the monthly forecast of the year 1994 calculated from the date of 68 stations are communicated in Table 1. in two variants (that time we have not yet calculated with a precipitation of 90 % of occurring probability). The national average of the effectively occurring PAI values became 8.01 °C/100 mm.

Table 1. The forecast values of the average of the drought index (PAI) of 1994, in Hungary

Time of the forecast issue	Forecast PAI (°C/100 mm) with a precipitation of	
	10 %	50 %
6th April 1994	7.97	5.16
4th May, 1994	7.67	4.92
6th June, 1994	7.43	5.71
4th July, 1994	8.30	6.16
4th August, 1994	8.98	7.60

In Table 2. we present the most recent results of drought forecast, issued on the 13th March, 1998 in the case of one of the measuring stations in the Great Hungarian Plain (Túrkeve). During the calculation which has been carried out much earlier than the usual, the precipitation sum between the 13th - 31st March was substituted by its many years' average - for the sake of simplicity - and was handled as a fact date.

It can be seen from the Table 2., that the formulation and strength of the drought - subsequently from the differing weight of the monthly precipitation are decisively determined by the precipitation quality of the period between April-August, therefore the fact data used for the spring forecast give very few footing by themselves for the judgement of the expectable drought situation. The weighted precipitation sum (80 mm) of the period between the 1st October, 1997 - the 31st March, 1998 means only the 75 % of the many years' average which reflects a rather dry winter semester, however the PAI even in the case of an additional precipitation of 50 % would not reach the 6.0 threshold value.

Table 2. The results of the drought forecast issued in March 1998, at the measuring station of Túrkeve municipality

The character of data		The supposed probability of the precipitation occurrence		
		10 %	50 %	90 %
Fact data:				
(P_{X-III})	80			
k_{gw}	1.0			
Statistical data:				
$P_{IV-VIII}$ (mm)		166	265	432
Derivated data:				
$t_{IV-VIII}$ (°C)		18.61	18.12	17.28
k_t		1.16	1.06	0.94
k_p		1.13	1.04	0.90
Forecast data:				
PAI (°C/100 mm)		9.92	5.79	2.86

The drought forecast are set together at the Lower Tisza District Authority (Szeged), by the help of the data providing services of the other District Water Authorities and by the utilisation of the data of the National Meteorological Service. The ready forecast is controlled at the National Water Authority from where the data are disseminated - by E-mail network system - to all of the 12 District Water Authorities. They prepare also the shortened form of the forecast and send it to the Hungarian News Service - which depending on the severity of the drought situation - publishes the forecast in the daily press further in the agricultural and water-related specialised reviews.

The presented forecasting method has not the aim to forecast the changes in the water resources of rivers, however on the basis of the results we may get an approximate information about it. In Figure 3. as an example, we present the connection which exits between the territorial average of the drought index referring to the Hungarian part of the Tisza River Basin and the summer medium water discharge of the Tisza River (between June-August monthly). It can be seen that

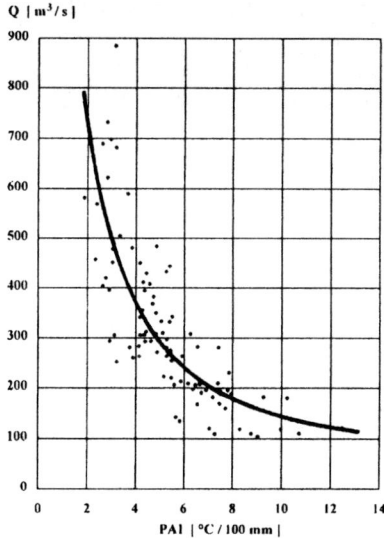

Fig. 3. The relationship between the territorial average of the drought index (PAI) in the Tisza valley (Hungarian part) and the summer medium water discharge of the Tisza River (Q) (water discharge recording section: Szeged).

stress producing contradiction which is know elsewhere too that the stronger the drought is the less is the water discharge of the river, from where irrigation demands have to be complied with. The water discharge of the Tisza River otherwise drives from the foreign part of the catchment area of the river, which increases the difficulties even move.

5 References

1. Pálfai, I.(1990): Description and forecasting of droughts in Hungary, in *ICID Fourteenth Congress*. Rio de Janeiro. Q.43.R.13.
2. Palmer, W.C.(1965): Meteorological drought. Research Paper, No.45, U.S. Weather Bureau, Washington D.C.
3. Pálfai, I. - Szilárd, Gy. - Váradi, J.(1997): Use of the drought index (PAI) and water management influence of the drought in Hungary: some practical examples, in *ICID 18th. European Regional Conference. Sustainable irrigation in areas of water scarcity and drought.* pp. 195-202, Oxford.

CHARACTERIZING AGRICULTURAL DROUGHTS IN EASTERN ROMANIA

C. CISMARU, I. BARTHA, I. COJOCARU, N. MARCOIE and V. GABOR
Land Reclamation Department, Technical University "Gh. Asachi" of Iasi, Romania

Abstract
In order to combat damages produced in agriculture by meteorological droughts its parameters, respectively intensity, frequency, duration have been analysed as local elements, and the extension area as well as the vulnerability, expressed by damages as elements of characterisation at regional scale. Intensity of droughts has been expressed by humidity index, proposed by Soroceanu, and by aridity index, defined by Pálfai. There has been analysed and analytically expressed the correlation between of the intensity of droughts and their duration. Using data from 15 meteorological stations, relatively uniform distributed on the territory, the average intensity of droughts in the subzone Galati - Tecuci (the "nucleus" of the droughts) have been correlated with their manifestation area. The agricultural damages due to drought have been established by simulation of the hydric regime dynamics of soil (assisted by ISAREG programme, for a representative soil - leached chernozem - and for three representative cultures (corn, sugar-beet and alfalfa). There is a close relationship between the relative damages of agricultural production and the two annual indexes of the intensity of droughts. Finally, there has been established a relationship which permits the estimation of the loss of production depending on the intensity of droughts in their nucleus, respectively in the south-eastern part of the region.
Keywords: agricultural droughts, drought duration, drought impacts, drought indices, drought intensity

1 Introduction

Drought has negative influences on economy, environment and, even, on social climate. Eastern part of Romania is characterised by a temperate climate with continental aspects, having an irregular distribution of rainfalls. That is due to general

Water and the Environment: Innovative Issues in Irrigation and Drainage. Edited by Luis S. Pereira and John W. Gowing. Published in 1998 by E & FN Spon. ISBN 0 419 23710 0

athmospherical circulation of currents, but also Carpathians, at the western part of the zone, are an obstacle in front of clouds brought by Atlantic currents. Although, the zone is opened to east, permitting the advance of steppe climate from the eastern part of Europe.

The number of droughty years, after N. Topor [1], represent 69% in Moldavia and 89% in Dobrogea, but in majority of these years, there are 1 or 2 droughty months. The multiannual duration average without rainfalls in the interval 15 June to 15 August is 22 days in the north (at Botosani) and, respectively, 26 days in the south (at Galati).

Dangerous factors of drought effects are: the soil erosion (which cause the reduction of capacity by retaining of soil water) and the pouring character of summer rains. The most part of agricultural lands are placed on hills and terraces and they do not benefit by water table. Control of droughts suppose to adopt measures in order to reduce their effects, which refer to the management of soil water, as well as, the management of water resources from lakes and reservoirs. All these measures are based on study and knowledge of drought characteristics.

From this point of view, this paper tries to improve the knowledge of drought parameters, such as: intensity, duration, frequency, affected area and vulnerability in the eastern zone of Romania.

2 Research and its results

Analyse of local parameters, such as: intensity, duration, and climatic frequency of drought, has been carried out for 15 meteorological stations, distributed almost uniform not only at the zone, but also at the hydrografic basins - Prut and Siret - of the zone, for a period of time along 16 to 106 year. There have been used data referring to the rainfalls and average monthly temperature of air, the relative air humidity, the wind velocity, the duration of sun's brightness, as well as yearly number of tropical days and maximum duration of periods without rainfalls in the interval 15 June to 15 August.

Intensity of drought has been expressed by humidity index, proposed by Soroceanu [2], and aridity index, defined by Pálfai [3]. The former index is established and verified in the studied zone and it is expressed by formula:

$$I_u = \frac{a\,q + Q}{etp} \tag{1}$$

where: "q" is the sum of precipitations in period of accumulation of soil humidity (November to February); "a" is the proportionality and storage coefficient of the water consumption from soil based on precipitation from the cold season; "Q" are the monthly precipitations in March through October; "etp" is the potential evapotranspiration in March through October.

The aridity index, proposed by Pálfai, has been calculated by formula:

$$PAI_0 = \frac{t_{IV-VIII}}{P_{X-VIII}} \cdot 100\,\% \tag{2}$$

where: "$t_{IV-VIII}$" are average monthly temperatures of the air in April to August (°C); "P_{X-VIII}" - monthly precipitations in October to August (mm), affected with weight coefficients (0.1 in October, 0.4 in November, 0.5 in December...April, 0.8 in May, 1.2 in June, 1.6 in July, 0.9 in August). Humidity monthly index I_u has been calculated on the period March to October, and for droughty intervals have been established the average values of this ($I_{u\,med}$). Aridity index PAI_0 has been calculated yearly.

There are four scales of intensity for appreciation of droughts for both two indexes respectively: slight, moderate, strong and extreme drought.

Duration of drought is considered to be the period in which the monthly values of indexes exceed critical, respectively: $I_u < 1$ and $PAI_0 > 2.5$ (value which corresponds to the conditions of analysed zone). Due to the strong correlation between the intensity and duration of droughts, the frequency refers only to the intensity expressed by $I_{u\,med}$.

Regional distribution of drought has been analysed in each year, for the period 1981 to 1996 being represented on maps of $I_{u\,med}$ value. There has been delimited zones having the same humidity index $I_{u\,med}$ and then calculated the area of these zones.

Agricultural damages have been established for a leached chernozem soil, which is representative for the analysed zone, and for three cultures: corn, sugar-beet and alfalfa, using ISAREG program [4]. The dynamics of soil water storage has been simulated annually for the period 1981-1996 at four meteorological stations (situated in the north, centre and south of the zone) being established - in nonirrigated conditions - the relative loss of productions which have been correlated with the intensity indexes: $I_{u\,med}$ and PAI_0 of the drought.

3 Intensity and duration of droughts

Intensity of droughts depends on the reduction of precipitations' rate and also on the evolution of air temperature, pedological characteristics, agricultural technologies, development of crops and crop sensibility for water deficit, genetic resistance to drought.

Annual and seasonal precipitations (from April to September), especially from July and August, have a great variability; this being the main element to determine droughts and other, secondary factors.

The annual average intensity of droughts has been correlated with their duration, resulting the regression in the fig.1.

Using these regressions, in which are introduced $I_{u\,med}$ for each drought interval (1 for slight drought, 0.75 for moderate drought, 0.5 for strong drought and 0.25 for extreme drought), the duration of drought correlated with $I_{u\,med}$ has been established (Tab 1)

It can be seen the increase of duration of drought with its intensity. The slight, moderate and a part of strong droughts have a duration which do not exceed one vegetation season, while the extremes and a part of those strong, are extended in the next vegetation season. During the first year the most strong and extreme droughts occur as slight and moderate droughts and they continue in the next year, when they run to strong and extreme droughts. There are years when they touch the maximum intensity in a single year, for e.g. 1894, which was an exceptional droughty year (nine droughty months). This situation imposes that drought management measures to be taken in real time, proportionally with its intensity.

Table 1. Duration of droughts for any intensity intervals (months) *

Intensity interval	Subzone North	Central	South
Slight drought	3-5	1-3.6 (1.1-4)	1.7-4.1 (2.3-4.0)
Moderate drought	4.4-6.3	3.6-6.3 (4.0-7.0)	3.35-5.4 (4.0-5.8)
Strong drought	6.3-8.2	6.3-9.0 (7.0-9.8)	5.4-7.7 (5.8-7.5)
Extreme drought	8.2-10.2	9.0-10.2 (9.8-12.8)	7.7-9.7 (7.5-9.3)

* Values from the brackets refer to the western alignment of the zone (Roman - Bacau - Adjud - Focsani).

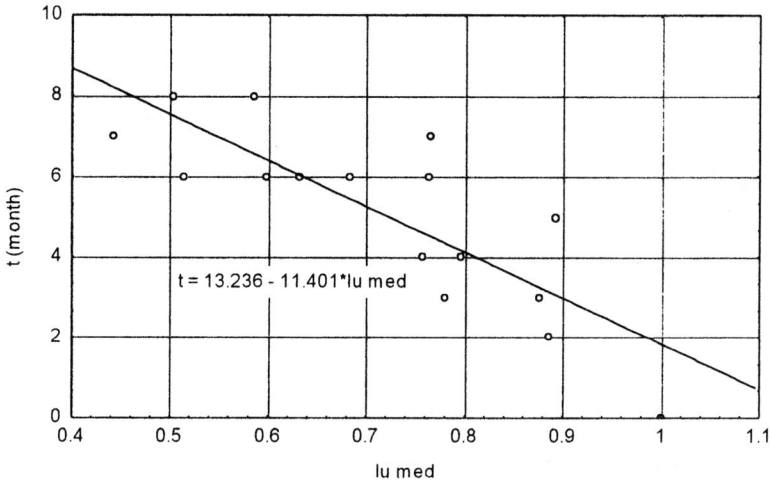

Fig. 1. Correlation between $I_{u\,med}$ and duration of drought at Iasi

Duration varies in the same way with intensity (not inversely as for pouring rains), and this determines a rapid evolution during the time, of severity of droughts as well as their consequences.

4 Dependence of frequency and intensity of droughts

Agricultural productions and farmers income, appreciated as a multiannual one, are influenced by frequency of drought (which depends on their intensity). Analysing frequency of I_{umed}, and PAI_0 values for the period 1981-1996 at mentioned meteorological stations, there has been calculated the values in Table 2.

Moderate droughts in the north and south part of the zone have the highest frequency in the studied period. It can be observed an increase of strong droughts frequency in south part of the zone (Galati Tecuci), which is considered to be the nucleus of droughts, and where occurs the first droughts.

An analysis for a long period (1891-1996), based on the precipitations from July and August at Iasi (situated in the central part) and for water consumption of 150 mm, for the spring cultures from this subzone (corn, sugar-beet and soya-beans), leades to the following frequency of droughts: (Table 3).

Table 2. Frequency of droughts for any intensity intervals (%) *

Intensity interval	Subzone North	Central	South
Slight drought	31.25 (31.25)	50.0 (12.5)	36.25 (18.75)
Moderate drought	62.5 (56.25)	37.5 (62.5)	62.5 (50.0)
Strong drought	6.25 (6.25)	6.25 (18.75)	31.25 (25)
Extreme drought	0 (0)	0 (0)	0 (0)
No droughts	0 (6.25)	0 (6.25)	0 (6.25)

* Values from the brackets refer to western alignment of the zone (Roman - Bacau - Adjud - Focsani).

Table 3. Frequency of droughts at Iasi (%)

Interval of droughts	July	August
Slight drought (100-75% from necessary)	9.8	4.9
Moderate drought (75-50% from necessary)	19.6	23.5
Strong drought (50-25% from necessary)	40.2	33.6
Extreme drought (25-0% from necessary)	24.5	33.3

These values show that for longer periods strong droughts have the highest frequency, followed by the extreme droughts.

5 The affected area depending on the intensity of droughts

Areas affected by droughts with different intensities in the period 1982-1996 are shown in the table together with $I_{u\,med}$ from the nucleus.

From the Table 4 and Fig. 2 it can be observed that the drought manifestation area grows at the same time with the intensity of droughts in the nucleus Galati-Tecuci. The maximum intensity, situated in the south-east, is continued to the north and west, on these areas having lower intensity. From the Fig. 2. it can be established the strong drought area as well as the affected area by moderate droughts in the zone.

Table 4. Affected areas by droughts and the $I_{u\,med}$ value, during 1982-1996

Year	Affected areas by droughts (km^2) strong	moderate	slight	$I_{u\,med}$ in nucleus	Localisation of nucleus
1982	-	27244	8028	0.550	Galati
1983	8876	26536	-	0.462	Galati-Tecuci
1984	-	4820	30588	0.628	Galati
1985	1620	15560	18232	0.486	Barlad
1986	33732	1680	-	0.342	Tecuci
1987	9500	25912	-	0.364	Galati-Tecuci
1988	-	7880	27532	0.512	Galati
1989	-	25496	9916	0.495	Focsani
1990	13296	22116	-	0.360	Galati
1991	-	-	5780	0.846	Galati
1992	-	35412	-	0.571	Tecuci
1993		27536	7876	0.582	Galati-Tecuci
1994	21176	14252	-	0.353	Barlad
1995	2760	31808	844	0.418	Tecuci
1996	1324	20460	13628	0.418	Galati

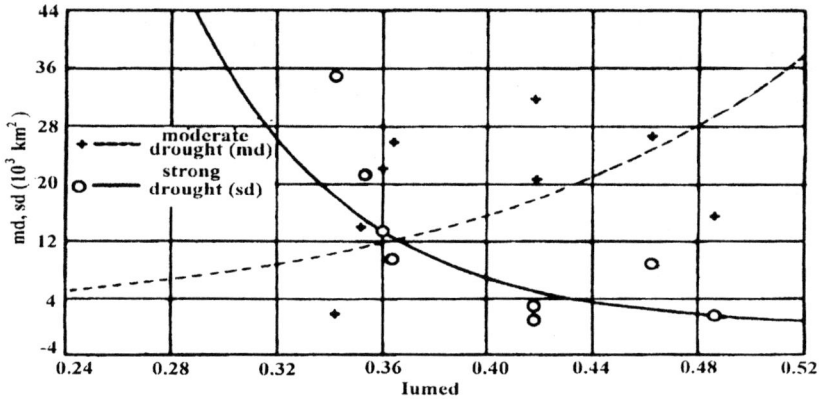

Fig. 2. Relationship between of $I_{u\,med}$ in nucleus and the affected areas by strong and moderate droughts

6 Relative loss of agricultural production due to droughts

Intensity of drought is the main element which, in any pedological and crop conditions, influence the amount of agricultural production (with respect the maximum potential production). Using the ISAREG programm [4] it has been simulated the dynamics of soil water storage and the production loss along 1981-1996, using data from four meteorological stations (Dorohoi, Iasi, Tecuci, Galati).

Water storage in sowing soil has been determined by the balance method in the period March 1st until the sowing time, considering that the water storage at the beginning of this period at the field capacity. Production loss due to drought has been correlated with average humidity index $I_{u\,med}$ and Pálfai aridity index PAI_0 (fig. 3).

It can be observed that the relative production loss has a strong relation with both indexes, and these correlations have a logarithmical form, which means that the production loss gradient will increase while the droughts intensity increase too.

The analysed crops have the production loss which depend on droughts interval as follows: 17-20% for slight drought, 36-40% for moderate drought, 60-77% for strong drought, and the extreme droughts could determine the total loss of production.

The logharitmical correlation between the relative production loss and the droughts intensity indexes show a non-linear biologically process of evolution, fact shown by Petrásovits, by the index AHP for the characterisation of agricultural droughts.

7 Vulnerability to droughts

The final aim of the droughts analysis is the estimation of their consequences, at regional scale, and their agricultural damages. Damages have to be expressed not only as annual values (for tactically management of drought), but also multiannual (for strategically management of drought by irrigations). For annual estimations are useful the relations between intensity-duration-affected area-production loss, while, for

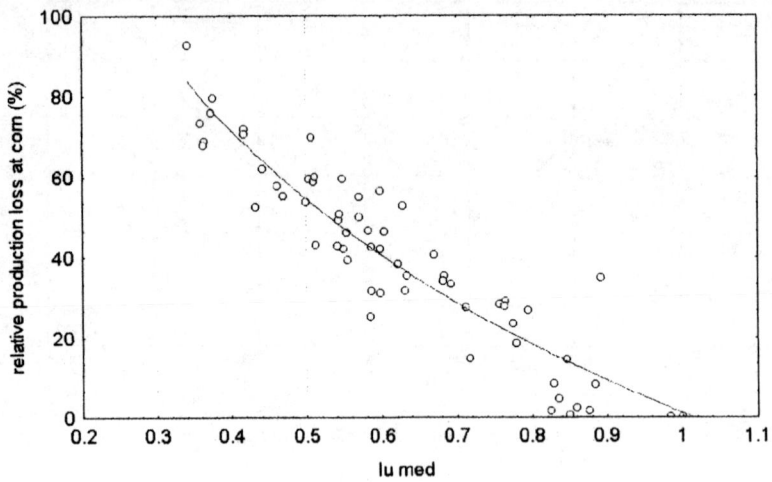

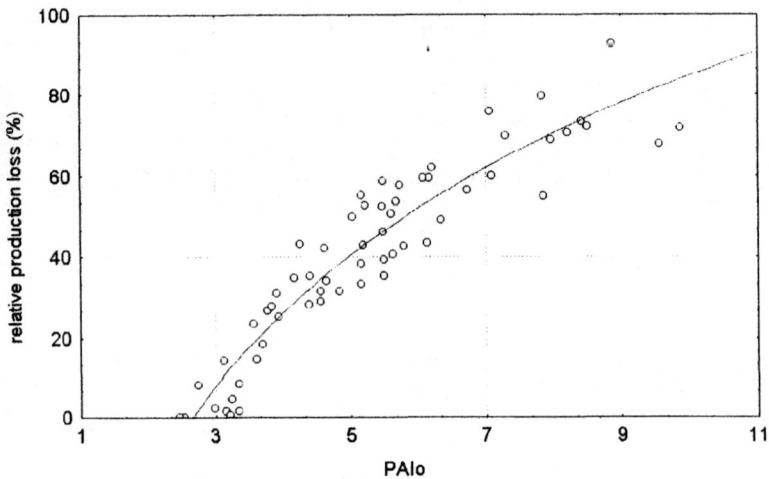

Fig. 3. Relative production loss for corn in case of a leached chernozem depending on: a) humidity index $I_{u\,med}$; b) Pálfai aridity index PAI_0.

multiannual estimations, the relations between intensity-frequency-affected area-production loss are important.

Using maps with territorial spreading of droughts intensity in each year, on the period 1982-1996 and the relationship between droughts intensity and relative production loss the average production loss in each year has been determined. Finally, a relationship between the intensity index in nucleus and the average production loss on the whole studied zone had been established. These relationships are the following:

corn: $RLP = -8.26 - 111.391 \log (I_{u\,med\,n})$;
sugar-beet: $RLP = -4.96 - 92.214 \log (I_{u\,med\,n})$;

alfalfa: $$RLP = -3.314 - 90.987 \log (I_{u\ med\ n}),$$

where: "RLP" is the relative loss of production expressed in % and "$I_{u\ med\ n}$"; - the average intensity index in Galati-Tecuci nucleus.

These relationships are useful for the operative evaluation of damages, practically in real time, together with the evolution of drought in the nucleus.

For the development of irrigation projects have been established multiannual average values of humidity index at 15 meteorological stations mentioned before and have been drowned izolines of humidity index. After that, using the relationship between $I_{u\ med}$ and relative production loss, the average multiannual damages and their territorial distribution have been established. (fig. 4).

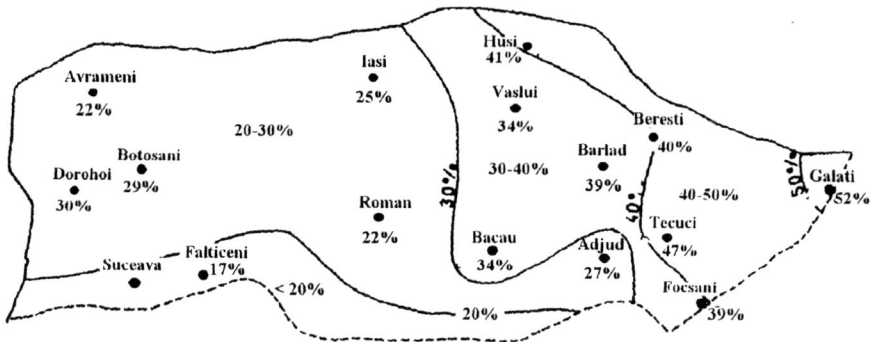

Fig. 4. Average multianual loss of production due to droughts

8 Conclusions

Droughts intensity is in strong relation with the duration: slight, moderate droughts and a part of those strong, do not exceed the period of a vegetation season, but extremes and a part of strongs are extended to the next agricultural season.

The affected area increase, while the droughts intensity increase too.

Using the simulation method of dynamics of soil water storage for a representative soil in the period 1981-1996 logarithmic correlations between droughts intensity and the relative production loss have been established.

There has been analysed the vulnerability to drought in real time and on a multiannual average scale on territory of Romanian Moldavia.

9 References

1. Topor, N. (1964) *Anii ploiosi si secetosi in R.P.Romana.*.C.S.A. Institutul meteorologic
2. Soroceanu, N. (1989) *Consideratii asupra conceptului si evaluarii fenomenului de seceta. Studii si cercetari meteorologice*, vol.3, INMH, Bucuresti.

3.   Palfăi, I., Szilárd, G.,Váradi, J. (1997) Use of the drought index (PAI) and water management influence of the drought in Hungary: some practical exemple, in *International Workshop "Sustainable Irrigation in Areas of Water Scarcity and Drought"*, Oxford, pp. 195-202.

4.   Teixeira, J.L. (1994) *Programa ISAREG - Guia do utilizador*, Dep de Engenharia Rural, Instituto Superior de Agronomia, Univ. Tecnica de Lisboa.

5.   Pálfai, I., Petrásovits I., Vermes L., (1995) Some methodological questions of the european drought-sensitivity map, in *International Workshop on Droughts in the Carpathians Region*, Budapest, Hungary, pp. 131-142.

MODELING THE OCCURRENCE OF DRY SPELLS AND THEIR IMPACT ON CROP YIELD

S.A.V. SOUSA
Department of Rural Engineering, ESALQ/USP-FAPESP, Piracicaba, SP, Brazil
J.A. FRIZZONE
Department of Rural Engineering, ESALQ/USP, Piracicaba, SP, Brazil

Abstract
A computational model, using Monte Carlo's Method, was development to simulate the occurrence of sequences of dry days during the cropping season and the consequent decrease in yield for a given crop. The simulation is based on series of frequency of sequence of dry days and evapotranspiration date. Valuation of the model was accomplished for corn production in the region of Piracicaba for two different crop periods: normal cropping season and cropping during the fall season ("safrinha"). Results show that in the normal cropping season the probability of substantial amount of lost is small. When cropping corn in Piracicaba during the fall season probabilities are higher of occurring substantial losses. For both cropping periods decreasing in yields was higher when dry days occurred during flowering stage. The stage of grain filling was the second most vulnerable crop stage to occurrence of dry days.
Keywords: Dry spells, irrigation planning, modeling, Monte Carlo method.

1 Introduction

Most of the Brazilian territory the agriculture is practiced without irrigation, that is to say, the crops are developed depending exclusively on the natural precipitation. The rainy period in Central-West, Southeast and South regions in Brazil is the summer, when most of the crops are grown up.

In some regions, specially that which natural vegetation is called "cerrados", the total rainfall of the rainy period is enough for the development of the agriculture, but is common the occurrence of sequence of dry days during the rainy season, what is known as dry spell. Depending on their duration and season occurrence, dry spells can

Water and the Environment: Innovative Issues in Irrigation and Drainage. Edited by Luis S. Pereira and John W. Gowing. Published in 1998 by E & FN Spon. ISBN 0 419 23710 0

affect in an accentuated way the crops development and, consequently, the final productivity.

The occurrence of long drought periods (dry spells) is common, mainly in Central and Central-West region of Brazil. The crop yield losses vary with the intensity and duration of the water stress, as well as the losses depend on the crop development stage [1].

The annual irregularity on the occurrence of dry spells makes corn a vulnerable crop to water deficit in any development stage, with visible damages its crop yield [2].

It was verified corn yield reductions up to 60%, when the water deficit occurred since the flowering stage to until the grain filling stages, and about 40% when it occurred during the flowering initiation [3]. The same authors observed that, for the experiment in that the drought periods occurred during whole crop reproductive process, or in a part of it , the supplemental irrigation allows practically to duplicate corn yield of the tested varieties.

The forecast of the occurrence of dry spells for a given region is fundamental, and added to the losses that these periods can cause, they are an important tool for the development of the agriculture with a smaller risk for the farmer.

This forecast can also be adopted as an auxiliary tool in the process of taking of decision for the elaboration of irrigation projects as form of minimizing the risks in the agriculture.   There are some techniques do the forecast of meteorological phenomenons. The simulation process can be adopted with this purpose, because it allows starting from historical data to simulate values of future occurrence.

The main objective, when developing a simulation model, is to obtain a realistic representation of a system behavior [4]. The choice of right model, deterministic or stochastic, is fundamental for reaching this objectify. In a deterministic model, the simulation result is function only of the incoming parameters, while in the stochastic models there are aleatory variable introduced, and each simulation leads to a different result, and the results must be statistically analyzed [5].

This work had the following objectives: to develop a computer program for simulation of occurrence of dry spells in a month for a region, and the respective crop yield decrease, corresponding to the evapotranspiration deficit in the different development stages. The model was used in the simulation of dry spells occurrence for two seasons of corn crop growing, at county of Piracicaba, State of São Paulo, Brazil: summer (corresponding to the months of January, February and March), and "safrinha" (corresponding to the months of May, June and July).

2 Material and methods

2.1 Software development

The simulation model allows a choice among three options. The first one, to calculate only the relative frequency of the dry spells occurrence for a region. This option requires only the data the observed frequency of sequence of days without rain, for a determined month, obtained of a historical series. It is recommended a series with a minimum of 20 years of observation for a good model adjustment.

The second option allows, besides the simulation of the dry spell, the simulation of the daily referential evapotranspiration during the dry spell duration, and calculates the daily value of the maximum evapotranspiration and real evapotranspiration for a crop.

It can also calculate the evapotranspiration deficit and its respective yield decrease for the referred crop, in each development stages. This option requires, besides the data of frequency of days without rain, the daily evapotranspiration data, the crop coefficients and coefficients to response to water in each stage, and the soil characteristics data.

The third option calculates the crop yield decrease for a dry spell with a know duration, occurred in each stage of development of a determined crop. The simulation done in a similar way to the second option, except for the dry spell duration, which that should be pre-established.

2.2 Method of simulation and results determination

It was used the method "Monte Carlo" of simulation [6].

Two functions for data adjustment are used on the dry spell simulation. The first bases on the empiric distribution of probabilities, that is to say, they are generated aleatory numbers that are compared with the accumulated frequency and starting from this, they are determined the different probabilities of durations of the dry spells.

In the second option the values are simulated with base in the first order on Markov's chain([7], [8], [9]).

The simulations are monthly, that is to say, the relative frequencies and the return periods are simulated, for each dry spell, of different durations, occurred in a month pre-established by the user.

The simulation of reference evapotranspiration (ETo) can be done with the use of two functions. The first one allows data adjustment by the normal function. In this option, the ETo data, obtained from a historical series, are divided into classes with interval of 1,0 mm, and the model adjusts the normal function in order to describe the data behavior. After the function adjustment, the model generates aleatory numbers and using these with the normal function adjusted, allows the obtaining the ETo values.

The second one simulates the ETo values using the triangular function. This function is used when there are a few available data. Its utilization requires the minimum, maximum and the most frequent expected value (mode). Aleatory numbers are generated and adjusted to the triangular function ,and starting from that, daily ETo values are simulated.

The other model calculations are deterministics.

The values of the maximum evapotranspiration (ETm), for the chosen crop, are determined for each day into an interval of dry spell duration, at different development stages. ETm is obtained from the multiplication of each ETo simulated value by the respective crop coefficient. The model also determines the real evapotranspiration of crop (ETr), in every day, in the different development stages. For the ETr determination it was considered in the model the methodology presented by [10].

It is determined the sum of maximum evapotranspiration and the real evapotranspiration for each consecutive sequence of days without rain, in the different stages of development of crop chosen. The model also determines the evapotranspiration deficit and the respective relative crop yield, given by the relation between the yield obtained due to the dry spell, and the maximum crop yield, for the chosen crop. These determinations are made using the methodology proposed by [11].

2.3 Model evaluation

In order to evaluate the model, it was studied the probability of dry spell occurrence in two crop growing seasons, at the region of Piracicaba (São Paulo, Brazil): summer (corresponding to the months of January, February and March) and the period of the "safrinha" (corresponding the months of May, June and July). It was simulated the yield decrease on corn crop, due of the dry spell occurred in each period.

It was used in the simulation a 20 years historical data series (1976-1995), obtained in Physics and Meteorology Department of ESALQ-USP (Superior School of Agriculture "Luiz de Queiroz", State University of São Paulo). It was determined from this data series, for each month considered on the work, the observed frequency of sequence of days without rain, since 1 up to 31 days, case occurred. It was defined a criterion for defining dry days. Dry days were that which the rainfall had less than 3,0 mm.

Daily values of ETo were obtained from the same series, for the respective months. The values were organized into classes of 1,0 mm, for normal function adjustment. The ETo values used on adjustment were determinedes for the equation of Thornthwaite, quoted by [12].

It was considered the corn crop at four development stages, that is to say, vegetative, flowering, grain filling and maturation. The respective crop coefficients were obtained with the methodology proposed by [13], adapted to the region of Piracicaba.

In order to calculate the relative yield, in function of the maximum yield, were used the corn coefficients of response to water (Ky), at the same development stages already quoted. These values were obtained on a table presented by [11].

Were used the soil data obtained by [14], in experiment conduced in the experimental area of the Rural Engineering Department of ESALQ.

3 Results and discussion

The Fig. 1 shown the values of the relative yield, simulated with the use of the software, in function of each dry spell, at corn crop development stage, for the tree months studies, at the two considered seasons: summer and "safrinha".

Observing the Fig. 1, it can be verified that, at every month studied, the yield decrease due to the dry spells is accentuated when these occur in the flowering stage, followed by grain filling stage. When the dry spell occur in the stage vegetative and maturation, the yield decreases are small, even with more intense dry spells. This was just hope to occur, because Ky values (response factors of the crop to the water) used, they show that the more sensitive stage of crop corn are flowering and grain filling, being also this observed by [1] and [2].

For the three months studied in the summer, the largest decreases in the yield, at each stage, were provoked of dry spells with duration corresponding to 20 days, occurred in February.

For the period of the "safrinha", the largest yield decreases were obtained for a dry spell of 30 days in May, although its occurrence probability is small (0,36%).

Comparing dry spells of corresponding duration, it could be observed that, for a same development stage, the dry spells occurred in February (summer) and May ("safrinha") was the ones that they provided the largest decreases in the yields.

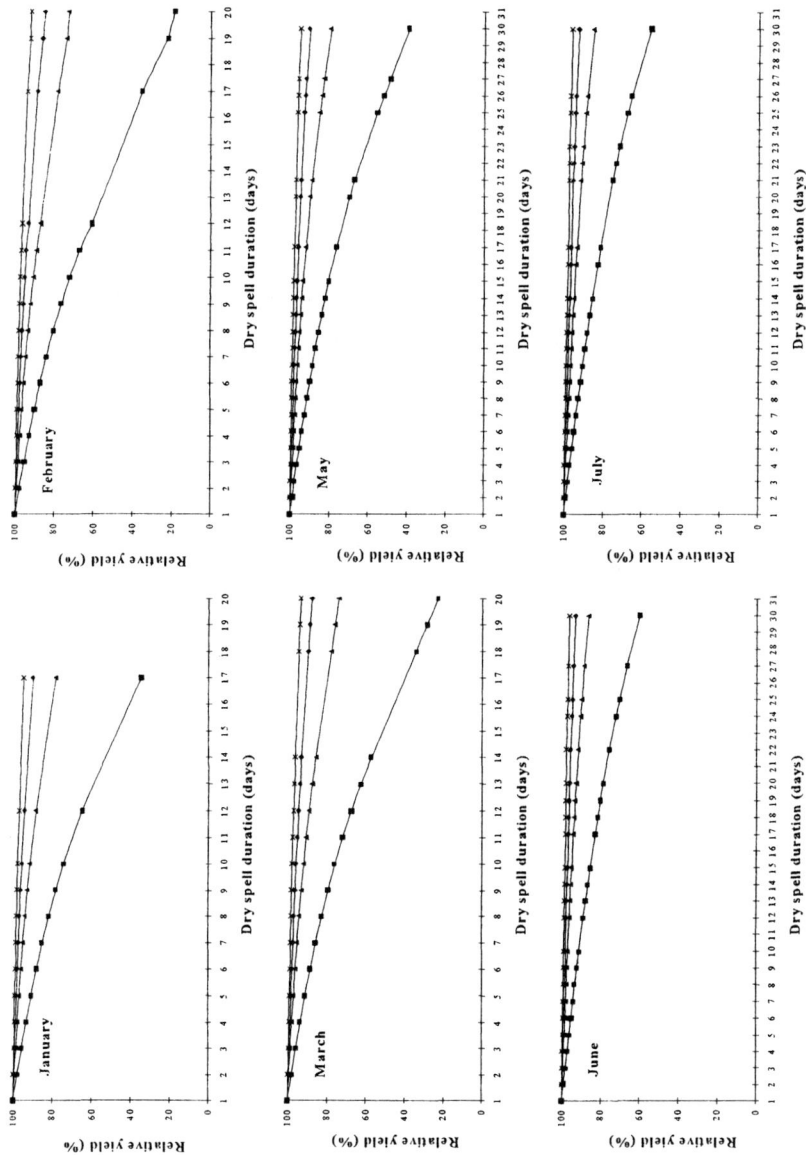

Fig. 1. Relative yield, in the different corn development stages, for different dry spells durations (vegetative periods: ◆ growth, ■ flowering, ▲ grain filling, × maturation).

This can be better observed in Fig. 2, that it presents the relative yield for dry spells occurred in the flowering stage, in every month, in the periods of summer and "safrinha".

In Fig. 2 are observed that, for a dry spell of same duration, the relative yields obtained in the summer, were smaller for dry spells occurred in February, followed by the occurred in January and March, although the differences between the decreases in the yields corresponding to a same duration are small. In the "safrinha", the months sequence whose dry spells provided smaller relative yields were May, July and June. In this period, the dry spells occurred in May, mainly the one of larger duration, provide very more decreases accentuated in the yield, when compared with occurred them in the other two months.

Opted for presenting the flowering stage, for this to be the most sensitive to the dry spells, us other stages, the sequences of the months went the same to both growing periods.

The Table 1 shows the simulated values of the relative frequency of the dry spells from different durations occurred in the region of Piracicaba - SP, for every month, in the two studied periods. It is verified that the probability of occurrence of dry spells of large duration is small in the summer, because the largest simulate frequencies, were obtained for dry spells of duration of up to 4 days.

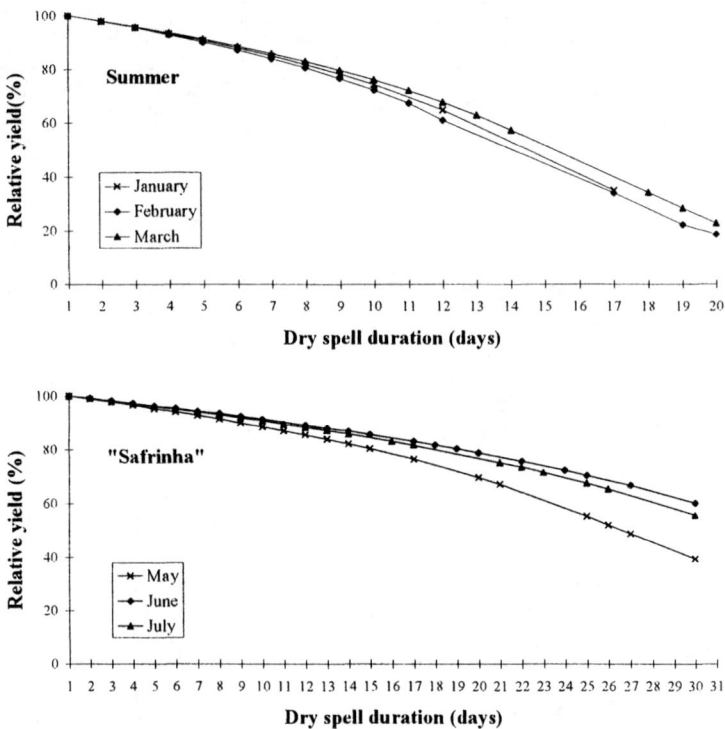

Fig. 2. Relative yield, in the corn flowering stage, for different dry spells durations.

In the "safrinha" there are larger probabilities of occurrence dry spells of more intense. In this period, the month of June is what presents larger probability of occurrence of dry spells of larger duration.

Comparing the values of Table 1 with Fig. 1 and 2, it is possible determinate which the probability of occurrence of dry spells of different durations and the respective relative yield for the corn, at each month studied for Piracicaba region. As an example, it is waited that the probability of occurrence of a dry spell with duration of 7 days in the month of January is of 5.52%. For this period is waited a relative yield of 97.0, 85.0, 95.0 and 99.0% of the maximum yield, if the referred dry spell occurs in the vegetative, flowering, grain filling and maturation stage, respectively.

Table 1. Relative frequency (%) of simulate dry spell at Piracicaba region

Dry spell duration	Months					
(days)	January	February	March	May	June	July
1	35.68	26.84	19.20	26.78	12.58	9.66
2	17.00	24.90	28.94	11.04	19.20	5.52
3	14.32	12.93	10.52	15.98	2.24	3.78
4	10.96	13.98	14.46	11.88	6.66	8.58
5	4.72	5.75	4.10	1.26	7.14	5.26
6	3.72	3.97	0.58	1.88	5.94	2.10
7	5.52	1.31	7.95	6.02	3.26	1.02
8	3.74	5.71	3.87	2.30	2.70	2.54
9	1.66	0.27	2.03	2.30	2.14	18.56
10	1.64	2.11	1.60	2.66	4.68	1.98
11	0.00	1.32	0.47	4.40	0.00	1.98
12	0.44	0.01	4.36	0.88	4.04	0.48
13	0.00	0.00	1.96	1.76	4.54	7.36
14	0.00	0.00	0.22	2.18	2.50	0.50
15	0.00	0.00	0.00	0.84	2.50	0.00
16	0.00	0.00	0.00	0.00	0.00	2.36
17	0.60	0.01	0.00	1.26	0.48	2.36
18	0.00	0.00	1.20	0.00	1.92	0.00
19	0.00	1.00	0.02	0.00	1.40	0.00
20	0.00	1.16	0.02	1.98	1.42	0.00
21	0.00	0.00	0.00	1.20	0.00	8.18
22	0.00	0.00	0.00	0.00	0.94	2.68
23	0.00	0.00	0.00	0.00	0.00	3.96
24	0.00	0.00	0.00	0.00	3.16	0.00
25	0.00	0.00	0.00	1.14	1.36	2.20
26	0.00	0.00	0.00	1.52	0.00	1.80
27	0.00	0.00	0.00	0.38	1.00	0.00
28	0.00	0.00	0.00	0.00	0.00	0.00
29	0.00	0.00	0.00	0.00	0.00	0.00
30	0.00	0.00	0.00	0.36	8.20	7.14

For the other months and dry spells durations the analysis can be made in a similar way. Coming this way, it can be determined which the probability of occurrence of the dry spells and the respective decrease of yields , being this an important methodology, so much for the planning of the agriculture, in the choices of the growing seasons that will provide smaller risks of yields decreases, as in the study of the viability of implantation of irrigation projects and its management.

4 Conclusions

1. The developed model allows simulating the occurrence of dry spells of different duration intensities, and the respective relative yield, being considered the region and the month of development of the crop.
2. For the region of Piracicaba, the probability of occurrence of dry spells of large duration in the summer is small. In the "safrinha" there are larger probabilities of occurrence of more intense dry spells.
3. For the two growing periods, the stages of flowering and grain filling are the most sensitive the dry spells occurrence. Dry spells, even if they are intense, in the stages vegetative and maturation they provoke small yield decreases.
4. The largest simulate yields decreases in the summer, are provoked of dry spells with duration of 20 days in February. In the "safrinha", the largest decreases are provoked of dry spells of 30 days occurred in May.
5. Considering a same duration and occurrence stage, in the summer, dry spells in February is the ones that they provide the largest decreases in the yield. In the "safrinha", the largest decreases are provoked of dry spells in May.

5 References

1. Couto, L.; Costa, E.F.; Viana, R.T. (1986) Avaliação e comportamento de cultivares de milho em diferentes condições de disponibilidade de água no solo, in *Relatório técnico anual do Centro Nacional de Pesquisa de Milho e Sorgo: 1980-1984.* Sete Lagoas, MG: Embrapa-CNPMS, p.77-78.
2. Barbosa, J.V.A. (1986) Efeito do veranico sobre a produção de cultivares de milho, in *Relatório técnico anual do Centro Nacional de Pesquisa de Milho e Sorgo: 1980-1984.* Sete Lagoas, MG: Embrapa-CNPMS, p.80-82.
3. Espinoza, W.; Azevedo, J.; Rocha, L.A. (1980) Densidade de plantio e irrigação suplementar na resposta de três variedades de milho ao déficit hídrico na região dos cerrados. *Pesquisa Agropecuária Brasileira*, v.15, n.1, p.85-95.
4. Emshoff, J.R.; Sisson, R.L. (1970) *Design and use of computer simulation models.* New York: Macmillan Publishing, 302 p.
5. Costa, M.H. (1991) Modelo de otimização dos recursos hídricos para irrigação, conforme a época de plantio. Viçosa, 111 p. Dissertação de Mestrado – Universidade Federal de Viçosa.
6. Hillier, F.S.; Lieberman, G.J. (1988) *Introdução à pesquisa operacional.* São Paulo: EDUSP, 805p.
7. Coe, R.; Stern, R.D. (1982) Fitting models todaily rainfall data. *Journal of Applied Meteorology*, V.21, p.1024-1031.
8. Katz, R.W. (1974) Computing probabilities associated with the Markov chain models for precipitation. *Journal of Applied Meteorology*, v.13, p.953-954.
9. Stern, R.D.; Coe, R. (1982) The use of rainfall models in agricultural planning. *Agricultural Meteorology*, v.26, p.35-50.

10. Bernardo, S. (1989) *Manual de Irrigação*. 5.ed. Viçosa, MG: UFV, 596p.
11. Doorenbos, J.; Kassam, A.H. (1979) *Yield response to water*. (Irrigation and Drainage Paper, 33). Rome: FAO, 193p.
12. Jensen, M.E. (1968) *Consumptive use of water and irrigation water requirements*. New York: American Society of Civil Engineers, 215p.
13. Doorenbos, J.; Pruitt, W.O. (1977) *Crop water requirements*. (Irrigation and Drainage Paper, 24). Rome: FAO, 144p.
14. Duarte, S.N. (1989) Efeitos do horário e da lâmina de irrigação na cultura da batata (Solanum tuberosum L.). Piracicaba, 148p. Dissertação de mestrado – Escola Superior de Agricultura Luiz de Quiroz -USP.

RECONNAISSANCE OPTIMAL SUSTAINABLE GROUNDWATER PUMPING STRATEGIES FOR THE LOWER GHAGGAR BASIN

A. KUMAR and R. SHYAM
Pant College of Technology, G.B. Pant University of Agriculture and Technology, Nanital, India
N.K. TYAGI
Central Soil Salinity Research Institute, Karnal, India
R.C PERALTA
Dept. Biological and Irrigation Engineering, Utah State University, Logan, UT, USA

Abstract
A steady-state groundwater optimization model is constructed to compute sustainable groundwater development strategies for part of Lower Ghaggar Basin in Haryana State (India) region and has flat topography. Basin lies in arid and semi-arid part. Recharge from huge canal network has resulted in rising trend of water table. The model developed represents steady-state groundwater flow response to assumed and computed optimal discharges and recharges. The model computes the spatially distributed maximum sustained yield pumping volumes for the project area. The optimal pumping rate is greater than current pumping which can be used to fulfill unmet demands for irrigation water. The difference in pumping results from an increase in recharge from the rivers and a decrease in groundwater evapotranspiration losses due to reversal of water table behaviour. The number of pumping wells needed to extract water at the optimal rates are also computed.
Keywords: Groundwater management, irrigation management, modeling, pumping, sustained yield.

1 Introduction

Planned groundwater development is necessary for ensuring sustained groundwater availability. Planning and management agencies are generally interested in determining the maximum sustainable pumping rates that will not cause unacceptable drawdowns or other flows. Determining the best acceptable pumping strategy for a region requires consideration of spatially variable hydrogeology and management goals. Use of mathematical optimization procedures can improve planning and development of

Water and the Environment: Innovative Issues in Irrigation and Drainage. Edited by Luis S. Pereira and John W. Gowing. Published in 1998 by E & FN Spon. ISBN 0 419 23710 0

sustainable irrigation systems. Simulation and optimization models are often used to achieve the above goals [1], [2], [3]. Development agencies can be relatively sure that groundwater system response will be in the acceptable range if procedures employed to develop the water management strategies are appropriate.

The present study aims at developing optimal groundater pumping strategies for the Lower Ghaggar River Basin, which is a part of the Indo-Gangetic quaternary basin in the northern Indian subcontinent (Fig. 1). The Ghaggar River flows through the northwestern part of the study area. The river is mostly an influent stream in the study area and contributes substantially to groundwater recharge. An extensive canal network distributes water for irrigation. Seventy percent of the land is irrigated with an average water allowance of 1.0 cumecs per 5000 ha.

Numerous exploratory bore hole investigations made by Haryana State Minor Irrigation and Tubewells corporation (HSMITC) and Central Ground Water Board (CGWB) indicate that the basement rocks are at a depth of 200-300 m on the southern side. The thickness of the alluvium decreases from northeast to southwest. The quality of the ground water varies both in vertical as well as lateral directions. Spatial variations are shown in Fig. 2. Electrical conductance (EC) generally increases with depth. Nearly 35 percent of the area is underlain by saline water (EC > 6 dS/m) [4]. The fresh water (EC < 2 dS/m) underlies about 12 percent of the area. The rest of the aquifer is marginally saline (EC:2-6 dS/m). Both salinity and sodicity problems exist. Rise of water table in saline groundwater zone which leads to development of waterlogging and the scarcity of water to meet the irrigation demands are the major problems.

2 Simulation and optimization models

In the present study development of optimal groundwater management strategies involves application of both simulation as well as optimization model. The simulation models used is basically [5] algorithm using finite difference technique. For optimization a steady state optimisation model is formulated on the basis of three dimensional partial equation of groundwater flow [6]. The linear programming model is translated into GAMS [7]. The objective function gives maximum sustainable pumping yield under well defined constraints and bounds. The study state is preferred over unsteady state because the basin lies in arid and semi-arid region where there are less chances of prolonged wet periods. The study area is discretized as shown in Fig. 1. It is the part of a larger unconfined aquifer. Artificial boundary conditions (constant-head/constrained flux cells) are used around the periphery. They permit maintenance of boundary heads while assuring that induced recharge across system boundaries will not disrupt regional flows outside the optimized area. The objective function maximises pumping from all variable head cells under well defined constraints and bounds [1].

An upper bound and a lower bound on head are fixed for each cell to prevent waterlogging and dewatering of saturated thickness of aquifer due to excessive pumping, respectively. An upper bound on groundwater extraction is imposed which is some multiple of current extraction rate. Lower bound is set to current extraction rate. Similarly recharge entering the study area through boundary cells is not allowed to

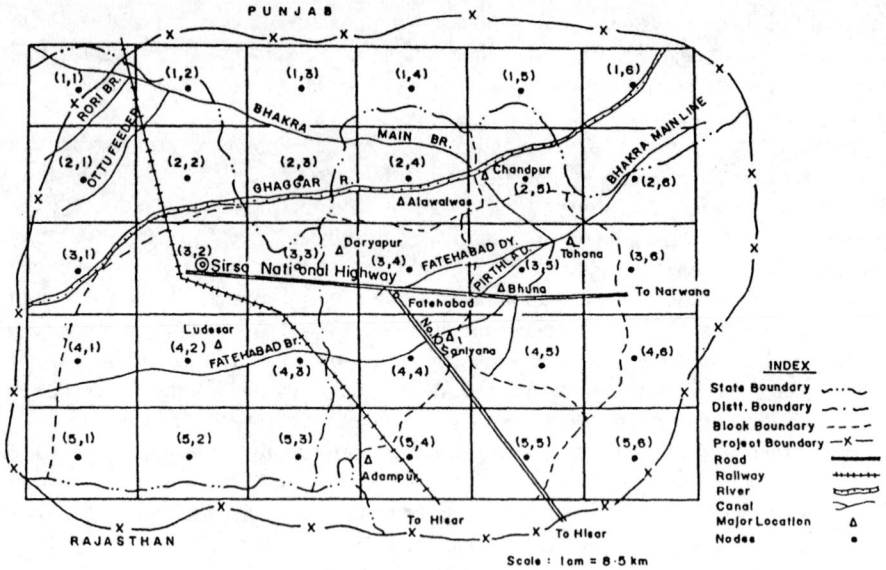

Fig. 1. Location map of the area

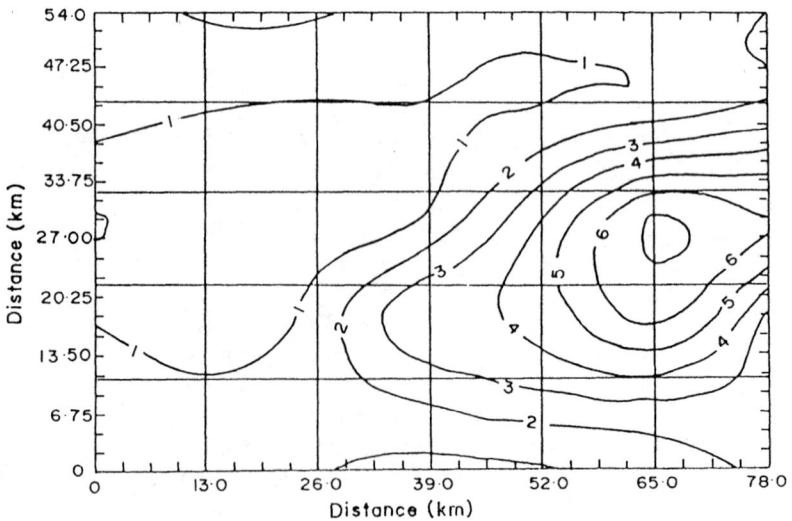

Fig. 2. Groundwater salinity contours (dS/m)

exceed current rates. Evapotranspiration on volume basis is assumed maximum when water table is above ET surface elevation. It ceases when depth of water table below ET surface exceeds specified interval. Between these limits evapotranspiration from water table varies linearly with water table elevation. The different flows in the model are shown in schematic diagram (Fig. 3). Detailed model description is given by [1].

Frequently, a simulation model is used to verify the consequences of implementation the optimal strategy [8]. The same approach is used to predict the time needed for attaining optimal steady state potentiometric surface in the area.

3 Model application

The groundwater simulation model is calibrated with three years of data from 1981 to 1983. During the calibration value of relaxation factor adopted in analysis is 0.80 and allowable error for the Gauss-Seidal iteration equation is taken as 0.5 Mm³/year. The calibrated model represents the best output that can be achieved with available data. The verification of model is done with water table data of 1984 and 1991. Thereafter, model is used for simulation.

The basic data required by optimisation model essentially includes information on topography, climatology, surface hydrology and geohydrology. The corrected and verified data (Table 1) from ground water simulation model is used in optimisation model as input.

4 Results

The basic data required for model essentially includes information on topography, climatology, surface hydrology and geohydrology. The corrected and verified data (Table 1) from groundwater simulation model are used in optimisation model. Computed optimal results are shown in Table 2. These include optimal pumping rates, stream-aquifer interflow, and potentiometric surface. Total optimal pumping is 44.1 cumecs which is more than twice the current pumping.

There is wide variation in optimal pumping rates between cells. The values range from 0.25 cumecs to 8.48 cumecs. The pumpable quantities of ground water depend largely on recharge opportunity. Areas near rivers and canals have greater opportunity for recharge than cells more distantly located. For example, river cells (2,1), (2,2), (2,3), (2,4), (2,5), (3,1) have pumping rates 4 to 20 times those of non-stream cells.

The total sustainable optimal pumping is significantly greater than current pumping. Implementation of the optimal strategy will prevent further water table rise. The optimal potentiometric surface lies 4 to 22 m beneath the ground surface (Fig. 4).

The calibrated simulation model is used to estimate how many years the optimal strategy would take for the potentiometric surface to evolve into the optimal surface (Table 2). River cells would attain their optimal potentiometric surfaces earlier than non-river cells.

Current flow from surface water bodies to the aquifer is 16.21 cumecs. The optimal strategy will cause this to increase to 26.1 cumecs. The increase in stream-aquifer interaction indicates the feasibility of generating more water resources from river flow,

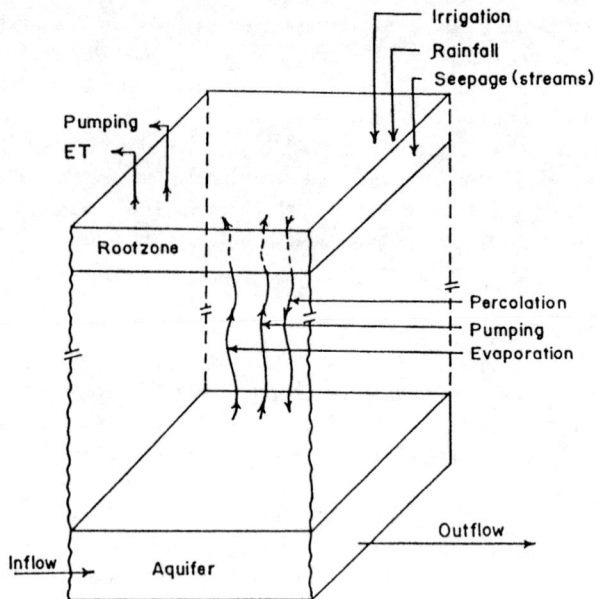

Fig. 3.   Groundwater system

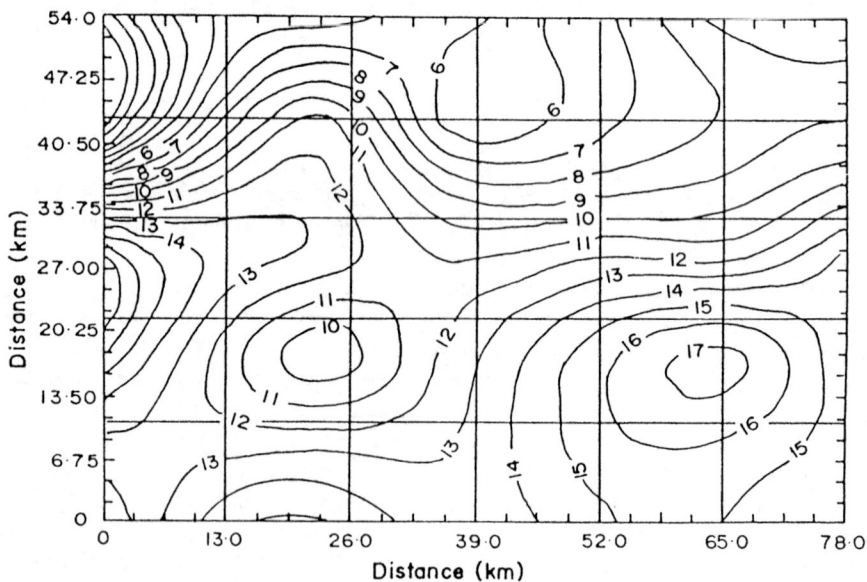

Fig. 4.   Depth to watertable contours (m)

which at present goes waste and creates waterlogging problems downstream in Rajasthan.

Table 1. Values of groundwater model inputs

Cell (Row, Column)	Hyd. conductivity (m/day)	Ground surface elevation (m above MSL)	Aquifer base elevation (m above MSL)	Initial hydraulic head (m above MSL)
1,1	12.45	203.82	35.66	200.06
1,2	14.98	209.87	91.60	204.33
1,3	16.23	212.57	100.93	206.34
1,4	18.35	221.56	124.38	214.91
1,5	18.23	226.83	86.78	217.41
1,6	18.09	230.05	76.98	220.15
2,1	14.56	201.25	88.88	196.81
2,2	18.23	204.68	92.08	192.61
2,3	19.98	209.53	100.30	203.96
2,4	16.92	217.98	117.03	211.26
2,5	15.24	223.15	95.21	214.98
2,6	14.60	226.80	80.93	217.75
3,1	15.32	198.86	64.39	184.56
3,2	19.62	204.28	83.50	181.16
3,3	19.98	210.23	101.32	199.71
3,4	16.92	213.59	95.63	202.27
3,5	13.80	217.42	90.68	206.53
3,6	11.79	220.60	94.43	206.52
4,1	14.13	196.70	50.60	183.01
4,2	13.92	207.10	73.90	197.96
4,3	13.92	209.00	80.05	196.72
4,4	15.86	210.90	86.19	195.48
4,5	11.62	219.83	79.59	197.43
4,6	10.82	219.72	98.46	205.60
5,1	11.95	202.67	55.40	189.88
5,2	11.40	208.32	79.13	194.47
5,3	11.70	210.93	87.70	197.89
5,4	11.44	213.89	63.73	198.60
5,5	10.84	217.12	78.94	201.80
5,6	10.24	217.12	103.54	202.82

5 Field implementation of the optimal pumping strategy

Implementation of optimal strategy requires the number of tubewells to be increased by about eight-fold, mostly in river cells. In this area many pumpsets are in operation for pumping groundwater from shallow tubewells. The technology for groundwater development through shallow tubewells is available at village level. Further, the water availability from canals is inadequate as it meets the requirement of only 35-40 percent area during each crop season and therefore, development of groundwater would be profitable and economically viable proposition.

Table 2. Simulation/Optimisation model output and years required to achieve the optimal heads

Internal nodes	Drawdown (m)	Saturated thickness (m)	Optimal head (m)	Optimal pumping (cumecs)	Years to achieve opt. head*
2,1	1.26	106.66	195.54	2.59	2
2,2	-8.00	108.53	200.61	7.93	2
2,3	1.33	102.32	202.62	8.48	2
2,4	0.67	93.55	210.58	8.10	2
2,5	1.00	118.76	213.97	2.39	4
3,1	8.00	112.17	176.56	2.42	3
3,2	-3.16	100.82	184.32	2.02	3
3,3	4.27	94.11	195.43	0.85	7
3,4	7.00	99.64	195.27	1.32	9
3,5	6.00	109.85	200.53	1.65	7
4,1	-2.80	135.21	185.81	1.90	8
4,2	-3.29	127.35	201.25	0.25	7
4,3	3.77	112.89	192.94	1.50	2
4,4	1.56	107.72	193.91	1.34	7
4,5	-0.35	118.19	197.78	1.39	8

*Determined via post-optimization simulation.

6 Summary

The management model that has been formulated to compute optimal sustained groundwater yield strategy helps in evolving the potentiometric surfaces. In Lower Ghaggar Basin, the current pumping rates are only half of the optimal pumping rates indicating the scope for further groundwater development. The difference in the current and optimal pumping rates is primarily due to increased flow from rivers to the aquifer and a reversal of water table behaviour under optimal pumping strategy.

7 References

1.   Kumar, Ambrish (1993). Developing optimal ground water pumping strategies: a case study of Lower Ghaggar Basin, unpublished thesis submitted to G.B. Pant University of Agriculture and Technology, Pant Nagar in partial fulfilment of the requirement of master's degree in Agricultural Engineering, p. 153.
2.   Peralta, R.C. and Aly, A.H. (1966). Software for optimizing groundwater or conjunctive water management, In proceedings sixth international conference on Computers in Agriculture, Cancum, Medico, June 1966, ASAE, St. Joseph, MI, pp. 1058-1066.
3.   Tyagi, N.K., Tyagi, K.C., Pillai, N.N. and Willardson, L.S. (1993). Decision support to irrigation system improvement in saline environment, *Agricultural Water Management*, 19(4):285-301.

4. Anonymous (1983). Groundwater Studies in the Ghaggar River Basin in Punjab, Haryana and Rajasthan. Final Technical Report, DP/UN/INP/7400 9/2, India, United Nations, New York.

5. Tyson, H.N. and Weber, E.M. (1964). Groundwater Management for the Nations Future. Computer Simulation of Groundwater Basin. *Jr. Hydraulics Division of ASCE* 90:(HV-4)59-78.

6. Mc Donald, M.G. and A.W. Harbaugh (1988). A Modular Three Dimensional Finite Difference Ground Water Flow Model. U.S. Geol. Surv. Tech. Water Resource. Inv., 6:A1.

7. Brook, A., Kendrik, D. and A. Meeraus (1988). *GAMS A User's Guide*, The Scientific Press, 651 Gateway Boulevard, Suite 100, South San Francisco, CA94080-7014.

8. Daubert, J.T., and R.A. Young (1982). Groundwater Development in Western River Basin: Large Economic Gains with unseen costs. *Groundwater*, 20(1):80-85.

IRRIGATION, SUSTAINABLE DEVELOPMENT AND LAW

M. MATHUS ESCORIHUELA
School of Law, Cuyo National University, Mendoza, Argentina

Abstract
The growing demand for irrigation water and the ever-increasing shortage of water resources claim for new legal approaches, for a water administration reform and for an active participation of water users. These changes, which aim at getting sustainable development, urge that the concept of "efficiency" be redefined. This is an ambiguous term and it admits different interpretations and uses in water policy, legislation, as well as in water administration. The purpose of this paper is to clear up the matter as regards its environmental dimension by considering the impact of new technologies, pollution control and water reuse and administration. The regulation of these aspects in Argentine legislation are analyzed, and some proposals are put forward.
Keywords: Ecosystem, efficiency, irrigation, law, sustainable development, use, water quality, water use.

1 Introduction

Law orders social reality and its function is to reach "goals" that the legislator considers valuable to achieve the common good. Therefore, the **efficacy** of the law lies in the adequate selection of the means to reach those goals.

All legal institutions should be geared to achieving **not only** the fair end of common good, **but also** the proper means to obtain that end. If the end is reached, the law as a means is efficient. At the same time, this concept is related to the effective enforcement of the law, and this happens when law is accepted and abided by society.

In this paper the concept of efficiency is considered in terms of water resources and one use in particular, irrigation, while considering the environmental aspect as well.

Water and the Environment: Innovative Issues in Irrigation and Drainage. Edited by Luis S. Pereira and John W. Gowing. Published in 1998 by E & FN Spon. ISBN 0 419 23710 0

2 Background

The Bruntland Report updated the Conservation principles applied to the use and conservation of natural resources and has allowed for tackling jointly environmental and socio-economic development problems.

Sustainable development aims at meeting the present basic social needs without affecting the environmental resources of future generations. The goal of intergenerational justice is conditioned by the resources themselves, by the use of new technologies, by the unequal socio-economic organization of the world, and by the restricted capacity of the biosphere to buffer the noxious effects of human activity.

Sustainable development requires, therefore, that population growth be related to the productive capacity of ecosystems in a dynamic process which can increase goods and services to redistribute wealth and overcome the inequality of today's global economic systems.

3 Relevance

In the context of the water crisis that characterizes the turn of the century, irrigation plays a crucial role in achieving sustainable development. The increase in irrigated agricultural areas and their esential production aim at enlarging the stock of food resources. Consequently, it is important to improve irrigation efficiency, which also entails preserving the quantity and quality of water resources.

4 Limitations

Several Laws, Codes, Covenants and International Agreements, have considered the problem of water contamination and reuse, but few legal regulations (latu sensu) have defined the concept of **efficiency** in water use. Nor have other issues related to efficiency in water use been analyzed which condition or hinder such efficiency. Therefore, it is necessary to define the concept in some fields of application:

1. Water is an esential component of the environment. If by "efficiency" we understand the capacity to achieve an effect, then the water resources of an ecosystem work efficiently when water acts as a dynamic interdependence factor of all the environmental components of that ecosystem, the preservation of the genetic diversity of species and the sustained use of its components.
2. Efficiency in the use of water is an effect resulting from different factors. It is not just achieved by the efficient action of the user whose intervention is a mere phase in a complex process. An efficient result in irrigation, as well as in other uses of water, derives from: the efficiency of the Water Policy, the Law, the Structure and Administration of the Water Management Institution, an adequate infrastructure of works, the active participation of users and the balance of the ecosystem involved. Thus, water is used efficiently when it is used at its fullest without degrading its quality and quantity.

Let us analize the factors that contribute to obtaining that result:

3. The Water Policy is a set of regulations and institutions that determine the goals to be achieved and the means to be used with respect to water at a certain time and place. It results from a rational plan, supplemented with   the cultural tradition of a certain society whose basic objective is the good   use of water with the optimal benefit of the water resources involved. Optimal benefits and society's concept of beneficial use are terms that must be defined by policy-makers according to the average values and the socio-economic needs of the population. Thus, concerning the water policy, the concept of beneficial use, which assumes an efficient use, is contingent and variable.

4. The law is a conditioning factor of efficient use. The law is efficient when it allows for solving cases and situations that the legislator foresaw or wished to solve. Laws may either hinder or maximize the use of water by the user. If the legislation is a "transplant" that does not respect the immemorial uses and customs of a community, it will hinder any beneficial change. If, on the contrary, it promotes the use of new technologies, the training of human resources and financial support, it will encourage the application of new, more efficient uses.

5. Water management-- carried out by public and private organizations connected to the management of water (in the sense of administration)-- is a complex activity. It includes the use (delivery of volume, use and control of affluents), hydraulic works (project, construction, operation and maintenance), defence against noxious effects, the taxing and financial system, user participation, etc. Efficiency in management is one of the factors to achieve the goals of the water policy and, like the law, it can either hinder or maximize its use, that is, efficiency in its use. It means counting on trained human resources, dynamic procedure regulations, adequate coordination among the participating offices and efficient user participation.

6. In some cases, seemingly efficient uses of water through the use of new technologies have altered the environmental balance. The increase in the irrigated area as well as in yield has damaged environmental components. Pressurized irrigation using subsurface water has brought about the salinization of soils, the alteration of the hydric balance of acquifers, and the disuse of drainage networks. On the contrary, traditional irrigation systems (using furrows and flooding), which are "inefficient" in terms of hydrological standards, are acceptably efficient taking into account the social, cultural and economic reality of poor irrigated areas which lack technology, financial resources and good users. Such uses are better adapted to the ethnic and social requirements of communities which have consolidated a true "irrigation culture" with normal production and profitability indexes.

5 Efficiency as a criterion

The efficient use of water, consequently, is not something technical belonging to hydric engineering, or specific to agricultural economy. It is a **criterion** of use, of valuing use, which is quite complex in nature. The criterion leads us to asserting-- within an irrigated ecosystem with its own cultural characteristics-- that in a certain time and place, the amount of water available for irrigation serves useful social purposes, preserving its quality, quantity, as well as the environment. These circumstances of value should be taken into account by policy-makers and should be reflected in irrigation legislation and management.

6 Other related issues

Water scarcity poses two interrelated problems: efficiency in its use and the reassignment of use rights. Inefficient uses of water in poor irrigation areas should be replaced by more profitable and productive uses which render a higher social benefit as a result of higher productivity, income and profit. This means reassigning use rights to the most efficient ones, but this poses at least two problems related to political security and social peace:

1. What authority - administrative or judiciary - will be in charge of judging when a certain use is more efficient and on the basis of what parameters or regulations.
2. What compensation will be given to the user who will be deprived of his acquired right to the use of water.

Laws should foresee these transient right situations. The reassignment of use rights presents yet another complexity: that of evaluating efficiencies among users and uses (who must receive water, what volume, to what end and of what level of quality). These are issues related to Use and Resource Planning which require a high degree of pre and post-information esential for evaluation and not always available.

7 Efficiency in Argentine legislation

When regulating use rights, the Argentine Civil Code establishes the regular and not the abusive exercise of rights in its articles 2513 and 2514, imposing the **functional use** of the faculties inherent to the use. This is a generic criterion of efficiency which prohibits, in the case of water, degrading or wasting it. It is a system analogous to the theory of reasonable use and to the prohibition of acts emulating the Angloamerican law.

The Water Code of Cordoba Province (Argentina), when legislating the concession for the use of public waters and their modalities, establishes norms that require the user to make an efficient use of irrigation water (Law 5589/73, articles 41, 45, 114, 117 and related articles). Besides establishing the functional use of the right, it also forbids the chemical or physical alteration of water and any variation in flow which can be prejudicial to ecosystems. Water delivery is decided on the basis of a **beneficial use rate** established by the authority according to spatial and temporal circumstances which indirectly imposes on the user the need to make an efficient use of water.

Few water laws take care of the reuse of water which, after adequate treatment, can be used for irrigation. Reuse maximizes use; it encourages the preservation of the resource and the environment, and it is an encouragement for the user as it gives him the possibility of having a higher volume of water. Reuse, then, is a technique that makes water use efficient. Laws should promote the use of new recovery technologies.

The already cited Water Code of Cordoba, in its article 115 addresses this situation and encourages the user by extending his use right with the recovered water at the goverment´s own expense.

8 Conclusions and proposals

1. Public and private actions to achieve Sustainable Development require an efficient use of water resources. Therefore, the Water Policy, Legislation and Water Management should set as their ultimate goal achieving the highest possible efficiency in water use.
2. Efficiency in water use results from a set of factors: the ends and means established by the water policy, a reasonable instrumentation of those ends in the legislation, an adequate structure of management organisms that allows for dynamic management, enough financial resources, infrastructure of adequate works, education, training and user participation and conservation of the environment. For these reasons, efficiency in use more than a standard concept is a **holistic criterion of valorization,** changing and contingent, applicable in a specific time and place, and whithin a certain socio-cultural framework.
3. The use of new irrigation technologies in themselves do not ensure an efficient use of water. The cultural idiosincracy, uses and customs of each community must be respected.
4. The law must foresee the reassignment of use rights as a means to achieve higher efficiency, ensuring juridical stability and the respect for acquired rights through adequate compensations.
5. Water legislation should address the use of recovery and reuse techniques, fostering their implementation.

9 References

1. UNESCO/ROSTLAC. (1994) *Efficient Water Use*, Montevideo.
2. Hunt, D., Johnson, C. (1996) *Environmental Management Systems, Principles and Practice*, Mc Graw Hill International (UK) Ltd. Madrid.
3. Moisset de Espanes, L., Lopez, J. (1977) *Derecho de Aguas. Regimen Transitorio y Normas de Conflicto*. CONFAGUA 14/23. Argentina.
4. Lopez, J. (1982) *Legislacion sobre los Conflictos entre Usos y Usuarios del Agua y su Resolucion*. MR CELA. Mendoza, Argentina.
5. Hotschewer, R.W. (1997) *Impacto de la Problematica Ambiental en el Derecho*. Law School, National Litoral University, Argentina.
6. International Workshop on Participatory Irrigation Management: Second Generation Problems. (1997) *Participatory Irrigation Management and Second Generation Problems*. IIMI. World Bank. Colombia.
7. *Codigo Civil de la Republica Argentina*. National Law # 340/ law 17711/1869 and 1968.
8. *Codigo de Aguas de la Provincia de Cordoba*. Argentina. Law 5589/73.

COMPREHENSIVE CRITERIA FOR WATER ECOSYSTEMS SUSTAINABILITY ASSESSMENT

V.I. SOKOLOV
Scientific-Information Center of the Interstate Commission for Water Coordination in the Aral Sea Basin, Tashkent, Uzbekistan
V.A. NIKOLAYENKO
Research Institute SANIIRI, Tashkent, Uzbekistan

Abstract
Environmental sustainability of natural water ecosystems in the Aral Sea Basin has been upset in the past 40-50 years. Lack of a transparent water resources management strategy has resulted in deterioration of the nature manifested in desertification, land salinization, waterlogging, formation of brackish lakes in depressions, disappearance of natural lakes in delta and a general deterioration of quality of water resources. To assess environmental sustainability of water sources of various genetic types it is necessary to identify the criteria characterizing their basic ecological qualities with due regard for securing an acceptable habitat within which the key hydrobionts (taxons) could exist. Control over the content of the established basic ecological criteria on the basis of the selected parameters will permit to take the necessary measures to sustain the water-related ecosystems.
Keywords: Aral Sea Basin, comprehensive criteria, environmental sustainability, hydrobionts, water ecosystem.

1 Introduction

Water is a strategic resource playing a vital role in the economic and social life of arid Central Asia. Over the last 40 years, intensive cotton farming and onesided agricultural development has diverted so muchwater from the two rivers which feed the Aral Sea that its shore line has retreated by more than 120 km in some places. Abandoned and derelict fishing boats in a landscape of salt and crusted sand first drew the world's attention to the human and ecological crisis facing the Aral Sea and its shore region. But degradation due to the nonsustainable use of water and related land

Water and the Environment: Innovative Issues in Irrigation and Drainage. Edited by Luis S. Pereira and John W. Gowing. Published in 1998 by E & FN Spon. ISBN 0 419 23710 0

resources - nonsustainable agriculture - extends far beyond the Sea and its shores. Over the last few decades, the upper Basin (flow formation zone) has lost about 50% of its forest cover. Soil erosion has intensified, not only reducing agricultural productivity but also silting storage reservoirs. The massive discharges of drainage water from irrigated land into rivers have resulted in a drastic increase of water salinity. Soil salinization and waterlogging is a serious problem throughout the Basin. These effects are directly linked to a decline in human health and agricultural productivity in the Basin.

In recent years, hydrological and ecological studies were carried out by specialists from the five countries of Central Asia within the framework of the Project "The Fundamental Provisions of Water Resources Management Strategy in the Aral Sea Basin" (this Project was initiated in 1995 by the International Fund for the Aral Sea Saving).

One of the results of this previous work is a better understanding of the problem setting, key issues and ecosystem approaches for water resources management in the Aral Sea Basin. As one of the critical problems, sustainability of water-related ecosystems can be mentioned.

Rehabilitation of ecosystems up to their original states (target - maximum) or their stabilization in terms of well-being ecological conditions (realistic target) will require a tremendous investments. In a given current economic conditions the Central Asian countries have not possibilities for a large-scale mitigating actions. That is why, the strategy for water resources management should be formulated with taking into account the economic ability - on the one hand, and the high priority socio-economic demands - on the other hand. From a such viewpoint, the strategy component addressed to the water resources quality management should focus on the complex measures to stabilize the environmental sustainability of water bodies and water-related ecosystems.

2 Approach to establishing the environmental sustainability criteria

The preservation of the given current ecological conditions or prevention of further degradation and restoration of the ecological balance in the water bodies - is a goal function. To support this task it is necessary to develop a set of criteria for assessment of the environmental sustainability of various genetic types of water sources: natural (e.g. rivers, lakes) and artificial (canals, reservoirs, drainage collectors, etc.). These water bodies are located in a different geographical zones and they differentiate in morphology, hydrology, physical and other characteristics. At the same time these water bodies have a different socio-economic value. The all offered water bodies are the elements of the environment. Thus, the water resources management activities should be based on the ecosystem approach.

To assess environmental sustainability of water sources of various genetic types it is necessary to identify the criteria characterizing their basic ecological qualities with due regard for securing an acceptable habitat within which the key hydrobionts (taxons) could exist. Control over the content of the established basic ecological criteria on the basis of the selected parameters will permit to take the necessary measures to sustain the water-related ecosystems.

A comprehensive classification should include water quality standards based on the salt content (mineralization and main ions) and hydro-ecological sanitary parameters ecompassing physical, chemical, hydro-biological, micro-biological and toxicological indicators. The Central Asian countries are widely using standards, which were generally accepted in previous Soviet times. For instance, these are:

1. Sanitary Norms and Rules for Surface Water Protection Against Pollution, Moscow, 1988.
2. State Standards "Drinking Water" (GOST - 2874-82). Moscow, 1982.
3. Rules for Water Quality in Water Bodies and Waterways Control, Moscow, 1982.
4. Sources for Centralized Domestic and Drinking Water Supply, Moscow, 1984.
5. Summarized List of Maximum Allowable Concentration and Approximate Safe Levels of Harmful Substances Impacts to Water Bodies Having Significance for Fishery, Moscow, 1990.

The initial milestone of proposed research work is a wide-known sustainability criteria: *"Nowadays human generation should remain to future generations such ecological conditions, which will be not worse than existing conditions"*. An assessment of the environmental sustainability should be based on comprehensive criteria in terms of quantity and quality of water resources, and also from viewpoint of life conditions for the key hydrobionts (taxons) in water bodies. The following types of criteria are suggested:

1. *Environment protection criteria* - indicators which take into account the preservation of integrity of water-related ecosystems with a normal functioning and reproduction of biogenesis formed in various genetic types of water sources (rivers, lakes, reservoirs, etc.).
2. *Ecological-sanitary criteria* - indicators for assessment of the limits of adaptability of hydrobionts to the environment and for assessment of a maximum of human-induced impacts which can be endured by the biogenesis without changing of it's structure and functional capacity.
3. *Economic criteria* - indicators for assessment of the water quality in water-related ecosystems from the viewpoint of water users constraints.

Development of criteria of environmental sustainability of water-related ecosystems in the Central Asian states should be conduced on the basis of the following principles:

- Recognition of equal importance of the environmental, social and economic components for the assessment of water-related ecosystems.
- Assumption of potentially permissible withdrawal of water without upsetting strongly the existing or perspective water balance, and without a considerable damage to any water user.
- Cautious interference in changing of various genetic types of water-related ecosystems, that is a gradual change of external factors of influence to water resources without upsetting sharply the stability of the existing biogenesis.

On the basis of the aforesaid principles the comprehensive criteria of environmental sustainability of water-related ecosystems can be singled out by referring each criteria to the components of the system 'water source - water user':

1. Criteria for assessment of safety of water body as an ecosystem:

   - sustainability criteria characterizing the ability of water body to preserve the existing mechanisms of formation and supporting their habitude;
   - resistance criteria characterizing the ability of hydrobionts to endure the deviations of the environment under different impacts.

2. Criteria for assessment of water quality as a natural resource:

   - criteria for assessment of water fitness for domestic and drinking water supply;
   - criteria for assessment of fishery safety;
   - criteria for assessment of water fitness for irrigation;
   - criteria for assessment of water fitness for industrial water consumption.

3 Policy addressed to the sustainable environment

The Aral Sea Basin Program has its origin not only in the concern for the extraordinary decline of the sea, but also in the degraded state of the environmental and related socio-economic conditions along the rivers, in the deltas, and the littoral zone of the sea. The decline in river flow and water quality (from both increased salinity and pollution mainly from agriculture) and the desiccation of the Aral Sea has resulted in three major impacts on the deltas and coastal zones: degradation and desiccation of valuable wetlands and grazing areas; a deterioration of the quality and availability of fresh water supply for drinking and irrigation; and a decline in fisheries not only in the sea but in the numerous lakes and reservoirs that were a major feature of the deltas.

The agreement on the objectives and content of the "Principal Provisions of the Concept for Improvement of the Environmental and Socio-economic Situation in the Aral Sea Basin" adopted by the Head of States in 1994 gives high priority to the restoration of the environmental conditions in the Aral Sea coastal and river deltas areas, improvement of the quality of life of the population, and restoring both natural and sustainable economic activity. The challenge of restoring sustainable environment conditions in these areas is to re-establish a hydrologic regime that emulates the natural regime in terms of the quantity, quality and variability of water flows through the complex pattern of channels and lakes that characterized the deltas. Annually and seasonally, a suitable quantity of water at appropriate times and with appropriate quality must be allocated and delivered to these areas, and managed. The pettern and level of flood flows is particularly significant as is the annual pattern of flow and water level, also the design and placement of structures needed to control and direct flows is important. The pattern of flow, including its annual variability, is as vital as the volume of water and its quality.

These objectives and concepts pose critical questions for the formulation of a policy, strategy, and action program for water and salinity management at regional and national level. Firstly, the most appropriate long-term hydrological regime (the quantities of water needed and their quality) must be determined along with criteria that could form the basis for transparent water allocation mechanisms. Secondly, the basin states should agree on these mechanisms for determining (annually and seasonally) the allocations of water to the delta areas and the sea, and make commitments that their individual actions (water diversion and use) will be consistent with this agreement. And thirdly, the states should agree on individual and collective strategies and measures to ensure that the agreed long-term hydrologic regime can be sustained. These measures will, no doubt, include greatly increased attention at the national level regarding water conservation and demand management and to the control of salinity in drainage flows and return flows.

The integrated regional and national policy, strategy, and action program could, for example, include the following:

- Future national water availability and demands, based on a equitable share of transboundary water resources and on reasonable effectivity criteria for water use.
- Standards for transboundary river salinity based on the economic impact that different river salinity levels have on the various water uses.
- Sustainable hydrologic regimes and related mechanisms for water allocation and control for satisfying the sanitary and ecological demands of the transboundary rivers and the delta areas (in accordance with the above mentioned classification).
- Screened measures, or combination of measures, in land management, water management and salt disposal, which can be implement on national level to arrive at a sustainable water and salt balance in the Amu Darya and Syr Darya basins that meets regional and ecological criteria; their likely impact, in terms of economic and social benefits \ costs, river and soil salinity reduction, ecological benefits, and the instruments them (incentive framework, institutional tools, economic priority setting, financial possibilities, priority setting). Where and what kind of national measures are technically, economically, managerially, most attractive to comply with the competing national and regional demands?
- Answer to, for example, the question if sustainable water and salt balances can be achieved by implementing the above national measures only, or are regional measures unavoidable from a point of view river salinity control and economic benefits? Screened measures for regional implementation for additional salinity control in terms of: (i) the reduction of salinity loads to be discharged into transboundary river water, (ii) the reduction of salinity levels of irrigated land and (iii) economic benefits and costs; and the recommended policies and mechanisms to recover the costs.
- An evaluation of alternative financing mechanisms for: (i) regional organizations in charge of water and salinity management at basin level, (ii) operation and maintenance and improvement of existing basin infrastructure, and (iii) the construction of new infrastructure, and (iii) the construction of new infrastructure to improve regional water and salinity management.
- An evaluation of alternative monitoring arrangements to assess the impact of all works and measures at national and regional level, and the compliance with agreed allocations from transboundary rivers and return flows (salt loads) into transboundary rivers.
- The mechanisms to deliver the strategy and action program «on the ground»

(communities get involved in identification of local natural resources issues; development and implementation of action based management plans for its locality; promotion and adoption of improved management practices; and communication to government of aspirations and concerns for management of natural resources at the local, national and basin-wide level).

4 Concluding remarks

The above discussed initial framework should be used to make several trial water balances to derive a range of future regional scenarios in terms of limits and constraints on water demands that can be used for the national and regional analysis of water conservation and the identification of options. In developing the initial framework, it is necessary to undertake and assess the water demands (quantity and quality) for different options for environmental sustainability in the Aral Sea Basin, including:

• Evaluate the present and future hydrologic and management to sustain and restore eco-systems under alternative development scenarios, to address the needs of fisheries, pastures, forests, wildlife and other natural and economic uses; together with an assessment of the sources and availability of water to meet these demands including surface water (both normal and flood flows), groundwater and return flows.
• Identify the hydrologic regime and its key characteristics and critical parameters required to sustain these activities and eco-systems (in accordance with above discussed classification of water sources).
• Assess the annual and multi-year planning implications of these requirements.
• Evaluate how these requirements are translated into water demands (quantity and quality) and how they should be accounted for in the overall water balance.

SUBJECT INDEX

—H—

herbicide, 12, 20, 21, 22, 23, 24, 25, 27, 28, 29, 30, 31, 32, 33, 34, 52, 53, 54, 56, 57
hydraulic conductivity, 3, 8, 30, 37, 47, 50, 66, 67, 68, 71, 95, 101, 104, 158, 160, 161, 390, 391, 392, 393, 394

—I—

infiltration, 20, 25, 28, 34, 37, 43, 44, 46, 49, 50, 84, 85, 86, 87, 89, 90, 101, 106, 120, 123, 125, 129, 140, 141, 142, 143, 144, 146, 147, 149, 150, 151, 152, 153, 154, 155, 156, 157, 158, 159, 160, 161, 341, 343, 348, 349, 350, 351, 355, 356, 357, 391, 394, 408, 409, 419
infrared surface temperature, 180, 183, 186
irrigated soils, 12
irrigation agronomy, 396
irrigation canal modernisation, 274
irrigation change, 235, 318, 320, 321, 322, 323
irrigation delivery scheduling, 239, 242, 243, 245
irrigation management, 141, 259, 262, 266, 267, 269, 272, 313, 318, 319, 320, 321, 322, 323, 324, 325, 326, 396, 399, 407, 448
irrigation planning, 223, 229, 422, 439
irrigation requirements, 126, 202, 257, 259, 293, 300, 303, 304, 306, 409, 412
irrigation scheduling, 112, 129, 141, 166, 180, 189, 190, 197, 198, 202, 203, 239, 240, 241, 242, 280, 300, 307, 407, 408, 409, 410, 412, 413
irrigation systems, 111, 112, 115, 116, 118, 140, 149, 156, 162, 163, 165, 166, 173, 174, 175, 177, 178, 179, 224, 226, 231, 232, 233, 234, 239, 241, 245, 246, 247, 248, 249, 251, 253, 254, 256, 258, 260, 262, 276, 283, 319, 322, 323, 324, 325, 331, 332, 334, 338, 349, 449, 458
irrigation technology, 131, 223, 229, 321
irrigation water demand minimum level, 292

—K—

kriging, 300

—L—

land reclamation, 75

leaching, 6, 7, 12, 17, 18, 20, 21, 24, 25, 43, 44, 47, 50, 54, 56, 62, 67, 75, 78, 79, 81, 101, 141, 166, 374
level basin irrigation, 43, 58, 60, 84, 85, 86, 87, 88, 89, 90, 121, 122, 125, 129, 131, 132, 139, 140, 141, 149, 154, 224
low-land areas, 20, 21
low pressure irrigation systems, 256

—M—

mathematical models, 100
modelling, 3, 6, 8, 41, 67, 78, 100, 120, 123, 125, 129, 140, 149, 156, 197, 314, 331, 356, 390, 391, 439, 448
monitoring, 23, 39, 54, 92, 122, 125, 240, 244, 266, 267, 268, 269, 270, 272, 273, 274, 275, 278, 280, 281, 288, 289, 384, 465
multicriteria analysis, 111, 112, 115

—N—

nitrate, 3, 5, 6, 7, 8, 9, 35, 36, 39, 42, 43, 44, 48, 50, 52, 53, 54, 57, 92, 100, 101, 104, 105, 106, 112, 368
nitrate leaching, 7, 43, 50
nitrate losses, 35, 36, 39, 44
nitrate management, 52
nitrogen, 6, 7, 35, 36, 53, 54, 57, 100, 101, 103, 105
no-tillage, 5, 6, 27, 28, 381, 382, 385, 386, 387, 388, 389
runoff, 3, 4, 5, 6, 8, 25, 27, 28, 29, 30, 31, 32, 33, 34, 36, 43, 53, 78, 85, 86, 87, 89, 115, 117, 131, 134, 136, 137, 139, 154, 155, 156, 157, 158, 159, 160, 161, 162, 163, 164, 172, 206, 331, 338, 340, 341342, 343, 344, 345, 346, 348, 349, 350, 351, 353, 356, 357

—O—

overland flow, 348, 349, 350, 351, 352, 353, 354
olive tree, 205, 206, 207, 211, 343, 390, 391
optimization, 140, 141, 142, 144, 146, 147, 173, 179, 256, 257, 260, 262, 299, 448, 449, 454
oxidation, 75, 76, 77, 78, 79, 80, 81, 82